CHROMOSOME MUTATION AND NEOPLASIA

ADVISORS TO THE EDITOR

John H. Edwards, *Oxford*
C. E. Ford, *Oxford*
Jean de Grouchy, *Paris*
Patricia A. Jacobs, *Honolulu*
Jérôme Lejeune, *Paris*
Orlando J. Miller, *New York*
*Klaus Patau, *Madison*
*Curt Stern, *Berkeley*

*The Editor and Advisors note, with sadness, the passing of two of their distinguished colleagues: Klaus Patau—born in Gelsenkirchen, Germany, 30 September 1908 and died in Madison, 30 November 1975; Curt Stern—born in Hamburg, 30 August 1902 and died near Berkeley, 23 October 1981.

CHROMOSOME MUTATION AND NEOPLASIA

JAMES GERMAN, Editor
The New York Blood Center
New York, New York

Alan R. Liss, Inc., New York

Address all Inquiries to the Publisher
Alan R. Liss, Inc., 150 Fifth Avenue, New York, NY 10011

Copyright © 1983 Alan R. Liss, Inc.

Printed in the United States of America.

Under the conditions stated below the owner of copyright for this book hereby grants permission to users to make photocopy reproductions of any part or all of its contents for personal or internal organizational use, or for personal or internal use of specific clients. This consent is given on the condition that the copier pay the stated per-copy fee through the Copyright Clearance Center, Incorporated, 21 Congress Street, Salem, MA 01970, as listed in the most current issue of "Permissions to Photocopy" (Publisher's Fee List, distributed by CCC, Inc.), for copying beyond that permitted by sections 107 or 108 of the US Copyright Law. This consent does not extend to other kinds of copying, such as copying for general distribution, for advertising or promotional purposes, for creating new collective works, or for resale.

Library of Congress Cataloging in Publication Data

Main entry under title:
Chromosome mutation and neoplasia.

Includes bibliographies and index.
1. Cancer—Genetic aspects. 2. Chromosome abnormalities. 3. Mutation (Biology) I. German, James. [DNLM: 1. Mutation. 2. Chromosome abnormalities. 3. Neoplasms—Etiology. 4. Clone cells. 5. Precancerous conditions. QZ 204 C557]
RC268.4.C48   1983   616.99'2071   82-21697
ISBN 0-8451-0220-6

CONTENTS

Contributors. ix
The Chromosomes Series. xiii
Introduction to *Chromosome Mutation and Neoplasia* xv
My Life in Cytology
   Sajiro Makino. xxvii

I. GENETICALLY DETERMINED HUMAN DISORDERS THAT FEATURE CHROMOSOME INSTABILITY, HYPERSENSITIVITY TO ENVIRONMENTAL CARCINOGENIC AGENTS, AND CANCER PRONENESS

A. The Clinical Disorders

Study of Familial Defects as a Way of Investigating the
Origin of Human Cancers
   John Cairns 3
Bloom's Syndrome
   Eberhard Passarge. 11
Ataxia-Telangiectasia: Search for a Central Hypothesis
   Richard A. Gatti and Kathleen Hall. 23
Long-Term Outcome in Fanconi's Anemia: Description of 26
Cases and Review of the Literature
   Blanche P. Alter and Nancy Upp Potter 43
Xeroderma Pigmentosum
   Alan D. Andrews. 63
Werner's Syndrome
   W. Ted Brown. 85

B. Commentaries on Selected Aspects of the Disorders

Patterns of Neoplasia Associated With the Chromosome-
Breakage Syndromes
   James German 97
The Cytogenetics of the "Chromosome-Breakage
Syndromes"
   James H. Ray and James German 135

Sister-Chromatid Exchange—The Phenomenon and Its
Relationship to Chromosome-Fragility Diseases
**Samuel A. Latt, Rhona R. Schreck, Charlotte P.
Dougherty, Karen M. Gustashaw, Lois A. Juergens, and
Tim N. Kaiser** 169

The Interrelationships in Ataxia-Telangiectasia of Immune
Deficiency, Chromosome Instability, and Cancer
Barbara Kaiser-McCaw and Frederick Hecht 193

Cellular Sensitivity to Mutagens and Carcinogens in the
Chromosome-Breakage and Other Cancer-Prone
Syndromes
**John A. Heddle, Alena B. Krepinsky, and
Richard R. Marshall**...................................... 203

Regulation of the Responses to DNA Damage in the
Hypersensitivity Diseases and Chromosome-Breakage
Syndromes
James E. Cleaver.. 235

II. GENOMIC ALTERATIONS, THE MOLECULAR TO THE MICROSCOPIC: THEIR RELEVANCE IN NEOPLASIA

Effects on Chromosomes of Carcinogenic Rays and
Chemicals
H.J. Evans.. 253

Molecular Alterations in DNA Associated With Mutation
and Chromosome Rearrangements
**Bernard Strauss, Kathleen Ayres, Kallol Bose, Peter
Moore, Samuel Rabkin, Robert Sklar, and
Valerie Lindgren**... 281

Viral Interactions With the Mammalian Genome Relevant
to Neoplasia
Warren W. Nichols...................................... 317

Genomic Rearrangements and the Origin of Cancer
Ruth Sager.. 333

Bloom's Syndrome. X. The Cancer Proneness Points to
Chromosome Mutation as a Crucial Event in Human
Neoplasia
James German 347

The Significance of Chromosome Change to Neoplastic
Development
R.S.K. Chaganti 359

Oncogenes, Chromosome Mutation, and the Development of
  Neoplasia
  R.S.K. Chaganti and Suresh C. Jhanwar 397
Tumor Progression and Clonal Evolution: The Role of
  Genetic Instability
  Peter C. Nowell............................. 413
Index 433

CONTRIBUTORS

Blanche P. Alter [43]
Departments of Medicine and Pediatrics, Polly Annenberg Levee Hematology Center, Mount Sinai School of Medicine, New York, NY 10029

Alan D. Andrews [63]
Department of Dermatology, Columbia University, New York, NY 10032

Kathleen Ayres [281]
Department of Microbiology, The University of Chicago, Chicago, IL 60637

Kallol Bose [281]
Department of Microbiology, The University of Chicago, Chicago, IL 60637

W. Ted Brown [85]
Department of Human Genetics, New York State Institute for Basic Research in Developmental Disabilities, Staten Island, NY 10314

John Cairns [3]
Department of Microbiology, Harvard School of Public Health, Boston, MA 02115

R.S.K. Chaganti [359, 397]
Laboratory of Cancer Genetics and Cytogenetics, Department of Pathology, Memorial Sloan-Kettering Cancer Center, New York, NY 10021

James E. Cleaver [235]
Laboratory of Radiobiology, University of California at San Francisco, San Francisco, CA 94143

Charlotte P. Dougherty [169]
Department of Pediatrics, Harvard Medical School, Boston, MA 02115

H.J. Evans [253]
Medical Research Council, Clinical and Population Cytogenetics Unit, Western General Hospital, Edinburgh, Scotland

Richard A. Gatti [23]
Department of Pathology, School of Medicine, University of California at Los Angeles, Los Angeles, CA 90024

James German [97, 135, 347]
The New York Blood Center, New York, NY 10021

Karen M. Gustashaw [169]
Department of Pediatrics, Harvard Medical School, Boston, MA 02115

Kathleen Hall [23]
Department of Pathology, School of Medicine, University of California at Los Angeles, Los Angeles, CA 90024

The boldface number in brackets following each contributor's name indicates the opening page number of that author's article.

Frederick Hecht [193]
The Genetics Center of Southwest Biomedical Research Institute, Tempe, AZ 85281

John A. Heddle [203]
Ludwig Institute for Cancer Research, Toronto, Ontario, Canada

Suresh C. Jhanwar [397]
Department of Pathology, Memorial Sloan-Kettering Cancer Center, New York, NY 10021

Lois A. Juergens [169]
Department of Pediatrics, Harvard Medical School, Boston, MA 02115

Tim N. Kaiser [169]
Harvard Medical School, Boston, MA 02115

Barbara Kaiser-McCaw [193]
The Genetics Center of Southwest Biomedical Research Institute, Tempe, AZ 85281

Alena B. Krepinsky [203]
Mutatech Incorporated, Department of Biology, York University, Downsview, Ontario, Canada

Samuel A. Latt [169]
Department of Pediatrics, Harvard Medical School, Boston, MA 02115

Valerie Lindgren [281]
Department of Microbiology, The University of Chicago, Chicago, IL 60637

Sajiro Makino [xxvii]
Chromosome Research Unit, Faculty of Science, Hokkaido University, Sapporo, Japan

Richard R. Marshall [203]
Department of Biology, York University, Downsview, Ontario, Canada

Peter Moore [281]
Department of Microbiology, The University of Chicago, Chicago, IL 60637

Warren W. Nichols [317]
Department of Cytogenetics, Institute for Medical Research, Camden, NJ 08103

Peter C. Nowell [413]
School of Medicine, University of Pennsylvania, Philadelphia, PA 19104

Eberhard Passarge [11]
Institut für Humangenetik, Universitätsklinikum, University of Essen, Federal Republic of Germany

Nancy Upp Potter [43]
Division of Epidemiology and Biostatistics, and of Pediatric Hematology and Oncology, Sidney Farber Cancer Institute, Boston, MA 02115

Samuel Rabkin [281]
Department of Microbiology, The University of Chicago, Chicago, IL 60637

James H. Ray [135]
The New York Blood Center, New York, NY 10021

Ruth Sager [333]
Division of Cancer Genetics, Sidney Farber Cancer Institute, Boston, MA 02115

Rhona R. Schreck [169]
Department of Pediatrics, Harvard Medical School, Boston, MA 02115

Robert Sklar [281]
Department of Microbiology, The University of Chicago, Chicago, IL 60637

Bernard Strauss [281]
Department of Microbiology, The University of Chicago, Chicago, IL 60637

THE CHROMOSOMES SERIES

Each volume in the Chromosomes series* is devoted to cytogenetic aspects of one broad subject, or to some aspect of cytogenetics itself. The chapters within a volume review various areas of knowledge pertaining to the given subject. Each chapter is authoritative and definitive, and each is written by an eminent scientist who himself has made important experimental contributions.

Authors have been given a free hand in style of writing and in approach. However, because the review paper plays an important role in contemporary science, a major objective in the preparation of this series is that the articles be comprehensible not only to those in cytogenetics or in the specific discipline of an author but also to those working in other branches of science.

I acknowledge with gratitude my Advisors, whose valuable views aid me both in the choice of major subject areas and possible chapter topics for review and in the selection of individual contributors.

—J. G.

---

*The first volume in the Chromosomes series was published in 1974 by John Wiley & Sons, Inc.

INTRODUCTION TO *CHROMOSOME MUTATION AND NEOPLASIA*

This is a book about cancer, but not cancer only. The term *cancer* is predominantly for clinical use, and this book is for biologists generally. The book's title and the following comments indicate my preference for viewing cancer as just one extreme of a spectrum; the name of the entire spectrum?—*neoplasia*. This view not only places in an appropriate perspective a process of immense importance to human health—cancer—but it also introduces a valuable dimension to the study of normal mammalian cell growth by contrasting the normal with the myriad variations from normal easily observable in neoplastic cells.

This book also is about clones, specifically, the populations of cells that develop a degree of proliferative autonomy as result of chromosome mutation.* Its contents derive heavily from observations made in five rare, genetically determined human disorders (to be mentioned again below) that predispose to the emergence of such clones, including clones that qualify for the diagnosis "cancer."

Why Is the Volume About Clones?

The acceptance of the notion that most cancers are populations of cells composed of the progeny of a single cell has developed gradually. Early in this century an authority wrote in the *Encyclopaedia Britannica* that "certain cells (emphasis added), which are apparently of a normal character and have previously performed normal functions, begin to grow and multiply in an abnormal way in some part of the body" [29]. However, by 1914 Boveri had concluded that "typically each tumor takes its origin from one and a single cell" [32]. In 1937 experimental leukemia was transmitted to 5 mice (out of 97 tries!) by the micromanipulation and intravenous injection into each mouse of a single leukemic cell [11]; this accomplishment provided evidence that "leukemia—which has hitherto been regarded

---

Chromosome mutation is defined as "any structural change involving the gain, loss or relocation of chromosome segments" [28]. Excluded are mutations that affect only one base pair in the DNA duplex. The term has been in use for at least 50 years, having been employed, for example, by Blakeslee and Davenport [6]. Although many chromosome mutations can be detected microscopically, some that cannot be will be demonstrable only by the techniques of molecular genetics, particularly deletions, duplications, and inversions that involve very short segments of a chromosome.

xv

by many workers as having a multicentric origin..." might itself be a disease of single cell origin.

By the 1950s newly improved cytogenetics techniques [21] revealed that the cells of certain "ascites tumors" that could be propagated in rodents by the transfer of intact cells had abnormal chromosome complements that included so-called marker chromosomes, i.e., chromosomes with morphologically detectable rearrangements; such observations led to the concept that a neoplastic cell population depends on the existence of a mutant stemline of cells, the stemline cell having a characteristic abnormal chromosome constitution [20]. In the mid-1950s, leukemias in rodents induced experimentally by X-irradiation were shown to consist of cell lineages with marker chromosome rearrangements [10, and discussed in 4]. (None of these observations indicated whether the visible genomic alterations were of etiological significance in the neoplasia or merely manifestations of it.)

In the mid-1960s certain benign tumors of the human uterus, the common leiomyomata, were shown by a biochemical method, electrophoretic protein separation, to be clones of cells; in women heterozygous (A/B) at the X-chromosome locus for glucose-6-phosphate dehydrogenase (G6PD), any one tumor was shown to be composed of cells of just one G6PD type, either A or B, rather than of a mixture of the two as were the non-tumorous myometria from which they had arisen [12]. (The Lyon hypothesis had been advanced early in that decade; by the Lyon effect, the G6PD locus on one of the two X chromosomes in a female (XX) cell undergoes selective inactivation, so that a cell and all its progeny express one type of the enzyme.) Subsequently, by the same approach, most human leukemias and "solid" cancers, and even the bone marrow in certain hematological disorders not generally classed as cancer but as precancerous [9], have been shown to be clones.

In a different type of experiment, monolayers of non-neoplastic cells proliferating in vitro that had been treated with some carcinogenic agent were found to contain foci of cells with altered growth patterns. That foci and not the whole monolayer of treated cells exhibited aberrant growth indicated that single cells had been "transformed," each giving rise to a colony that manifested features characteristic of neoplastic cells in vitro.

During the past two decades, cytogenetic evidence, viz., the presence of the identical marker chromosome(s) in each cell, has accumulated to indicate that several human leukemias and lymphomata, many solid tumors, and at least one benign type of tumor consist of clones of cells. (Subclones in the neoplastic populations have been observed also, and these are believed to be important in the evolution and changing clinical character of the cancers [24,25].) Even before banding techniques came into general use in the early 1970s, a unique marker chromosome in each cell of a tumor population, with or without additional gains or losses of chromosomes or rearrangements, had identified many human cancers of diverse types as descendants of single cells, the progenitor cells themselves necessarily having either produced or inherited the unique markers in each case. (Again, such studies did not bear on the possibility that the progenitor cell itself had been a member of a population that had inherited some alteration even before a chromosome mutation occurred in it, i.e., that in some obscure way it had become what has been termed "preneoplastic" [16].)

Chromosome banding techniques have permitted the recognition of more marker chromosomes and the better characterization of the rearrangements. Such studies are disclosing a striking degree of specificity between the breakpoint locations in the rearrangement and the types of cancer. (It is noteworthy in this respect that in 1981 the sixth of the international workshops convened at intervals since 1973 to summarize knowledge of human-chromosome mapping made an official tabulation not made by previous groups, of specific breakpoints of human cancers [14].) The first example to have been recognized of specificity of a chromosome rearrangement in a neoplasm is the translocation affecting the Nos. 22 and 9 that gives rise to the marker known as the Philadelphia chromosome [23] in chronic granulocytic leukemia. Subsequently, other examples have been recognized, in some of the acute leukemias, Burkitt's lymphoma, and (benign) meningioma. Examples of specificity of breakpoints have been identified also in certain malignant murine lymphoid neoplasms [16,17]; this information, in conjunction with the extensive mapping of genes to mammalian chromosomes that has been accomplished since 1968 (when the first gene was assigned to a human autosome [7]), makes apparent the important fact that in the mouse, as also in the human Burkitt's lymphoma, some of the breakpoints in malignant neoplasms are at or near structural loci for immunoglobulin genes, loci active in the type tissue in which the neoplasms develop [16,17]. Furthermore, the normal sites of so-called cellular oncogenes have in several cases been found to coincide with the specific chromosome breakpoints, with translocation of major portions of the oncogenes to new (abnormal) positions in the genome.

Not all clones with a visible chromosome mutation in their genome are neoplastic. A totally new observation was made soon after modern cytogenetics techniques began to be applied to the study of populations of human cells. Subpopulations of cells identified as clones by microscopically distinctive mutations in their chromosome complements are to be found growing among cells with normal complements, clones that by no pathologist's criteria would be classified malignant. First, some liveborn humans were found with mosaicism in multiple tissues, cells bearing a normal complement coexisting with cells bearing an abnormal one, e.g., translocation or deletion of a portion of some chromosome. That normal cells are present in such persons implies that the zygote was normal and that a chromosome mutation had occurred in a cell of the early conceptus and was transmitted thereafter by that cell's progeny. Although such major abnormal populations in mosaic individuals ordinarily are not thought of as clones, they in fact are. If their genome is unbalanced, the abnormal population may interfere with embryonic development. At other times mosaicism is detected in normally developed adults only after the clinical cytogeneticist has demonstrated that an occult abnormal clone had been responsible for the production of a gamete with an unbalanced genome; in such cases the mutant clone often comprises only a small proportion of body cells.

Clones of cells bearing chromosome rearrangements sometimes are found in cultures of fibroblasts derived from minute fragments of tissue (usually skin) taken from completely normal people [3]. Members of the clones are in the minority in such cultures, and it usually remains undetermined whether they were present in vivo or arose in vitro; however, in cell lines derived from tissue that has been X-irradiated in vivo, they may be present in abundance [8], indicating an in vivo origin at least under that unusual circumstance. Such mutant clones detected in fibroblast cultures usually have no clinical significance. Their proliferative capabilities in comparison to that of the fibroblasts with non-mutated complements have not been studied, but general observation suggests that they have no impressive growth advantage. Thus, mutation of the genome can occur in cells at various post-zygotic stages of life and give rise to non-neoplastic clones.

(I must exclude from my discourse here non-neoplastic clones in the immune system that arise in conjunction with a chromosome mutation that will permit a specific response to an antigen, because those rearrangements are not detectable microscopically. Time will tell what other non-neoplastic systems of immense interest I, for the same reason, have excluded unknowingly, for they have yet to be discovered!)

Introduction xix

If mutant clones exist in the circulating blood from normal people, they usually are not detected. Suggesting that very small clones may exist in vivo is the occasional observation made in many laboratories of single cells that have undergone a balanced translocation affecting specific points on a No. 7 and a No. 14 chromosome (discussed in [4]). (It also suggests that in T lymphocytes, the cells that can be brought into metaphase by phytohemagglutinin (PHA), certain specific chromosome regions are predisposed to rearrangement. It seems reasonable to speculate that loci so identified are regions undergoing active transcription as result of cell differentiation, being the loci concerned with specialized cell products and, or, with cell cycling in that particular cell type.)

In contrast to this apparent specificity of breakpoints and to the infrequency of occurrence of clones in blood from normal populations, mutant clones with various rearrangements apparently lacking specificity with respect to the chromosomes affected have been detected with no great difficulty in the blood of members of populations that had been exposed excessively to ionizing radiation, e.g., persons who have received roentgen therapy and survivors of the atomic blasts. In these populations, such mutant clones of T lymphocytes have exhibited no malignant potential, nor any clinical effect, although the exposed human populations from which the blood samples had been taken were strongly predisposed to cancer. That the clones are detectable at all, however, and that they persist and can be demonstrated at serial samplings of the blood as long as 35 years post-irradiation [1,2] suggest that from their inception they enjoy a small proliferative advantage over non-mutated lymphocytes, and, in this sense, a degree of autonomy with respect to cell cycling. (Note that the peak incidence of clinical chronic granulocytic leukemia in survivors of the atomic blasts in Japan occurred in 1953 [15], eight years after the events occurred that gave rise directly or indirectly to formation of a Philadelphia chromosome in a cell in each of the to-be-leukemic persons. This indicates that that well-known rearrangement of the genome endows the cell and its progeny with only a slight growth advantage—but a highly significant one. It also emphasizes the dependence on the passage of time, sometimes a long time, for the clinical "surfacing" of a malignant neoplasm.)

Finally, one small and heterogeneous, cancer-prone human population is known in which mutant clones of cells can be detected in various tissues relatively easily even without excessive exposure to exogenous clastogens. These are the persons with one of the exceedingly rare genetic disorders that, for our purposes here, may be grouped as "the chromosome-breakage syndromes," because, although the syndromes are dissimilar clinically, cells from affected persons present evidence that their genomes are unusually mutable, either spontaneously or following some environmental insult, or both [19,26,27]. Thus, in Fanconi's anemia, mutant clones have been found in fresh bone marrow aspirates and in circulating blood lymphocytes; in ataxia-telangiectasia, in circulating lymphocytes, and here a specificity as to the chromosomes affected has been detected, chromosomes Nos. 7, 14, or both predominating in the rearrangements; and in Werner's syndrome, in skin fibroblasts and B lymphocytes proliferating in vitro. Sometimes a very few cells will constitute the only evidence of the presence of a mutant clone—but, note that the finding of just 1–2% of T lymphocytes with the same rearrangement in a sample drawn from the blood of an adult reflects the existence at the moment of sampling of millions of members of the clone circulating in his blood, and many more stationed in lymphoid tissues elsewhere. In other cases the entire population of PHA-responding cells circulating in the blood will display the same mutated genome, and even then clinical evidence of leukemia may be lacking. In people with one of these syndromes, frank cancers will develop more often than expected in the general population, and marker-chromosome rearrangements have been found in most of the few cancers that have become available for cytogenetic study (e.g., Fig. 1), similar in type to those we have become familiar with in the benign clones. As cells and cell lineages from persons with these rare genetic constitutions are scrutinized, difficulty is encountered in deciding whether a clone proliferating excessively has graduated from benign to malignant status. In fact, the concept neoplasia itself becomes hazy here, and we begin to discern the spectrum of disturbed growth to which I referred at the outset. "Few if any areas of biology exist with a greater potential for elucidation of the step or steps taken as cells transform from 'normal' to 'neoplastic' and from 'benign' to 'malignant' " [26]. Since the early 1960s, the strikingly increased cancer occurrence in persons with genetically determined chromosome instability has been known. This has been a steady signal to many students of human cancer that pointed to a crucial role—or roles—for chromosome mutation in the etiology and, or, progression of neoplasia. It is for this reason that this particular volume on cancer—on neoplasia and on clones—takes as its point of departure the chromosome-breakage syndromes.

Introduction   xxi

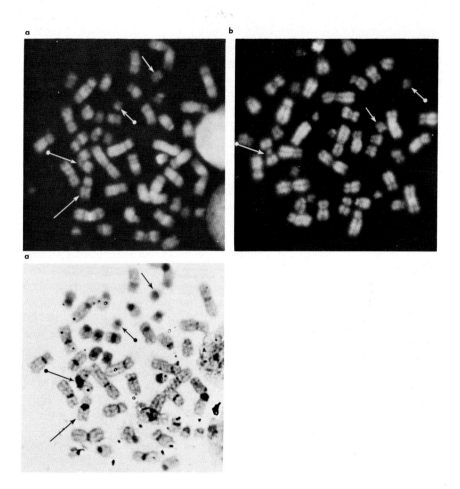

Fig. 1. Two metaphases from the leukemic clone populating the bone marrow of a young man with Bloom's syndrome and acute lymphocytic leukemia. Upper panels, Q-banding; lower panel, C-banding of the cell at upper left. Longer arrows indicate chromosome Nos. 9, the shorter ones the Nos. 20. The arrows marked with terminal dots point to structurally aberrant chromosomes, either an isochromosome for the long arm of No. 9—iso(9q)— or deletion of the long arm of No. 20—20q−. (Unpublished observation, with R.S.K. Chaganti.)

The Matters Covered in the Volume—and Some Not Covered

Among the several important matters addressed by the distinguished contributors to the volume are the following: the aberrant responses made to DNA-damaging agents by cells of persons with the rare syndromes; the mechanisms by which chromosome mutations may come about; the cytological and molecular nature of chromosome mutations; the cellular mechanism(s) by which they affect a cell's proliferative capacity; and their significance in relation to neoplasia. Each chapter in the book is an independent treatise, but they have been arranged into two sections. Section I consists of two parts, first, descriptions of the clinical disorders (IA), and then, reviews of special observational or experimental information either pertaining to the disorders specifically or derived from their study (IB); Section II consists of chapters on selected topics pertaining to genomic instability more generally and to changes in the genome associated with neoplastic transformation and progression.

This seemed to be enough for a single volume; therefore, I shall explain below the basis for the intentional exclusion of a large number of chapters on tumor virology, after an introductory comment. In the first volume in the series, entitled Chromosomes and Cancer (1974) [13], the geneticist H.L.K. Whitehouse was invited to consider possible ways by which the then-mounting body of information which quite clearly was pointing to a role for viruses in the etiology of cancer might be integrated with the equally formidable body of information pointing to but never proving chromosome mutation's etiological role. Previously, Whitehouse had not written on cancer, but he was asked to undertake this difficult task because of his earlier theoretical and experimental concern with the molecular nature (i.e., at the DNA-strand level) of matters such as genetic recombination, gene conversion, and chromatid exchange. What he accomplished [31] enhanced the book by pointing surprisingly clearly and with accurate foresight to the way things were to fall into place, as they rapidly are doing today, almost a decade after he wrote. In his "open-replicon hypothesis of carcinogenesis" he observed that host-DNA replication was needed for viral integration to take place, and that many viruses contain genes that can initiate such replication in the host. He proposed that because viral integration into a genome is advantageous to both the parasite and the host, selection will have occurred for genes in the host that will hold in check the DNA synthesis otherwise induced by the integrated virus and thereby permit survival of the host cell. Thereafter, the occasional loss from a cell of the activity of such suppressor loci, as by their mutation or deletion, could disrupt the balanced system that had been selected for and be responsible for triggering host-DNA replication, with an associated breakdown of the cell's replication control, i.e., could be responsible for the induction of neoplasia. Klein expressed the idea recently that "host-cell feed back regulations are delicately balanced and may be disrupted by relatively small changes" [17].

Some highly relevant basic information has accumulated during the decade since Whitehouse considered the matter, concerning (*i*) the specificity of chromosome breakpoints in different human and murine neoplasms and (*ii*) the exact localization on the chromosomes of these two species of several hundred genetic determinants—chromosome mapping—including very recently the known oncogenes. Also, (*iii*) some understanding has been obtained of the roles of viral and cellular oncogenes in neoplasia and the ways in which oncogenes can be activated abnormally. The availability of this information makes the present a period of intellectual fulfillment for students of neoplasia, particularly those who for many years have been concerned with chromosomes and the etiological role of chromosome mutation. Camps seemingly separated by a formidable chasm, namely tumor virologists and tumor cytogeneticists, find with the clearing of the mist not only no chasm but, in fact, a single camp. The present volume easily could have incorporated perhaps a dozen chapters not just on viral integration and transformation but, more broadly, on the remarkable instability that eukaryotic genomes can exhibit; but it does not, for at least three reasons, as follow. The volume's size would have been swelled unsuitably if the appropriate amount of space had been allocated to subjects such as transposable genetic elements, genomic changes associated with differentiation, the unscheduled activation of cellular oncogenes, oncogenic viruses and their integration into mammalian chromosomes, and transfection of cells with tumor DNA, all of which fall under the rubric "chromosome mutation." Second, because new information presently is accruing in these areas at such a rapid rate, definitive chapters on them very possibly cannot be prepared as yet, and definitiveness is one of the desiderata for inclusion of a paper in volumes in the *Chromosomes* series. Several chapters in Section II do address these matters briefly but authoritatively. Finally, molecular matters pertaining to eukaryotic genomic instability are being covered extensively in admirable fashion in several other current books, and these may be consulted as supplementary to *Chromosome Mutation and Neoplasia* (e.g., Refs. [5,18,22,30]). I do hope that definitive chapters on molecular aspects of chromosome instability as it pertains to neoplasia, as yet unwritten, will enrich future volumes in this series.

Acknowledgments. The symposium at which precursors of the chapters in this volume were presented orally ("Chromosome Breakage and Neoplasia," 11–12 December 1980, The Rockefeller University, New York City) was sponsored and underwritten by the National Foundation for Jewish Genetic Diseases. I acknowledge with warm gratitude the support of the Foundation and its president, Mr. George Crohn, Jr.

I acknowledge also Dr. Vivien B. Shelanski for assistance in editing portions of the volume and Ms. Rainelle Peters for assistance in compiling the various chapters for publication. Finally, I thank three of my colleagues for advice in developing the agenda for the symposium and the contents of this volume, namely, Drs. R.S.K. Chaganti, Eberhard Passarge, and James H. Ray.—J.G.

LITERATURE CITED

1. AWA, A.A.: Cytogenetic and oncogenic effects of the ionizing radiations of the atomic bombs. *In*: Chromosomes and Cancer (ed. J. German), John Wiley and Sons, N.Y., 1974, pp. 637–674.
2. AWA, A.A. Personal communication.
3. BENN, P.: Specific break points in chromosomally abnormal fibroblast subpopulations. *Cytogenet. Cell Genet.* 19:118–135, 1977.
4. CHAGANTI, R.S.K.: The significance of chromosome change to neoplastic development. (This volume)
5. COHN, W.E. (editor): Proceedings of symposium "Genetic Mechanisms of Carcinogenesis," Gatlinburg, Ten., 11–15 April 1982. *In*: Progress in Nucleic Acid Research and Molecular Biology, Vol. 29. (To be published.)
6. DAVENPORT, C.B. *In*: Yearbook, Carnegie Institution of Washington, Volume 27, page 35, 1927–28.
7. DONAHUE, R.P., BIAS, W.B., RENWICK, J.H., MCKUSICK, V.A.: Probable assignment of the Duffy blood group locus to chromosome 1 in man. *Proc. Natl. Acad. Sci. U.S.A.* 61:949–955, 1968.
8. ENGEL, E., FLEXNER, J.M., ENGEL-DE-MONTMOLLIN, M.L., FRANK, H.E.: Blood and skin chromosomal alterations of a clonal type in a leukemic man previously irradiated for lung carcinoma. *Cytogenetics.* 3:228–251, 1964.
9. FIALKOW, P.J.: Clonal origin and stem cell evolution of human tumors. *In:* "Genetics of Human Cancer" (ed. J.J. Mulvihill, R.W. Miller, and J.F. Fraumeni, Jr.), Raven Press, N.Y., 1977, pp. 439–453.
10. FORD, C.E., CLARKE, C.M.: Cytogenetic evidence of clonal proliferation in primary reticular neoplams. *In*: Canadian Cancer Conferences. (Academic Press, N.Y.) 5:129–146, 1963.
11. FURTH, J., KAHN, M.C.: The transmission of leukemia of mice with a single cell. *Am. J. Cancer* 31:276–282, 1937.
12. GARTLER, S.M.: Utilization of mosaic systems in the study of the origin and progression of tumors. *In*: Chromosomes and Cancer (ed. J. German), John Wiley and Sons, N.Y., 1974, pp. 313–334.
13. GERMAN, J. (editor): Chromosomes and Cancer, John Wiley and Sons, N.Y., 1974, 756 pages.
14. Human Gene Mapping 6. *Cytogenet. Cell Genet.* vol. 32, 1982 pp. 205–207.

15. KAMADA, N., UCHINO, H.: Chronologic sequence in appearance of clinical and laboratory findings characteristic of chronic myelocytic leukemia. *Blood* 51:843-850, 1978.
16. KLEIN, G.: Lymphoma development in mice and humans: Diversity of initiation is followed by convergent cytogenetic evolution. *Proc. Natl. Acad. Sci. U.S.A* 76:2442-2446, 1979.
17. KLEIN, G.: The role of gene dosage and genetic transpositions in carcinogenesis. *Nature* 294:313-318, 1981.
18. KLEIN, G. (editor): Advances in Viral Oncology, Vol. I, Raven Press, N.Y., 1982, 262 pages.
19. LATT, S.A., SHRECK, R.R., DOUGHERTY, C.P. GUSTASHAW, K.M., JUERGENS, L.A., and KAISER, T.N.: Sister-chromatid exchange—The phenomenon and its relationship to chromosome-fragility diseases. (This volume.)
20. MAKINO, S.: The chromosome cytology of the ascites tumors of the rat, with special reference to the concept of the stemline cell. *Int. Rev. Cytol.* 6:25-84, 1957.
21. MAKINO, S.: My life in cytology. (This volume.)
22. Movable Genetic Elements. Cold Spring Harbor Symp. Quant. Biol. Vol. 45, Cold Spring Harbor Laboratory, Cold Spring Harbor, N.Y., 1981, 1025 pages.
23. NOWELL, P.A., HUNGERFORD, D.A.: A minute chromosome in human chronic granulocytic leukemia. *Science* 132:1497, 1960.
24. NOWELL, P.C.: Chromosome changes and the clonal evolution of cancer. *In:* Chromosomes and Cancer (ed. J. German), John Wiley and Sons, N.Y., 1974, pp. 267-285.
25. NOWELL, P.C.: Tumor progression and clonal evolution: The role of genetic instability. (This volume.)
26. RAY, J.H., GERMAN, J.: The chromosome changes in Bloom's syndrome, ataxia-telangiectasia, and Fanconi's anemia. *In:* "Genes, Chromosomes, and Neoplasia" (edit. by F.E. Arrighi, P.N. Rao, and E. Stubblefield), Raven Press, N.Y., 1981, pp. 351-378.
27. RAY, J.H. GERMAN, J.: The cytogenetics of the "chromosome-breakage syndromes." (This volume.)
28. RIEGER, R., MICHAELIS, A., GREEN, M.M.: "Glossary of Genetics and Cytogenetics" Fourth Edition. Springer-Verlag, New York, 1976, pp. 98-102.
29. SHADWELL, A.: *In:* Encyclopaedia Britannica, 11th Edition, 1910-11, Vol. 5, p. 176.
30. WEISS, R., TEICH, N., VARMUS, H., COFFIN, J.: "RNA Tumor Viruses" Cold Spring Harbor Symp. Quant. Biol., Vol. 47, 1983, Cold Spring Harbor Laboratory, Cold Spring Harbor, N.Y., 1396 pages.
31. WHITEHOUSE, H.L.K.: Chromosome integration of viral DNA: The open replicon hypothesis of carcinogenesis. *In:* Chromosomes and Cancer (ed. J. German), John Wiley and Sons, N.Y., 1974, pp. 41-76.
32. WOLF, U.: Theodor Boveri and his book "On the Problem of the Origin of Malignant Tumors." *In:* Chromosomes and Cancer (ed. J. German), John Wiley and Sons, N.Y., 1974, pp. 3-20.

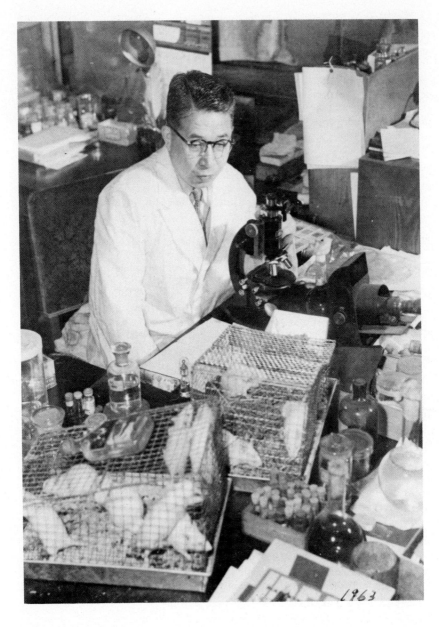

Professor Sajiro Makino engaged in cytogenetic studies of rodent neoplasia. Sapporo, 1963.

Editor's Note: *In each volume of* Chromosomes *an outstanding geneticist will be honored and invited to compose an autobiographical sketch. It is with pleasure that I select Professor Sajiro Makino for this second volume in the series. (The section headings below were editorial additions to Professor Makino's manuscript.)*—J.G.

MY LIFE IN CYTOLOGY

The following is a brief account of my career in cytological research, which began in 1930.

I was born, a fraternal twin, in Chiba prefecture near Tokyo in 1906. At the age of 12, when I was a student in middle school, I became curious about twin births and asked my natural-science teacher how they arose. He explained, in some detail, how multiple ovulation and fertilization led to twin births. His explanation aroused my interest in biology and undoubtedly was a contributing factor in my choice of biology as a career.

Early Influences Leading to a Career in Cytogenetics

I attended Hokkaido Imperial University as a student in the elementary course, and after completing it I was admitted to the Faculty of Agriculture at the same institution. My objective upon entering the Hokkaido Imperial University had been to learn chicken breeding so I could develop improved breeds of hens in Japan. Dr. Jinshin Yamane, a professor of animal breeding, informed me that I would need a basic knowledge of genetics in order to achieve my goal, and he introduced me to Professor Kan Oguma, a cytology professor, so that I might acquire that knowledge. This introduction, in fact, destined me to study animal cytogenetics as my life's work.

It was under the guidance of Professor Oguma (1885–1971) that I first studied chromosomes, initially those of insects, urodelans, and anurans. In 1930 I was graduated from the Department of Biology, Faculty of Agriculture. In the same year the Faculty of Science was opened at the Hokkaido Imperial University, and Professor Oguma was appointed to head the Department of Zoology. I joined the department as Professor Oguma's research associate. In 1935 I was promoted to Assistant Professor, and in 1940 obtained the Degree of Science (comparable to the Ph.D. degree in the United States). For my doctoral research, I chose to study mouse cytogenetics. The study included the comparative morphology of *Mus musculus, M. molossinus,* and *M. caroli* chromosomes as well as the maturation and fertilization of *M. musculus* eggs [1]. In 1947 I was appointed to the Chair of Zoology. I taught zoology and animal cytogenetics, along the way encouraging students and colleagues to engage in collaborative research with me in the area of animal cytogenetics.[1] Prior to my retirement in 1970, I established a new laboratory, the Chromosome Research Unit, which was independent from the Zoological Institute of the Faculty of Science. I also set up the Makino Library in the unit and donated some 270,000 scientific reprints that I had collected since 1928, along with numerous Japanese and foreign journals.

The Many Species Studied...

Beginning with *Bufo* chromosomes in 1930, over the years I have studied chromosomes from a wide variety of animal species. Fixed and sectioned testicular tissues were used in all the work. At first, the study of insect chromosomes predominated; various species of *Orthoptera, Lepidoptera, Odonata, Neuroptera,* and *Coleoptera* interested me deeply. In 1932 my interest turned to the chromosomes of vertebrates; I studied the chromosomes of various species of sharks, rays, goldfish, carp, salamanders, frogs, lizards, snakes, turtles, and birds. After 1940 I investigated the chromosomes of numerous mammalian species, including those of rodents such as squirrels, rats, and laboratory and field mice and ungulates such as horses, asses, cattle, swine, buffalo, reindeer, sheep, and goats. These data, along with those from other laboratories, pertaining to some 3,300 species, were compiled in *An Atlas of Chromosome Numbers in Animals,* which I published in 1951 [3]. I revised the Atlas in 1956, to incorporate data on a total of 4,850 species of animals, including man [5].

---

[1]Students who have worked under my direction include the following: H. Niiyama (1934); H. Hirai (Kichijo) (1934); E. Momma (1941); Y. Ojima (1942); S. Nogusa (1943); Y. Sun-Kang (1943); T.H. Yosida (1944); Y. Takenouchi (1946); J. Kitada (1948); K. Kano-Tanaka (1950); N. Nakamura (1950); K. Kawamura (1951); T. Tanaka (1951); K. Saitoh (1951); H. Nakahara (1951); K. Maeki (1952); T. Ishihara (1953); A. Tonomura (1953); T. Okada (1953); Y. Nakanishi (1953); Y. Ohnuki (1954); S. Koiwai (1954); Y. Hisada (1954); M. Sasaki (1954); H. Hori (1955); A.A. Awa (1956); T. Okigaki (1956); Y. Tonomura (1956); H. Hayashi (1956); H. Okumura (1956); S. Takayama (1956); S. Muramatsu (1956); T. Matsumoto (1956); K. Utsumi (1957); J. Kobayashi (1957); Y. Matano (1957); N. Inui (1958); Y. Kikuchi (1958); H. Oishi (1959); T. Seto (1960); T. Takayanagi (1960); H. Kato (1960); M.S. Sasaki (1961); K. Yamada (1961); Y. Toyakuni (1961); M. Mizutani (1961); N. Takagi (1962); K. Yamamoto (1962); M. Yoshida (1962); T. Sofuni (1963); T. Makita (1964); J. Muramoto (1964); H. Shimba (1965); R. Shoji (1965); T. Aya (1966); T. Ikeuchi (1966); S. Matsui (1966); Y. Obara (1967); S. Kono (1969); M. Okada (1970); I. Hayata (1970); S. Sonta (1970); M. Itoh (1970).

...With Some Serendipity Along the Way

During and after World War II, Japan faced serious financial difficulties. Chemical agents that were important for cytological studies, particularly osmic acid, were unavailable. I attempted to find other fixatives that could serve as suitable substitutes, but did not succeed. One day in the summer of 1948, during my search for a good fixative, I fortuitously discovered a method for obtaining well-spread chromosome preparations. I left the grasshopper testicular specimens that I was using in my search in a wet basin while I went out to buy some food for lunch from the farmers. Upon returning to the laboratory a half-hour later I took the testicular specimens and squashed them on the slides. On examining the slides I found to my surprise a beautiful metaphase figure that had well-spread, non-overlapping chromosomes. This prompted me to test the water pretreatment method on testicular specimens from more than 50 species of animals, including both invertebrates and vertebrates, over a two-year period [10]. In almost all cases the water pretreatment method proved satisfactory for the *in situ* study of chromosomes. This method probably led to the development of the hypotonic treatment of tissue culture material that is conventionally applied at present.

Another fortuitous discovery was the observation that *Drosophila virilis* somatic chromosomes could be induced to pair artificially [2]. Esophageal ganglia of third instar larvae grown at 24°–25° C contain many neurocytes in mitosis. Unintentionally, I left some vials containing such larvae at 5° C for 2–5 days, so that cell division in the ganglion cells was retarded. The vials then were returned to room temperature for a few hours before esophageal ganglion slide preparations were made. Microscopic examination of the slides revealed unusual metaphase figures in the ganglion cells: rather than the usual 12 (2n) chromosomes, there were 6 (n); the chromosomes had a remarkable appearance, like ring-shaped bivalents. Suppression of mitosis in larval ganglion cells by exposure to low temperature apparently had permitted homologous chromosome pairs in the non-mitotic nucleus to become unusually tightly associated. This implies that mitosis, when it is retarded, is interconvertible with meiosis. Some plant cytological studies of which I am aware also have indicated that meiotic activity is convertible to mitotic activity when cell activity is accelerated. Although the true mechanism of homologous chromosome pairing in meiosis remains to be elucidated, both the plant studies and those with *D. virilis* suggest that a lengthening of the cell cycle may, in part, be responsible.

Cancer Cytogenetics

In 1951 I became interested in the cytogenetics of cancer, particularly of the ascites tumors of rodents. At first I worked with the Yoshida ascites sarcoma of the rat. Later, several tumors produced experimentally by azo-dye treatment (e.g., the MTK-sarcoma) also were studied. Mitoses and chromosome behavior were analyzed in living ascites tumor cells with the aid of phase contrast microscopy, under both normal and experimental conditions. It was on the basis of these experimental observations and data that I proposed the theory of cancer stem lines as principal contributors to the growth and differentiation of tumors [4,6]. The cytological studies of ascites tumor cells, mostly of rats, all conducted in collaboration with my colleagues, were reported between 1951 and 1961 in a series of approximately 50 papers.

After 1959 my cytogenetic studies were extended to include canine and human tumors. Articles published in 1959 and 1964 detailed the chromosomes of more than 80 primary tumors that had arisen in various human organs [11]. Cytogenetic data from the study of moles showed that chorioepithelioma originated from chorionic villi and evolved through hydatidiform and destructive moles. I interpreted the observed chromosomal changes to indicate that normal villi, hydatidiform moles, destructive moles, and chorioepithelioma represent successive stages in malignant progression [12].

The effects of viruses on chromosomes interested me also. This was approached by studying chromosomes from persons with viral infections.

In 1961 our cytogenetic study of canine venereal tumors showed that dogs from widely separated localities in Japan had tumors with identically abnormal karyotypes. Then, similar findings were reported by cytogeneticists from other geographic areas, including the United States, France, and Jamaica. Throughout the world this tumor is characterized by the same drastically rearranged karyotype [7].

Human Cytogenetics

Since 1957 I have worked mainly with human chromosomes. We have investigated the chromosome complements from both normal subjects and those with any of a wide variety of congenital disorders and have published many articles on karyotype-phenotype relationships in persons with developmental defects. The major data were collected in a book, *Human Chromosomes*, which was published in 1975 [8]. In that book I reviewed and critically discussed both our cytogenetic data from persons with developmental defects and data obtained by other investigators.

While the final part of *Human Chromosomes* was being prepared, new staining techniques—chromosome banding—were developed, which enabled chromosomes to be identified easily and accurately. These techniques also made it possible to easily characterize the structurally altered chromosomes that occur in association with some congenital anomalies and certain pathological states. In 1979 I compiled the chromosomal data obtained using the new banding techniques in a book entitled *Chromosomes* [9].

Travel, New Interests, Approbation—and More Chromosome Studies

My career has provided me with numerous opportunities to travel. I have made several trips to the United States and also have been to Turkey, India, Switzerland, Italy, Australia, and England. During these travels I met with many cytogeneticists who influenced me greatly in my cytogenetic work. The most notable of these was my association with Dr. T.C. Hsu, in the winter of 1953 when he was in the Department of Anatomy of the University of Texas at Galveston; it was there that I had the opportunity to meet Dr. C.M. Pomerat, a man who became my life-long friend. In the Galveston laboratory I was introduced to the use of chicken embryo extract in tissue culture medium for the investigation of mammalian chromosomes, and this became the basis for the development of the study of human chromosomes in Japan.

In 1932 I happened to meet Dr. J.F. McClendon, a geneticist from the United States who was invited to our University to give a special lecture on physiological genetics. Dr. McClendon had brought with him two pairs of pure albino mice of a Carnegie Institute strain. He explained to me the importance of inbreeding animals for use in biological and medical experiments and told me that inbreeding should be possible merely by means of sister-brother matings. Before World War II, most Japanese scientists engaged in medical research paid little attention to the advantage of using pure strains of animals in their experiments. After the war, foreign scientists, primarily from the United States, stressed the importance and significance of using genetically pure-bred animals in experiments. To meet the need for such animals, I founded a small department in 1943 for this purpose where I, with the aid of my students and colleagues, began to inbreed mice and rats by the sister-brother mating system. On a trip to the United States in 1952 and 1953 I visited the Wistar

Institute (Philadelphia), the National Cancer Institute (Bethesda), the Sloan-Kettering Cancer Institute (New York), the City of Hope Medical Center (California), and other research institutes and was able to collect approximately 15 inbred strains of mice and rats. On my return to Japan, I brought them with me and bred them by means of the sister-brother mating system. At the present time we maintain 13 strains of inbred rats and 20 strains of mice, all of which are internationally registered. These strains are the principal source of inbred mice and rats in Japan, and they are supplied to many laboratories on request.

I have reviewed what I have accomplished in my chosen field of research—the study of animal chromosomes. Throughout my career my studies have been possible mainly because I enjoyed the collaboration of colleagues and co-workers who have assisted me scientifically in various ways[2]. I am happy that these scientific achievements have been recognized. Several organizations and scientific societies have presented me with awards[3], and I was appointed a member of the Japan Academy in 1972. Now, in emeritus status, I choose not to abandon research on chromosomes, and I shall continue to pursue those studies in the future. It seems natural for me to be engaged in chromosome cytology.

Sajiro Makino, Professor Emeritus
Chromosome Research Unit, Faculty of Science
Hokkaido University, Sapporo
7 December 1981

---

[2]Collaborators associated with me: K. Oguma; J.J. Asana; I. Nishimura; Y. Matsui; S. Ohno; T. Kajii; M. Hikita; Y. Kuroki; H. Kobayashi; T. Fukuschima; T. Kadotani; K. Oikawa; T. Ishikawa; K. Ohama; H. Takahara; D.A. Hungerford; Y. Yamashina.

[3]Included among the awards are the following: 1944—prize from the Genetics Society of Japan for cytological studies of the mouse; 1953—prize from Hokkaido Press for basic research on cancer; 1954—prize from Zoological Society of Japan for karyological studies of tumors; 1958—prize from the Japan Academy for chromosome studies in animals; 1962—prize from Toyo Rayon Company for chromosome studies in mammals and man; 1963—award at the Second International Conference on Congenital Malformations, New York; 1967—prize from the Japan Society of Human Genetics for chromosome studies of normal and congenital disorders of man; 1977—prize from the Princess Takamatsu Cancer Research Foundation for cytogenetic studies of cancer; 1979—prize from the Japanese Society of Fertility and Sterility for advances in clinical cytology.

LITERATURE CITED

1. MAKINO, S. Studies of the murine chromosomes. I. Cytological investigations of mice, included in the genus *Mus.. J. Fac. Sci. Hokkaido* Ser. VI, 7:305–380, 1941.
2. MAKINO, S. Artificial induction of meiotic chromosome pairing in the somatic cell of *Drosophila virilis. Cytologia* 12:179–186, 1942.
3. MAKINO, S. An Atlas of the Chromosome Numbers in Animals. Ames, Iowa: Iowa State College Press, 1951, pp. 1–290.
4. MAKINO, S. A cytological study of the Yoshida sarcoma, an ascites tumor of white rats. *Chromosoma* 4:649–674, 1952.
5. MAKINO, S. A Review of the Chromosome Numbers in Animals. Tokyo: Hokuryukan, 1956.
6. MAKINO, S. The chromosome cytology of the ascites tumors in rats, with special reference to the concept of the stemline cell. *Int. Rev. Cytol.* 6:25–84, 1957.
7. MAKINO, S. Cytogenetics of canine venereal tumors: Worldwide distribution and a common karyotype. In *Chromosomes and Cancer* (J. German, ed.), New York:John Wiley and Sons, 335–372.
8. MAKINO, S. *Human Chromosomes*. Tokyo: Igaku Shoin, 1975.
9. MAKINO, S. *Chromosomes*. (In Japanese) Tokyo: Igaku Shoin, 1979.
10. MAKINO, S., NISHIMIRA, I. Water-pretreatment squash technique: A new and simple method for the chromosome study of animals. *Stain Technol.* 27:1–7, 1952.
11. MAKINO, S., SASAKI, M.S., TONOMURA, A. Cytological studies of tumors, XL. Chromosome studies in fifty-two human tumors. *J. Natl. Cancer Institute* 32:741–777, 1964.
12. MAKINO, S., SASAKI, M.S., FUKUSCHIMA, T. Cytological studies of tumors. XLI. Chromosomal instability in human chorionic lesions. *Okajimas Fol. Anat. Jap.* 40:439–465, 1965.

I. GENETICALLY DETERMINED HUMAN DISORDERS THAT FEATURE CHROMOSOME INSTABILITY, HYPERSENSITIVITY TO ENVIRONMENTAL CARCINOGENIC AGENTS, AND CANCER PRONENESS

A. The Clinical Disorders

STUDY OF FAMILIAL DEFECTS AS A WAY OF INVESTIGATING THE ORIGIN OF HUMAN CANCERS

JOHN CAIRNS

INTRODUCTION

In the last 30 years a great change has come about in the ambitions and expectations of biologists. Before the 1950s biology was in many respects a pessimistic discipline; for example, I can remember that even though the chemistry of intermediary metabolism was thought to be potentially understandable in its entirety, the underlying strategy of biological systems was judged by all but a few to be essentially unknowable. This changed with the coming of molecular biology. Rightly or wrongly, everything now appears to be potentially within our grasp and we expect, sooner or later, to come up with all the answers. This newly found confidence extends into cancer research. We now really expect to be able to identify the fundamental defect or defects of the cancer cell, and we are confident that this knowledge will eventually lead to a solution to the cancer problem. Even though most cancer research quite properly is conducted at a rather esoteric level, its ultimate objective is a practical one: to lighten the human burden. In this chapter, therefore, I shall outline what I believe is the right way to view the interaction and impact of epidemiology and laboratory research and the way they bear upon the cancer problem.

By now it is abundantly clear that the incidence of the common human cancers is determined by various controllable, external factors. This is surely the single most important fact to come out of all cancer research; for, it means that cancer is a preventable disease. Unfortunately, there has been a strange reluctance on the part of most people to accept the consequences of this discovery; it seems that it is more exciting to try to produce something like interferon, which, if it works as expected, should save a few thousand lives each year in the U.S., than it is to battle against cigarette smoking—a battle which, in the end, could save more than 100,000 lives every year. So cancer research has come to be divided into two separate encampments. On one side, the laboratory worker is concerned with such things as the chemistry and biology of carcinogenesis, assays of carcinogenicity and mutagenicity, and the search for chemotherapeutic agents; these are activities that can be pursued with little or no attention to the human condition. On the other side, the epidemiologist seeks out the actual causes of human cancer. Thus far, this is something he has been able to do without having to consider the mechanism of carcinogenesis. For example, the two most common cancers in the Western world are those of the skin and the lung, and each happens to be so strongly dependent on a single factor (sunlight and smoking) that its cause could be identified without the need for any understanding of the biological basis of carcinogenesis. (This was also true for most of the known industrial carcinogens.)

So far, the epidemiologist has identified causative factors for about a third of the lethal cancers that occur in such countries as the United States [7]. Because at least 80% of cancers are thought to be preventable, this leaves at least 50% for which no precise cause has been found. Many of these—in particular, cancer of the stomach, large intestine, and breast—seem to be determined partly by features of diet, although it has not yet been possible to pin down exactly which components of the diet are responsible. The prime candidates are an excess of animal fats, a deficiency of fiber, and a shortage of certain vitamins. But it is probably going to be difficult to make a choice among these factors, both because they are not really independent variables and because it is rather difficult to monitor what each person is eating from one year to the next (much harder, for example, than to determine whether someone smokes or is exposed to asbestos).

Some of these questions about the role of diet might disappear if we had a precise understanding of the process of carcinogenesis. For example, if the total nucleotide sequences for several of these diet-associated cancers were known, and if it had been shown that the cells of the common human cancers characteristically contained a number of frame-shift mutations that were producing gross alterations in gene expression, then we could be fairly confident that we should be looking for frame-shift mutagens in the appropriate tissues and tracing these back to their source in the diet. However, it is going to be some time before such definite information about the DNA sequences and messenger-RNA composition of the human cancers becomes available, and until then, our choice of what to look for has to depend on circumstantial evidence.

This chapter deals with some of the pieces of circumstantial evidence. In it, I wish to make two points about the interaction between laboratory research and epidemiology. The first concerns the proper interpretation of what is known about experimental carcinogenesis in animals. The second concerns the study of people with certain genetic defects in order to learn more about the nature of the principal carcinogenic hazards of the human condition.

THE PRODUCTION OF CANCERS IN ANIMALS

Although different species often show marked differences in their susceptibility to particular carcinogens and each has its own particular spectrum of "spontaneous" tumors, it seems reasonable to assume that some animal cancers will be found that are close facsimiles for each of the major human cancers. In other words, we ought to be able to learn something about the causes of human cancer by studying carcinogenesis in animals. To an outsider, however, perhaps the most conspicuous feature of all the animal experiments is the wide variety of treatments and agents that are known to affect (positively or negatively) the incidence of cancer. We are faced with an embarrassment of riches, and our problem is going to be one of choice.

The most widely studied method for producing cancer in animals is exposure to repeated doses of a physical or chemical agent that is known to damage DNA and cause mutations. In general, the carcinogenic potency of such agents is correlated rather closely with their mutagenicity [1,2,15]. Furthermore, in some instances it has been possible to establish that the cancers are indeed arising as the consequence of DNA damage. For example, pyrimidine dimers in DNA must be the direct or indirect cause of certain cancers produced by ultraviolet light, because the cancers can be prevented by photoreversal with visible light (a reaction that is believed to be specific for such dimers) [11]. But the exact interpretation of most forms of chemical carcinogenesis has been complicated by the observation that these cancers often appear to be the result of a sequence of steps, some of which are being catalyzed by agents ("promoters") that are not themselves mutagenic [3,9]. Nevertheless, the marked carcinogenicity of nearly all mutagens has led, quite illogically, to the belief that most human cancers will prove to originate in the kinds of local change in base sequence that are being monitored in the various tests for mutagenicity.

In fact, there are many other ways of producing cancers in animals that apparently are not dependent on localized changes in the coding sequences of genes. Some of these, for example the production of the reversible change of embryonic cells into teratocarcinomas by appropriate transplantation [4,17], are rather bizarre and so are probably not good models for any of the common human cancers. But other methods are not so exotic. For example, about a quarter of all the known viruses of vertebrates have been shown to be capable of causing cancer in some animal [8]. Many of these viruses probably are carcinogenic because they lead to novel juxtapositions of genes from the virus and the host, and it is perhaps for this reason that they tend to cause particular kinds of cancers (lymphomas, leukemias, or sarcomas) rather than to act as nonspecific carcinogens. Certainly there is no evidence that they function as general mutagens. But the most conspicuous (and neglected) group of cancers, that are due to a treatment which is not obviously mutagenic, are those that arise from overfeeding—i.e., the cancers that can be prevented by quite minor reductions in food intake or (what is perhaps equivalent) by chronic infection with certain parasites [6,12,18]. These effects of nutrition on the incidence of several kinds of animal cancer are very similar to some of the correlations between diet and cancer observed by the epidemiologists—correlations that involve most of the major forms of cancer apart from those due to smoking.

The matter has been recently put very clearly by Doll and Peto [7]:

> Possibly, misleading conclusions are being drawn from overemphasis on the spectrum of chemicals found active in mutagenicity tests and in chronic carcinogenicity studies in rodents. For example, the authors of official guidelines on how to do long-term tests usually emphasize the importance of concurrent controls and the need for strictly identical diet, handling, heat, light, stress and infection in the treated and control animals. Why? Can minor details of the lifestyle of the animals really be important determinants of the animals' "spontaneous" tumor yields? And, if so, might not the same also be true for humans? In experimental animals, quite minor details of the total quantity of food and of the vitamins, fats and carbohydrates in the food can certainly be enormously important determinants of spontaneous tumor yields. But, because the mechanisms are not understood, short-term tests for the possible relevance to humans of these processes do not exist, and these phenomena are typically viewed as a potential nuisance to serious investigators who want to study the carcinogenicity of some trace environmental contaminant rather than as being themselves a potentially fruitful area of inquiry.

Once we accept the idea that cancer can be produced by a wide variety of factors, which seem to have little or nothing in common, somehow we have to decide which of these factors are most likely to be the rate-limiting variables that determine the incidence of the major human cancers. Certainly, in the absence of information about the underlying molecular biology of most human and experimental cancers, it is hard to see how the choice could be made solely on the basis of further experiments. What we need is additional information about human cancers that could help us select the right experimental model. And this, I believe, is where the effects on human cancer incidence of certain genetic defects will prove to be of crucial importance.

THE GENETICS OF SUSCEPTIBILITY TO CANCER

Certain inherited diseases are known to affect the incidence of certain kinds of cancer. By observing which abnormalities influence which cancers and, perhaps more importantly, which abnormalities are without an effect, we might be able to determine the classes of events that tend to be rate-limiting for the common cancers.

For example, inherited defects affecting the excision-repair pathway that handles all bulky DNA adducts (in bacteria, this is called the *uvr* pathway) lead to the disease xeroderma pigmentosum (XP). Patients with this disease show extreme skin sensitivity to ultraviolet light and have a high incidence of skin cancer. Although none of the tissues appear to be capable of excising bulky DNA adducts and thus are very sensitive to the lethal and mutagenic effects of most mutagens, these patients apparently do not have an abnormally high rate of death from the common internal cancers [5]. This suggests that the common fatal cancers are not caused by the kind of lesions of DNA that XP patients (and bacterial *uvr* mutants) are specifically unable to repair. This conclusion is potentially so important, and so much at variance with the conventional wisdom, that a great effort should now be made to collect more data on XP patients. It should not be too difficult to set up a systematic study of the fate of patients whose disease was diagnosed and reported many years ago. Such a study should establish whether the production of bulky lesions in our DNA is a rate-limiting step in the formation of most of the common cancers and, therefore, whether protection from exposure to most mutagens is likely to have an appreciable effect on cancer incidence.

Some information about certain of the rate-limiting steps in the production of the common human cancers comes from the study of Bloom's syndrome. Patients with this rare disease show a high frequency of chromosomal aberrations and exchanges [10] and an increased incidence of a wide variety of cancers. Although, strictly speaking, we do not know that these cancers arise by the same mechanisms and are of exactly the same kind as the cancers of normal subjects, it seems reasonable to suppose that the defect in Bloom's syndrome is raising the frequency of certain steps in "normal" carcinogenesis. If so, these steps are more likely to be the kind of large-scale genetic rearrangements seen in Bloom's syndrome than local changes of one or two bases, (such as those that presumably occur in XP). This conclusion fits well with the observation that many cancers show characteristic changes in karyotype (usually particular translocations).

A more precise picture of the causes of the common human cancers might be obtained if we could identify several different congenital defects that had a wide-spread effect on cancer incidence. Apart from Bloom's syndrome, the other obvious candidate is ataxia-telangiectasia (AT). This condition, like Bloom's syndrome, shows evidence of chromosomal instability and is associated with a rather general increase in cancer incidence; but, as in the case of XP, the published clinical reports of AT are not in a suitable form to yield sound estimates of age-specific death rates; here, too, a proper study should be undertaken.

In our search for congenital defects that illuminate the common steps in carcinogenesis, we cannot expect to get much information from the most severe congenital diseases, e.g., Fanconi's anemia, which seldom permit survival into adult life. But we do not have to confine our attention to diseases affecting DNA metabolism and chromosomal mechanics, or, if it comes to that, to diseases that are already known to affect cancer incidence. For example, the hypothesis has been gaining ground recently that one of the rate-limiting promoting steps in carcinogenesis is related to lipid oxidation in cell membranes. If that hypothesis is correct, we should expect to see an increased incidence of cancer in people with certain congenital defects in porphyrin metabolism because these patients show a characteristic photosensitivity due to a porphyrin-mediated oxidative reaction affecting cell surfaces (in particular, the surface of erythrocytes). But, as far as I know, no such epidemiological investigation has ever been proposed.

The other inherited conditions that could prove highly informative are those that affect the metabolism of certain groups of potential carcinogens. Mutations that block the induction of aryl hydrocarbon hydroxylase make mice more sensitive to the short-term toxicity and less sensitive to the long-term carcinogenicity of certain polycyclic aromatic hydrocarbons [13]. If we could show that similar mutations in humans lower the incidence of any of the common cancers, this finding would suggest that these compounds are important in human carcinogenesis. So far, however, such studies in human populations have been inconclusive.

I have deliberately excluded the familial conditions that are associated with a high incidence of particular cancers because these do not necessarily provide any information about the common cancers. For example, the high incidence of reticulum cell sarcomas in people with inherited or acquired immunodeficiency may tell us something about the origin of "spontaneous" reticulum cell sarcomas, but it has no obvious bearing on the origin of the major common cancers (and, of course, should never have been taken as evidence for the operation of a generalized system for immune surveillance).

Human populations are probably rather heterogeneous in susceptibility to each of the common cancers. This is certainly true for breast cancer [14] and may well be true for most other varieties [16]. Study of the basis for such variation could conceivably throw light on the mechanisms of carcinogenesis for each of the major cancers, which in turn could help epidemiologists identify the particular and potentially preventable causes of these cancers.

SUMMARY

The different kinds of human cancer vary in frequency from one population to another, which suggests that there are rather few factors in our environment and lifestyle that are determining the incidence of each variety of cancer. In contrast, cancers can be produced in experimental animals by a wide variety of agents which, in many cases, seem to have little in common and, therefore, presumably operate by different mechanisms. We must find some way of choosing which of these experimental systems is the best model for the common human cancers. Until that is known, it may be difficult to determine what we can do to prevent the occurrence of most forms of cancer.

Short of some method for distinguishing the cancers that arise from each possible class of carcinogenic sequence, the easiest way to identify what class or classes of agent tend to be rate-limiting for the common human cancers may be to determine which familial abnormalities raise (or lower) the incidence of cancer. Once we understand the nature of these defects, we might then be able to deduce the usual mechanism for human carcinogenesis. For example, the fact that XP does not appear to affect the frequency of the common lethal cancers suggests that these cancers are not due to mutagenesis by any of the agents that give rise to large DNA adducts.

For this reason, there is a pressing need for proper long-term follow-up studies of patients with such diseases as XP, ataxia-telangiectasia, Bloom's syndrome, and certain of the congenital porphyrias. In effect, these patients are human "tester" strains—and their experience may be the most accessible source of information about the causes of human cancer.

LITERATURE CITED

1. AMES, B.N., HOOPER, K.: Does carcinogenic potency correlate with mutagenic potency in the Ames assay? *Nature* 274:19-20, 1978.
2. BARTSCH, H., MALAVEILLE, C., CAMUS A.M., MARTEL-PLANCHE, G., BRUN, G., HAUTEFEUILLE, A., SABADIE, N., BARBIN, A., KUROKI, T., DREVON, C., PICCOLI, C., MONTESANO, R.: Bacterial and mammalian mutagenicity tests: Validation and comparative studies on 180 chemicals. *In* Molecular and Cellular Aspects of Carcinogen Screening Tests (Montesano, R., Bartsch, H., Tomatis, L., eds.), Lyons, France: IARC, 179-241, 1980.
3. BERENBLUM, L.: Carcinogenesis as a Biological Problem. New York: American Elsevier, 1974.
4. BRINSTER, R.L.: The effect of cells transferred into the mouse blastocyst on subsequent development. *J. Exp. Med.* 140:1049-1056, 1974.
5. CAIRNS, J.: The origin of human cancers. *Nature* 289:353-357, 1981.
6. CARROLL, K.K.: Experimental evidence of dietary factors and hormone-dependent cancers. *Cancer Res.* 35:3374-3383, 1975.
7. DOLL, R., PETO, R.: Quantitative estimates of avoidable risks of cancer in America today. *J. Natl. Cancer Inst.* 66:1197-1312, 1981.
8. FENNER, F., McAUSLAN, B.R., MIMS, C.A., SAMBROOK, J., WHITE, D.O.: The Biology of Animal Viruses. New York: Academic Press, 1974.
9. FRIEDEWALD, W.F., ROUS, P.: The initiating and promoting elements in tumor production. *J. Exp. Med.* 80:101-144, 1944.
10. GERMAN, J., ARCHIBALD, R., BLOOM, D.: Chromosomal breakage in a rare and probably genetically determined syndrome of man. *Science* 148:506-507, 1965.
11. HART, R.W., SETLOW, R.B., WOODHEAD, A.D.: Evidence that pyrimidine dimers in DNA can give rise to tumors. *Proc. Natl. Acad. Sci. USA* 74:5574-5578, 1977.
12. JOSE, D.G.: Dietary deficiency of protein, aminoacids and total calories on development and growth of cancer. *Nutr. Cancer* 1:58-63, 1979.
13. KOURI, R.E., RATRIE, H., WHITMIRE, C.E.: Evidence of a genetic relationship between susceptibility to 3-methylcholanthrene-induced subcutaneous tumors and inducibility of aryl hydrocarbon hydroxylase. *J. Natl. Cancer Inst.* 51:197-200, 1973.
14. MACKLIN, M.T.: Comparison of the number of breast-cancer deaths observed in relatives of breast-cancer patients, and the number expected on the basis of mortality rates. *J. Natl. Cancer Inst.* 22:927-951, 1959.
15. MESELSON, M., RUSSELL, K.: Comparisons of carcinogenic and mutagenic potency. *In* Origins of Human Cancer (Hiatt, H.H., Watson, J.D., Winsten, J.A., eds.), Cold Spring Harbor, N.Y.: 1473-1481, 1977.
16. PETO, J.: Genetic predisposition to cancer. *In* Cancer Incidence in Defined Populations. Cold Spring Harbor, N.Y.: Cold Spring Harbor Lab, Banbury Report 4, 203-213, 1980.
17. STEVENS, L.C.: The development of transplantable teratocarcinomas from intratesticular grafts of pre- and postimplantation mouse embryos. *Dev. Biol.* 21:364-382, 1970.
18. TANNENBAUM, A., SILVERSTONE, H.: Nutrition in relation to cancer. *Adv. Cancer Res.* 1:451-501, 1953.

BLOOM'S SYNDROME

EBERHARD PASSARGE

Bloom's syndrome (BS) represents the clearest example known of a hereditary disorder that features chromosome instability and an increased risk of malignant tumor formation. These features, along with its pre- and postnatal growth deficiency and variable immune deficiency, make this disorder a model for the study of growth, carcinogenesis, mechanisms of chromosome exchange, and immune function. My survey here is derived from the accumulated clinical data of the 98 persons with BS that have been recognized over the past 20 years in different parts of the world. These 98 persons comprise the Bloom's Syndrome Registry [12,13,16,17].

DISCOVERY OF THE SYNDROME

In 1941, the pediatrics department of New York University referred a 9-year-old boy to Dr. David Bloom in the dermatology clinic at the same institution, as a possible case of lupus erythematosus. The boy was abnormally small and had a telangiectatic erythema of the face which resembled lupus. It seemed to Dr. Bloom that the association of the skin lesion and the stunted growth did not permit a definite diagnosis, and the child's mother refused to have a biopsy made to confirm the possibility of lupus. He decided to follow the child's clinical course and arranged for him to make regular visits to the dermatology clinic.

Twelve years later, in 1953, two patients were presented at the New York Academy of Medicine with features very similar to those of the patient Dr. Bloom had observed since 1941: In March a boy, from the dermatology department of Columbia University College of Physicians and

Surgeons; in October a girl, from The New York Hospital. It then became evident to Dr. Bloom that the skin eruption and stunted growth represented a definite but previously unrecognized clinical entity. As a name for the new entity, he suggested "congenital telangiectatic erythema resembling lupus erythematosus in dwarfs" [6]. In 1954, he published a report describing the new disorder [5] (in the Bloom's Syndrome Registry those first three patients are identified as 1[GeSo], 2[SuBu], and 3[HoCo]) [5,12,13,16], and in 1966, a second report [7]. The condition now is referred to as Bloom's syndrome (although for reasons discussed elsewhere [22] I suggest that "Bloom's disease" is a more appropriate name).

CLINICAL FEATURES

Small stature resulting from severe pre- and postnatal growth deficiency, dolichocephaly (a long narrow head), a narrow and characteristic face, and sun-sensitive telangiectatic erythema of the face are characteristic features of the syndrome, a phenotype that can easily be recognized in most persons with BS [12,13,14].

Facial Appearance

In affected children, a narrow face is almost invariably present (Fig. 1); hypoplasia of the malar areas, a prominent nose, and a small mandible are characteristic. In adults, the characteristic facial appearance becomes less impressive.

Skin Lesion

In most patients, a telangiectatic erythema appears during the first or second summer of life, after exposure to sun. Its characteristic distribution over the upper portions of the cheeks, often in the butterfly configuration, is responsible for the erroneous diagnosis of lupus erythematosus with stunted growth that was made in the first patients (as mentioned above). The telangiectatic erythema may extend to the upper part of the ears, especially between the helix and antihelix. Frequently the eyelids, lips, and dorsa of the hands and forearms are involved. The legs and trunk are not affected in spite of exposure to sunlight. In some patients the telangiectatic erythema is severe and becomes inflamed, whereas in at least one and possibly two patients, skin lesions have never been observed. The severity of the skin lesion varies considerably from patient to patient, but for a given patient it tends to remain unchanged for several years. The skin lesion is worse in the summer than in the winter, and during the second and third summers of life it often is worse than it was during the first. With increasing age, especially after the ages of 6–9 years, the skin lesion tends to fade and may gradually disappear. In

Fig. 1. A boy with Bloom's syndrome – 30(MaKa) in the Bloom's Syndrome Registry – at (a) 10 years and (b) 16 years of age.

general, the skin lesion is more prominent in males than in females. Protection from the sun and the use of ultraviolet-absorbing creams are helpful measures in reducing the severity of the skin lesion.

Growth Deficiency

The average birth weight after a normal 40-wk gestation period is 1,960–2,020 gm for males and 1,880 gm for females. No prenatal growth data are available, but it is assumed that BS fetal growth lags behind normal at all stages of pregnancy. Birth length is reduced to about 40–45 cm. Growth is below the third percentile throughout childhood, but the rate of growth is comparable to normal (Fig. 2); however, a growth spurt before puberty is lacking or minimal in BS individuals. The average adult height is about 150 cm for males and 145 cm for females. Body proportions are normal.

Other Clinical Manifestations

Patients with BS are predisposed to respiratory tract infections, including otitis media; gastrointestinal infections are frequent during infancy. These infections are life-threatening and require prompt treatment. They

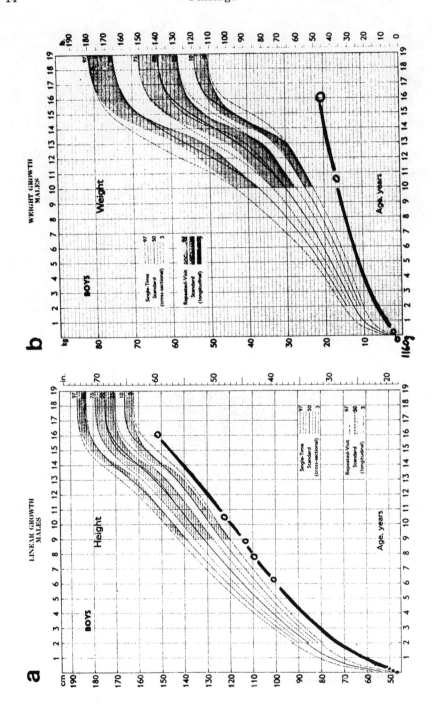

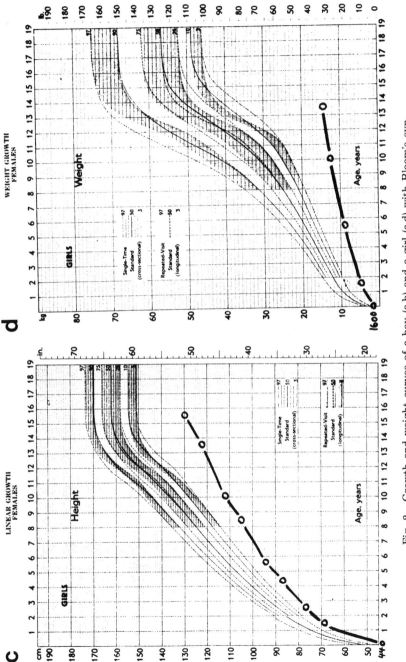

Fig. 2. Growth and weight curves of a boy (a,b) and a girl (c,d) with Bloom's syndrome – entries 30 (MaKa) and 35 (EvNe) in the Bloom's Syndrome Registry.

may be accompanied by a nonspecific and variable derangement of immunoglobulin concentrations in the blood. German [14] has pointed out that the predisposition to infections must have caused a high infant mortality in the past and quite possibly delayed BS's description until antibiotic therapy was introduced into medicine.

Congenital malformations occur in BS with a slightly higher frequency than expected, but no recognizable pattern has emerged. Malformations that have been observed include congenital heart disease, annular pancreas, abnormal thumb, absent toe, and cryptorchidism. Depigmented spots are quite common on the extremities or trunk. Malocclusion of the teeth, missing teeth, high arched palate, nasal septal deviation, kyphoscoliosis, mild syndactyly, clinodactyly, and other minor mal- and deformations have been noted.

Patients may have a high-pitched, squeaky voice with poor speech development. Some consonants are not articulated; e.g., in one patient, the lack of a pronounced k or c (replaced by a t) and other deviations were noted.

Adult males appear to be infertile, probably as a result of testicular hypogonadism. Azoospermia has been observed in the few males who have been tested. In females, menarche occurs at about the expected time or may be delayed a few years. One young woman with BS conceived and carried a pregnancy to term [21].

Mental development usually is normal. However, many patients do less well in school than their unaffected sibs and can be considered mentally less well developed. It is not clear just how small stature and recurrent infectious illnesses contribute to the mild intellectual retardation, but they certainly contribute to some degree.

Many patients with BS display characteristic behavioral patterns. They tend to be restless, unconcentrated, and irritable, but they usually are well integrated into their family. They may try to dominate their younger but taller sibs by coercion and violence. They move in a graceful, age-corresponding manner that contrasts sharply with their small stature.

OCCURRENCE OF CANCER

Because of the high frequency of malignant tumor development [9,15] BS has become a model for studying the relationship between chromosomal breakage and neoplasia. Of the 98 recorded patients, 25 of them are known to have developed at least one malignant tumor. The tumors that have been observed are shown in Table I. Leukemias occur in BS at a higher incidence than in the normal population, whereas other tumors

TABLE I. Occurrence of Cancer in Persons with Bloom's Syndrome

Type of cancer	Age at diagnosis	No. of patients
One primary site		
Leukemia, acute nonlymphocytic	4, 13, 23, 23, 25	5
Leukemia, acute lymphocytic	4, 9, 15, 15	4
Lymphosarcoma, abdominal	4	1
Lymphosarcoma, cervical	12	1
Lymphoma	12, 13, 25	3
Lymphoma, diffuse histiocytic	31	1
Hodgkin's disease	16	1
Carcinoma, squamous cell, tongue	30	1
Carcinoma, adeno, transverse colon	31	1
Wilms tumor	8	1
More than one primary site		
Carcinoma, squamous cell, epiglottis	30	1
Lymphoma	30	
Carcinoma, adeno-, sigmoid colon	39	1
Carcinoma, squamous cell, esophagus	39	
Carcinoma, adeno-, sigmoid colon	37	1
Carcinoma, squamous, gastro-esophageal junction	44	

occur at a much earlier age but not necessarily at an increased rate. Most patients develop tumors by the age of 30. At least three patients have developed more than one primary tumor (Table I).

The choice of a cancer treatment regime in BS, especially for leukemia, must take into account the fact that the bone marrow is extremely sensitive to most antileukemic drugs. Bone marrow failure occurs more readily than usual.

CYTOGENETIC FEATURES

Only the principal cytogenetic features of BS are mentioned here since they are described in detail elsewhere in this volume. They include a spontaneously increased rate of sister-chromatid exchange (SCE) to 60-80 SCEs per 46 chromosomes in lymphocyte [9,10,15] and fibroblast cultures [2], the rate being higher in lymphocytes in short-term cultures than in fibroblasts. Signs of chromosomal breakage and reunion are evident in 5-15% of the metaphases of cultured lymphocytes or fibroblasts. A quadriradial exchange figure between homologous chromosomes present in about 1% of the metaphases [12,13,19,23] is the cytological hallmark of the disease.

DIAGNOSIS

Any individual with pre- and postnatal growth deficiency and a sun-sensitive telangiectatic erythema should be suspected of having BS. An increased rate of SCE and increased chromosomal instability should serve as confirmation of the diagnosis. It should be noted that in some metaphases the rate of SCE is not increased [20]. Chromosomal breaks are evident in about 15% of metaphases.

An absence of growth deficiency, the characteristic facial appearance, or the cytological features makes the BS diagnosis highly unlikely. Facial lesions, on the other hand, may be absent entirely. Thus, minimal diagnostic criteria would be (1) pre- and postnatal growth deficiency, (2) the characteristic facial appearance, and (3) cytogenetic confirmation.

BS probably can be identified in utero. We have had experience with several pregnancies at risk for this disease. By chance, all have resulted in the birth of unaffected children. The best approach for prenatal diagnosis would be (1) precise ultrasonographic fetal size determinations between wk 16 and 24 of pregnancy (the time until a pregnancy may be terminated legally), and (2) determination of the frequency of SCE and chromosomal aberrations in cultured aminotic fluid cells. If these parameters are normal, an unaffected fetus can be predicted quite safely. The opposite has yet to be observed, but we feel that both parameters would be abnormal in an affected fetus.

Whether BS is a disease where antenatal diagnosis and pregnancy termination are indicated is debatable. However, one family that I know that lost a beloved child from BS with a malignancy expressed quite clearly the desirability of antenatal testing, emphasizing that they would rather try a new pregnancy than have another affected child.

MANAGEMENT

No effective therapy for the growth deficiency is available. Nevertheless, patients and their families need sound medical support. This includes early antibiotic therapy for infections, immunoglobulin administration in some cases, dietary measures for poor feeding in infancy and childhood, advice concerning protection of the face from the sun, and vocational support. It must be determined individually to what extent the risk for cancer should be discussed with the family of an affected child. As is often the case with genetic disorders, the more familiar the family is with the disease, the better they are able to adjust to it. Neoplastic lesions occurring anywhere in the body should be considered malignant until proven otherwise.

Prophylactic bone marrow cryopreservation should be considered for later possible use in autologous bone marrow transplantation. In one patient, 32(MiKo), this may have proven to be a useful measure. The patient developed acute lymphocytic leukemia at age 15; after chemotherapy, bone marrow failure developed. The infusion of autologous bone marrow cryopreserved eight years earlier was followed by the return of normal marrow function [J. German, unpublished].

GENETICS

An autosomal recessive mode of inheritance has been demonstrated [12–14,18]; thus, offspring from a marriage between heterozygotes for the mutant gene have a risk of 25% of having BS. It has been observed in different parts of the world, but the gene is quite rare except in Ashkenazi Jewish populations. There, a heterozygote frequency of greater than 1:120 has been determined [22]. Increased parental consanguinity is present in affected non-Jewish families, reflecting the low gene frequency. No evidence for genetic heterogeneity has been forthcoming. Two Jewish/non-Jewish marriages have resulted in the typical disease, which suggests that the same allele was present in both populations. German has discussed the possible reasons for the relatively high frequency among Ashkenazim and explained it by founder effect in fluctuating populations [14].

Heterozygotes cannot be identified by laboratory tests at present.

STUDIES IN VITRO AND SEARCH FOR THE MOLECULAR DEFECT

The molecular basis for BS is unknown. German and Schonberg [19] reviewed the many cytological, biochemical, and immunological studies made of BS cells in vitro. These have included the analysis of chromosome aberrations and SCEs, cell survival after UV-irradiation and exposure to chemicals (e.g., ethyl methanesulfonate and methyl methanesulfonate), DNA repair, DNA-polymerase activities, DNA-fork movement and chain maturation, proliferative response of lymphocytes to mitogens in mixed cultures, and others.

Of particular interest have been studies employing the co-cultivation of BS and normal cells. One such study showed that BS cells can increase the rate of SCE in normal cells. However, in several other studies the opposite effect was observed; i.e., normal cells reduced the rate of SCE in

BS cells by co-cultivation [3,4,25,26,27] and by cell fusion [1,8]. Attention has been drawn to the possible existence of a "Bloom corrective factor" produced by normal cells that are capable of reducing the increased rate of SCE and chromosomal breaks in BS cells [3,24,25]. According to these studies, BS cells lack this "factor." On the other hand, a "breakage factor" has been described by other authors [11]. At present it is unclear whether or how these factors are related and what their significance is. A recent study shows that heterozygous BS cells are less capable than normal cells of reducing the rate of SCE exchange in BS cells during co-cultivation [4].

CONCLUSIONS

BS is the prototype of hereditary disorders associated with chromosomal breakage and neoplasia. It is characterized by a distinct clinical and cellular phenotype, consisting of severe pre- and postnatal growth deficiency, a characteristic facial configuration, an increased risk of neoplasia, and an increased chromosome instability.

Acknowledgments. This investigation was supported in part by research grants from the Deutsche Forschungsgemeinschaft. It is a pleasure to acknowledge the cooperation with Drs. James L. German and David Bloom of New York in studying patients with this disorder during the past 15 years.

LITERATURE CITED

1. ALHADEFF, B., VELIVASAKIS, M., PAGAN-CHARRY, I., WRIGHT, W.C., SINISCALCO, M,: High rate of sister chromatid exchanges of Bloom's syndrome chromosomes is corrected in rodent human somatic cell hybrids. *Cytogenet.Cell Genet.* 27:8–23, 1980.
2. BARTRAM, C.R., KOSKE-WESTPHAL, T., PASSARGE, E.: Chromatid exchanges in ataxia telangiectatisia, Bloom's syndrome, Werner syndrome, and xeroderma pigmentosum. *Ann.Hum.Genet.* 40:79–86, 1976.
3. BARTRAM, C.R., RUDIGER, H.W., PASSARGE, E.: Frequency of sister chromatid exchanges in Bloom syndrome fibroblasts reduced by co-cultivation with normal cells. *Hum.Genet.* 46:331–334, 1979.
4. BARTRAM, C.R., RUDIGER, H.W., SCHMIDT-PREUSS, U., PASSARGE, E.: Functional deficiency of fibroblasts heterozygous for Bloom syndrome as specific manifestation of the primary defect. *Am.J.Hum.Genet.* 33:928–934, 1981.
5. BLOOM, D..: Congenital telangiectatic erythema resembling lupus erythematosus in dwarfs. Probably a syndrome entity. *Am.J.Dis.Child.* 88:754–758, 1954.
6. BLOOM, D.: Discussion of a presentation on primordial dwarfism: Discoid lupus erythematosus, by D.P. Torre and J. Cramer. *Arch.Dis.Dermatol.* 69:512, 1954.
7. BLOOM,. D.: The syndrome of congenital telangiectatic erythema and stunted growth. *J.Pediatr.* 68:103–113, 1966.
8. BRYANT, E.M., HOEHN, H., MARTIN, G.M.: Normalization of sister chromatid exchange frequencies in Bloom syndrome fibroblasts via euploid cell hybridization. *Nature* 279:795–796, 1979.
9. CAIRNS, J.: The origin of human cancers. *Nature* 289:353–357, 1981.
10. CHAGANTI, R.S.K., SCHONBERG, S., GERMAN, J.: A manyfold increase in sister

chromatid exchanges in Bloom's syndrome lymphocytes. *Proc.Natl.Acad.Sci. USA* 71:4508-4512, 1974.
11. EMERIT, I., CERUTTI, P.: Clastogenic activity from Bloom syndrome fibroblast cultures, *Proc.Natl.Acad.Sci.* 78:1868-1872, 1982.
12. GERMAN, J.:Bloom's syndrome. I. Genetical and clinical observations in the first twenty-seven patients. *Am.J.Hum.Genet.* 21:196-227, 1969.
13. GERMAN, J.: Bloom's syndrome. II. The prototype of human genetic disorders predisposing to chromosome instability and cancer. *In* Chromosomes and Cancer. (German, J., ed.) New York: John Wiley and Sons, Inc., pp 601-617, 1974.
14. GERMAN, J.: Bloom's syndrome. VIII. Review of clinical and genetic aspects. *In* Genetic Diseases Among Ashkenazi Jews. (Goodman, R.M., Motulsky, A.G., eds.) New York: Raven Press, pp. 121-139, 1979.
15. GERMAN, J.: Chromosome-breakage syndromes; Different genes, different treatments, different cancers. *In* DNA Repair and Mutagenesis in Eukaryotes. (Generoso, W.M., Shelby, M.D., de Serres, F.J., eds.) New York: Plenum Press, pp. 429-439, 1980.
16. GERMAN, J., BLOOM D., PASSARGE, E.: Bloom's syndrome. V. Surveillance for cancer in affected families. *Clin.Genet.* 12:162-168, 1977.
17. GERMAN, J., BLOOM, D., PASSARGE, E.: Bloom's syndrome. VII. Progress report for 1978. *Clin.Genet.* 15:361-367, 1979.
18. GERMAN, J., BLOOM, D., PASSARGE, E., FRIED, K., GOODMAN, R.M., KATZENELLEN-BOGEN, I., LARON, Z., LEGUM, C., LEVIN, S., WAHRMAN, J.: Bloom's syndrome. VI. The disorder in Israel and an estimation of the gene frequency in the Ashkenazim. *Am.J.Hum.Genet.* 29:553-562, 1977.
19. GERMAN, J., SCHONBERG, S.: Bloom's syndrome. IX. Review of cytological and biochemical aspects. *In* Genetic and Environmental Factors in Experimental and Human Cancer. (Gelboin, H.V., et al., eds.) Tokyo: Japan Sci. Soc. Press, pp. 175-186, 1980.
20. GERMAN, J., SCHONBERG, S., LOUIE, E., CHAGANTI, R.S.K.: Bloom's syndrome. IV. Sister-chromatid exchanges in lymphocytes. *Am.J.Hum.Genet.* 29:248-255, 1977.
21. MULCAHY, M.T., FRENCH, M.: Pregnancy in Bloom's syndrome. *Clin.Genet.* 19:156-158, 1981.
22. PASSARGE, E.: Elemente der Klinischen Genetik. Grundlagen und Anwendung der Humangenetik in Studium und Praxis. Stuttgart: G. Fischer Verlag, 1979.
23. RAY, J.H., GERMAN, J.: The chromosome changes in Bloom's syndrome, ataxia-telangiectasia, and Fanconi's anemia. *In* Genes, Chromosomes, and Neoplasia. (Arrighi, E., Rao, P.N., Stubblefield, E., eds.) New York: Raven Press, pp. 351-378, 1981.
24. RUDIGER H.W.: Cytogenetic demonstration of a corrective factor in Bloom syndrome. *In* Host Factors in Carcinogenesis. Lyon: Int. Agency for Research on Cancer, in press.
25. RUDIGER, H.W., BARTRAM, C.R., HARDER, W., PASSARGE, E.: Rate of sister chromatid exchanges in Bloom syndrome fibroblasts reduced by co-cultivation with normal fibroblasts. *Am.J.Hum.Genet.* 32:150-157, 1980.
26. TICE, R., WINDLER, G., RARY, J.M.: Effect of co-cultivation on sister chromatid exchange frequencies in Bloom's syndrome and normal fibroblast cells. *Nature* 273:538-540, 1978.
27. VAN BUUL, P.P.W., NATARAJAN, A.T., VERDEGAAL-IMMERZEEL, E.A.M.: Suppression of the frequencies of sister chromatid exchanges in Bloom's syndrome fibroblasts by co-cultivation with Chinese hamster cells. *Hum.Genet.* 44:187-189, 1978.

ATAXIA-TELANGIECTASIA: SEARCH FOR A CENTRAL HYPOTHESIS

RICHARD A. GATTI AND KATHLEEN HALL

BACKGROUND

Most investigators in the field of cytogenetics are acquainted with the general clinical and laboratory aspects of ataxia-telangiectasia (AT) from various reviews [4,6,40,60]. That few have ever seen a child affected with this disease stems from the fact that such patients are rarely encountered even in pediatric practice. It is estimated that the frequency is less than 1/30,000.

The many abnormalities that now have been described in AT patients encompass more biology than any group of investigators can competently engage. What is most needed at this writing is a unifying hypothesis or hypotheses that, once tested and established, would explain the basis for an underlying disease process which involves these many facets. One would hope such understanding would also lead to a rational approach to therapy.

Briefly described, AT is an inherited disorder characterized by a progressive cerebellar ataxia which appears in early childhood, accompanied by a typical skin lesion which is most simply described as a permanent dilation of blood vessels (i.e., telangiectases) over the eyes and ears and a history of repeated infections of the sinuses and lungs. These children are normal at birth and meet all developmental landmarks on time until ataxia presents as an unsteady gait, usually between 3 and 6 years of age. The presence of a progressive cerebellar ataxia in a child with telangiectases distinguishes this diagnosis from other forms of progressive ataxia of childhood, of which there are actually very few. Approximately 15% of such children develop malignancies, usually lymphoid [23,56,65]. The disease is not limited to any particular ethnic group. Death typically occurs during the second decade of life, either from overwhelming pulmonary infection or cancer.

Neurologic findings include progressive cerebellar ataxia, apraxia of eye movements, choreoathetosis, and dysarthria. When conjugate gaze is attempted, the movements are initiated very slowly. At postmortem examination, Purkinje cells are severely depleted in the cerebellum with no accompanying reactive gliosis. Anterior horn cells and neurones also are reduced in number in the spinal cord. Neuroaxonal atrophy has been noted in the cerebellar cortex, pons, and medulla as well. Denervation atrophy of muscle also has been reported. Variations in the sizes of nuclei are seen in many tissues, notably myocardium, endocrine organs, liver, and central nervous system.

In 1972, Waldmann et al. [76] noted that serum levels for alphafetoprotein (AFP) were elevated in all but a few AT patients. Such elevations also are associated with hepatic carcinoma, pregnancy, and, in the amniotic fluid, an open neural tube defect in the fetus. AFP is synthesized primarily in the liver, especially in the immature liver. Since elevations of liver enzymes such as serum glutamic-oxaloacetic transaminase, alkaline phosphatase, and lactic acid dehydrogenase are commonly noted in AT patients, despite normal bilirubin levels and liver clearance studies, it would appear that elevations of AFP in these patients are just another reflection of immature hepatic cells.

Immunodeficiency

Immunologic aberrations encompass both humoral and cell-mediated compartments. The early postmortem descriptions of Boder and Sedgwick [5] and Peterson et al. [54] documented that the thymus was almost always grossly abnormal, usually appearing vestigial or embryonic and lacking corticomedullary differentiation, Hassal's corpuscles, and well-developed vasculature. Lymph nodes display depletion of lymphocytes in the deep cortex or T-dependent areas while germinal follicles appear intact. The tonsils are usually quite hypoplastic and often reveal poor germinal follicles and secondary crypt formation. In the blood, T-cell proportions and absolute numbers are decreased in most patients, apparently unrelated to the length of illness, and B-cell proportions are often elevated [21]. T-helper-cell proportions and absolute numbers are diminished [74].

Functional assays reveal that T-lymphocytes are defective in many ways [15,75]. They respond poorly to stimulation by mitogens, antigens, and allogeneic cells. They do not respond to autologous stimulation [75]. Lymphocytotoxic assays using various targets are deficient. Nelson et al. [45] showed that influenza-infected fibroblasts are not killed appropriately by Class I (Major Histocompatibility Complex antigens)-restricted lymphocytes from these patients. T-helper function is deficient, as shown when lymphocytes from AT patients are added to purified B cells from normal persons and in vitro immunoglobulin (Ig) synthesis is measured. T-suppressor activity seems to be grossly intact in most patients tested [21].

IgA deficiency in AT patients was first noted by Thieffry et al. [72]. Most studies have concluded that approximately 60% of patients have decreased IgA levels. A recent report by Griscelli and co-workers [57] suggests that of the remaining 40%, normal IgA1 concentrations mask marked IgA2 deficiencies in almost all AT patients. Oxelius and coworkers report that IgG2 is deficient in most patients [21,49] and that these levels may be closely associated with recurrent infections. While IgG4 levels are described as deficient in two recent reports [49,57], we have not found this in a study of 12 patients whose IgG4 levels were measured independently in three laboratories [21]. IgG1 and IgG3 levels have been found to be generally normal in several studies. IgE deficiencies tend to accompany IgA deficiency although elevated IgE levels are seen occasionally in some patients. IgM levels usually are normal despite the fact that AT patients have primarily an 8S-type IgM in their serum rather than the typical 19S form, and this can lead to an artifactual report of elevated IgM levels.

Delayed hypersensitivity skin testing usually reveals profound anergy [54] and confirms laboratory findings of T-cell abnormalities. Recurrent infections in these patients are associated more with bacterial than with viral organisms; problems with progressive vaccinia or viral encephalitides have not been observed in this group of patients. Of course, smallpox vaccination usually has been given in infancy at a time when such patients remain undiagnosed and *may* be immunologically intact for most of the above-described functions. It is unclear whether the myriad of immunologic aberrations is acquired or congenital. Nor is it clear whether the thymic abnormalities are acquired or congenital. A common bias is that these abnormalities exist even before the ataxia manifests itself (see below).

Radiation Sensitivity

Gotoff et al. [26] noted when treating an AT patient with lymphosarcoma that conventional radiation doses applied by standard techniques resulted in severe unexpected sequellae. This observation was confirmed by Morgan et al. [44]. Further similar untoward reactions have been described following cyclophosphamide therapy [16].

Cells from many patients with AT are sensitive to gamma-induced radiation damage, as measured by cell survival and DNA replication [7,51,70,73]. It was theorized that this enhanced sensitivity represents a defect in DNA repair of the induced damage. However, not all patients show abnormal results in all assay systems. In fact, most of the data are based on experiments performed on a handful of fibroblastoid and lymphoblastoid cell lines which were derived from a few patients with AT. Most of these experiments do not include age-matched, control cells. Further, recent new insights into the mechanisms of DNA repair indicate that following radiation damage, the viability of cells, the efficiency of colony formation, and the uptake of tritiated thymidine reflect more than just DNA-repair mechanisms.

Hall and Gatti [to be published] studied repair of unscheduled DNA synthesis following gamma radiation-induced damage, as measured by tritiated thymidine uptake. Eleven AT patients and all members of their immediate families were tested; age-matched controls were included. As can be seen in Figure 1, little difference was noted between the patient group (AT) and the age-matched control group (<14 years). An unexpected finding was that the adult control group had a wide range of values, and when either the patient group or the parent group was compared to it, significant differences could be calculated. We therefore ex-

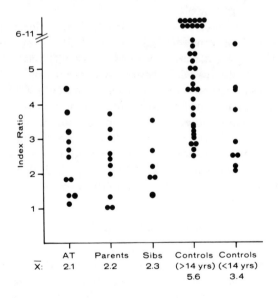

Fig. 1. Thymidine uptake of peripheral blood lymphocytes following 30 K rads of gamma radiation and 6-h incubation in the presence of hydroxyurea.

panded our studies of controls at various ages and found that this response to gamma radiation-induced damage may not be homogeneous throughout a normal population: (1) The majority of subjects have ratios between 2.4 and 6 (standard error = 0.2); (2) a small, potential, subset of subjects have much higher ratios, and we are retesting them at intervals to rule out the possibility that their higher readings may have reflected an infection or some other transient stimulation of their immune system; and (3) all prepubertal subjects (comparable to the age of most AT patients when studied) had ratios in the 2.5 to 6 range. Further, evidence from mice [27] suggests that levels of DNA-repair activity may vary according to genetic influences. (These influences may play a role in lifespan potentials as well, a concept that may be relevant to the reported premature aging of AT patients. It is even conceivable that a gene for DNA-repair activity might be closely linked to the AT gene but not otherwise related to it or to the pathogenesis of the disorder, i.e., an epistatic effect.) Thus, high-repair and low-repair groups may exist, making it difficult to evaluate the results of studies of unscheduled DNA synthesis following gamma-radiation damage.

Experiments performed by Painter and Young [50] on fibroblast lines from two AT patients led these workers to suggest that "the cause of increased radiosensitivity in AT cells may be a defect in their ability to respond to DNA damage rather than a defect in their ability to repair it. Doses of x-radiation that markedly inhibited the rate of DNA synthesis in normal human cells caused almost no inhibition in AT cells." They suggest that AT cells are defective in a system that in normal cells prolongs the time during which strand breaks can be repaired, replicon initiation inhibited, and chromatin structure normalized. The result in AT patients would be chromatid aberrations in daughter cells. They tested this by producing radiation damage in replicating cells (as monitored by ^{14}C-thymidine labeling) while the rate of postradiation replication was monitored by ^{3}H-thymidine labeling. Radiation-induced changes in the rate of replication could thus be evaluated by comparing the ^{3}H to ^{14}C ratio. We and others [34] have extended these observations, using fresh lymphocytes from AT patients, and the data essentially confirm the findings of Painter and Young, although the differences between patients and controls are minimal (Fig. 2). Lymphocytes from parents (i.e., obligate carriers) do not allow reliable identification of an intermediate pattern.

Paterson and colleagues [28] recently have reevaluated enzyme repair of x-ray-induced DNA strand breaks by an alkaline elution method. They compared repair (i.e., ligation) in three AT fibroblast culture strains with that seen in a strain from a clinically normal host of similar age. No dif-

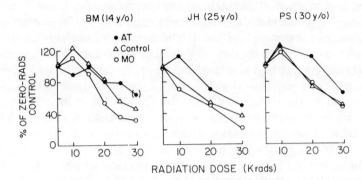

Fig. 2. DNA synthesis following gamma irradiation of replicating lymphocytes. See text for details. Note that response of AT patients to radiation damage is less impaired than controls, suggesting that such cells do not respond appropriately to radiation damage.

ferences were found. The three AT fibroblast strains used belonged to each of three complementation groups: exr⁻ type A, exr⁻ type B, and exr⁺ [52]. Sheridan and Huang [61] had looked earlier at endonuclease S1 activity in DNA strand breakage and rejoining and found it to be essentially normal. Fornace and Little [18] measured repair of DNA single-strand breakage in diploid fibroblasts isolated from normal individuals and AT patients by the alkaline elution method. No differences were observed between these cell strains with respect to the rate of rejoining DNA-strand breaks by low and moderate doses of x-rays or by treatment with bleomycin. However, the latter investigators induced radiation damage in nondividing cells, whereas Paterson and colleagues tested the damage-inducing effects on fibroblasts in the exponential phase of the growth cycle. Taken together, it now seems unlikely that the radiation sensitivity of AT cells and patients reflects decreased *physical* repair of DNA-strand scissions. Whether the *rate* or *accuracy* of repair is normal remains to be established.

Poly (ADP-ribose) polymerase is a eukaryotic chromosomal enzyme which utilizes ADP-ribose to synthesize the nucleic acid homopolymer (ADP-ribose)$_n$. While the precise function has not been established, it is known to modify chromosomal proteins of this polymer by ADP-ribosylation. Normal patients show an increase in the synthesis of the polymer along with depressed DNA synthesis following gamma-irradiation. Irradiation of lymphoblastoid cells from AT patients does not result in depression of DNA synthesis or in an elevation in the levels of (ADP-ribose) [12]. The synthesis of (ADP-ribose)$_n$ has been implicated in the regulation of DNA synthesis and suppression of strand breaks following DNA damage [2,8]. The latter mechanism would provide an explanation for the multiple breaks which are characteristic of AT karyotypes. (ADP-ribose)$_n$ also fails to increase in xeroderma pigmentosum patients after UV radiation [3].

Chromosomal Defects

As discussed elsewhere in this volume, cytogenetic studies by McCaw et al. [39] identified a tandem translocation on the long arm of chromosome 14 in seven of eight patients. In one AT patient studied at UCLA, a leukemic clone of lymphocytes manifested the 14q translocation [59,64]. However, this recurring aberration has not been identified in the majority of AT patients who have been karyotyped worldwide. Also, the frequency of 14q translocations in the normal population remains unclear. Further, the fact that many other aberrations are observed in the karyotypes of these patients suggests that the major cytogenetic lesion in these patients involves a broader mechanism of chromosomal repair, replication, and, or segregation and not just a 14q translocation.

MODELS OF PATHOGENESIS
A Single Rare Autosomal Recessive Gene

The classical genetic model for AT is that of a simple autosomal recessive trait [17,41,69]. This is based on (1) phenotypically normal parents, and (2) a frequency of affecteds in over 60 families which approached the expected 25% for an autosomal recessive disorder. Other genetic models have been considered and rejected.

Upon close scrutiny, however, the phenotypically normal parents do manifest some immunologic abnormalities in the laboratory: intermediate levels (i.e., between normal values and those seen in affected patients) have been noted for T-cell levels, Con A capping, cyclic GMP in T cells, DNA repair, and neutrophil chemotaxis [21]. Further, the frequency of cancer in AT parents appears to be increased when compared to various control populations [67,68].

Although the above facts do not seriously challenge the concept of an autosomal recessive model, a more disturbing observation is that, with the exception of the Turkish families [15], very few AT patients are the product of consanguineous marriages, an otherwise typical characteristic of an automosal recessive disorder. Swift suggests [personal communication] that this could result if the AT gene were, in fact, much more common than presently appreciated and that most affecteds die in utero while the living patients represent only a small subset (perhaps even an atypical one). The consanguinity rate among 20 Turkish familes was 85% [15].

Heterogeneity of the AT syndrome also challenges a simple genetic model. For example, the typical age of death in some families extends well beyond the second decade [1], despite typical patterns of onset and progression of disease in these patients and characteristically elevated AFP levels, IgA deficiency, and reduced T-cell numbers. Further, within several of our families, the AFP levels are normal in all affected members (as well as in unaffected members, of course). Again, no discernible clinical differences exist, although some of these patients are probably not of normal intelligence (the typical normal I.Q. of most AT patients is difficult to document by standard testing due to the very sluggish responses of such patients and the cerebellar speech). Finally, several studies have documented complementarity for DNA-repair defects among cell lines from different patients [52]. What is lacking in such studies is a clear demonstration that the same pattern of complementarity holds true in all affected members of any single family. If this finding could be confirmed, one would be forced to accept subsets or complementarity groups among AT patients, similar to the situation in xeroderma pigmentosum [58]. On the other hand, as discussed above, the three complementation groups of AT all are deficient in rejoining strand breaks [42], suggesting a *common* repair defect rather than subsets. Thus, while it is probable that most AT represents one disease, it is hazardous to exclude other genetic models. While the syndrome might result from several different genetic defects which form part of a common pathway to phenotypy, it could equally be considered to arise from a single developmental abnormality which, depending upon the timing and severity of the abnormality, could produce a spectrum of phenotypes (see below).

The very recent introduction of a mouse model for AT, the *wasted* mouse [62], promises to be of great assistance in analyses of this disorder. This model already provides several crucial insights into previous findings. First and foremost, it establishes that a single genetic point mutation can produce the phenotype of (1) neurologic deterioration leading to premature death, (2) immunodeficiency, (3) susceptibility to gamma-radiation damage, and (4) increased frequency of chromosomal breaks. Second, it again links the neurological deterioration to Purkinje cell changes and the immunodeficiency to thymic hypoplasia and lymphocyte depletion. Linkage studies in the *wasted* mice have so far failed to map the mutation to areas on chromosomes 1–11, 13–15, 17, and 19. On the other hand, the gene could still be located on any of those chromosomes at greater distances from the markers tested than linkage analyses would detect. Most provocative is the fact that chromosome 12 has not yet been evaluated for linkage since this is where the Ig heavy chain genes are located in the mouse and this region is homologous with the area of tandem translocations (i.e., 14q) in the AT patients.

We are embarked upon genetic studies which assume a single rare autosomal recessive gene model and attempt to link the AT gene to other genetic markers so as to map it within the human genome. Using nine multiplex AT families, we found no evidence for linkage to HLA [31]. We presently are focusing on markers for the long arm of chromosome 14 (see Fig. 3). These include D14S1 [11,79], Pi [9,53] (alpha-l-antitrypsin), Gm [10,63], and possibly DNA fragment length polymorphisms that may be revealed with probes for IGH and Pi. Considering the pattern of immunoglobulin deficiencies seen in AT patients, it seems plausible to consider a genetic defect that influences the expression of the IGH genes which map downstream from Gamma 3 (i.e., Gamma 2, Gamma 4, Alpha 1, Epsilon, and Alpha 2) on chromosome 14q. On the other hand, it is difficult at the present writing to imagine how such a defect would also result in the progressive neurologic deterioration.

Waldman [75] and Wall [personal communication] and co-workers have documented that the mu, gamma, and alpha structural genes are present in the genome of AT patients and in their proper arrangements. This is not surprising since B cells in these patients express each of the major immunoglobulin classes. T-cell defects in AT patients might reflect a 14q-related problem in that there is much circumstantial evi-

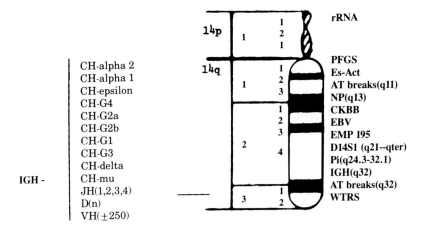

Fig. 3. Map of genetic loci on chromosome 14.

dence that places a gene or gene family for T-cell Ig-like receptors downstream (i.e., proximal) from the IGH region on 14q [33].

DNA-Related Enzyme Defect

If the defect in expression of IGH or T-cell receptor genes were secondary to an enzyme involved in a switching mechanism, there is no guarantee that the gene for that enzyme would map to 14q. Thus, this hypothesis will not be tested by such linkage studies but will require a more direct analysis of DNA-related enzymes and the coding mechanisms responsible for switching.

Little is known about the enzymes which are involved in rearrangement (i.e., switching/splicing) of the Ig heavy chain genes. At least two types of enzymes would appear to be necessary: a nuclease for splicing out the unwanted gene(s), and a ligase for restoring the continuity of the DNA chains. Similar types of enzymes are involved in repair of DNA damage induced by radiation. We have already mentioned Edwards and Taylor's [12] finding that the level of ADP ribose polymer does not increase appropriately in AT patients following x-radiation damage, suggesting a polymerase deficiency. uv endonuclease activity is inferred to be intact in AT patients since UV damage is repaired normally. The complementarity of "DNA repair" studies [52] suggests that a pathway of enzymes may exist which is abnormal in different ways in different subsets of AT.

Thus, the basic defect would be the formation of an abnormal enzyme(s) which normally plays an important role in maintaining the integrity of the DNA itself. Such a basic perturbation of cell function could impact upon the DNA replication necessary for clonal expansion of lymphoid compartments in immune responses as well as upon DNA repair and maintenance. Thus, this area is a prime candidate for providing a unifying model for the pathogenesis of AT.

An Immunoregulatory Defect

A favored model of pathogenesis in AT has been that of an underlying immunoregulatory defect. This is supported by an abundance of evidence for impaired thymic development and function of the T-cell compartment [21,45,54,74,75]. This would reasonably be expected to result in humoral immunodeficiencies as well if, for example, T-cell help and, or suppression were impaired. The high incidence of cancer in these patients also could be due to a primary immunodeficiency in this disease, since despite the controversial role of immunosurveillance as a major factor in oncogenesis, many examples exist of increased cancer frequency in immunosuppressed hosts [23].

An important limitation to the hypothesis of an immunoregulatory defect is that it is difficult to design a series of clinically feasible experiments which would directly test this hypothesis in AT patients. Indeed, it seems premature and misguided to apply each new immunological technique to the study of these children under the guise of a broad but vaguely defined immunoregulatory hypothesis. More than a dozen in vitro immunological parameters are abnormal in a significant proportion of these patients [21,75,81]. Even when considered as a group, these findings do not constitute a general theme and do little to explain the karyotypic aberrations, the DNA repair/replication defects, or the progressive neurologic deteriorations seen in these children. It appears to us that most of the immunological findings in AT patients represent epiphenomena of more central defects, the investigations of which should take priority in the difficult conquest of this fatal disorder. Perhaps the *wasted* mouse model will provide a better opportunity to dissect this facet of the disease process.

An immunoregulatory defect that would impact upon humoral immunodeficiencies would probably be mediated via class-specific T-helper or T-suppressor cells. While there is little evidence for any profound abnormalities of T-suppressor cells in AT patients, abundant evidence points to T-helper dysfunction. With the recent availability of monoclonal antibodies against Ig classes and subclasses, perhaps laboratory technology for evaluating class- and subclass-specific T-helper functions in these patients may become available. Whether the T-cell receptors for such cells utilize the IGH gene family on 14q is also unknown [33].

A Microtubular Defect

Concanavalin-A (Con-A) molecules are capped to completion by a greater proportion of peripheral blood lymphocytes in AT patients than that seen with normal lymphocytes under comparable conditions (e.g., 37°C for 90 min) [21]. Since capping is dependent upon depolymerization of microtubules [48], it seemed appropriate to examine other parameters which interrelate with microtubular function. Cell motility is one of these. We found that chemotaxis of neutrophils to a synthetic tripeptide in Boyden chambers was significantly impaired [21]. Microtubular integrity is dependent upon cAMP to provide energy for repolymerization of tubulin. Levels of cyclic AMP in E-rosette-enriched T-cell preparations were markedly increased over adult and age-matched controls in 11 of 12 AT patients [21]. Cyclic GMP levels also were elevated in half of these patients. These results neither prove nor even test the hypothesis that a microtubular defect might underlie the pathogenesis of this disease. However, they alerted us to the possibility that a defect of microtubular integrity, whether a first-degree structural defect or one secondary to a biochemical defect in energy metabolism, might produce a syndrome similar to that of AT: frequent chromosomal breaks (secondary to disrupted spindle formation and mitosis), atrophy of Purkinje cells (secondary to inadequate synaptic connections, one function of microtubular structure), and immunologic deficiencies (as a result of inappropriate cell-surface-receptor mobility and defective clonal expansion resulting from defective mitosis). Our finding of enhanced Con-A capping is similar to those of Gelfand et al. of enhanced Con-A capping of lymphocytes in a patient with severe combined immunodeficiency disease [24,48].

Perhaps the major weakness of a cytoskeletal hypothesis in AT is that one might expect such a far-reaching defect to produce more devastating results. Recall, however, the suspicion of Swift that the lack of consanguinity among parents of these patients could be interpreted to indicate that the gene frequency for AT is much more common than realized and that the homozygous state may be a lethal lesion in most situations [67].

An Inborn Nutritional Deficiency

A trace-element or vitamin deficiency is difficult to imagine in children receiving a well-balanced diet, unless the pathway(s) for the absorption or utilization of that dietary factor is genetically defective. This model assumes that such patients would be born normal and develop their disease secondary to a deficiency of some poorly metabolized nutrient. Pursuing this logic, we recently addressed the problem of whether the Purkinje cells which are notably depleted at postmortem might ever have been present in normal numbers or were absent at birth in AT patients. This can be estimated by determining whether "empty" basket cells are present, using a Bielschowsky stain, since basket cells form around existing Purkinje cells. Resectioning of cerebellum specimens from five former cases revealed numerous empty basket cells. A similar observation was published by Amromin et al. [1] in one patient and now the *wasted* mice confirm a similar ongoing degeneration of Purkinje cells [62]. Thus, it appears likely that the depleted Purkinje cell layer seen in virtually all AT patients at postmortem examination represents an *acquired* problem, one which probably correlates with the neurologic deterioration that begins in the early years of the disease. One must now ask how many of the other abnormalities noted in this disease are acquired and what common factor could lead to such acquired defects. As mentioned above, a model of inborn nutritional deficiency would allow for differing phenotypes, depending upon either (1) the stage of development at which the deficiency was manifested, or (2) blocks in different stages of a common metabolic pathway. It also provides hope for a therapeutic approach which, if initiated early after diagnosis, might reverse the ataxia.

Other Hypotheses

Other hypotheses have been proposed over the past 20 years. A slow virus infection occurring in an immunodeficient host might be responsible for the syndrome [22]. A recent report suggests that cellular cAMP levels, abnormally high in AT T cells [21], may determine the shift between acute and persistent viral infections [43]. A viral hypothesis is supported further by the similarity of AT to a parvovirus-induced ataxia of cats [38].

An autoimmune mechanism might underlie the neurologic deterioration and explain some of the endocrinological abnormalities noted in some of these patients [13,32,71]. A neuroendocrine dysfunction could involve some basic pituitary irregularity which might impact upon thymic as well as neurologic development. This hypothesis may be too far ahead of current understanding of neuroendocrine-immune interactions to be rigorously tested [20]. Thymosin-alpha-1 serum concentrations of our patients were within normal ranges [21]. An early model suggested that a primary mesenchymal defect might prevent normal differentiation of tissue [54], and evidence exists that the liver, thymus, gonads, and cerebellum do not mature normally.

EPILOGUE

A Model for the Cancer Susceptibility of AT Patients

The answer to why one in eight of these young patients develops cancer is unclear at this writing. AT patients typically develop lymphoid cancers [23], usually of the T-cell type. A T-cell chronic lymphatic leukemia was characterized in detail by Saxon and co-workers [59].

Very recent methodologic advances combining chromosomal banding, mouse-man cell fusion and molecular hybridization using cloned cDNA probes have revealed new genetic relationships and concepts which may be relevant to the question of why AT patients are cancer-prone. These advances are exemplified by the 20-year long attempt to explain the role of the Philadelphia chromosome, a non-random chromosomal aberration which has been clearly associated with malignancy:

1960. Discovery of the Philadelphia chromosome (Ph1) in chronic myelogenous leukemia [46].

1973. Characterization of Ph1 as a deleted chromosome 22 which can be found translocated on chromosome 9 [55]. Realization that the deleted piece of chromosome 22 in CML occasionally translocates to another chromosome instead, but still results in leukemia [29,36].

1982. Demonstration by studying mouse-man cell fusion hybrids that the oncogene, *abl*, which normally maps to chromosome 9 is found in association with the Philadelphia chromosome in CML [Bootsma and co-workers, personal communication].

A similar chronology has been witnessed in the studies of Burkitt's lymphoma:

1972. Description of a non-random chromosomal aberration which involves a reciprocal translocation between the long arms of chromosomes 8 and 14 [37].

1978. Rearrangement of Ig genes in normal persons by DNA splicing so that V genes are joined to constant regions during the ontogeny of antibody responses [25].

1979. Description of translocations in mouse plasmacytomas between chromosome 12 and chromosome 15 and between chromosome 6 and chromosome 15 [35,47,77]. The region of chromosome 15 which is translocated is regularly duplicated in both B- and T-cell leukemias of viral or chemical origin [66,78]. A similar chromosomal duplication has been described in man at chromosome 12q13-22, which is associated with B-cell chronic lymphocytic leukemia [19].

1979–1981. Mapping of the Ig genes: heavy chains on chromosome 14q in man and 12 in mouse, kappa light chains on chromosome 2 in man and 6 in mouse, and lambda chains on chromosome 22 in man [10,11,30,63].

1981. Realization that a subset of patients with Burkitt's lymphoma have translocations between chromosome 8 and chromosome 2 or 22, instead of 14. These tumors express Ig light chains instead of heavy chains [35]. Taken together, these observations nicely parallel the mouse plasmacytoma findings.

1982. Demonstration that the translocated fragment from chromosome 14q found on chromosome 8 in Burkitt's lymphoma cells often contains the VH gene [14].

1982. Demonstration that the translocated fragment from chromosome 8q found on chromosome 14q in a subset of Burkitt's lymphoma patients contains the *myc* oncogene inserted next to a *mu* gene in a 5' to 5' orientation [Croce and co-workers, personal communication].

While these investigations have reached a point of great excitement and provoke much speculation, the common denominator remains unclear. Is it the V genes being translocated to a new site that promote oncogenesis or is it the translocation of the oncogenes that do this? When

V genes are translocated from either chromosomes 14, 2 or 22 onto 8q, Burkitt's lymphoma results. Do the V genes activate CML-inducing genes when translocated, instead, to chromosome 9? It may be that V genes resemble oncogenes by functioning as promoters to activate DNA sequences downstream from new splice sites, their natural role in Ig production. Equally plausible, the reciprocally translocated cellular oncogenes may activate normal sequences, but without appropriate feedback controls for later reversing this process. The term "cellular oncogene" itself may be misleading in that these genes can function without inducing cancer; they are probably best viewed as *differentiation* genes. What is emerging, then, is a model of oncogenesis dependent upon chromosomal rearrangements which place cellular oncogenes and/or promoter genes in inappropriate sites.

If one accepts (1) that AT chromosomes are prone to breakage, (2) that breakage increases the frequency of translocations, especially the tandem translocation of 14q in these patients, (3) that frequent translocations increase the number of possible chances for a "cancer-producing" translocation to occur, and (4) that V genes are among the most common promoter-like candidates in the normal genome which involve rearrangement, one would then expect lymphoid malignancy to be one of the most common forms of cancer in the general population and, also, in any cancer-prone population. Lymphoid malignancy accounts for about half of all pediatric cancer [80]. Thus, the absence of nonlymphoid cancers in AT patients may simply be masked by the predominance of lymphoid tumors for the above reasons. The frequent malignancies would then seem to relate primarily to the chromosomal instability seen in this disease.

CONCLUSIONS AND SUMMARY

The rarity of this syndrome and the limitations of performing extensive analyses on small children provide a formidable challenge to the clinical investigator. Only by being prepared to maximize the information that can be derived from each affected individual, each tissue biopsy, each tumor, and each family, will we eventually arrive at an understanding that will allow a rational and effective treatment of this disorder. The syndrome can result from a single point mutation. The ataxia appears to be a *degenerative* process which might be reversible if appropriate therapy were found and instituted early after diagnosis (or even before that if the diagnosis could be made preclinically). The basic abnormality appears to be subcellular, most likely a molecular one which results in the formation of an abnormal protein. That protein must play an important role either in maintaining the integrity of the DNA itself or in transcribing and, or translating genomic messages into other proteins.

LITERATURE CITED

1. AMROMIN, G.D., BODER, E., TEPLITZ, R.: Ataxia-telangiectasia with a 32 year survival. A clinicopathologic report. *J. Neuropathol. Exp. Neurol.* 38:621–643, 1979.
2. BERGER, N.A., PETZOLD, S.J., BERGER, S.J.: Association of poly(ADP-ribo) synthesis with cessation of DNA synthesis and DNA fragmentation. *Biochem. Biophys. Acta* 564:90–104, 1979.
3. BERGER, N.A., SIDORSKI, G.W., PETZOLD, S.J., KUROHARA, K.D.: Defective poly (adenosine diphosphoribase) synthesis in xeroderma pigmentosum. *Biochemistry* 19:289–293, 1980.
4. BODER, E., SEDGWICK, R.P.: Ataxia-telangiectasia, a familial syndrome of progressive cerebellar ataxia, oculocutaneous telangiectasia and frequent pulmonary infection. *Pediatrics* 21:526–554, 1958.
5. BODER, E., SEDGWICK, R.P.: Ataxia-telangiectasia. A review of 150 cases. *Proc. Intnl. Copenhagen Congress on the Scientific Study of Mental Retardation.* Denmark, Aug. 1964.
6. BRIDGES, B.A., HARNDEN, D.G.: Untangling ataxia-telangiectasia. *Nature* 289:223–224, 1981.
7. CHEN, P.C., LAVIN, M.F., KIDSON, C., MOSS, D.: Identification of ataxia-telangiectasia heterozygotes, a cancer prone population. *Nature* 274:484–486, 1978.
8. CLAYCOMB, W.B.: Poly (Adenosine Diphosphate Ribose) polymerase activity and nicotinamide adenine dinucleotide in differentiating cardiac muscle. *Biochem. J.* 154:387–393, 1976.
9. COX, D.W., MARKOVIC, V.D., TESHIMA, I.E.: Genes for immunoglobulin heavy chains and for alpha-1-antitrypsin are localized to specific regions of chromosome 14q. *Nature* 297:428–430, 1982.
10. CROCE, C.M., SHANDER, M., MARTINIS, J., CIRCUREL, L., D'ANCONA, G.G., DOLBY, T.W. KOPROWSKI, H.: Chromosomal location of the genes for human immunoglobulin heavy chains. *Proc. Natl. Acad. Sci. USA* 76:3416–3420, 1979.
11. DE MARTINVILLE, B., WYMAN, A., WHITE, R., FRANKE, U.: Assignment of the first highly polymorphic DNA marker locus to a human chromosome region. *Am. J. Hum. Genet.* 34:216–226, 1982.
12. EDWARDS, M.J., TAYLOR, A.M.R.: Unusual levels of (ADP-ribose)n and DNA synthesis in ataxia telangiectasia cells following γ-ray irradiation. *Nature* 287:745–747, 1980.
13. EISEN, A.H., KARPATI, G., LASZLO, T., ANDERMANN, F., ROBB, J.P., BACAL, H.L.: Immunologic deficiency in ataxia-telangiectasia. *N. Engl. J. Med.* 272:18–22, 1965.
14. ERIKSON, J., FINAN, J., NOWELL, P.C. AND CROCE, C.M.: Translocation of Immunoglobulin VH genes in Burkitt lymphoma. *Proc. Natl. Acad. Sci. USA*, 79:5611–5615, 1982.
15. ERSOY, F. BERKEL, A.I.: Clinical and immunological studies in twenty families with ataxia-telangiectasia. *Turk. J. Pediatr.* 16:145–160, 1974.
16. FEIGIN, R.D., VIETTI, T.J., WYATT, R.G., KAUFMAN, D.G., SMITH, C.H., JR.: Ataxia telangiectasia with granulocytopenia. *J. Pediatr.* 77:431–438, 1970.
17. FERAK, V., BENKO, J., CAJKOVA, E.: Genetic aspects of Louis-Bar syndrome (ataxia-telangiectasia). *Cs. Neurol.* 5:319–327, 1968.
18. FORNACE, A.J., LITTLE, J.B.: Normal repair of DNA single-strand breaks in patients with ataxia telangiectasia. *Biochem. Biophys. Acta* 607:432–437, 1980.
19. GAHRTON, G., ROBERT, K.H., FRIBERG, X.K., JULIUSSON, G., BIBERFELD, P., ZECH, L.: Cytogenetic mapping of the duplicated segment of chromosome 12 in lymphoproliferative disorders. *Nature* 297:513–514, 1982.
20. GATTI, R.A.: Ataxia-telangiectasia: A neuro-endocrine-immune disease? Alternative models of pathogenesis. *In* Immunoregulation (Fabris, N., Garaci, E., eds.), New York: Plenum Press, in press.

21. GATTI, R.A., BICK, M., TAM, C.F., MEDICI, M.A., OXELIUS, V., HOLLAND, M., GOLDSTEIN, A.L., BODER, E.: Ataxia-telangiectasia: A multiparameter analysis of eight families. *Clin. Immunol. Immunopathol.* 23:501-516, 1982.
22. GATTI, R.A., GOOD, R.A.: The immunological deficiency diseases. *Med. Clin. North Am.* 54:281-307, 1970.
23. GATTI, R.A., GOOD, R.A.: Occurrence of malignancy in immunodeficiency disease. *Cancer* 28:89-98, 1971.
24. GELFAND, E.W., OLIVER, J.M., SCHOORMAN, R.K., MATHESON, D.S., DOSCH, A.M.: Abnormal lymphocyte capping in a patient with severe combined immunodeficiency disease. *N. Engl. J. Med.* 301:1245-1249, 1979.
25. GILMORE-HERBERT, M., WALL, R.: Immunoglobulin light chain in RNA is processed from large nuclear RNA. *Proc. Natl. Acad. Sci. USA* 75:342-345, 1978.
26. GOTOFF, S.P., AMIRMOKRI, E., LIEBNER, E.J.: Ataxia-telangiectasia: Neoplasia, untoward response to x-irradiation and tuberous sclerosis. *Am. J. Dis. Child* 114:617-618, 1967.
27. HALL, K.Y., BERGMANN, K., WALFORD, R.L.: DNA repair, H-2 and aging in NZB and CBA mice. *Tissue Antigens* 16:104-110, 1981.
28. HARIHARAN, P.V., ELECZKO, S., SMITH, B.P., PATERSON, M.: Normal rejoining of DNA strand breaks in ataxia telangiectasia fibroblast lines after low x-ray exposure. *Radiat. Res.* 86:589-597, 1981.
29. HAYATA, I., KAKATI, S., SANDBERG, A.A.: A new translocation relative to the Philadelphia chromosome. *Lancet* 2:1384, 1973.
30. HENGARTNER, J., MEO, T., MULLER, D.: Assigment of genes for immunoglobulin K and heavy chains to chromosomes 6 and 12 in mouse. *Proc. Natl. Acad. Sci. USA* 75: 4494-4498, 1978.
31. HODGE, S., BERKEL, I., GATTI, R.A., BODER, E., SPENCE, M.A.: Ataxia-telangiectasia and xeroderma pigmentosum: No evidence of linkage to HLA. *Tissue Antigens* 15: 313-317, 1980.
32. HONG, R., AMMANN, A.J.: Ataxia-telangiectasia. *N. Engl. J. Med.* 283:660, 1970.
33. JANEWAY, C.A., CONE, R.E., ROSENSTEIN, R.W.: T-cell receptors. *Immunol. Today* 3: 83-86, 1982.
34. JASPERS, N.G.J., SCHERES, J.M.J.C., DEWIT, J., BOOTSMA, D.: Rapid diagnostic test for ataxia telangiectasia. *Lancet* 2:473, 1981.
35. KLEIN, G.: The role of gene dosage and genetic transpositions in carcinogenesis. *Nature* 294:313-318, 1981.
36. LAWLER, S.D.: The cytogenetics of chronic granulocytic leukaemia. *Clin. Haematol.* 6:55-75, 1977.
37. MANOLOV, G., MANOLOVA, Y.: Marker band in one chromosome 14 from Burkitt lymphoma. *Nature* 237:33-34, 1972.
38. MARGOLIS, G., KILHAM, L., JOHNSON, R.H.: The parvoviruses and replicating cells: Insights into the pathogenesis of cerebellar hypoplasia. *Prog. Neuropathol.* 1:173-210, 1971.
39. MCCAW, B., HECHT, F., HARNDEN, D.G., TEPLITZ, R.L.: Somatic rearrangement of chromosome 14 in human lymphocytes. *Proc. Natl. Acad. Sci. USA* 72:2071-2075, 1975.
40. MCFARLIN, D.E., STROBER, W., WALDMANN, T.A.: Ataxia-telangiectasia. *Medicine* 51:281-314, 1972.
41. MCKUSICK, V.A., CROSS, H.E.: Ataxia-telangiectasia and Swiss-type agammaglobulinemia. *J. A. M. A.* 195:739-745, 1966.
42. MCKUSICK, V.: Mendelian Inheritance in Man. Baltimore: Johns Hopkins University Press, pp. 702-703, 1978.
43. MILLER, C.A., CARRIGAN, D.R.: Reversible regression and activation of measles

virus infection in neural cells. *Proc. Natl. Acad. Sci. USA* 79:1629-1633, 1982.
44. MORGAN, J.L., HOLCOMB, T.M., MORRISSEY, R.W.: Radiation reaction in ataxia telangiectasia. *Am. J. Dis. Child.* 116:557-558, 1968.
45. NELSON, D.L., BIDDISON, W.E., BUNDY, B.M., SHAW, S.: Influenza virus specific HLA restricted cytotoxic T-cell responses in humans – heterogeneity of responsiveness in immunodeficiency patients. *Proc. Fourth International Congress of Immunology*, Abstr. No. 14.2.20, Paris, France, July 21-26, 1980.
46. NOWELL, P.C., HUGERFORD, D.A.: Minute chromosome in human granulocytic leukemia. *Science* 132:1497, 1960.
47. OHNO, S., BABONITS, M., WIENER, F., SPIRA, J. KLEIN, G.: Nonrandom chromosome changes involving the Ig gene-carrying chromosomes 12 and 6 in pristane-induced mouse plasmacytomas. *Cell* 18:1001-1007, 1979.
48. OLIVER, J.M., GELFAND, E.W., PEARSON, C.B., PFEIFFER, J.P., DOSCH, H.-M.: Microtubule assembly and concanavalin A capping in lymphocytes: Reappraisal using normal and abnormal human peripheral blood cells. *Proc. Natl. Acad. Sci. USA* 77:3499-3503, 1980.
49. OXELIUS, V.-A., BERKEL, I., HANSON, L.A.: IgG2 deficiency in ataxia-telangiectasia. *New Engl. J. Med.* 306:515-517, 1982.
50. PAINTER, R.B., YOUNG, B.R.: Radiosensitivity in ataxia-telangiectasia: A new explanation. *Proc. Natl. Acad. Sci. USA* 77:7315-7317, 1980.
51. PATERSON, M.C., SMITH, B.P., LOHMAN, P.H.M., ANDERSON, A., FISHMAN, L.: Defective excision repair of gamma-ray-damaged DNA in human (ataxia telangiectasia) fibroblasts. *Nature* 260:444-446, 1976.
52. PATERSON, M.C., SMITH, P.J.: Ataxia telangiectasia: An inherited human disorder involving hypersensitivity to ionizing radiation and related DNA-damaging chemicals. *Ann. Rev. Genet.* 13:291-318, 1979.
53. PEARSON, S., TETRI, P., FRANCKE, U.: Chromosome 14 codes for human alpha 1-antitrypsin (PI) expression in rat hepatoma x human fetal liver cell hybrids. *Am. J. Hum. Genet.* 33:148A, 1981.
54. PETERSON, R.D.A., BLAW, M., GOOD, R.A.: Ataxia-telangiectasia: A possible clinical counterpart of the animals rendered immunologically incompetent by thymectomy. *J. Pediatr.* 63:701-703, 1963.
55. ROWLEY, J.D.: A new consistent abnormality in chronic myelogenous leukemia identified by quinacrine fluorescence and Giemsa staining. *Nature* 243:290-293, 1973.
56. REED, W.B., EPSTEIN, W.L., BODER E., SEDGWICK R.P.: Cutaneous manifestations of ataxia-telangiectasia. *J. A. M. A.* 195:746-753, 1966.
57. RIVAT-PERAN, L., BURIOT, D., SALIER, J.-P., RIVAT, C., DUMITRESCO, S.-M., GRISCELLI, C.: Immunoglobulins in ataxia-telangiectasia: Evidence for IgG4 and IgA2 subclass deficiencies. *Clin. Immunol. Immunopathol.* 20:99-110, 1981.
58. ROBBINS, J.H., FRAEMER, K.H., LUTZNER, M.A., FESTOFF, B.W., COON, H.G.: Xeroderma pigmentosum: An inherited disease with sun sensitivity, multiple cutaneous neoplasms and abnormal DNA repair. *Ann. Intern. Med.* 80:221-248, 1974.
59. SAXON, A., STEVENS, R.H., GOLDE, D.W.: Ataxia-telangiectasia and chronic lymphocytic leukemia: Evaluation of a T lymphocyte clone with differentiation to both helper and suppressor T lymphocytes. *New Engl. J. Med.* 300:700-704, 1979.
60. SEDGWICK, R.P., BODER, E.: Ataxia-telangiectasia. *In* Handbook of Clinical Neurology (Vinken, P.J., Bruyn, G.W., eds.) Amsterdam: North-Holland Publ. Co., pp. 267-339, 1972.
61. SHERIDAN, R.B., HUANG, P.C.: Further considerations of the evidence for single strand break repair. *Mutat. Res.* 61:415-417, 1979.
62. SHULTZ, L.D., SWEET, H.O., DAVISSON, M.T., COMAN, D.R.: "Wasted", a new mutant

of the mouse with abnormalities characteristic of ataxia-telangiectasia. *Nature* 297: 402-404, 1982.
63. SOLOMON, E., GOODFELLOW, P., CHAMBERS, S., SPURR, N., HOBART, M.J., RABBITS, T.H., POVEY, S.: Confirmation of the assignment of immunoglobulin heavy chain genes to chromosome 14, using cloned DNA as molecular probes. *Human Gene Mapping Workshop* IV. Abstr. 96, 1981.
64. SPARKES, R.S., COMO, R., GOLDE, D.W.: Cytogenetic abnormalities in ataxia-telangiectasia with T-cell chronic lymphocytic leukemia. *Cancer Genet. Cytogenet.* 1:329-336, 1980.
65. SPECTOR, B., PERRY, G., KERSEY, J.: Genetically-determined immunodeficiency diseases (GDID) and malignancy. Report from the Immunodeficiency-Cancer registry. *Clin. Immunol. Immunopathol.* 11:12-29, 1978.
66. SPIRA, J., BABONITS, M., WIENER, F., OHNO, S., WIRSCHUBSKI, Z., HARAN-GHERA, N., KLEIN, G.: Nonrandom chromosomal changes in Thy-1-positive and Thy-1-negative lymphomas induced by 7,12-dimethylbenzanthracene in SJL mice. *Cancer Res.* 40:2609-2616, 1980.
67. SWIFT M.: Disease predisposition of ataxia-telangiectasia heterozygotes. *In* Ataxia-telangiectasia – A Cellular and Molecular Link Between Cancer, Neuropathology and Immune Deficiency (Bridges, B.A., Harnden, D.G., eds.) London: John Wiley and Sons Ltd., pp. 355-361, 1982.
68. SWIFT, M., SHOLMAN, L., PERRY, M., CHASE, C.: Malignant neoplasms in the families of patients with ataxia-telangiectasia. *Cancer Res.* 36:209-215, 1976.
69. TADJOEDIN, M.K., FRASER, D.C.: Heredity of ataxia-telangiectasia (Louis-Bar syndrome). *Am. J. Dis. Child* 110:64-68, 1965.
70. TAYLOR, A.M.R., HARNDEN, D.G., ARLETT, C.F., HARCOURT, S., LEHMANN, A.R., STEVENS, S., BRIDGES, B.A.: Ataxia-telangiectasia: A human mutation with abnormal radiation sensitivity. *Nature* 258:427-429, 1975.
71. TEPLITZ, R.L.: Ataxia-telangiectasia. *Arch. Neurol.* 35:553-554, 1978.
72. THIEFFRY, S., ARTHUIS, M., AICARDI, J., LYON, G.: L'ataxie-telangiectasie (7 observation personnelles). *Rev. Neurol.* (Paris) 105:390-405, 1961.
73. TOLMACH, L.J., JONES, R.W.: Dependence of the rate of DNA synthesis in X-irradiated HeLa S_3 cells on dose and time after exposure. *Radiat. Res.* 69:117-133, 1977.
74. TROMPETER, R.S., LAYWARD, L., HAYWARD, A.R.: Primary and secondary abnormalities of T cell subpopulations. *Clin. Exp. Immunol.* 34:388-392, 1978.
75. WALDMAN, T.A.: Immunological abnormalities in ataxia-telangiectasia. *In* Ataxia-telangiectasia – A Cellular and Molecular Link Between Cancer, Neuropathology and Immune Deficiency (Bridges, B., Harnden, D.G., eds.). London: John Wiley and Sons, Ltd., pp. 37-51, 1982.
76. WALDMANN, T.A., MCINTIRE, K.R.: Serum-alpha-fetoprotein levels in patients with ataxia-telangiectasia, *Lancet* 2:1112-1115, 1972.
77. WIENER, F., BABONITS, M., SPIRA, J., KLEIN, G., POTTER, M.: Cytogenetic studies on IgA/Lambda-producing murine plasmacytomas: Regular occurrence of a T(12;15) translocation. *Somatic Cell Genet.* 6:731-738, 1980.
78. WIENER, F., OHNO, S., SPIRA, J., HARAN-GHERA, N., KLEIN, G.: Cytogenetic mapping of the trisomic segment of chromosome 15 in murine T-cell leukamia. *Nature* 275:658-660, 1978.
79. WYMAN, A.R., WHITE, R.: A highly polymorphic locus in human DNA. *Proc. Natl. Acad. Sci. USA* 77:6754-6758, 1980.
80. YOUNG, J.L., MILLER, R.W.: Incidence of malignant tumors in U.S. children. *J. Pediatr.* 86:254-258, 1975.
81. YOUNT, W.J.: IgG2 deficiency and ataxia-telangiectasia. *N. Engl. J. Med.* 306: 541-543, 1982.

LONG-TERM OUTCOME IN FANCONI'S ANEMIA: DESCRIPTION OF 26 CASES AND REVIEW OF THE LITERATURE

BLANCHE P. ALTER AND NANCY UPP POTTER

INTRODUCTION

The association of pancytopenia and physical abnormalities was first described by Fanconi in 1927 [26]. Since then, more than 300 cases have been reported. These are reviewed with detailed references elsewhere [2]. This chapter will summarize that review and provide data on 26 previously unpublished cases seen at the Children's Hospital Medical Center in Boston (CHMC) from 1960 to 1980. The major complications of Fanconi's anemia (FA) will be described.

FA is the most frequent form of constitutional (inherited or congenital) aplastic anemia. From 1958 to 1977, 134 children with aplastic anemia were seen at CHMC [3]. Forty of these (30%) had constitutional disease, and 26 of the 40 (65%) had clear-cut FA. Thus, approximately 20% of cases of childhood aplastic anemia have FA. Ten of our cases were thought to be familial non-FA. However, in retrospect two of those probably did have FA. In four of the 40 constitutional cases, aplasia developed following amegakaryocytic thrombocytopenia. The ages at diagnosis of bone marrow failure in the constitutional aplastics at CHMC are shown in Figure 1. Figure 2 shows the ages at diagnosis of FA in 303 cases in the literature, separated according to sex. It has been stated that the onset of pancytopenia in FA is slightly earlier in males than in females; this is confirmed in Figure 3. Half the males were diagnosed by age 6, while the median age for diagnosis in females was 7.5 years. Figure 2 shows that the peak for males was 6 years, and for females, 9 years; the oldest female was 35 at diagnosis, the oldest male, 27. Ninety percent of males were diagnosed by 12 and 90% of females by 14 years of age. In the 26 cases seen at CHMC, the range was from 1 month to 13 years; the median age for boys was 5 years, and for girls, 6 years.

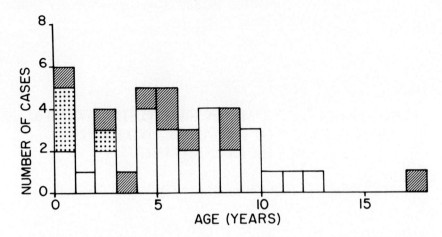

Fig. 1. Age at diagnosis of constitutional aplastic anemia in 40 pediatric patients seen at Children's Hospital Medical Center (CHMC), Boston, in 1958-1977. White = 26 patients with FA. Stippled = four patients with amegakaryocytic thrombocytopenia, who developed aplastic anemia. Hatched = ten patients in six families, with familial non-FA bone marrow failure (from Alter et al. [3]).

The diagnosis of FA is usually made in a child with aplastic anemia (or perhaps leukemia, see below) who has one or more characteristic physical anomalies and,or, chromosomal instability. Cytogenetic study was not made in earlier cases and on occasion may be normal in bona fide cases. The male:female ratio is 174:129 (or 1.4:1) in the literature cases and 17:9 (or 1.9:1) in our 26 cases.

The inheritance pattern of FA is apparently autosomal recessive [55]. In our 26 cases, two were adopted, one was the product of incest, and two were the products of consanguineous marriages; seven families had more than one affected child (two families of two brothers each are included, and in five families we saw only one of the affected sibs). Phenotypic expression is variable: Li and Potter [39] reported recently that a family of five children with aplasia, described by Estren and Dameshek [24] as familial aplastic anemia without physical anomalies, in fact carried the FA gene; a child with classic FA had parents who were second cousins to each other and also cousins to the original family described by Estren and Dameshek. The frequency of FA heterozygotes has been estimated at between 1/300 and 1/600 [63].

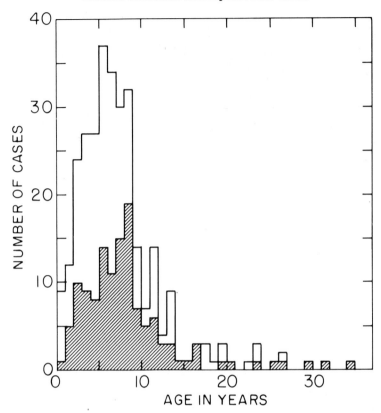

Fig. 2. Age at diagnosis of bone marrow failure in 303 published cases with FA. White=174 males. Hatched=129 females. Cases summarized from references (which are given in [2]).

PRESENTATION

The typical patient may have had orthopedic problems from birth, such as abnormal thumbs. He may have been small, and grew poorly during childhood. Between 5 and 10 years of age, he may appear pale, but medical help may not be sought until bruises and petechiae appear. At this time, he will be found to be anemic, thrombocytopenic, and leukopenic. The parents will realize in retrospect that he has been pale. It is very unusual for infections to be a problem during the onset of pancytopenia.

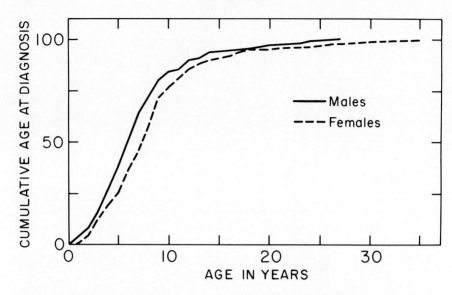

Fig. 3. Cumulative percentage distribution of age at diagnosis for 303 cases of FA (references in [2]).

PHYSICAL FINDINGS

Table I compares the physical abnormalities that have been reported in the literature with those described in our 26 cases. Five of our cases had café au lait spots, six were hyperpigmented, and eight had both skin findings. Thumb anomalies occurred in 14 patients: two had an extra digit, six lacked thumbs bilaterally, and six had abnormally small thumbs. In four of those lacking thumbs, radii also were absent or hypoplastic. Flat thenar eminences were noted in an additional six. Three patients had scoliosis, and one had a Sprengel deformity. Twenty-two of the children were abnormally short, and ten had microcephaly. Abnormally low birth weights were reported in seven of 22 cases, although only two were premature by dates. Structural renal abnormalities occurred in 13: four lacked one kidney, three had ectopic and three pelvic kidneys, and one had a horseshoe kidney; two had double collecting systems without other anomalies, and three had double collecting systems plus other anomalies. Two males had undescended testes, one had abnormally small testes, and one had an unusually small penis. Eye findings were common, with seven cases of strabismus and nine of microophthalmia (usually small and, or, close-set eyes). Hyperreflexia was noted in seven patients and mental retardation in six. Deafness was found in eight cases

TABLE I. Physical Abnormalities in Fanconi's Anemia

Abnormality	Literature*		CHMC	
	Number	Percentage	Number	Percentage
Number of patients	129	100	26	100
Hyperpigmentation and, or, café au lait spots	99	77	19	73
Thumb anomalies	48	37	14	54
Other skeletal anomalies	37	29	7	15
Microsomy (small stature)	78	60	22	85
Low birth weight	20/36†	56	7/22†	32
Microcephaly	51	40	10	38
Renal anomalies	36	28	13	50
Hypogenitalism	26M,5F	24	4M	16
Strabismus	28	22	7	27
Microophthalmia	20	16	9	35
Hyperreflexia	24	19	7	27
Mental deficiency	22	17	6	23
Ear anomalies and, or, deafness	9	7	10	38
Congenital heart disease	8	6	3	12

*Modified in [2] from [29].
†Denominator = number of cases where data available.

and was usually conductive, due to anomalies of bones of the middle ear. Five had abnormal-appearing ears. Three patients had heart disease due to patent ductus arteriosis. Three patients had congenital hip dislocation and three had high-arched palates. The proportions of the anomalies in our small group of patients were similar to those reported in the literature. Of particular note is the fact that all our patients had at least one abnormal physical finding. The photograph of a representative patient is shown in Figure 4. He was unusually short, mentally deficient, deaf, hyperpigmented, and lacked thumbs and radii.

LABORATORY STUDIES

The onset of pancytopenia is gradual. Although the presenting sign is often anemia, thrombocytopenia and leukopenia may be found at the same time. The anemia is often macrocytic, with mild poikilo- and anisocytosis (Fig. 5) and reticulocytopenia. The bone marrow is usually hypocellular, although foci of hypercellularity may be found early in the evolution of marrow failure. Marrow biopsy will provide the best evaluation of cellularity.

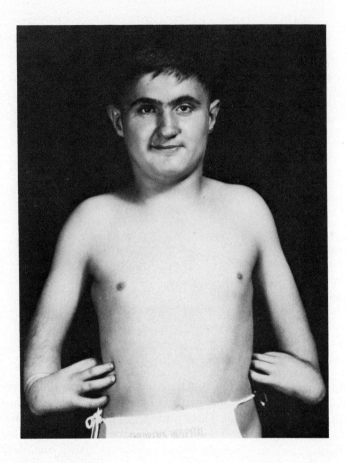

Fig. 4. Twelve-year-old patient with FA. Note absence of radii and thumbs (from [2]).

The evolution of marrow failure and, or, the response to treatment (see below) are accompanied by "stress erythropoiesis," in which red cells are produced that have "fetal" characteristics [1]. These cells are macrocytes, with fetal hemoglobin (distributed heterogeneously) and i antigen. Figure 6 shows the incidence of these fetal-like erythroid characteristics in some of our patients. Macrocytes and, or, increased fetal hemoglobin have been observed in patients prior to the appearance of overt pancytopenia [57].

The most specific laboratory findings come from studies of metaphase chromosomes in peripheral blood lymphocytes or, less commonly, cells in

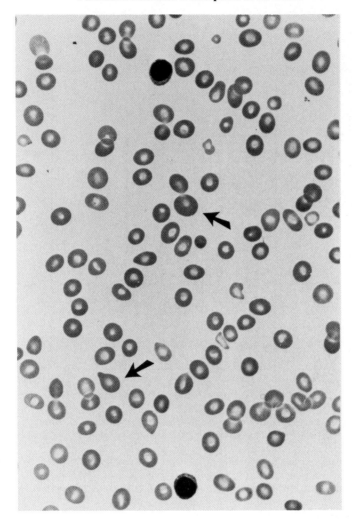

Fig. 5. Peripheral blood from a patient with FA. Note anisocytosis and occasional macrocytes (arrows), thrombocytopenia, and neutropenia. Photograph courtesy of Dr. Gail Wolfe (reproduced from [2]).

the marrow. These studies show chromatid breaks or gaps, rearrangements, chromatid interchange figures, and endoreduplication in 10-75% of the cells [13]. These abnormalities have been reported prior to pancytopenia as well as during remission. The proportion of abnormal metaphases does not correlate with the clinical status, and cytogenetic study

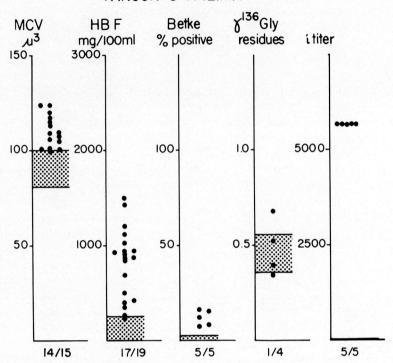

Fig. 6. Parameters of fetal-like erythropoiesis in patients with FA. Numbers indicate the number that were fetal-like in the number tested. Stippled area indicates the normal range (from [1]). MCV = mean corpuscular volume. Hb F = fetal hemoglobin, determined by alkali denaturation. Betke = acid-resistant cells, according to method of Kleihauer and Betke. γ^{136}Gly residues = proportion of gamma chain of fetal hemoglobin that has Glycine at position 136, compared to gamma with Glycine plus gamma with Alanine at this position. Titer of i = dilution of antiserum to i which agglutinates patient red cells. High values in all of these parameters indicate cells with fetal-like characteristics.

may be normal on one occasion and abnormal on another [53]. The proportion of abnormal cells is greater in peripheral blood lymphocytes stimulated to divide by phytohemagglutinin than in direct unstimulated marrow [10]. In our group of patients, chromosome studies were available in 22 and abnormal in 19. The three patients not showing increased chromosome abnormalities were the following: (1) a 10-year-old pancytopenic boy with an abnormal and small left hand, in whom an examination of 50 cells showed two with a break and seven with gaps; (2) a 1-year-old pancytopenic boy who was short and had bilateral double renal collecting systems, whose chromosome study showed a rare trisomic cell (+F or +G) and occasional ring chromosomes; and, (3) a 14-year-old girl with pancytopenia, short stature, hyperpigmentation, microophthalmia, and a double collecting system, in whom two cells out of the 60 examined had breaks.

Culture of patients' cells in the presence of mitomycin C increased the number of breaks but produced fewer sister-chromatid exchanges than in cells from normal persons. Latt et al. [37] reported similar results in four of our patients and the authors suggested that the patients have a defect in DNA repair. Other laboratories have reported increased in vitro transformation of FA fibroblasts by SV40 virus [66] and increased rates of chromosome damage following culture with specific carcinogens [5]. Inasmuch as pancytopenia does not appear until childhood or later, it may reflect the combination of an abnormality of DNA repair coupled to an environmental insult (toxic or viral). Genetic heterogeneity of FA was suggested recently by complementation studies [67]; somatic cell hybrids of SV40-transformed FA cells from one patient corrected the high mitomycin-C-induced aberration rate in untransformed cells from a second but not from a third FA patient.

Cultures of bone marrow cells reveal markedly decreased or absent hematopoietic progenitor cells (CFU-C, CFU-E, and BFU-E) [17,20,40,52]. These progenitors have been reported to be absent even in patients who were not anemic when studied [20]. Thus, the pluripotent stem cell may be defective. The rare successful engraftment with donor marrow (see below) supports the concept of a stem-cell disorder.

TREATMENT

Treatment of patients with FA involves support with specific transfusions of red cells, white cells, and platelets. In the past, when only blood transfusions were available, it was said that 80% of patients died within 2 years of diagnosis and almost all within 4 years. Figure 7 shows the cumulative survival for 280 cases reported in the literature. Half had androgen therapy (see below) and half either had not or treatment was not stated. The median survival was 4.5 years overall: 2.5 years in the group that apparently did not receive androgens and 8 years in the androgen-treated group. Eighty percent mortality occurred at 12 years overall, but at 8 years in the untreated and 14 years in the treated patients. Thus, the androgen-treated group did survive longer, by approximately 6 years. This apparent prolongation of survival may reflect selective reporting as well as improved supportive care in recent years. Overall, there were 64

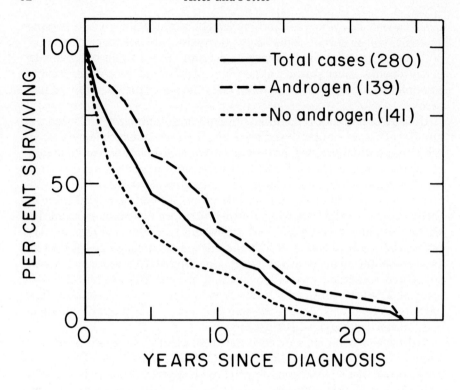

Fig. 7. Cumulative survival of published cases with FA, 1927–1980 (references in [2]).

deaths reported among 139 treated patients (46%) and 92 among 141 untreated patients (65%). Thus, androgen treatment both prolonged life and led to an apparently lower overall mortality rate; however, further information is needed concerning the subsequent status of patients who were alive when reported.

Figure 8 shows the cumulative survival analysis of the CHMC patients. Fifty percent survived 4 years and 20% 12 years. All patients did receive androgens; despite this, the survival curve of our patients is no better than the overall survival in the literature cases (Fig. 7). The five current survivors are 3–18 years from diagnosis (mean, 10 years).

Androgen treatment was first reported in 1959 by Shahidi and Diamond [56]. Review of the literature suggests that more than 50% of patients may respond to such treatment initially [2] by improvement of blood counts. Fourteen of our 26 patients (54%) responded. The course of one such patient (an 11-year-old girl) is shown in Figure 9. The reticulocyte count increased initially, with a lag before the hemoglobin rose. The responses of the white cell count and platelet count were more gradual. A decrease in androgen dosage resulted eventually in a decrease in the platelet count, which then rose when the androgen was increased again [data not shown]. Our current therapeutic regimen consists of prednisone (10 mg every other day) for vascular stability, along with oxymetholone (2-5 mg/kg/d orally) or 19-nortestosterone decanoate (1-2 mg/kg per week intramuscularly). If the patient responds, the androgen dose is decreased, but therapy usually is not discontinued entirely.

COMPLICATIONS

The major complications reported in FA have been hepatic disease and malignancies (Table II, Fig. 10). Hepatocellular carcinomas were reported in six cases [32,34,36,44,50,58]; hepatomas in eight [12,16,25,33, 46,47,51,62]; adenomas in three [18,27,45]; peliosis hepatis in three [6,36];

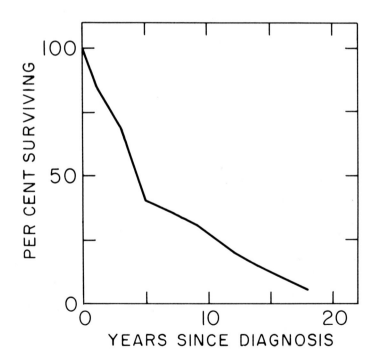

Fig. 8. Cumulative survival of 26 CHMC patients, followed until 1980.

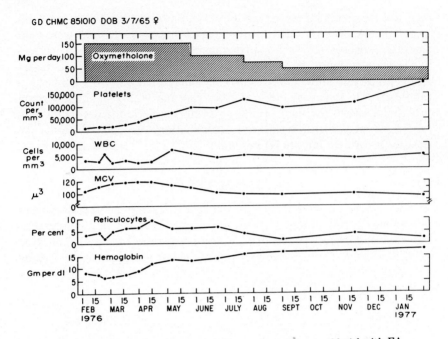

Fig. 9. Course of response to androgen treatment in an 11-year-old girl with FA.

and hepatic coma in one [55]. Although androgens have been implicated in the development of hepatic tumors, the cause-and-effect relationship remains to be proven. One hepatoma developed in a patient who had received no androgens [16], while the others had received this treatment. Three patients with hepatoma also had acute leukemia [46,47,51]. In our own group of patients, one 3-year-old girl had peliosis hepatitis, but no case of liver tumor was seen. Oxymetholone may cause cholestatic jaundice and necessitate a change to a different form of androgen.

Acute leukemia is reputed to be the terminal event in 5–10% of FA patients. In fact, 22 examples of leukemia in persons with FA are mentioned or described in the literature. This includes three of our patients, two of whom were mentioned but not described in detail. Five of those reported had acute myelocytic leukemia [22,30,46,59,68]; seven myelomonocytic [14,19,22,47,51,54,55,59]; three monocytic [13,38]; four erythroleukemia [7,10,43,49]; and in three the type of leukemia was not described [11,31,55]. Our patients are cited [in references 13 and 22]. Five of the reported patients with leukemia apparently had had no androgen therapy [13,19,31,55]. Three leukemic patients also had hepatomas [46,47,51]. One

TABLE II. Complications in Fanconi's Anemia*

		Number of cases		
		Androgen therapy	No androgen therapy or not stated	Total
Liver disease		19	2	21
Hepatocellular carcinoma	6			
Adenoma	3			
Hepatoma (three had leukemia)	8			
Peliosis hepatitis	3			
Hepatic coma	1			
Acute leukemia		17	5	22
Myeloblastic	5			
Myelomonocytic	7			
Monocytic	3			
Erythroleukemia	4			
Type unspecified	3			
Squamous cell carcinomas†		5	1	6
Bone marrow transplantation		11	7	18
Dead (usually from GVH reaction)	14			
Alive	4			
Death from aplastic anemia		19	81	100
Total deaths		64	92	156

*See text for references. Cases are from literature.
†Esophagus; gingiva (2); anus; anus and vulva; tongue and hepatocellular carcinoma. Two still alive.

had not had pancytopenia preceding the diagnosis of leukemia [19]. In the others, the interval from pancytopenia to leukemia ranged from less than 1 year in two patients, to 14 years in one patient. The median was 5 years, similar to the overall median survival shown in Figure 7. Death from leukemia usually ensued soon after diagnosis (less than 1 month in 13 instances, 1-3 months in five, and less than 1 year in the rest). Only five patients received specific chemotherapy, and only one had even a partial response. (Since the writing of this chapter, one additional case of acute myelomonocytic leukemia was reported in a boy who had received oxymetholone for 3 years [J Kunze: Estren-Dämeshek-Anämie mit myelomonocytärer Leukämie (Subtyp der Fanconi-Anämie?), Klinische Genetik in der Pädiatrie, 2. Symposon in Mainz, Georg Thieme Verlag, Stuttgart, 1980].)

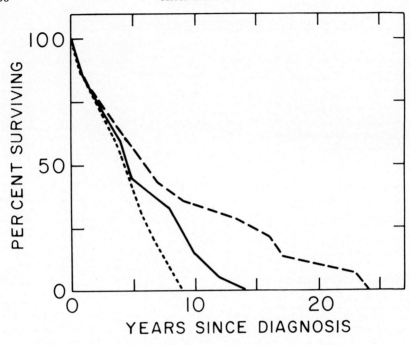

Fig. 10. Cumulative percentage survival from onset of pancytopenia to death from complications. Solid line, 22 cases with leukemia; dashed line, 14 cases with liver tumors; dotted line, 18 cases who had bone marrow transplantation. See text for references.

Other malignancies that have been reported are six cases with squamous cell carcinomas [23,32,42,51,65]. Five of the six patients were being treated with androgens prior to this complication. One patient also had hepatocellular carcinoma [32]. Four of the six patients died. These six were all older than 20 when they developed the tumors, and only two had pancytopenia before 20 years of age.

Androgen therapy may permit the persons with FA to survive long enough to manifest their genetic predisposition to leukemia or other malignancies. However, the survival from the onset of pancytopenia to death from leukemia was no better than that of the entire group of FA cases (Fig. 10). The median survival was 5 years, and all patients had died by 12 years following the onset of pancytopenia. Thus, the role of androgen therapy as a possible cause of leukemia is unclear. In contrast, approximately half of the patients who developed liver tumors survived longer than expected. The median survival of those who developed liver tumors was 6 years, but five patients survived more than 13 years from the diagnosis of pancytopenia; however, the number of patients with these tumors is very small. Either this long survival led to expression of the tendency for tumors or prolonged androgen treatment played a role in tumor development. Because almost all patients do receive androgen treatment now, the controversy about its tumorigenic potential may be difficult to clarify.

It was at one time suggested by Swift [63] that the FA gene itself was associated with malignancy. He reported an increased incidence of cancer in presumed heterozygotes. However, he subsequently studied 25 extended families of FA probands and found no excess of cancers or cancer deaths among obligate heterozygotes [64], thus superseding his earlier report. A similar study of 15 of the CHMC families (including six of the 25 families studied by Swift) also indicates no increased incidence of cancer in FA heterozygotes [48]. Thus, the association of the FA gene and malignancy may require homozygosity.

Bone marrow transplantation has now been reported in 18 persons with FA [4,8,15,21,28,35,41,60,61], only four of whom were still alive at the time of the reports. One of our patients was so transplanted but died of sepsis before engraftment. All marrow-transplanted patients had graft-versus-host disease. Berger et al. [9] observed that the cells of FA patients were unusually sensitive to a metabolite of cytoxan, the main agent used to prepare all the patients for marrow transplantation. Thus, rapidly dividing cells (e.g., intestinal mucosa, liver, skin) of FA patients may be vulnerable to DNA damage by the alkylating agents used in marrow transplantation (as well as in chemotherapy). Another potential problem with marrow transplantation is that a sibling to be used as a donor may be an undiagnosed FA homozygote, or at least an FA heterozygote, whose cells may have a partial defect in DNA repair.

The actual outcome of all reported persons with FA cannot be determined from the literature because follow-up reports are not always available. Review of all the 139 patients who received androgens indicates that at least 75 (54%) responded initially. Of these 139 patients, 17 (12%) developed leukemia and 20 (14%) developed liver disease; 13 of the liver problems were fatal tumors. Eventually most patients failed to respond to androgens, and only seven succeeded in discontinuing the medications for 1-8 years. The final outcomes of the latter group are not reported. Twenty-one of our 26 cases have died (Table III), 18 from complications of aplastic anemia and three from leukemia. Our five survivors are 3, 5, 7, 16, and 18 years postdiagnosis. One of these, a 21-year-old man, is a poor androgen responder but was maintained for 7 years by transfusions. This patient has now responded to the experimental use of etiocholanolone, despite failure to respond to oxymetholone or testosterone. The other

TABLE III. Causes of Death in Fanconi's Anemia at CHMC

	Number of cases	Years between diagnosis of FA and death	
		Range	Median
Aplastic anemia	18	1–19	4
Leukemia	3	1– 5	
Total	21		

four responded to the usual types of androgens, and one has apparently received no treatment for 12 years.

Although our group of patients is small (26 cases), it is quite representative of the disorder and has the advantage of providing complete information from a single center. Despite the advent of androgen treatment, the prognosis of FA remains poor, because most patients eventually do fail to respond to androgens.

SUMMARY

Fanconi's anemia is an autosomal recessive disorder. Its biochemical basis is unknown. Associated chromosome instability is present, which suggests the possibility of defective DNA repair. Most patients have characteristic congenital physical abnormalities. The onset of aplastic anemia usually is in middle childhood. Many patients respond to androgen therapy although aplastic anemia reappears ultimately. The prognosis is poor, and despite the use of androgen therapy, chemotherapy, and bone marrow transplantation, death usually results from complications of pancytopenia. The incidence of malignancies such as leukemia and liver tumors is increased.

Acknowledgments. We are grateful to David Nathan, Frederick Li, James German, Arleen Auerbach, and Paul Berk for helpful discussions and advice, and to Judith Goldberg for statistical consultation. This work was supported in part by National Institutes of Health grant HL26132, an Irma T. Hirschl Career Scientist Award to B.P.A., and the Amy Potter Memorial Fund.

LITERATURE CITED

1. Alter, B.P.: Fetal erythropoiesis in bone marrow failure syndromes. In Cellular and Molecular Regulation of Hemoglobin Switching (Stamatoyannopoulos, G., Nienhuis, A.W., eds.), New York: Grune & Stratton, 87–105, 1979.
2. Alter, B.P., Parkman, R., Rappeport, J.M.: Bone marrow failure syndromes. In Hematology of Infancy and Childhood (Nathan, D.G., Oski, F.A., eds.), 2nd ed., Philadelphia: W.B. Saunders, 168–249, 1981.

3. ALTER, B.P., POTTER, N.U., LI, F.P.: Classification and aetiology of the aplastic anemias. *Clin. Haematol.* 7:431-465, 1978.
4. AUERBACH, A.D., ADLER, B., CHAGANTI, R.S.K.: Prenatal and postnatal diagnosis and carrier detection of Fanconi's anemia by a cytogenetic method. *Pediatrics* 67:128-139, 1981.
5. AUERBACH, A.D., WOLMAN, S.R.: Susceptibility of Fanconi's anaemia fibroblasts to chromosome damage by carcinogens. *Nature* 261:494-496, 1976.
6. BANK, J.I., LYKKEBO, D., HAGERSTRAND, I.: Peliosis hepatis in a child. *Acta Paediatr. Scand.* 67:105-107, 1978.
7. BARGMAN, G.J., SHAHIDI, N.T., GILBERT, E.F., OPITZ, J.M.: Studies of malformation syndromes of man. XLVII: Disappearance of spermatogonia in the Fanconi anemia syndrome. *Eur. J. Pediatr.* 125:162-168, 1977.
8. BARRETT, A.J., BRIGDEN, W.D., HOBBS, J.R., HUGH-JONES, K., HUMBLE, J.G., JAMES, D.C.O., RETSAS, S., ROGERS, T.R.F., SELWYN, S., SNEATH, P., WATSON, J.G.: Successful bone marrow transplant for Fanconi's anaemia. *Br. Med. J.* 1:420-422, 1977.
9. BERGER, R., BERNHEIM, A., GLUCKMAN, E., GISSELBRECHT, C.: In vitro effect of cyclophosphamide metabolites on chromosomes of Fanconi anaemia patients. *Br. J. Haematol.* 45:565-568, 1980.
10. BERGER, R., BERNHEIM, A., LE CONIAT, M., VECCHIONE, D., SCHAISON, G.: Chromosomal studies of leukemic and preleukemic Fanconi's anemia patients. *Hum. Genet.* 56:59-62, 1980.
11. BERGER, R., BUSSEL, A., SCHENMETZLER, C.: Somatic segregation and Fanconi anemia. *Clin. Genet.* 11:409-415, 1977.
12. BERNSTEIN, M.S., HUNTER, R.L., YACHNIN, S.: Hepatoma and peliosis hepatis developing in a patient with Fanconi's anemia. *N. Engl. J. Med.* 284:1135-1136, 1971.
13. BLOOM, G.E., WARNER, S., GERALD, P.S., DIAMOND, L.K.: Chromosome abnormalities in constitutional aplastic anemia. *N. Engl. J. Med.* 274:8-14, 1966.
14. BOURGEOIS, C.A., HILL, F.G.H.: Fanconi anemia leading to acute myelomonocytic leukemia. Cytogenetic studies. *Cancer* 39:1163-1167, 1977.
15. CAMITTA, B.M.: Histocompatible bone marrow transplantation for hereditary and acquired marrow aplasias in children. *In* Aplastic Anemia (Hibino, S., Takaku, F., Shahidi, N.T., eds.), Tokyo: University of Tokyo Press, 429-440, 1978.
16. CATTAN, D., KALIFAT, R., WAUTIER, J-L., MEIGNAN, S., VESIN, P., PIET, R.: Maladie de Fanconi et cancer du foie. *Arch. Fr. Mal. App. Dig.* 63:41-48, 1974.
17. CHU, J-Y.: Granulopoiesis in Fanconi's aplastic anemia. *Proc. Soc. Exp. Biol. Med.* 161:609-612, 1979.
18. CORBERAND, J., PRIS, J., DUTAU, G., RUMEAU, J-L., REGNIER, C.: Association d'une maladie de Fanconi et d'une tumeur hépatique chez une malade soumise à un traitement androgénique au long cours. *Arch. Fr. Pediatr.* 32:275-283, 1975.
19. COWDELL, R.H., PHIZACKERLEY, P.J.R., PYKE, D.A.: Constitutional anemia (Fanconi's syndrome) and leukemia in two brothers. *Blood* 10:788-801, 1955.
20. DANESHBOD-SKIBBA, G., MARTIN, J., SHAHIDI, N.T.: Myeloid and erythroid colony growth in non-anaemic patients with Fanconi's anaemia. *Br. J. Haematol.* 44:33-38, 1980.
21. DOOREN, L.J., KAMPHUIS, R.P., DE KONING, J., VOSSEN, J.M.: Bone marrow transplantation in children. *Semin. Hematol.* 11:369-382, 1974.
22. DOSIK, H., HSU, L.Y., TODARO, G.J., LEE, S. L., HIRSCHHORN, K., SELIRIO, E.S., ALTER, A.A.: Leukemia in Fanconi's anemia: Cytogenetic and tumor virus susceptibility studies. *Blood* 36:341-351, 1970.
23. ESPARZA, A., THOMPSON, W.R.: Familial hypoplastic anemia with multiple congeni-

tal anomalies (Fanconi's syndrome) – report of three cases. *R.I. Med. J.* 49:103-110, 1966.
24. ESTREN, E., DAMESHEK, W.: Familial hypoplastic anemia of childhood. *Am. J. Dis. Child.* 73:671-687, 1947.
25. EVANS, D.I.K.: Aplastic anaemia in childhood. *In* Aplastic Anaemia (Geary, C.G., ed.), London: Baillière Tindall, 161-194, 1979.
26. FANCONI, G.: Familiäre infantile perniziosaartige Anämie (perniziöses Blutbild und Konstitution). *Z. Kinderheilkd.* 117:257-280, 1927.
27. FARRELL, G.C.: Fanconi's familial hypoplastic anaemia with some unusual features. *Med. J. Aust.* 1:116-118, 1976.
28. GLUCKMAN, E., DEVERGIE, E., SCHAISON, G., BUSSEL, A., BERGER, R., SOHIER, J., BERNARD, J.: Bone marrow transplantation in Fanconi anaemia. *Br. J. Haematol.* 45:557-564, 1980.
29. GMYREK, D., SYLLM-RAPOPORT, L.: Zur Fanconi Anämie (FA). Analyse von 129 beschriebenen Fällen. *Z. Kinderheilkd.* 91:297-337, 1964.
30. GMYREK, D., WITKOWSKI, R., SYLLM-RAPOPORT, I., JACOBASCH, G.: Chromosomal aberrations and abnormalities of red-cell metabolism in a case of Fanconi's anaemia before and after development of leukaemia. *Ger. Med. Mon.* 13:105-111, 1968.
31. GOTZ, M., HANDEL, H., PICHLER, E., WEIPPL, G.: Fanconi-Anämie. Übergang in Leukämie und Erythrozytenstoffwechselbefunde. *Kinderärtzl. Prax.* 42:69-74, 1974.
32. GUY, J.T., AUSLANDER, M.O.: Androgenic steroids and hepatocellular carcinoma. *Lancet* 1:148, 1973.
33. HOLDER, L.E., GNARRA, D.J., LAMPKIN, B.C., NISHIYAMA, H., PERKINS, P.: Hepatoma associated with anabolic steroid therapy. *Am. J. Roengentol. Radium Ther. Nucl. Med.* 124:638-642, 1975.
34. JOHNSON, F L., FEAGLER, J.R., LERNER, K.G., MAJERUS, P.W., SIEGEL, M., HARTMANN, J.R., THOMAS, E.D.: Association of androgenic-anabolic steroid therapy with development of hepatocellular carcinoma. *Lancet* 2:1273-1276, 1972.
35. KALLENBERG, C.G.M., MULDER, N.H., THE, T.H., SPECK, B.: Defective stimulating capacity of leukocytes in mixed leukocyte culture in constitutional aplastic anemia caused by suppressor T cells. A case study. *Acta Haematol.* 63:81-87, 1980.
36. KEW, M.C., VAN COLLER, B., PROWSE, C.M. SKIKNE, B., WOLFSDORF, J.I., ISDALE, J., KRAWITZ, S., ALTMAN, H., LEVIN, S.E., BOTHWELL, T.H.: Occurrence of primary hepatocellular cancer and peliosis hepatis after treatment with androgenic steroids. *S. Afr. Med. J.* 50:1233-1237, 1976.
37. LATT, S.A., STETTEN, G., JUERGENS, L.A., BUCHANAN, G.R., GERALD, P.S.: Induction by aklylating agents of sister chromatid exchanges and chromatid breaks in Fanconi's anemia. *Proc. Natl. Acad. Sci. USA* 72:4066-4070, 1975.
38. LEVY, J.M., STOLL, C., KORN, R.: Sur un cas de leucémie aiguë chez une filiette atteinte d'anémie de Fanconi. Revue de la littérature. *Nouv. Rev. Fr. Hematol.* 14:713-720, 1974.
39. LI, F.P., POTTER, N.U.: Classical Fanconi anemia in a family with hypoplastic anemia. *J. Pediatr.* 92:943-944, 1978.
40. LUI, V.K., RAGAB, A.H., FINDLEY, H.S., FRAUEN, B.J.: Bone marrow cultures in children with Fanconi anemia and the TAR syndrome. *J. Pediatr.* 91:952-954, 1977.
41. MARMONT, A.M., VAN LINT, M.T., AVANZI, G., REALI, G., ADAMI, R., SOLDA, A., STRADA, P., BARBANTI, M., MINGARI, M.C., SORO, O., GRAZI, G., PEDULLA, D., CERRI, R., ROSSI, E., GIORDANO, D., SANTINI, G., CARELLA, A., RISSO, M., VIMERCATI, R., PIAGGIO, G., RAFFO, M.R., LIBRACE, E., VITALE, V., BACIGALUPO, A.: Bone marrow transplantation for severe aplastic anemia. A report of 9 cases. *Acta Haematol.*

62:121-127, 1979.
42. McDonough, E.R.: Fanconi anemia syndrome. *Arch. Otolaryngol.* 92:284-285, 1970.
43. Meisner, L.F., Taher, A., Shahidi, N.T.: Chromosome changes and leukemic transformation in Fanconi's anemia. *In* Aplastic Anemia (Hibino, S., Takaku, F., Shahidi, N.T., eds.), Tokyo: University of Tokyo Press, 253-271, 1978.
44. Mokrohisky, S.T., Ambruso, D.R., Hathaway, W.E.: Fulminant hepatic neoplasia after androgen therapy. *N. Engl. J. Med.* 296:1411-1412, 1977.
45. Mulvihill, J.J., Ridolfi, R.L., Schultz, F.R., Borzy, M.S., Haughton, P.B.T.: Hepatic adenoma in Fanconi anemia treated with oxymetholone. *J. Pediatr.* 87:122-124, 1975.
46. Obeid, D.A., Hill, F.G.H., Harnden, D., Mann, J.R., Wood, B.S.B.: Fanconi anemia. Oxymetholone hepatic tumors, and chromosome aberrations associated with leukemic transition. *Cancer* 46:1401-1404, 1980.
47. Perrimond, H., Juhan-Vague, I., Thevenieau, D., Bayle, J., Muratore, R., Orsini, A.: Evolution médullaire et hépatique après androgénothérapie prolongée d'une anémie de Fanconi. *Nouv. Rev. Fr. Hematol.* 18:228, 1977.
48. Potter, N.U., Sarmousakis, C., Li, F.P.: Cancer in relatives of patients with aplastic anemia. Submitted.
49. Prindull, G., Jentsch, E., Hansmann, I.: Fanconi's anaemia developing erythroleukaemia. *Scand. J. Haematol.* 23:59-63, 1979.
50. Recant, L., Lacy, P.: Fanconi's anemia and hepatic cirrhosis. *Am. J. Med.* 39:464-475, 1965.
51. Sarna, G., Tomasulo, P., Lotz, J., Bubinak, J.F., Shulman, N.R.: Multiple neoplasms in two siblings with a variant form of Fanconi's anemia. *Cancer* 36:1029-1033, 1975.
52. Saunders, E.F., Freedman, M.H.: Constitutional aplastic anaemia. Defective haematopoietic stem cell growth in vitro. *Br. J. Haematol.* 40:277-287, 1978.
53. Schroeder, T.M., Drings, P., Beilner, P., Buchinger, G.: Clinical and cytogenetic observations during a six-year period in an adult with Fanconi's anemia. *Blut* 34:119-132, 1976.
54. Schroeder, T.M., Pohler, E., Hufnagl, H.D., Stahl-Mauge, C.H.: Fanconi's anemia: Terminal leukemia and "forme fruste" in one family. *Clin. Genet.* 16:260-268, 1979.
55. Schroeder, T.M., Tilgen, D., Kruger, J., Vogel, F.: Formal genetics of Fanconi's anemia. *Hum. Genet.* 32:257-288, 1976.
56. Shahidi, N.T., Diamond, L.K.: Testosterone-induced remission in aplastic anemia. *Am. J. Dis. Child.* 98:293-302, 1959.
57. Shahidi, N.T., Gerald, P.S., Diamond, L.K.: Alkali-resistant hemoglobin in aplastic anemia of both acquired and congenital types. *N. Engl. J. Med.* 266:117-120, 1962.
58. Shapiro, P., Ikeda, R.M., Ruebner, B.N., Connors, M.H., Halsted, C.C., Abildgaard, C.F.: Multiple hepatic tumors and peliosis hepatis in Fanconi's anemia treated with androgens. *Am. J. Dis. Child.* 131:1104-1106, 1977.
59. Skikne, B.S., Lynch, S.R., Bezwoda, W.R., Bothwell, T.H., Bernstein, R., Katz, J., Kramer, S., Zucker, M.: Fanconi's anaemia, with special reference to erythrokinetic features. *S. Afr. Med. J.* 53:43-50, 1978.
60. Storb, R., Thomas, E.D., Buckner, C.D., Clift, R.A., Deeg, H.J., Fefer, A., Goodell, B.W., Sale, G.E., Sanders, J.E., Singer, J., Stewart, P., Weiden, P.L.: Marrow transplantation in thirty "untransfused" patients with severe aplastic anemia. *Ann. Intern. Med.* 92:30-36, 1980.

61. Storb, R., Thomas, E.D., Weiden, P.L., Buckner, C.D., Clift, R.A., Fefer, A., Fernando, L.P., Giblett, E.R., Goodell, B.W., Johnson, F.L., Lerner, K.G., Neiman, P.E., Sanders, J.E.: Aplastic anemia treated by allogeneic bone marrow transplantation: A report on 49 new cases from Seattle. *Blood* 48:817–841, 1976.
62. Sweeney, E.C., Evans, D.J.: Hepatic lesions in patients treated with synthetic anabolic steroids. *J. Clin. Pathol.* 29:626–633, 1976.
63. Swift, M.: Fanconi's anemia in the genetics of neoplasia. *Nature* 230:370–373, 1971.
64. Swift, M., Caldwell, R.J., Chase, C.: Reassessment of cancer predisposition of Fanconi anemia heterozygotes. *J. Natl. Cancer Inst.* 65:863–867, 1980.
65. Swift, M., Zimmerman, D., McDonough, E.R.: Squamous cell carcinomas in Fanconi's anemia. *J.A.M.A.* 216:325–326, 1971.
66. Todaro, G.J., Green, H., Swift, M.R.: Susceptibility of human diploid fibroblast strains to transformation by SV40 virus. *Science* 153:1252–1254, 1966.
67. Zakrzewski, S., Sperling, K.: Genetic heterogeneity of Fanconi's anemia demonstrated by somatic cell hybrids. *Hum. Genet.* 56:81–84, 1980.
68. Zawartka, M., Restorff-Libiszowska, H., Kowalski, R., Litwiniszyn-Krzewicka,K., Schejbal, J.: Bialaczka szpikowa w przebiegu anemii Fanconiego (AF). *Wiad. Lek.* 29:145–150, 1976.

XERODERMA PIGMENTOSUM

ALAN D. ANDREWS

INTRODUCTION

Xeroderma pigmentosum (XP) is a rare hereditary syndrome manifested most notably by an increased susceptibility to sunlight-induced cutaneous damage. Common clinical manifestations include acute sun sensitivity, abnormal pigmentation, atrophy of the skin, and cutaneous tumors, both benign and malignant. Some persons with XP also manifest severe progressive neurological abnormalities.

Biochemically, the hallmark of the syndrome is a reduced capacity for repair of ultraviolet (UV)-induced damage to cellular DNA. Through *in vitro* studies of their cells' DNA repair capacity, XP patients have been divided into eight different genetic subgroups, designated as complementation groups A, B, C, D, E, F, and G, and the "variant" group [7,14, 17,40,44]. These genetically distinct subgroups will be described later.

Although the phenotype of XP is recognized easily in persons presenting with well-developed clinical abnormalities, the age of onset and the degree of clinical abnormalities vary widely among affected individuals. This variability results from the numerous XP genotypes, from the wide differences in cumulative sunlight exposure among XP patients, and, undoubtedly, from other genetic and environmental factors.

This paper will describe in detail the clinical abnormalities of XP, and point out the correlations between clinical manifestations and genetic or environmental factors which are known.

EPIDEMIOLOGY

While xeroderma pigmentosum has been reported in all races and virtually all areas of the world, the estimated prevalence of the syndrome has varied widely. It has been estimated that, in the United States and

Europe, XP occurs in 1 out of 250,000 persons [69], whereas in Japan the estimate is 1 out of 40,000 persons [60]. The pattern of inheritance for the majority of cases is autosomal recessive. Males and females are affected equally and, as would be expected, the rate of parental consanguinity in families with affected individuals is high [27,69,74]. In areas of the world where consanguineous marriage is a common cultural pattern and sunlight is abundant, as in Egypt, a large number of XP cases have been reported [28,36]. There is one report of a family that may suffer from a dominantly inherited form of XP [3], and it is possible that one of the uncommon genetic forms of XP, such as group B (see below), which has been identified in only one individual, could be inherited in a manner different from that of the more common genetic forms.

ACUTE SUN SENSITIVITY

Beginning in infancy, some, but not all, XP patients experience an acute sunburnlike erythema after minimal-to-moderate sun exposure [69]. In the more severe reactions the erythema may be accompanied by blisters and subsequent desquamation. The reaction is confined to skin surfaces actually exposed to sunlight and so most often is reported on the face and arms, but there is no evidence that any one area of the body is more susceptible than another. Many patients who experience this acute sun sensitivity also report that the apparent severity of reactions diminishes gradually over the first 10–20 years of life. Clinical phototesting of XP patients using broad band or monochromatic UV sources has been reported in only a few cases, and the results have not been correlated clearly with the presence or absence of a positive history for acute reactions after natural sun exposure [42,62]. Nevertheless, the testing suggests that at least some XP patients may be sensitive to ultraviolet light of longer wavelengths than those that cause sunburn in normal individuals [38]; that the time between irradiation and attainment of the peak (most intense) erythema may be abnormally long in XP patients; and, that occasionally the morphological response of XP patients may be significantly different from that of normal persons [42,62].

In those XP patients who are subject to acute sun sensitivity, those reactions are often the first clinical indication of XP to appear. Current evidence suggests that this symptom occurs most often in XP patients who have the more severe DNA-repair deficits, at least to the extent that those deficits are reflected by studies in vitro of cellular UV sensitivity [4,10]. XP patients with a history of acute sun sensitivity are also more likely to develop XP-associated neurological abnormalities than are XP patients without such a history [10].

CHRONIC CUTANEOUS CHANGES

Virtually all XP patients, whether they experience acute sun-sensitivity reactions or not, eventually develop certain chronic skin changes in areas of repeated sun exposure. When these changes, along with the ocular changes discussed in the next section, occur in the absence of neurologic abnormalities, they constitute what is sometimes referred to as the "classical" form of XP.

The chronic skin changes occur gradually over a period of years, but the age of onset and the rate of progression vary widely. In general, a progression through several overlapping stages occurs. In the first or pigmentary stage, sun-exposed areas develop multiple hyperpigmented macules, identical in appearance to normal freckles, i.e., small, flat, uniformly pigmented lesions, generally light brown in color. The only suggestion of abnormality at this stage may be the unusually early appearance of such freckling, sometimes within the first year of life. Eventually, however, the pigmented lesions can be distinguished from ordinary freckles by their more varied size, shape, and color. Individual lesions vary in color from light brown to black; in size, from a millimeter or two to several centimeters; and in shape from round or oval to angular and markedly irregular. Hypopigmented macules appear, interspersed randomly with the hyperpigmented lesions. They, too, often begin as small, freckle-sized spots and later develop into larger more irregular patches. A characteristic stippled "salt and pepper" pattern results from the mixture of small hyper- and hypopigmented macules in early cases, whereas a more bizarre, patchwork appearance results from the addition of larger and more irregular lesions. In some patients the more irregular pigmentary changes are seen on one part of the body (e.g., face or arms) while less bizarre patterns are seen in other, usually more sun-protected, areas. The percentage of the skin surface affected varies widely among XP patients, due to differences in customs or habits of sun exposure. Most patients, however, show relative sparing of skin in the axillae and on the buttocks. Occasionally patients will exhibit small hyperpigmented macules of the palms or soles and of the tongue [43]. Pigmentary changes of the lips, especially the lower, are common, and similar to those of other sun-exposed skin.

Although pigmentary changes often occur before the age of 2, some patients report a later onset, and in persons in whom early diagnosis is followed by stringent measures to avoid sun exposure most of the pigmentary changes (and other chronic skin changes of XP) may be prevented [50].

The second, or atrophic and telangiectatic, stage of XP begins after freckling has appeared but progresses in concert with and in the same areas of skin as the most severe of the pigmentary changes. This stage is characterized by the development of dry, scaly, finely wrinkled skin with multiple small, red, venous dilatations, sometimes appearing as small "spider"-like angiomas or matted tangles of vessels. The most severely affected areas show marked atrophy of the skin, often associated with depigmentation; despite the appearance of fine wrinkling, the skin of these areas may be very tightly stretched, making it difficult to pinch the skin up into a fold. This atrophy and tightness of the skin is especially common on the face, where it may lead to ectropion of the lower eyelids and to difficulty in fully opening the mouth. The nose, ears, and mouth may all appear to be compressed by this atrophic process. The subcutaneous fat generally is normal despite severe atrophy of the overlying skin. Some patients, however, especially those with severe growth deficiency and neurologic abnormalities (see below), may show loss of subcutaneous fat. This is usually a generalized process, extending to all areas of the skin, whether sun-exposed or not, and, therefore, probably is not linked to sun damage as are the other abnormalities described.

The final, or neoplastic, stage is characterized by the appearance of multiple skin tumors, and usually begins with the development of actinic keratoses and other benign warty or papillomatous growths. The first appearance of actinic keratoses may be as early as 2 years of age, and the number of such tumors is greatest in the areas that show the greatest degree of pigmentary and atrophic change. The actinic keratoses occurring in XP patients are similar to those seen in normal individuals with severely sun-damaged skin; they consist of small, irregularly shaped plaques or papules with dense, adherent, "horny" scales on the surface. Their color generally varies from light yellow-brown to dark brown, and the surrounding skin is sometimes reddened due to inflammation or telangiectasia or both. Often the keratosis on the skin surface is most readily located by touch, as its hard, warty surface is more easily appreciated by the fingertip than is its appearance by the eye, especially as these occur in areas of already checkered pigmentation.

Other benign lesions, such as keratoacanthomas, angiomas, angiofibromas, and neurofibromas occur, but usually in much smaller numbers than actinic kertoses. At least one XP patient, however, has been reported who developed a large number of angiomatous lesions [32].

Malignant skin tumors, which have appeared as early as 2 and 3 years of age, more commonly develop several years after the onset of the other severe skin changes. In XP patients, basal cell carcinomas are the most common malignant tumors, but squamous cell carcinomas are only slightly less frequent. Both types of tumors occur most commonly in the areas of greatest sunlight-induced skin change. In many cases, the squamous cell tumors develop from preexisting actinic keratoses, as they often do in normal individuals. The clinical appearances of basal and squamous cell carcinomas in XP patients are as varied, and as characteristic, as similar tumors in normal persons. Thus, basal cell carcinomas may begin as small shiny, pearly papules, as pink, red brown, or black nodules, or as sclerotic plaques. As they enlarge, they often become increasingly irregular in shape and in color, often ulcerate in the center, and if left untreated, eventually will invade and destroy tissues beneath the skin. Squamous cell carcinomas frequently begin as actinic keratoses, which then show progressive thickening and expansion and an increased inflammatory reaction in the surrounding skin. In other cases squamous cell carcinomas arise as new nodules or plaques that may ulcerate as they enlarge and which may be impossible to differentiate clinically from a nodular basal cell carcinoma. Like basal cell carcinomas, squamous cell carcinomas also will invade deeper tissues if not treated early.

There are no well-documented cases of metastatic basal or squamous cell tumors in XP patients. Similar tumors that arise in sun-damaged skin of normal individuals very rarely metastasize, or, if they do metastasize, do so very late in the course of development. Thus, it seems that the biological behavior of these tumors in XP patients is similar to that of the corresponding tumors in normal individuals. In both groups, the tumors do their greatest harm by invading locally and destroying tissues, especially when they occur on the face, where destruction of the nasal or other facial bones, the eye socket, the ear, the sinuses, and even the cranium has occurred [43].

Malignant melanoma is the next most common malignant neoplasm in XP patients, but its reported frequency varies from zero, in a Japanese series of over 50 patients [74], to about 50%, in one series from the United States [69]. All histologic forms of malignant melanoma have been found in XP patients, and these tumors have a very high frequency of early metastasis to distant sites. Spontaneous regression of metastatic melanoma has been reported in XP patients [49], but more commonly metastatic disease is progressive and eventually fatal [69,75]. However, vigilant medical follow-up of patients with XP can lead to long-term survival, even of patients with multiple primary melanomas [69,71], provided that the individual lesions are excised before distant metastasis has occurred.

Sarcomas are the only other malignant skin tumors so far reported in XP patients [33], but apparently they are very uncommon.

OCULAR ABNORMALITIES

Almost all XP patients describe at least a mild photophobia, and most patients develop some degree of chronic inflammation of the bulbar or palpebral conjuctivae or both. As mentioned earlier, atrophy of the skin of the face and of the eyelids themselves can result in ectropion; this in turn can lead to exposure keratitis. In some patients such drying of the cornea may be aggravated further by decreased lacrimation. Corneal clouding and ulceration can lead to blindness; vascularization of the cornea, which often accompanies the other processes, may render the patient a poor candidate for corneal transplantation. Other complications of conjunctival inflammation include pterygia, pinguecula, and symblepharon [43]. Inflammation of the iris is less common but can lead to the development of synechia [43,69]. Retinal changes do not occur in XP, presumably because most or all of the harmful wavelengths of ultraviolet light in sunlight are absorbed by the anterior elements of the eye.

Benign and malignant tumors of the same types that occur elsewhere on the skin of XP patients occur also on the eyelids. In addition, tumors of the conjunctivae, of the cornea, and, less commonly, of the iris can occur. The most frequent conjunctival tumors are intraepithelial epitheliomas and squamous cell carcinomas; and, although they most often arise at the corneal-scleral lumbus [69], they may arise from any part of the conjunctiva. Primary tumors of the cornea include epitheliomas, squamous cell carcinomas, sarcomas, and even melanomas [57,67].

NEUROLOGICAL ABNORMALITIES

The first description of XP patients who had severe neurological abnormalities in addition to the typical cutaneous and ocular findings of classical XP was that of three siblings, reported in 1932 by De Sanctis and Cacchione [21]. Since that report, the combination of classical XP with neurological abnormalities usually has been referred to as the De Sanctis-Cacchione syndrome. This term can be misleading, however; the patients described in 1932 had almost all of the neurological abnormalities associated with XP, as well as severe growth deficiency and immature sexual development, whereas many XP patients who manifest neurologic abnormalities have only a portion of this full De Sanctis-Cacchione syndrome [69]. The abnormalities can include microcephaly, low intelligence, progressive mental deterioration, hypo- or areflexia, progressive sensorineural deafness, choreoathetosis, ataxia, spasticity, and, rarely, epilepsy. Markedly deficient growth and immature sexual development have been reported mainly in patients with the most severe neurological abnormalities, but primary or secondary amenorrhea sometimes occurs in the absence of neurological abnormalities [69].

Pathological reports indicate that the principal abnormality in XP patients with neurological symptoms is the simple loss of neurons, primarily from the cerebral cortex and the cerebellum [21,69,77]. This loss of neurons probably is an ongoing process inasmuch as affected patients show a progressive clinical deterioration, gradually becoming less and less able to walk, to hear, to speak, or, in extreme cases, to take care of basic personal needs. In the most severe cases, death in childhood may result from progressive cachexia or infection, or both [21,66]. In other cases, however, onset of neurological abnormalities may not be apparent clinically until the second decade of life, the abnormalities may be few, and progression may be relatively slow. There is evidence from electromyography and muscle biopsies that affected patients suffer a progressive lower motor neuron degeneration, which may account in part for the hyporeflexia that occurs so frequently in these patients [69]. Electroencephalographic abnormalities have also been reported in some patients [69]. Computed tomography scans of severely affected patients have demonstrated atrophy of the cerebral cortex and ventricular dilatation [35].

OTHER CLINICAL ABNORMALITIES

Neoplasms other than those of the skin and eye occasionally have been reported in XP patients: at least two cases of primary brain tumors [43,62], two of leukemia [11,65], one of testicular sarcoma [56], and one of breast carcinoma [75]. However, whether or not internal malignancies are abnormally frequent in XP is an unanswered question. Squamous cell carcinomas of the tip of the tongue have been reported [43], but as these could be attributed to sunlight exposure, it does not seem appropriate to include them in computing the incidence of "internal" neoplasms in XP.

Deficient growth and sexual development occur primarily in patients who experience the early onset of severe neurological abnormalities. No specific laboratory evidence of endocrine or metabolic abnormality has been identified consistently among such patients, but occasional reports of abnormalities in isolated patients have appeared [25,26,29,59], including reports of amino aciduria, low blood glutathione, and low 17-keto- and 17-hydroxysteroids.

Case reports of XP patients occasionally mention infection as part of the clinical picture, especially late in the course of severely debilitated patients, but evidence of primary immunodeficiency is lacking in most such reports. A few specific abnormalities have been identified, however, including a deficiency in the C8 component of the complement cascade [30], circulating factors inhibitory to lymphocyte proliferation [45], and defects in standard tests of cellular immune function [24,72].

One XP patient in complementation group C (see below) has not only the typical changes of XP but also the signs and symptoms of systemic lupus erythematosus, including arthritis, anemia, and a positive antinuclear antibody test [34].

Another patient had both XP and Cockayne's syndrome, another rare disorder that is usually inherited as an autosomal recessive trait [69]. Cockayne's syndrome consists of cachectic dwarfism, deafness, mental deficiency, pigmentary retinal degeneration, optic atrophy, normal pressure hydrocephalus, hyperreflexia, diminished peripheral nerve conduction velocities, and photosensitivity [13]. This particular patient had all of these abnormalities; although some are similar to abnormalities seen in other XP patients, the retinal changes, diminished nerve conduction velocities, and hyperreflexia distinguished Cockayne's syndrome from the De Sanctis-Cacchione syndrome. This patient died at age 33 of acute hypertension, a disorder reported as the cause of death in other patients with Cockayne's syndrome [61]. She is the only known patient in XP complementation group B (see below).

COMPLEMENTATION GROUPS

DNA-excision repair of ultraviolet light (UV) damage to DNA can be shown to be below normal in skin fibroblasts and other cells cultured from most persons with XP. An assay widely used to demonstrate this finding is the autoradiographic determination of unscheduled DNA synthesis (UDS). In this assay the relative rate of incorporation of tritium-labeled thymidine into non-S-phase nuclei of ultraviolet-irradiated cells is assessed by counting silver grains which appear over such nuclei in the developed autoradiograms. This same assay can be used to demonstrate genetic complementation between cells from certain pairs of DNA excision repair-deficient XP patients. Complementation, in this instance, is defined as the appearance of normal UV-induced UDS in some of the binuclear cells that are generated by fusion of fibroblasts from two different XP patients. If the fused cells of two excision repair-deficient XP patients show complementation, they are said to belong to different complementation groups. To date, in fusion experiments involving many, but not all, of the existing XP tissue culture cell lines, seven complementation groups have been defined. These groups have been designated by letters A through G [7,40,47].

Cells from patients in each of these groups have been found to have reduced rates not only of UV-induced UDS, but also of removal of UV-induced pyrimidine-dimer sites [78], which suggests that the defect in each of these groups is located at or before the endonucleolytic step of the classical excision-repair scheme. Numerous biological consequences of these defects have been measured in XP cells, including reduced post-UV colony-forming ability of XP fibroblasts [4], reduced post-UV proliferation of XP lymphoblasts [6,58], reduced host-cell reactivation of UV-irradiated DNA viruses [19,51], increased frequencies of post-UV chromosome abnormalities [37,54], increased frequencies of post-UV sister-chromatid exchanges [15,22], and increased UV-induced mutation rates [31,52]. The magnitude of such biological consequences varies among the complementation groups and, in some instances, within the groups. Some interesting correlations between the clinical characteristic of XP patients and the in vitro sensitivities of their cells have been found and will be discussed below. With the exceptions to be noted, each of the complementation groups has certain characteristic clinical and, or, laboratory findings that distinguish it, at least partially, from the other groups (Table I).

Group A

Of the patients classified according to complementation group, approximately one-third of those from the United States, Europe, and Egypt and over half of those from Japan have been found to be in group A [43]. This group is comprised largely of patients who have XP with severe neurologic abnormalities. These abnormalities generally begin in the first few years of life, almost always are evident by 6 or 7 years of age, and progress relatively rapidly. Thus, these patients closely resemble those reported by De Sanctis and Cacchione [21].

In studies of post-UV colony-forming ability of XP fibroblasts [4], the cells from group A patients with severe neurologic abnormalities were the most UV-sensitive of all XP cells tested. On the other hand, the fibroblast strains from three group A patients who, unlike most, had minimal or no neurologic abnormalities, had less UV-sensitivity in the post-UV colony-forming assay than did the typical group A strains. In fact, these three "atypical" group A strains (XP12BE, XP1LO, and XP8LO) had less UV-sensitivity than did strains from any patients with XP-associated neurologic abnormalities, including those of groups D [4] and G [10]. Thus, a correlation exists between the presence of severe XP-associated neurologic abnormalities and the degree of UV-sensitivity of the patient's fibroblasts in the post-UV colony-forming assay. This finding supports the suggestion [4] that the DNA repair defects in XP are causally related to the neurologic abnormalities of XP. Nevertheless, that is not to imply that UV-induced DNA damage is responsible for the neurologic abnormalities of XP. It is more likely that some other form of DNA damage, e.g., "spontaneous" damage (that is, damage resulting from endoge-

TABLE I. Clinical and Laboratory Characteristics of the Xeroderma Pigmentosum Complementation Groups

Group	Rate of UV-induced unscheduled DNA synthesis (% of normal)	D_0 for fibroblast post-UV colony-forming ability (J/M^2)*	History of acute sun sensitivity†	Neurological abnormalities‡
A	<2§	0.25–0.35‖	+ and −#	++++, +, and −¶
B	3–7	(0.60–0.70)**	(+)**	(++++)**
C	10–25	0.55–1.10	+ and −††	++ and −‡‡
D	25–50	0.35–0.50	+	+++
E	40–65	2.80–3.00	Unknown	−
F	10–20	1.60–1.90	Unknown	−
G	<2	0.30–0.45	+	++++ and +++
Variant	90–100	2.50–2.80	−	−

*The D_0 is the UV dose required to reduce colony-forming ability from any given value to 37% of that value; the D_0s given in this table were calculated from the principal portions of the colony-forming curves and do not reflect the size or shape of "shoulders," i.e., the portion of the curve at low UV dose which sometimes has a slope different from that of the rest of the curve. The D_0 for normal human fibroblasts is 3.0–4.5 J/M^2. Values were calculated from data in 4, 9, and 10.

†+, present; −, absent.

‡++++, multiple severe neurological abnormalities present by 7 years of age.

+++, multiple severe neurological abnormalities occurring between 7 and 12 years of age.

++, microcephaly and mental deficiency without other neurological abnormalities.

+, hyporeflexia, without other neurological abnormalities.

−, no neurological abnormalities.

§Cells from one group A patient (XP8LO) differ from those of all other group A patients by having a rate of UV-induced unscheduled DNA synthesis of 30% [23].

‖Cells from three "atypical" group A patients have higher post-UV colony-forming ability than those of other group A patients: XP12BE, $D_0 = .70$ J/M^2; XP1LO, $D_0 = .88$ J/M^2; and, XP8LO, $D_0 = 1.74$ J/M^2. These three patients also differ clinically from the other group A patients in that they have few or no neurological abnormalities and no history of acute sun sensitivity.

#With the exception of the atypical group A patients mentioned in the previous footnote, all group A patients have had acute sun sensitivity.

¶All but the three atypical group A patients have severe (++++) neurological abnormalities; XP12BE has only hyporeflexia; XP1LO and XP8LO have no neurological abnormalities.

**This group is composed of only one patient, and she had both XP and Cockayne's syndrome; the extent is unknown, therefore, to which her clinical sun sensitivity and neurological abnormalities or the reduced post-UV colony-forming ability of her fibroblasts can be attributed to her XP.

††In general, the cells of group C patients who experienced acute sun sensitivity had lower post-UV colony-forming ability than those of group C patients who did not experience acute sun sensitivity.

‡‡Only one patient in group C has had significant XP-associated neurological abnormalities (++); that patient's cells had lower post-UV colony-forming ability than the cells of any other group C patient.

nous metabolic processes) or damage from certain exogenous chemicals, the repair of which requires, at least in part, the same mechanisms utilized to repair UV damage, leads to premature death of cells in the central nervous system. In fact, XP cells have been found to be sensitive to the killing effects of numerous DNA-damaging chemicals [43,69].

As a rule, the clinical findings in siblings with XP are similar, and studies of repair capacity and ultraviolet sensitivity with cells from XP siblings give identical results. A notable exception to this rule was recently reported from Italy [73]. A brother and sister with XP, both in complementation group A, had markedly different clinical findings. The brother presented the De Sanctis-Cacchione type of XP, with severe neurological abnormalities, immature sexual development, growth deficiency, and many skin cancers. He died of metastatic melanoma at age 22. The younger sister, at age 16, had had only one skin cancer, was of normal intelligence, and had undergone normal sexual development. The sister also had a higher level of UV-induced UDS than her brother, in studies with both leukocytes and skin fibroblasts. The latter studies revealed evidence among the sister's cells of two populations of fibroblasts, one repair-proficient, the other repair-deficient; this led the authors to suggest that unstable regulatory factors, which might allow normal repair in some cells and not others, could be involved. Furthermore, they speculated that inasmuch as the three aforementioned atypical group A patients with minimal or no neurologic abnormalities were, like this patient, female, the hypothetical regulatory gene(s) for repair might be located on the X-chromosome. An additional point of interest in this regard is that post-UV colony-forming assays with two of the atypical group A fibroblast strains (XP12BE and XP1LO) also show evidence that each strain may consist of two populations of cells. Specifically, the curve of post-UV colony-forming ability vs. UV dose for each of these strains shows two distinct components of different slope [4]. Attempts by this author to isolate clones from either XP12BE or XP1LO that have only one of the two components of the "parental" post-UV colony-forming curve have been unsuccessful.

Most group A fibroblast lines have extremely low rates of unscheduled DNA synthesis, generally less than 2% of the normal rates [63]. Of the atypical group A strains, XP12BE and XP1LO also have UDS rates less than 2% of normal, but XP8LO has a UDS rate about 30% of normal [23]. Cells from the female sibling in the Italian report (above) were mixed, some with normal USD, some with very low UDS (<10%).

Judging by those cases in which it is known whether or not a patient exhibited acute sun sensitivity, it appears that the atypical group A patients did not [4,23]. In the unusual Italian sibship, the brother with severe neurologic abnormalities did have acute sun sensitivity, whereas the sister had neither neurological abnormalities nor acute sun sensitivity.

Group B

To date only one XP patient in complementation group B has been reported. This is the woman mentioned earlier who had both XP and Cockayne's syndrome. The rate of UV-induced UDS in that patient's cells is between 3 and 7% of normal [43]. By contrast, the rate of UV-induced UDS is normal in cells from patients who have Cockayne's syndrome without XP [5].

It is possible but very unlikely that the group B patient was homozygous for two independently inherited and extremely rare recessive traits, i.e., that she inherited one set of genes for Cockayne's syndrome and another set for the seemingly unique group B form of XP. Conservatively, the probability of such an occurrence would be at least $10^{-6} \times 10^{-6}$, or one chance in 10^{12}. It would seem more probable that this patient's clinical picture was the result of homozygosity for a single recessive trait or, perhaps, of a single new dominant mutation. In this regard it is noteworthy that, in addition to the clinical overlap between Cockayne's syndrome and the De Sanctis-Cacchione form of XP, both the XP and Cockayne's syndrome fibroblasts are sensitive to the same group of chemical and physical DNA-damaging agents [2,76]. Also, after UV irradiation [48], Cockayne's syndrome cells (like XP cells) recover their ability to replicate DNA more slowly than do normal cells. In light of these facts, even though no specific defect(s) in the known repair pathways has been identified in cells from patients with Cockayne's syndrome, it seems reasonable to hypothesize that DNA-repair defects of some kind (perhaps in an as yet undiscovered repair process) may exist in such patients. For the sake of this argument the concept of "repair" would include any process that leads to recovery of a cell's ability to perform DNA replication following chemical or physical damage to the cell's DNA. Applying this hypothesis to the case of the XP patient in group B, one can imagine that a gene may be common to both the excision repair pathway and to some other repair pathway (i.e., the hypothetical pathway affected in Cockayne's syndrome), be it a regulatory gene or one coding for an enzyme necessary to both repair pathways, and that a mutation in this gene could result in the clinical findings of both XP and Cockayne's syndrome in the same individual.

It is not possible to determine whether the group B patient had any XP-associated neurological abnormalities, inasmuch as all of her neurological abnormalities might have been the result of her Cockayne's syndrome defect. Therefore, the level of post-UV colony-forming ability of her fibroblasts is not relevant to the aforementioned correlation between post-UV colony-forming ability and the presence of XP-associated neurological abnormalities. It is interesting to note, however, that the colony-forming ability of her fibroblasts after 254-nm ultraviolet radiation is lower than that of all other strains of Cockayne's syndrome fibroblasts tested so far [5].

Group C

About 40% of the XP patients who have been classified as to complementation group in the United States, Europe, and Egypt have been in group C [43]. But in Japan, only one sibship out of 40 has been found to be in group C. Almost all the patients in this group have the classical form of XP; i.e., they have the cutaneous and ocular findings of XP but lack XP-associated neurological abnormalities. Two exceptions have been described. One patient, XP10BE, has an IQ of 81, with no other neurological abnormalities [69]; it is possible that her slightly subnormal IQ is coincidental and unrelated to her XP. The other patient, mentioned earlier, has both XP and the autoimmune disorder, systemic lupus erythematosus. She also has microcephaly and severe mental deficiency, but no other neurological abnormalities [34]. It seems worth noting that her fibroblasts have lower post-UV colony-forming ability than any other group C fibroblasts tested [68].

Except for the patient just described, the post-UV colony-forming abilities of group C fibroblast lines, while below normal, are higher in all cases than those of fibroblast lines from patients with XP-associated neurological abnormalities. The post-UV colony-forming abilities of strains from affected members of any one group C kindred have been identical, but between members of different group C kindreds significant differences in UV sensitivity have been found [4]. These findings suggest that heterogeneity of the repair defects exists within group C, much as within group A.

Most patients in group C have not reported acute sun sensitivity as a clinical problem; however, the fibroblasts from those who have had this problem have generally had lower post-UV colony-forming ability than those from group C patients without the problem. The rate of UV-induced UDS in group C fibroblasts falls between 10 and 25% of normal (Table I) [43,44].

Group D

Of the XP patients from the United States, Europe, and Egypt whose group has been determined, approximately 10% are in group D. In Japan, as with Group C, only one kindred out of 40 has been found to be in group D [43]. Patients in this group all have had XP-associated neurological abnormalities, but have generally developed abnormalities later than typical group A patients; some group D patients have not manifested significant neurological abnormalities until well into their second decade of life [62]. Growth and sexual maturation generally are normal in group D patients.

Rates of UV-induced unscheduled DNA synthesis in group D range between 25 and 50% of normal. Post-UV colony-forming ability of group D fibroblasts is lower than that of group C fibroblasts and higher than that of typical group A fibroblasts [4]. All group D patients for whom relevant information is available have had acute sun sensitivity.

Group E

Group E has been diagnosed in only two persons, related Europeans who presented a relatively mild form of classical XP [43]. Their fibroblasts perform UV-induced unscheduled DNA synthesis at 40 to 60% of the normal rate [12,44] and have post-UV colony-forming ability just at the lower limits of the normal range [4,10]. Host-cell reactivation studies with UV-irradiated viruses have demonstrated convincingly the abnormal sensitivity of group E fibroblasts to UV-induced damage [1,19].

Group F

Group F has been reported in only three kindreds, all in Japan [43]. The patients have had no neurological abnormalities. The rate of UV-induced UDS in their cells is 10–20% of normal, and the post-UV colony-forming ability of their fibroblasts is as high as or slightly higher than that of the most UV-resistant group C fibroblasts [7,10]. Thus, these patients are very similar to the less severely affected members of group C, both clinically and in terms of cellular responses to UV light.

Group G

Group G consists of two unrelated individuals in Europe [9,40]. Both have severe XP-associated neurological abnormalities, but their case histories suggest that one of them, XP3BR [9], had a significantly earlier onset of such abnormalities than did the other, XP2BI [16]. Also, patient XP3BR had profound growth deficiency not reported for patient XP2BI. It is interesting that, although cells of both patients have very low (<2%) rates of UV-induced UDS, subtle but detectable differences appear to exist between their cell lines in post-UV colony-forming ability. Specifically, cells from patient XP3BR, the patient with the earlier and more profound neurological abnormalities, showed slightly lower post-UV colony-forming ability than did cells from XP2BI; i.e., the post-UV colony-forming curve for XP3BR was nearly coincident with curves of typical group A fibroblasts [9], whereas that for XP2BI, in studies by two different laboratories [9,10], was coincident with the slightly higher curves of group D fibroblasts. Furthermore, cells of patient XP3BR were found to be abnormally sensitive to gamma radiation, in contrast to the normal gamma radiation sensitivity of XP2BI [9] and of all other XP patients tested [9,70].

Acute sun sensitivity was reported for both patients in group G. In neither case, however, was malignant skin tumor reported. This is somewhat surprising in view of the fact that the older of the two, XP2BI, was 17 years of age when her case was reported [16]. One possible explanation is that early diagnosis followed by careful, consistent sun protection has resulted in a much below-average cumulative UV exposure of the skin in these patients. This possibility, however, would be difficult to confirm.

XP VARIANTS

An eighth XP group, designated the "variant" group, is comprised of patients with the classical clinical form of XP (i.e., none has been found to have neurological abnormalities) whose cells, in contrast to those of the XP complementation groups described above, have normal levels of UV-induced UDS [14,69,70]. Cells from XP variants have an abnormality in DNA repair which is manifested by a reduced rate of increase in the molecular weight of daughter-strand DNA that is synthesized on UV-damaged parental strands [47]. The normal repair process, whereby unusually low molecular weight DNA that is synthesized immediately after UV exposure is converted to larger molecular weight DNA, has been termed "postreplication" repair, but its identity with the processes of similar names which have been studied in bacteria is uncertain. All XP variant cell lines tested have shown not only a marked reduction in the rate of such postreplication repair but also a marked sensitivity to caffeine inhibition of that process [46,47]. Cells from individuals with certain excision-repair-deficient forms of XP, namely groups A, B, C, D, and G, also have abnormally slow postreplication repair [9,46]. In these patients, however, such repair is not as slow or as caffeine-sensitive as it is in XP variants. Group E cells have normal postreplication repair [46], and results for group F cells have not yet been reported.

Jung [39] described a group of patients who were similar clinically to patients with relatively mild classical XP except that skin changes appeared at a later age than is usual. These patients had normal excision repair, and, in the belief that they represented a group clinically distinct from XP, Jung coined the term "pigmented xerodermoid" to describe their condition. Recent studies, however, suggest that cells from patients with pigmented xerodermoid are indistinguishable, in terms of post-UV DNA synthesis and repair characteristics, from cells of XP variants [18]. It is likely, therefore, that patients originally described as pigmented xerodermoid are, in fact, XP variants who merely had a later clinical onset. Thus, significant clinical variability appears to exist within the XP variant group, analogous to that in several of the excision-repair-deficient XP complementation groups. Of course, this could be the result of differences among patients in the precise nature of their genetic defect or in their exposure to UV radiation.

Cells of XP variants are below normal in both post-UV colony-forming ability [4] and host-cell reactivation of UV-irradiated DNA viruses [20]; their relative performance in each assay, however, is better (i.e., closer to normal) than that of cells from group C XP patients. Like the excision-repair-deficient XP fibroblasts that have been tested, XP variant fibroblasts are more highly mutable by UV than are normal fibroblasts [53].

XP HETEROZYGOTES

Persons who are heterozygous for XP (i.e., parents of affected individuals) cannot be distinguished from noncarriers on clinical grounds. As a group, they may have a slightly higher than normal incidence of skin cancer [55], but even if so, this would not be useful diagnostically in individual cases. Most in vitro studies of UV sensitivity and DNA-repair capacity with cells from XP heterozygotes have revealed either no abnormalities or abnormalities in some but not all of the cell lines tested [12,19,41]. Recently, however, Rainbow [64] reported that by using a very sensitive host-cell reactivation assay that utilizes immunofluorescent detection of viral structural antigens, he detected an abnormally slow repair of UV-irradiated adenovirus in each of the cell strains which he tested from persons heterozygous for excision-repair-deficient forms of XP (groups A, C, and D). Whether heterozygotes for other forms of XP, or even whether all the group A, C, and D heterozygotes can be detected with this assay, must await further studies. For the present, no proven, practical method exists for detecting XP heterozygotes.

LITERATURE CITED

1. ABRAHAMS, P. J., VAN DER EB, A. J.: Host-cell reactivation of ultraviolet-irradiated SV-40 DNA in five complementation groups of xeroderma pigmentosum. *Mutat. Res.* 35:13–22, 1976.
2. AHMED, F. E., SETLOW, R. B.: Excision repair in ataxia telangiectasia, Fanconi's anemia, Cockayne's syndrome, and Bloom's syndrome after treatment with ultra-

violet radiation and N-acetoxy-2-acetylaminofluorene. *Biochim. Biophys. Acta* 521: 805-817, 1978.
3. ANDERSON, T., BEGG, M.: Xeroderma pigmentosum of mild type. *Br. J. Dermatol.* 62:402-407, 1950.
4. ANDREWS, A. D., BARRETT, S. F., ROBBINS, J. H.: Xeroderma pigmentosum neurological abnormalities correlate with colony-forming ability after ultraviolet radiation. *Proc. Natl. Acad. Sci. USA* 75:1984-1988, 1978.
5. ANDREWS, A. D., BARRETT, S. F., YODER, F. W., ROBBINS, J. H.: Cockayne's syndrome fibroblasts have increased sensitivity to ultraviolet light but normal rates of unscheduled DNA synthesis. *J. Invest. Dermatol.* 70:237-239, 1978.
6. ANDREWS, A. D., ROBBINS, J. H., KRAEMER, K. H., BUELL, D. N.: Xeroderma pigmentosum long-term lymphoid lines with increased UV sensitivity. *J. Natl. Cancer Inst.* 53:691-693, 1974.
7. ARASE, S., KOZUKA, T., TANAKA, K., IKENAGA, M., TAKEBE, H.: A sixth complementation group in xeroderma pigmentosum. *Mutat. Res.* 59:143-146, 1979.
8. ARLETT, C. F., HARCOURT, S. A.: Survey of radiosensitivity in a variety of human cell strains. *Cancer Res.* 40:926-932, 1980.
9. ARLETT, C. F., HARCOURT, S. A., LEHMANN, A. R., STEVENS, S., FERGUSON-SMITH, M. A., MORLEY, W. N.: Studies on a new case of xeroderma pigmentosum (XP3BR) from complementation group G with cellular sensitivity to ionizing radiation. *Carcinogenesis* 1:745-751, 1980.
10. BARRETT, S. F., TARONE, R. E., MOSHELL, A. N., GANGES, M. B., ROBBINS, J. H.: The post-UV colony-forming ability of normal fibroblast strains and of the xeroderma pigmentosum group G strain. *J. Invest. Dermatol.* 76:59-62, 1981.
11. BERLIN, C., TAGER, A.: Xeroderma pigmentosum, report of eight cases of mild to moderate type and course: A study of response to various irradiations. *Dermatologica* 116:27-42, 1958.
12. BOOTSMA, D., MULDER, M. P., POT, F., COHEN, J. A.: Different inherited levels of DNA repair replication in xeroderma pigmentosum cell strains after exposure to ultraviolet irradiation. *Mutat. Res.* 9:507-516, 1970.
13. BRUMBACK, R. A., YODER, F. W., ANDREWS, A. D., PECK, G. L., ROBBINS, J. H.: Normal pressure hydrocephalus: recognition and relationship to neurological abnormalities in Cockayne's syndrome. *Arch. Neurol.* 35:337-345, 1978.
14. BURK, P. G., LUTZNER, M A., CLARKE, D. D: Ultraviolet stimulated thymidine incorporation in xeroderma pigmentosum lymphocytes. *J. Lab. Clin. Med.* 77:759-767, 1971.
15. CHANG, W. S., TARONE, R. E., ANDREWS, A. D., WHANG-PENG, J. S., ROBBINS, J. H.: Ultraviolet light-induced sister chromatid exchanges in xeroderma pigmentosum and in Cockayne's syndrome lymphocyte cell lines. *Cancer Res.* 38:1601-1609, 1978.
16. CHEESBROUGH, M. J., KINMONT, P. D. S.: Xeroderma pigmentosum – a unique variant with neurological involvement. *Br. J. Dermatol.* 99 (Suppl. 16):61, 1978.
17. CLEAVER, J. E.: Xeroderma pigmentosum: Variants with normal DNA repair and normal sensitivity to ultraviolet light. *J. Invest. Dermatol.* 58:124-128, 1972.
18. CLEAVER, J. E., ARUTYUNYAN, R. M., SARKISIAN, T., KAUFMANN, W. K., GREENE, A. E., CORIELL, L.: Similar defects in DNA repair and replication in the pigmented xerodermoid and the xeroderma pigmentosum variants. *Carcinogenesis* 1:647-655, 1980.
19. DAY, R. S.: Studies on repair of adenovirus 2 by human fibroblasts using normal, xeroderma pigmentosum, and xeroderma pigmentosum heterozygous strains. *Cancer Res.* 34:1965-1970, 1974.

20. DAY, R. S.: Xeroderma pigmentosum variants have decreased repair of ultraviolet-damaged DNA. *Nature* 253:748-749, 1975.
21. DE SANCTIS, C., CACCHIONE, A.: L'idiozia xerodermica. *Riv. Sper. Freniatr.* 56: 269-292, 1932.
22. DEWEERD-KASTELEIN, E. A., KEIJZER, W., RAINALDI, P.: Introduction of sister chromatid exchanges in xeroderma pigmentosum cells following UV exposure. *Mutat. Res.* 46:163, 1977.
23. DEWEERD-KASTELEIN, E. A., KEIJZER, W., SABOUR, M., PARRINGTON, J. M., BOOTSMA, D.: A xeroderma pigmentosum patient having a high residual activity of unscheduled DNA synthesis after UV is assigned to complementation group A. *Mutat. Res.* 37:307-312, 1976.
24. DUPUY, J. M., LAFFORET, D.: A defect of cellular immunity in xeroderma pigmentosum. *Clin. Immunol. Immunopathol.* 3:52-58, 1974.
25. EL-HEFNAWI, H., EL-HAWARY, M. F. S.: Chromatographic studies of amino acids in sera and urine of patients with xeroderma pigmentosum and their normal relatives. *Br. J. Dermatol.* 75:235-240, 1963.
26. EL-HEFNAWI, H., EL-HAWARY, M. F. S., EL-KONY, H. M., RASHEED, A.: Xeroderma pigmentosum: V. Studies of 17-ketosteroids and total 17-hydroxycorticosteroids. *Br. J. Dermatol.* 75:484-489, 1963.
27. EL-HAFNAWI, H., RASHEED, A.: Xeroderma pigmentosum: A further clinical study of 12 Egyptian cases. *J. Egypt. Med. Assoc.* 45:106-141, 1962.
28. EL-HEFNAWI, H., SMITH, S. M., PENROSE, L. S.: Xeroderma pigmentosum, its inheritance and relationship to the ABO groups in the U.A.R. *Br. J. Dermatol.* 77:35, 1965.
29. FERUERSTEIN, M., LANGHOFF, H.: Free amino acids in the urine and the amino acid content of the hairs of xeroderma pigmentosum. *Arch. Klin. Exp. Dermatol.* 220:486-487, 1964.
30. GIRALDO, G., DEGOS, L., BETH, E., SASPORTES, M., MARCELLI, A., GHARBI, R., DAY, N. K.: C8 deficiency in a family with xeroderma pigmentosum; Lack of linkage to the HLA region. *Clin. Immunol. Immunopathol.* 8:377-384, 1977.
31. GLOVER, T. W., CHANG, C. C., TROSKI, J. E., LI, S. S.: Ultraviolet light induction of diphtheria toxin-resistant mutants of normal and xeroderma pigmentosum human fibroblasts. *Proc. Natl. Acad. Sci.* 76:3982-3986, 1979.
32. GOLDMAN, L. RICHFIELD, D. F., LOUTZENHISER, J., HANISZKO, G.: Features uncommon to xeroderma pigmentosum: Case report with a study of 92 biopsy specimens. *Arch. Dermatol.* 33:272-278, 1961.
33. HADIDA, E., MARILL, F. G., SAYAG, J.: Xeroderma pigmentosum (A propos de 48 observations personnelles). *Ann. Dermatol. Syphiligr.* 90:467-496, 1963.
34. HANANIAN, J., CLEAVER, J. E.: Xeroderma pigmentosum exhibiting neurological disorders and systemic lupus erythematosus. *Clin. Genet.* 17:39-45, 1980.
35. HANDA, J., NAKANO, Y., AKIGUCHI, I.: Cranial computed tomography findings in xeroderma pigmentosum with neurologic manifestations (DeSanctis-Cacchione syndrome). *J. Comput. Assist. Tomogr.* 2:456-459, 1978.
36. HASHEM, N., BOOTSMA, D., KEIJZER, W., GREENE, A., CORIELL, L., THOMAS, G., CLEAVER, J. E.: Clinical characteristics, DNA repair, and complementation groups in xeroderma pigmentosum patients from Egypt. *Cancer Res.* 40:13-18, 1980.
37. HUANG, C. C., BENERJEE, A., HOU, Y.: Chromosomal instability in cell lines derived from patients with xeroderma pigmentosum. *Proc. Soc. Exp. Biol. Med.* 148: 1244-1248, 1975.
38. ICHIHASHI, M., FUJIWARA, Y.: Clinical and photobiologic characteristics in Japa-

nese xeroderma pigmentosum variants. Abstracts of the Eighth International Congress of Photobiology, Abstract 159, 1980.
39. JUNG, E. G.: New form of molecular defect in xeroderma pigmentosum. *Nature* 228: 361-362, 1970.
40. KEIJZER, W., JASPERS, N.G., ABRAHAMS, P. J., TAYLOR, A. M., ARLETT, C. F., ZELLE, B., TAKEBE, H., KINMONT, P. D., BOOTSMA, D.: A seventh complementation group in excision-deficient xeroderma pigmentosum. *Mutat. Res.* 62:183-190, 1979.
41. KLEIJER, W. J., DEWEERD-KASTELEIN, E. A., SLUYTER, M. L., KEIJER, W., DEWIT, J., BOOTSMA, D.: UV-induced DNA repair synthesis in cells of patients with different forms of xeroderma pigmentosum and of heterozygotes. *Mutat. Res.* 20: 417-428, 1974.
42. KOBZA, A., GIANNELLI, F.: Xeroderma pigmentosum. *Proc. R. Soc. Med.* 65:9-10, 1972.
43. KRAEMER, K. H.: Xeroderma pigmentosum. *In* Clinical Dermatology (Demis, D. J., Dobson, R. L. McGuire, J., eds.), Vol. 4, Hagerstown: Harper & Row, unit 19-7, 1-33, revised Edition 1980.
44. KRAEMER, K. H., DEWEERD-KASTELEIN, E. A., ROBBINS, J. H., KEIJZER, W., BARRETT, S. F., PETINGA, R. A., BOOTSMA, D.: Five complementation groups in xeroderma pigmentosum. *Mutat. Res.* 33:327-340, 1975.
45. LAFFORET, D., DUPUY, J. M.: Inhibitory factors of lymphocyte proliferation in serum from patients with xeroderma pigmentosum. *Clin. Immunol. Immunopathol.* 4: 165-173, 1975.
46. LEHMANN, A., KIRK-BELL, S., ARLETT, C., HARCOURT, S. A., DEWEERD-KASTELEIN, E. A., KEIJZER, W., HALL-SMITH, P.: Repair of ultraviolet light damage in a variety of human fibroblast cell strains. *Cancer Res.* 37:904-910, 1977.
47. LEHMANN, A. R., KIRK-BELL, S., ARLETT, C. F., PATERSON, M. C., LOHMAN, P. H. M., DEWEERD-KASTELEIN, E. A., BOOTSMA, D.: Xeroderma pigmentosum cells with normal levels of excision repair have a defect in DNA synthesis after UV-irradiation. *Proc. Natl. Acad. Sci.* 72:219-223, 1975.
48. LEHMANN, A. R., KIRK-BELL, S., MAYNE, L.: Abnormal kinetics of DNA synthesis in ultraviolet light-irradiated cells from patients with Cockayne's syndrome. *Cancer Res.* 39:4237-4241, 1979.
49. LYNCH, H. T., FRICHOT, B. C., FISHER, J., SMITH, J. L. JR., LYNCH, J. F.: Spontaneous regression of metastatic malignant melanoma in two sibs with xeroderma pigmentosum. *J. Med. Genet.* 15:357-362, 1978.
50. LYNCH, H. T., FRICHOT, B. C., LYNCH, J. F.: Cancer control in xeroderma pigmentosum. *Arch. Dermatol.* 113:193-195, 1977.
51. LYTLE, C. D., AARONSON, S. A., HARVEY, E.: Host-cell reactivation in mammalian cells: II. Survival of herpes simplex virus and vaccinia virus in normal human and xeroderma pigmentosum cells. *Int. J. Rad. Biol.* 22:159-165, 1972.
52. MAHER, V. M., MCCORMICK, J. J.: Effect of DNA repair on the cytotoxicity and mutagenicity of UV irradiation and of chemical carcinogens in normal and xeroderma pigmentosum cells. *In* Biology of Radiation Carcinogenesis (Yuhas, J. M., et al., eds.), New York: Raven Press, 129-145, 1976.
53. MAHER, V., OULLETTE, L. M., CURREN, R. D., MCCORMICK, J. J.: Frequency of ultraviolet light-induced mutations is higher in xeroderma pigmentosum variant cells than in normal human cells. *Nature* 261:593-595, 1976.
54. MARSHALL, R. R., SCOTT, D.: The relationship between chromosome damage and cell killing in UV-irradiated normal and xeroderma pigmentosum cells. *Mutat. Res.* 36:397-400, 1976.

55. MARX, J. L.: New clues to carcinogenesis. *Science* 200:518-521, 1978.
56. MILLER, R. W.: Childhood cancer and congenital defects: A study of US death certificates during the period 1960-1966. *Pediatr. Res.* 3:389-397, 1969.
57. MORTADA, A.: Incidence of lids, conjunctival and orbital malignant tumors in xeroderma pigmentosum in Egypt. *Bull. Ophthalmol. Soc. Egypt* 61:231-236, 1968.
58. MOSHELL, A. N., TARONE, R. E., NEWFIELD, S. A., ANDREWS, A. D., ROBBINS, J. H.: A simple and rapid method for evaluating the survival of xeroderma pigmentosum lymphoid lines after irradiation with ultraviolet light. *In Vitro* 17:299-307, 1981.
59. MOSS, H. V., JR.: Xeroderma pigmentosum: Report of two cases with metabolic studies. *Arch. Dermatol.* 92:638-644, 1965.
60. NEEL, J.V., KODAI, M., BREWER, R., ANDERSON, R. C.: The incidence of consanguineous matings in Japan: With remarks on the estimation of comparative gene frequencies and the expected rate of appearance of induced recessive mutations. *Am. J. Hum. Genet.* 1:156-178, 1949.
61. OHNO, T., HIROOKA, M.: Renal lesions in Cockayne's syndrome. *Tohuku J. Exp. Med.* 89:151-166, 1966.
62. PAWSEY, S. A., MAGNUS, I. A., RAMSAY, C. A., BENSON, P. F., GIANELLI, F.: Clinical, genetic and DNA repair studies on a consecutive series of patients with xeroderma pigmentosum. *Q. J. Med.* (new series) 48:179-210, 1979.
63. PETINGA, R. A., ANDREWS, A. D., TARONE, R. E., ROBBINS, J. H.: Typical xeroderma pigmentosum complementation group A fibroblasts have detectable ultraviolet light-induced unscheduled DNA synthesis. *Biochim. Biophys. Acta* 479:400-410, 1977.
64. RAINBOW, A. J.: Reduced capacity to repair irradiated adenovirus in fibroblasts from xeroderma pigmentosum heterozygotes. *Cancer Res.* 40:3945-3949, 1980.
65. REED, W. B., MAY, S. B., NICKEL, W. R.: Xeroderma pigmentosum with neurological complications: The DeSanctis-Cacchione syndrome. *Arch. Dermatol.* 91:224-226, 1965.
66. REED, W. B., SUGARMAN, G. I., MATHIS, R. A.: DeSanctis-Cacchione syndrome. A case report with autopsy findings. *Arch. Dermatol.* 113:1561-1563, 1977.
67. REESE, A., WILBER, L.: The eye manifestations of xeroderma pigmentosum. *Am. J. Ophthalmol.* 26:901-911, 1943.
68. ROBBINS, J. H.: Significance of repair of human DNA: Evidence from studies of xeroderma pigmentosum. *J. Natl. Cancer Inst.* 61:645-656, 1978.
69. ROBBINS, J. H., KRAEMER, K. H., LUTZNER, M. A., FESTOFF, B. W., COON, H. G.: Xeroderma pigmentosum – an inherited disease with sun sensitivity, multiple cutaneous neoplasms and abnormal DNA repair. *Ann. Intern. Med.* 80:221-248, 1974.
70. ROBBINS, J. R., LEVIS, W. R., MILLER, A. E.: Xeroderma pigmentosum epidermal cells with normal ultraviolet-induced thymidine incorporation. *J. Invest. Dermatol.* 59:402-408, 1972.
71. ROBBINS, J. H., MOSHELL, A. N.: DNA repair processes protect human beings from premature solar skin damage: Evidence from studies on xeroderma pigmentosum. *J. Invest. Dermatol.* 73:102-107, 1979.
72. SALAMON, T., STOJAKOVIC, M., BOGDANOVIC, B.: Delayed hypersensitivity in xeroderma pigmentosum. *Arch. Dermatol. Forsch.* 251:277-280, 1975.
73. STEFANINI, M., KEIJZER, W., DALPRA, L., ELLI, R., PORRO, M. N., NICOLETTI, B., NUZZO, F.: Differences in the levels of UV repair and in clinical symptoms in two sibs affected by xeroderma pigmentosum. *Hum. Genet.* 54:177-182, 1980.
74. TAKEBE, H., MIKI, Y., KOZUKA, T., JUN-ICHI, F., TANAKA, K., SASAKI, M., FIJIWARA,

Y., AKIBA, H.: DNA repair characteristics and skin cancers of xeroderma pigmentosum patients in Japan. *Cancer Res.* 37:490-495, 1977.
75. VAN PATTER, H. T., DRUMMOND, J. A.: Malignant melanoma occurring in xeroderma pigmentosum. *Cancer* 6:942-947, 1953.
76. WADE, M. H., CHU, E. H. Y.: Effects of DNA damaging agents on cultured fibroblasts derived from patients with Cockayne's syndrome. *Mutat. Res.* 59:49-60, 1979.
77. YANO, K.: Xeroderma pigmentosum with changes in the C.N.S.: A histological study. *Folia Psychiatr. Neurol. Jpn.* 4:143, 1950.
78. ZELLE, B., LOHMAN, P. H.: Repair of UV-endonuclease-susceptible sites in the seven complementation groups of xeroderma pigmentosum A through G. *Mutat. Res.* 62: 363-368, 1979.

WERNER'S SYNDROME

W. TED BROWN

INTRODUCTION

An unusual syndrome was first reported in 1904 by Otto Werner as "cataract in combination with scleroderma" [37]. Four brothers and sisters between the ages of 30 and 40 each had a prematurely aged appearance, short stature, premature graying of hair, early onset of cataracts, skin changes with an appearance like scleroderma, and hyperkeratotic lesions of the soles of the feet. The distinction between Werner's syndrome (WS) and a syndrome with juvenile cataracts and atrophic skin changes, described in 1868 by Rothmund [26], was unclear until 1934 when Oppenheimer and Kugel [24] published a paper that distinguished the two syndromes. In 1945, Thannhauser [35] presented a classic study in which he distinguished the skin changes seen in WS from those of both scleroderma and Rothmund's syndrome. In 1966, Epstein et al. [7] published a comprehensive study of WS in which they contrasted it with the process of natural aging.

Because the remarkable phenotypic features of WS in many ways resemble an acceleration of the aging process, some suggest that the syndrome may be a disease model for studies of aging. In addition, an increase in chromosome rearrangement appears to be associated with the disease and may underly a predisposition to malignancy. In this regard, WS may be similar to the other genetic syndromes considered in this volume, all of which feature increased chromosome instability.

CLINICAL FEATURES

The most striking feature of patients with WS are growth deficiency and severe degenerative changes that affect a variety of tissues and organs. In their review of 125 documented patients, Epstein et al. [7] noted that the mean age of the patients at diagnosis of the disease was

38.7 years and the mean age of death of 25 patients was 47.0 years. A remarkably high frequency of sarcomas in WS patients has been reported. Of the nine patients for whom the cause of death was noted, three died of myocardial infarction and six died of malignancies. The disease is inherited in an autosomal recessive fashion and the frequency of consanguineous marriages is high. Based upon the frequency of parental consanguinity, Epstein et al. estimated that there are between one and 22 cases per million population in the United States [7]. This implies that in the United States several hundred cases should be recognizable. Many cases may not be diagnosed, however, due to the late age of recognition of the disease.

Generally the patients grow normally during childhood but cease to grow by their early teenage years. The mean height and weight for males are about 157 cm (5 ft, 1 in) and 45 kg (99 lbs); for females, the means are 146 cm (4 ft, 9 in) and 40 kg (88 lbs) [7]. The patients have a low weight relative even to their short stature. The trunk is often stocky and the abdomen protuberant, whereas the arms and legs are exceptionally thin and spindly. Premature graying of the hair is an early sign of the disease, with onset usually before age 20. The graying or whitening of the hair may be abrupt or may occur gradually over a period of 5–20 years. Hair loss from the head can range from minimal changes to total alopecia. The hair also becomes thin over the pubic, axillary, eyebrow, and eyelash areas.

The development of cataracts is a characteristic feature of the disease. Cataract formation and the need for opthalmological surgery is often the major presenting symptom that leads to medical attention and diagnosis. The onset of cataracts is generally between 20 and 30 years of age, and affects the posterior cortex and subcapsule of the lens in a homogeneous and striate fashion. Cataracts usually occur bilaterally with slight variation in the degree of development. Cataract extraction is usually successful; however, in a review of the ophthalmological features of the disease, Petrohelos [25] cautioned that eye surgery has a guarded prognosis, as indicated by reports of postoperative complications such as bullous keratopathy and degenerative corneal changes. In addition, other eye changes, including retinitis pigmentosum, retinal macular degeneration, and blue sclerae [7] have been associated with isolated cases.

Frequently the patients have a voice that is described as high-pitched or thin. The voice abnormalities may be present from birth or develop during adolescence. About half of the patients examined have some degree of vocal cord atrophy. Often, the muscles of the arms and legs show a considerable degree of atrophy and muscle wasting, even though the trunk may retain a stocky appearance. No specific muscle abnormalities have been noted, and tests of serum muscle enzyme levels have generally been within the normal range.

Ever since the first descriptions by Werner, it has been noted that the striking skin changes have a superficial resemblance to scleroderma. The major sites of skin involvement are the face, hands, and feet, where there appears to be a thinning and atropy of the skin. The skin appears to be adherent to the underlying connective tissue. In a histologic study of the skin changes in WS, Fleischmajer and Nedwich [8] reported a remarkable replacement of underlying subcutaneous tissue by connective tissue which looks like an extension of the dermis. A characteristic feature of the disease is the development of a marked degree of hyperkeratosis and callous formation over the soles of the feet and ankles. These callouses may ulcerate and heal poorly, and they frequently lead to complications that necessitate amputation.

In addition to the skin changes of the extremities, severe peripheral vascular disease frequently develops and may contribute to the ulceration of the ankles. Arteriosclerosis of the Monckeberg type (medial calcinosis) and vessel calcification of the legs are common. Metastatic calcifications in the soft tissue overlying the tendons of the knees, ankles, and elbows are detectable by x-ray in a moderate proportion of patients. Widespread atherosclerosis affecting the aorta, major vessels, and coronary arteries is present in nearly all cases at autopsy. The mitral and aortic valve leaflets show striking degrees of calcification.

Diabetes mellitus has been recognized in about half of the reported cases. The diabetes appears to show peripheral resistance of a type similar to that seen in lipoatropic diabetes. In three patients studied [21] there was a hyperresponsive diabetic profile. Endocrine and hormone studies have indicated no specific abnormalities. Resistance to human chorionic gonadotropic hormone and luteinizing hormone-releasing hormone have been reported [20]. Hypogonadism is a prominent feature of WS. All male patients studied histologically in the series by Epstein et al. [7] had moderate to severe testicular atrophy, with complete hyalinization of the seminiferous tubules. Females may have irregular menses and usually terminate the menses at an early age. One ovary examined was described as atrophic. The degree of female hypogonadism that develops appears to be variable.

X-ray examinations often reveal generalized osteoporosis, which is particularly severe in the hands, feet, and hips [38]. Goto et al. [15] noted very flat feet in each of 15 Japanese patients studied. Hypoplasia of the nasal cartilage, along with the skin changes of the face, gives the nose a beaked or pinched appearance.

Chromosome Abnormalities

Unusual chromosome changes in WS patients have been noted in both fibroblasts and lymphocytes. Hoehn et al. [17] and Norwood et al. [23] examined the chromosomes in cultured skin fibroblasts from a patient with WS. They found multiple, variable, and predominantly stable translocations that were clonal in nature, and used the term "variegated translocation mosaicism" (VTM) to describe this unusual phenomenon. Salk et al. [28] has confirmed the finding of VTM in 92% of 1,538 metaphases from 29 independent cell lines derived from five WS patients, including the original patient of Hoehn et al. [17]. The distribution of chromosomes involved in the translocations appeared to be nearly random, although chromosomes 2 and 3 were slightly underrepresented and 10 and X were somewhat overrepresented. Controls showed about 5% of metaphases with VTM in eight of 95 cultures, and one culture derived from a trisomy 21 patients with 90-100% VTM. According to Salk et al. [28], a high frequency of translocations was confirmed in another unrelated patient by Dr. M. Fraccaro. Fibroblast cultures from two other patients have been deposited in the Aging Cell Repository at the Institute for Medical Research, Camden, New Jersey; chromosomal analysis of these has also shown a variety of translocations [32]. Salk et al. [29] observed that VTMs can be used to identify individual cellular clones and study their growth patterns. They observed a generally reproducible pattern of clonal succession during the repeated *in vitro* aging of several mass cultures of one strain. In two parallel derivative cultures, they noted rapidly growing clones that were not observed in the parental cultures. They suggested that clonal succession and clonal alteration may be occurring in the mass cultures.

In contrast to fibroblasts, blood lymphocyte chromosomes do not have the structural rearrangements of chromosomes. Cytogenetic analyses using unbanded preparations of lymphocytes from 19 patients have been reported as normal [3,20,21,34]. However, Nordenson [22] reported an increase in chromosome breakage: an average of 0.231 breaks per cell (including gaps, chromatid, or chromosome aberrations) in lymphocytes from four WS patients as compared to 0.016 in controls, a 14-fold increase. Additions of superoxide dismutase, or catalase or both to the cell culture media (Parker's 199) were tested for their effect on breakage frequency. The addition of either enzyme alone reduced the number to 0.14 breaks per cell, whereas the simultaneous addition of both lowered the frequency to 0.10. Nordenson suggested that these enzymes, by acting as free-radical scavengers, would decrease the *in vitro* environmental levels of free radicals which might be affecting chromosome breakage.

Salk et al. [30] treated two WS fibroblast strains with superoxide dismutase and catalase and observed no effect on *in vitro* life-span and VTM frequency.

We examined the sister-chromatid exchange frequencies in lymphocytes from three WS patients [5]. No consistent increase was detected. Addition of mitomycin C to the cultures induced an *increase* greater than controls in two out of three of the patients. A normal frequency of sister-chromatid exchange in untreated lymphocytes was found also by Bartram et al. [1], and in fibroblasts by Salk et al. [28].

We have established lymphoblastoid cultures from the two WS patients previously reported [4], with the assistance of Dr. Earl Henderson and Dr. Arthur Greene. These cultures (AG3829 and AG3364A) were deposited at the Aging Cell Repository at the Institute for Medical Research, Camden, N.J., and are available for general distribution. We found no specific chromosome abnormality in these lines. D. Salk also examined these lines and noted a high frequency of dicentric chromosomes in some harvests [D. Salk, personal communication]. Further studies are needed to characterize these lines.

Other Abnormalities

Cultured fibroblasts from patients with WS show a greatly reduced potential for *in vitro* growth and a reduced life-span. Whereas 50–80 generations is typical for fibroblastlike cells in culture [19], WS cells generally have a growth potential of only ten to 20 generations [9,19]. Salk et al. [31] studied 20 separate cell cultures derived from three WS patients. They found reduced *in vitro* life-spans which averaged 27% of normal. Cocultivation with normals did not influence their lower population growth rates. Somatic cell hybrids with normal cells resulted in reduced growth potentials of the hybrids to that typical of WS. Ultrastructural changes that accompany *in vitro* senescence of normal fibroblast strains were found to be present in early passage WS fibroblasts [2]. In addition, fibrous material with a high subunit molecular weight, probably a form of incompletely processed procollagen, was produced in large amounts both by early passage WS cells and senescing normal fibroblasts [2].

The reduced growth potential of WS fibroblasts may be a reflection of a slower rate of DNA replication. Fujiwara et al. [9] reported that DNA chain elongation, as estimated by molecular weight increase on alkaline sucrose gradients, was significantly reduced in cultured fibroblasts from four WS patients and similar to that observed in Bloom's syndrome [16]. Tanaka et al. [33] examined the rate of DNA synthesis in heterodikaryons between WS and either normal diploids or HeLa cells. When the fusion was with HeLa, the rate of DNA-chain elongation in WS nuclei was greatly increased, and when it was with normal cells, the rate was moderately increased. This is consistent with the notion that the slower rate of DNA synthesis in WS cells has properties of a recessive characteristic which can be corrected by hybridization with normal or neoplastic cells, and that the missing factor or factors needed for normal rates of DNA synthesis may be supplied.

Normally, people excrete in the urine small amounts of acid glycosaminoglycans (AGAG), of which a trace amount is in the form of hyaluronic acid (HA). Tokunaga et al. [33] reported that the urine of five persons with WS contained normal total amounts of AGAG but excessive amounts of HA. The HA in these patients comprised a mean of 7.1% (range, 1.7–16.7%) of the AGAG compared to less than 0.1% in normal individuals. They proposed that WS may be a disease involving a metabolic disorder of HA and suggested that it could represent a variant type of mucopolysaccaridosis. Goto et al. [15] also found excessive levels of HA in 13 WS patients [13]. The pooled urine from two patients was found to contain 16.4% of total AGAG as HA. We have also quantitated HA in two separate 24-hr collections from a WS patient [W.T. Brown and J. Distler, unpublished] and found HA to comprise an average of 61.4% of the uronic acid, a measure of total AGAG. Our results thus confirm the findings of Tokunaga et al. and Goto et al. Whether this reflects a primary defect of the disease or a secondary effect is unclear. Perhaps a recessive deficiency of an enzyme, which alters metabolism of AGAG, could have a generalized effect on the basal ground substance of tissues and lead to the clinical features of WS.

It has been suggested that the proportion of abnormally heat-labile enzymes in WS is unusually high. Holliday et al. [18] found that about 20% of the glucose-6-phosphate dehydrogenase (G6PD) from the cultured fibroblasts of a patient with WS were abnormally heat labile, whereas control cultures had about 5% heat lability. They regarded this as evidence in favor of the hypothesis that errors in protein synthesis occur in WS cells and lead to accelerated aging. Goldstein and Singel [12] and Goldstein and Moerman [10] reported an increased proportion with elevated thermolability of G6PD, 6-phosphogluconate dehydrogenase (6PGD), and hypoxanthine-guanine phosphoribosyltransferase. Levels of 20–50% were found in WS fibroblasts, compared to 0–7% in controls. Subsequently they have suggested that increased levels of thermolabile enzymes also are found in erythrocytes or fibroblasts in patients with Hutchinson-Gilford progeria syndrome, Down's syndrome, or diabetes, and in patients who are in intensive care units [2,12, and personal communication]. We tested levels of abnormal thermolabile G6PD and

6PGD in erythrocytes from two WS patients [4] and found no evidence of an increased fraction of abnormal enzyme. At present, the existence, molecular basis, specificity, and possible significance of thermolabile enzymes in WS or other syndromes are unclear.

To date, no specific abnormalities of immunofunction have been described in WS, but several reports have suggested that some type of immune abnormality may be present. Goto et al. [14] reported that purified T-lymphocytes from seven patients with WS syndrome showed a decreased intensity of cell-surface staining when examined with a naturally occurring autoantibody to T-lymphocytes found in sera of patients with systemic lupus erythematosis or with a heterologous antiserum against human brain tissue. A natural T-lymphocyte-toxic autoantibody was found in six of seven WS patients. These results were interpreted to suggest that in WS, an increase in an autoantibody to T-lymphocytes may lead to a change in T-lymphocyte subpopulations. Djawari et al. [6] reported a reduced lymphocyte response to phytohemagglutinin in one patient, and Salamon et al. [27] reported a reduced number of T-lymphocytes in another patient. As yet, however, there is no clear indication that a decrease in T-lymphocytes or in classes of T-lymphocytes, or specific increase in autoimmunity are typical in WS patients.

SUMMARY

Many features of WS resemble accelerated aging. Further, WS patients have a markedly reduced life-span. A high frequency of sarcomas is found. Also, WS fibroblasts *in vitro* have a severely reduced life-span. In culture, WS fibroblasts exhibit variegated translocation mosaicism. Although most patients studied have shown no chromosome instability in blood lymphocytes, a 14-fold increase in the frequency of chromosome gaps and breaks in lymphocytes has been reported by one laboratory. The presence of elevated concentrations of urinary hyaluronic acid may suggest that WS has an underlying metabolic defect that leads to widespread abnormalities in mesenchymal tissue development. A specific molecular or immune abnormality that might underlie unusual genetic phenotype has yet to be elucidated.

LITERATURE CITED

1. BARTRAM, C.R., KOSKE-WESTPHAL, T., PASSARGE, E.: Chromatid exchanges in Ataxia telangiectasia, Bloom's syndrome, Werner's syndrome and xeroderma pigmentosum. Ann. Hum. Genet. 40:79–86, 1976.
2. BASLER, J.W., DAVID, J.D., AGRIS, P.F.: Deteriorating collagen synthesis and cell ul-

trastructure accompanying senescence of human normal and Werner's syndrome fibroblast cell strains. *Exp. Cell Res.* 118:73-84, 1979.
3. BEADLE, G.F., MACKAY, I.R., WHITTINGHAM, S., TAGGART, G., HARRIS, A.W., HARRISON, L.C.: Werner's syndrome, a model of premature aging? *J. Med.* 9:377-403, 1978.
4. BROWN, W.T., DARLINGTON, G.J.: Thermolabile enzymes in progeria and Werner syndrome: Evidence contrary to the protein error hypothesis. *Am. J. Hum. Genet.* 32: 614-619, 1980.
5. DARLINGTON, G.J., DUTKOWSKI, R., BROWN, W.T.: Sister chromatid exchange frequencies in progeria and Werner syndrome patients. *Am. J. Hum. Genet.* 33: 762-766, 1981.
6. DJAWARI, D., LUKASCHEK, E., JECHT, E.: Altered cellular immunity in Werner's syndrome. *Dermatologica* 161:233-237, 1980.
7. EPSTEIN, C.J., MARTIN, G.M., SCHULTZ, A.L., MOTULSKY, A.G.: Werner's syndrome: A review of its symptomatology, natural history, pathologic features, genetics and relationship to the natural aging process. *Medicine (Baltimore)* 45:177-221, 1966.
8. FLEISCHMAJER, R., NEDWICH, A.: Werner's syndrome. *Am. J. Med.* 54:111-118, 1973.
9. FUJIWARA, Y., HIGASHIKAWA, T., TATSUMI, M.: A retarded rate of DNA replication and normal level of DNA repair in Werner's syndrome. Fibroblasts in culture. *J. Cell. Physiol.* 92:365-374, 1977.
10. GOLDSTEIN, S., MOERMAN, E.J.: Heat-labile enzymes in Werner's syndrome fibroblasts. *Nature* 255:159, 1975.
11. GOLDSTEIN, S., MOERMAN, E.: Heat-labile enzymes in skin fibroblasts from subjects with progeria. *New Engl. J. Med.* 292:1305-1309, 1975.
12. GOLDSTEIN, S., SINGAL, D.P.: Alteration of fibroblast gene products in vitro from a subject with Werner's syndrome. *Nature* 251:719-721, 1974.
13. GOTO, M., MURATA, K.: Urinary excretion of macromolecular acidic glycosaminoglycans in Werner's syndrome. *Clin. Chim. Acta* 85:101-106, 1978.
14. GOTO, M., HORIUCHI, Y., OKUMURA, K. TADA, T.: Immunological abnormalities of aging: An analysis of lymphocyte subpopulations of Werner's syndrome. *J. Clin. Invest.* 64:695-699, 1979.
15. GOTO, M., HORIUCHI, Y., TANIMOTO, K., ISHI, T., NAKASHIMA, H.: Werner's syndrome: Analysis of 15 cases with a review of the Japanese literature. *J. Am. Geriatr. Soc.* 26:341-347, 1978.
16. HAND, R., GERMAN, J.: A retarded rate of DNA chain growth in Bloom's syndrome. *Proc. Natl. Acad. Sci. USA* 72:758-762, 1975.
17. HOEHN, H., BRYANT, E.M., AU, K., NORWOOD, T.H., BOMAN, H., MARTIN, G.M.: Variegated translocation mosaicism in human skin fibroblast cultures. *Cytogenet. Cell Genet.* 15:282-298, 1975.
18. HOLLIDAY, R., PORTERFIELD, J.S., GIBBS, D.D.: Premature ageing and occurence of altered enzyme in Werner's syndrome fibroblasts. *Nature* 248:762-763, 1974.
19. MARTIN, G.M., SPRAGUE, C.A., EPSTEIN, C.J.: Replicative lifespan of cultivated human cells: Effects of donor's age, tissue, and genotype. *Lab. Invest.* 23:86-92, 1970.
20. MCKUSICK, V.A.: Medical Genetics 1962. (paragraph 792). *J. Chron. Dis.* 16: 599-603, 1963.
21. NAKAO, Y., KISHIHARA, M., YOSHIMI, H., INOUE, Y., TANAKA, K., SAKAMOTO, N., MATSUKURA, S., IMURA, H., ICHIHASHI, M., FUJIWARA, Y.: Werner's syndrome: In vivo and in vitro characteristics as a model of aging. *Am. J. Med.* 65:919-932, 1978.
22. NORDENSON, I.: Chromosome breaks in Werner's syndrome and their prevention in

vitro by radical-scavenging enzymes. *Hereditas* 87:151–154, 1977.
23. NORWOOK, T.H., HOEHN, H., SALK, D., MARTIN, G.M.: Cellular aging in Werner's syndrome: A unique phenotype? *J. Invest. Dermatol.* 73:92–96, 1979.
24. OPPENHEIMER, B.S., KUGEL, V.H.: Werner's syndrome—a heredo-familial disorder with scleroderma, bilateral juvenile cataract, precocious graying of hair and endocrine stigmatization. *Trans. Assoc. Am. Physicians* 49:358, 1934.
25. PETROHELOS, M.: Werner's syndrome: A survey of three cases, with review of the literature. *Am. J. Ophthalmol.* 56:941–953, 1963.
26. ROTHMUND, A.: Über Kataract in Verbindung mit einer eigentümlichen Hautdegeneration. *Arch. Ophthalmol.* 14:158, 1868.
27. SALAMON, T., BOGDANOVIC, B., LAZOVIC-TEPAVAC, O., BLAZEVIC, A., MACANOVIC-BOGNER, K.: Werner's syndrome and cellular immune reactions. *Acta Derm-Vener,* (ed). (Stoch.) 58:543–544, 1978.
28. SALK, D., AU, K., HOEHN, H., MARTIN, G.M.: Cytogenetics of Werner syndrome cultured skin fibroblasts: Variegated translocation mosaicism. *Cytogenet. Cell. Genet.* 30:92–107, 1981.
29. SALK, D., AU, K., HOEHN, H., STENCHEVER, M.R., MARTIN, G.M.: Evidence of clonal attenuation, clonal succession, and clonal expansion in mass cultures of aging Werner's syndrome skin fibroblasts. *Cytogenet. Cell Genet.* 30:108–117, 1981.
30. SALK, D., AU, K., HOEHN, H., MARTIN, G.M.: Effects of Radical-scavenging enzymes and reduced oxygen exposure on growth and chromosome abnormalities of Werner's Syndrome cultured skin fibroblasts. *Hum. Genet.* 57:269–275, 1981.
31. SALK, D., BYANT, E., AU, K., HOEHA, H., MARTIN, G.M.: Systemic growth studies, cocultivation, and cell hybridization studies of Werner Syndrome cultured skin fibroblasts. *Hum. Genet.* 58:310–316, 1981.
32. SCHONBERG, S.A., HENDERSON, E., NIERMEIJER, M.F., GERMAN, J.: Werner's syndrome: Preferential proliferation of clones with translocations. *Am. J. Hum. Genet.* 33:363A, 1981.
33. TANAKA, K., NAKAZAWA, T., OKADA, Y., KUMAHARA, Y.: Increase in DNA synthesis in Werner's syndrome cells by hybridization with normal human diploid and HeLa cells. *Exp. Cell Res.* 123:261–267, 1979.
34. TAO, L.C., STECKER, E., GARDNER, H.A.: Werner's syndrome and acute myeloid leukemia. *Can. Med. Assoc. J.* 105:952–954, 1971.
35. THANNHAUSER, S.J.: Werner's syndrome (progeria of the adult) and Rothmund's syndrome: Two types of closely related heredo-familial atrophic dermatosis with juvenile cataracts and endocrine features. A critical study with five new cases. *Ann. Intern. Med.* 23:559, 1945.
36. TOKUNAGA, M., FUTAMI, T., MAKAMATSU, E., YOSIZAWA, Z.: Werner's syndrome as "hyaluronuria." *Clin. Chim. Acta* 62:89–96, 1975.
37. WERNER, O.: Über Katarakt in Verbindung mit Sklerodermie. (Doctoral dissertation, Kiel University). Kiel: Schmidt & Klaunig, 1904.
38. ZUKER-FRANKLIN, D., RIFKIN, H., JACOBSON, H.G.: Werner's syndrome: An analysis of ten cases. *Geriatrics* 23:123–138, 1968.

B. Commentaries on Selected Aspects of the Disorders

PATTERNS OF NEOPLASIA ASSOCIATED WITH THE CHROMOSOME-BREAKAGE SYNDROMES

JAMES GERMAN

INTRODUCTION

Certain persons are more predisposed to develop clinical neoplasia within a given period of time than persons of comparable age in the general population. A person is identified as cancer-prone (i) if he can be shown to be either heterozygous for certain rare but well-known genes (e.g., those of retinoblastoma or Gardner's syndrome) or homozygous for certain others (e.g., those of Bloom's syndrome or ataxia-telangiectasia), or (ii) if, though genetically not so predisposed, he has been exposed excessively to some environmental agent known to be a tumor initiator or promoter.

Identification of a person as cancer-prone does not imply necessarily that he is at increased risk of developing neoplasms of the usual types or at the sites most commonly affected in the general population. Rather, a degree of specificity as to cancer type and site may be imparted by the particular pro-neoplasia gene inherited or by the particular environmental agent to which he has been exposed. An understanding of the bases for the specificities probably will provide insight into the mechanisms by which cells become neoplastic. The beginning point, however, is an accurate tabulation of just which neoplasms occur with increased frequencies in persons bearing the various cancer-predisposing genes and in persons exposed excessively to the various classes of environmental agents; hence, the tabulations I present here for several genetic disorders.

Dominant Genes

With respect to genes that predispose to neoplasia when inherited in a single dose, or when acquired as result of mutation, some knowledge of the type and site of neoplasms has become available simultaneously with the recognition of the several genes themselves, for the following reason. Such dominant genes often produce recognizable clinical syndromes by interfering with the development or function of a particular type tissue. Familiar examples of this include Gardner's syndrome, neurofibromatosis, and the basal cell nevus syndrome. Pedigree analyses of affected families indicate an autosomal dominant transmission for the clinical entities. The neoplasms that emerge in such persons are of a specific type and affect the particular tissue affected in the syndrome. Consequently, a specific neoplasm is often thought of as an integral part of the clinical syndrome itself. In the case of a few dominant genes, a neoplasm of a particular tissue is the first if not the only clinical evidence of the gene's presence; e.g., retinoblastoma, polyposis coli. A tabulation of the type and age of onset of neoplasms associated with these dominant genes, although outside the scope of the present report, should prove informative. Only relatively recently, for example, has it been recognized that persons inheriting the retinoblastoma gene are predisposed also to tumor formation in tissues other than retina.

Recessive Genes

Genes that require homozygosity if they are to predispose significantly to neoplasia are discovered to be segregating in a kindred only after the birth, to normal-appearing parents, of a child with a particular abnormal phenotype, one known to physicians as a specific clinical entity. That is, each of these rare clinical disorders shows recessive inheritance on formal genetic analysis, usually autosomal (e.g., xeroderma pigmentosum, Bloom's syndrome) but in some instances X-linked (e.g., dyskeratosis congenita, the Wiskott-Aldrich syndrome). In the recessive cancer-predisposing disorders, as in the dominant, the types and sites of neoplasms are different from those in the general population. Until recently, however, the rarity of the disorders has precluded a determination of either the distributions of tumor types or their ages of onset. In 1979, on the occasion of the Sixth International Congress of Radiation Research in Tokyo, I undertook a tabulation of the types of malignant neoplasms that occur in one group of recessively inherited, cancer-predisposing conditions, the so-called chromosome-breakage syndromes [50]; subsequently, the tabulations were presented for discussion at a workshop on DNA repair and mutagenesis [51]. The tabulation revealed that an impressively distinctive distribution of cancer types exists for each of these rare conditions, and the differences raised some intriguing questions [e.g., 22].

Here, again on the occasion of a symposium devoted to a consideration of chromosome breakage in relation to the etiology of neoplasia, I have chosen to tabulate neoplasms that occur in several rare, recessively transmitted human disorders, viz., the five conditions described in the preceding chapters of this volume: Bloom's syndrome (BS), ataxia-telangiectasia (AT), Fanconi's anemia (FA), xeroderma pigmentosum (XP), and Werner's syndrome (WS). Several approaches have been used in compiling the data, including the study of certain disease registries, extensive review of published papers reporting both series and single cases of the disorders, and personal contact with students of the disorders. The information tabulated here for BS is reliable because of the maintenance since the early 1960s of the Bloom's Syndrome Registry [47–49]. The distribution of types of neoplasms in AT is reliably extracted from data in the Immunodeficiency-Cancer Registry (ICR), an international registry maintained since the early 1970s of immunodeficient persons who develop cancer, [69,70,128,129]. The tabulations for FA, XP, and WS must be considered tentative, the literature having been my main source of information.

The periods of time that the five syndromes have been known in clinical medicine vary from over a century for XP to just over a quarter-century for BS (Fig. 1); thus, even though they all are rare entities, the sizes of the populations of affected persons that have been observed by physicians, and from which those developing cancer may or may not have been reported, differ considerably. This should be taken into consideration during examination of Tables I, III, IV, and V in which an attempt was made to list all of the neoplasms that have been reported for four of the five entities—BS, FA, XP, and WS. With respect to the estimation of the risk of neoplasia in a person with one of these rare disorders, it will be apparent from Figure 1 that the denominator will be larger the longer an entity has been known in clinical medicine. Also to be taken into consideration, if risk estimates are being attempted, is the tendency of physicians (and editors) not to report patients after the condition becomes well known unless a particular patient presents some unusual feature—such as cancer.

A final introductory comment concerns genetic heterogeneity. This has been recognized in AT, FA, and XP using complementation analysis of cells in culture. Thus, homozygosity for more than one mutant gene, possibly at more than one locus, can produce an array of abnormalities that is recognized clinically as a single syndrome. It is too early to tabulate the cancers in these rare syndromes by complementation group, but this certainly will be desirable at some future time.

NEOPLASIA IN BLOOM'S SYNDROME

BS was described in 1954. Study of the condition was begun in my laboratory six years later, in 1960, when I saw one of the first three patients Dr. David Bloom had reported. The child's death from leukemia in 1963 and my discovery at that time that Dr. Bloom's very first patient also had died of leukemia suggested that persons with the syndrome might be cancer-prone. This, coupled with the interesting cytogenetic features of BS cells, impelled me to study as many affected persons as possible.

Information concerning neoplasia occurrence in BS has been taken from the files maintained in the Laboratory of Human Genetics of The New York Blood Center on each person ever known to have had BS. These 99 persons comprise the Bloom's Syndrome Registry [47-49]. Contacts with affected persons and their families by mail, personal visits, or telephone conversations have provided a steady stream of information

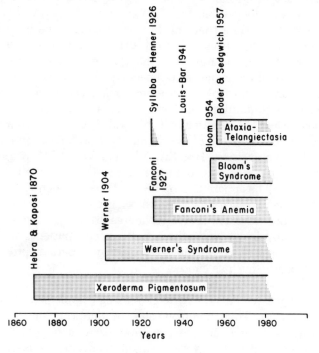

Fig. 1. Durations of clinical observations of xeroderma pigmentosum, Werner's syndrome, Fanconi's anemia, Bloom's syndrome, and ataxia-telangiectasia. (References to the original descriptions of these rare syndromes appear in "Literature Cited" [62,146,36,140,77,17] with the exception of that for Bloom's syndrome, which is the following: Am. J. Dis. Child. 88:754-758, 1954).

TABLE I. Neoplasia Diagnosed in the 99 Persons in the Bloom's Syndrome Registry, Indicating Ages of Onset (This Table Defines the Neoplasms Represented by Boxes in Figure 1)

Type of neoplasm		Age of onset
Malignant		
Leukemia, lymphoid	ALL*	4
	ALL	9
	ALL	15
	ALL	15
Lymphoid tissue, not leukemia	Lymphosarcoma, abdomen	4
	Lymphoma, abdomen	12
	Lymphosarcoma, cervical area	12
	Lymphoma, lymphoblastic type, diffuse, epipharynx	13
	Lymphoma, mixed histiocytic-lymphocytic	25
	Lymphoma, disseminated	30
	Lymphoma, diffuse histiocytic	31
Leukemia, not lymphoid	ANLL†	4
	ANLL	13
	ANLL	23
	ANLL	23
	ANLL	25
Carcinoma		
Squamous cell	Tongue, base of	30
	Epiglottis	30
	Esophagus, mid-	39
Adeno-	Colon, transverse	31
	Colon, sigmoid	37
	Colon, rectosigmoid	39
	Esophago-gastric junction	44
Metastatic‡	To liver	30
Other	Wilms's tumor	8
	Hodgkin's disease	16
Benign		
	Meningioma, temporal area	9
	Fibroma, subepidermal, thigh	39
	Polyps, colon and jejunum	39–44

*ALL, acute lymphocytic leukemia.
†ANLL, acute nonlymphocytic leukemia.
‡Tail of the pancreas suspected as primary site, but histological proof lacking.

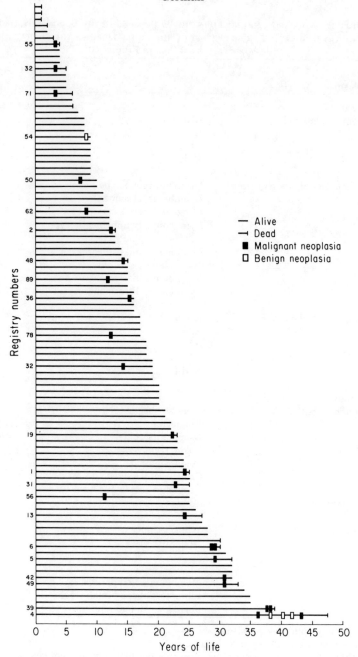

Fig. 2. Neoplasia in Bloom's syndrome. Representation by lines commensurate with years of their lives of persons in the Bloom's Syndrome Registry. Malignant and benign neoplasms are indicated by solid and open boxes, respectively, positioned on the life lines at ages of diagnosis. (Numbers at the origin of lines of persons with a neoplasm are the identification numbers of those persons in the Registry.)

regarding the general health and development of the affected persons and the occurrence of neoplasia in them and their close relatives. Consequently, information regarding the incidence of various types of neoplasms, their age of onset, and the year when they were diagnosed (Table I, Figs. 2, 3) can be considered reliable. The initial report that the BS gene may in the homozygous state predispose to cancer appeared in 1965, a decade after the description of the syndrome itself had been published. A program of cancer surveillance of affected persons throughout the world was put into effect at that time, and the predisposition to cancer was confirmed. That the cancer predisposition is enormous rapidly became apparent. Approximately a fourth of the 96 persons accessioned to the Bloom's Syndrome Registry who survived infancy have developed some form of neoplasia.

Each person in the Registry is depicted in Figure 2; it and Table I are complementary. Cancer in BS is impressive in three ways: by the frequency of its occurrence; by the diversity of type and tissue distribution; and, for the carcinomata, by the unusually early age of onset. Acute leukemia is the most common single type of malignancy that has occurred, acute lymphocytic leukemia (ALL), the most common malignancy of childhood, having been diagnosed four times, acute non-lymphocytic leukemia (ANLL) five times. Lymphoid tissue often has been affected: the ALL just mentioned, lymphosarcoma in each of two sibs, and lymphoma in five other persons. Hodgkin's disease has occurred once, as has Wilms's tumor. Epithelial cancer has been diagnosed seven times, at the following sites: base of the tongue, epiglottis, esophagus, colon. In addition, metastatic liver cancer was the cause of death of one individual (identified by double dagger in Table I), but histological proof and identification of the primary site of the cancer are lacking. These cancers that emerged in epithelium in BS did so at extraordinarily early ages in comparison to the general population. The mean age of onset in BS of all types of carcinoma combined was 36 (range 30–44),

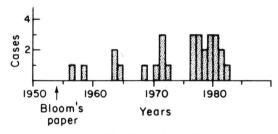

Fig. 3. Neoplasms in Bloom's syndrome, by year of occurrence.

whereas the mean age of onset of these particular tumors in the American population would have been 66 (range 63-68) (i.e., they occurred approximately three decades prematurely). Their occurrence at these anatomic sites at such young ages is to be expected in only ≤ 1-2% of the general population (estimated by Dr. N.R. Schneider from Third National Cancer Survey data [30]).

Figure 2 shows that 23 of the 99 individuals known to have had BS have developed at least one neoplasm; 18 of the 71 who lived 10 years did so, as did 11 of the 35 who lived 20 years and 6 of the 13 who lived 30 years. The only person with the syndrome known to have reached age 40 already had developed one cancer by that age, and he died of a second one at age 47. Attrition of living entries from the Registry through premature death from cancer is partially responsible for the low average age—16.8 years—of the 78 individuals alive today. The mean age at death from cancer is 23. Cairns [22] estimated that the death rate from cancer in each age group of the BS population is approximately 100-fold greater than that observed in the general population.

Three persons developed cancers at two sites: a cigarette smoker of 30 (No. 6 in Fig. 2) who developed squamous cell carcinoma of the epiglottis was found during surgery for the carcinoma to have lymphoma as well; a man of 39 (No. 39 in Fig. 2) who sought medical attention because of dysphagia caused by esophageal carcinoma also was found to have an asymptomatic sigmoid colon carcinoma; and, a man (mentioned above, No. 4 in Fig. 2) who was cured of sigmoid colon carcinoma at age 37 developed esophago-gastric junction carcinoma at age 44.

The etiology and genetics of benign neoplasms have attracted little scientific attention. It is interesting to note, therefore, that two persons with BS have developed such neoplasms: a boy of 9 (No. 54 in Fig. 2) developed symptoms from a large meningioma in the left temporal area, exceptionally early for this tumor; and, the man mentioned above (no. 4 in Fig. 2) who developed the sigmoid and esophageal carcinomata at ages 37 and 44 had to have a subepidermal fibroma of the thigh removed at age 39 and several polyps of the jejunum and colon between ages 39 and 44.

In summary, the risk that a person with BS will develop any of a variety of neoplasms is strikingly increased in comparison to people of the same age in the general population. With the exception of xeroderma pigmentosum with its myriad lesions that affect the sunlight-exposed integument, the cancer risk of BS probably is greater than in any of the other recessively transmitted disorders that predispose to cancer.

NEOPLASIA IN ATAXIA-TELANGIECTASIA

It is widely believed that the first genetically determined immunodeficiency (GDID) reported in the literature was X-linked agammaglobulinemia, which was described in 1952 by Bruton [21]. In the succeeding decades, more than a dozen other clinical immunodeficiency states, each rare, were described, several of which are known to follow recessive inheritance. In fact, a major GDID, AT, first was described in 1926 [140] (Fig. 1), and again in 1941 [77]; however, not until a third report in 1957 [17] — after the advent of antibiotic therapy — did AT take on significance as a clinical entity and become known generally among physicians.

The immunodeficiency states most commonly encountered in clinical medicine are the following, in decreasing order of frequency of occurrence: selective decrease or absence of one serum immunoglobulin, IgA being the most commonly affected; acquired agammaglobulinemia, also known as variable immunodeficiency; immunodeficiency accompanied by ataxia and telangiectasia of the conjunctiva and skin (AT); severe combined-system immunodeficiency (SCID); and, immunodeficiency accompanied by thrombocytopenia and eczema (the Wiskott-Aldrich syndrome, WAS). Among the least frequently encountered of the disorders is immunodeficiency accompanied by growth deficiency and sun-sensitivity, a combination of features clinically referred to as BS. Genetic heterogeneity exists in some of the GDIDs; it has not been demonstrated as yet in Bruton's agammaglobulinemia, WAS, and BS. Neither of the two most commonly encountered entities, selective IgA (or IgM) deficiency and acquired agammaglobulinemia, follows simple genetics; although familial clustering of cases of these is known, and increased numbers of relatives of those affected may show immunological disturbances, these two entities as yet cannot with certainty be classed as GDIDs.

By 1963, reports of malignant neoplasia in persons with a primary immunodeficiency had begun to appear [90]. Dr. E. Boder, who had learned of three neoplasms in persons with AT, emphasized the relationship at a major national meeting of academic pediatricians [18]. Actually, one of the seven patients in the first series of cases of AT to be published "died reportedly of sarcoma of the mastoid at age 9 years" [17]. In 1971, an extensive report about immunodeficiency states made by a WHO committee stated that Dr. R. A. Good, then at the University of Minnesota, would organize a registry "to comprise cases of malignant tumor found in patients with primary immunodeficiency . . ." [40]. That same year, Gatti and Good published an extensive tabulation of cancers reported to have occurred in the various immunodeficiency states [43], a tabulation that can be considered to be the precursor of the Immunodeficiency-Cancer Registry (ICR) developed and maintained at the University of Minnesota; 38 neoplasms were listed for AT at that time [43], and it was estimated that at least 10% of persons with AT die with malignancy. By 1973 [69] 151 neoplasms in 145 patients with some immunodeficiency had been registered by the ICR; 52 of these neoplasms were in persons with AT.

Even though persons with AT comprise only about 8% of the clinical cases from which come the immunodeficient persons with cancer accessioned to the ICR, AT consistently has been the major contributor of cancers to the ICR; 90 of the 267 cancers in the ICR in 1978 (34%) were in persons with AT [128]. Although data are unavailable to determine the true risk that persons with AT will develop cancer, Spector et al. [128] made a guarded estimate of 11.7%, based on selected ICR data; of the other immunodeficiency states, only WAS was estimated to be higher, at 15.4%.

The types of neoplasms that occur in persons with primary immunodeficiency are strikingly different from that in the general population matched for sex and age; e.g., lymphoreticular neoplasms predominate (non-Hodgkin's disease lymphoma and lymphoid leukemia) and constitute 59% of all cases in the ICR (158 of 267). In addition, different patterns of malignancy also exist among the different clinical immunodeficiencies. Further, the types of neoplasms in the various clinical entities differ from those seen in another important population of immunodeficient persons, i.e., those made so for medical reasons by the therapeutic regimens accompanying the transplantation of kidneys, bone marrow, and livers. (The major source of information concerning cancer in homograft recipients is the Denver Transplant Tumor Registry (DTTR), maintained since 1968 by Dr. I. Penn [92–94], which by 1980 contained data concerning 1,023 patients who had developed cancer following transplantation.) The incidence of cancers following kidney transplantation is from 2 to 13%. In kidney homograft recipients, 39% of the cancers are of the skin or lip (two-thirds of which are squamous cell carcinomata, in contrast to the predominance of basal cell in the general population), and 18% are lymphomata (predominantly reticulum cell and Kaposi's sarcomata). Thus, as in the primary immunodeficiencies, neoplasms commonly observed in the general population are under-represented, i.e., those of the lung, prostate, colon, breast, and cervix uteri. Noteworthy is the high incidence of lymphoma in immunodeficient persons in both the ICR and the DTTR – of neoplasia that originates in a cell of the immune system itself—in view of its low incidence in the general population (i.e., about 4% of all malignancies). However, the type of lymphoma that predominates varies with the type of immunodeficiency.

In both the primary and secondary immunodeficiencies, multiple primary neoplasms in one person occur more commonly than is the case generally in persons who develop cancer.

Spector et al. [129] have analyzed the data in the ICR with respect specifically to neoplasia in AT. Table II has been prepared on the basis of those data. Most neoplasms occurred in children under 16 years of age. Although carcinomata are unusually common in AT, non-Hodgkin's disease lymphomata and leukemias made up 44% and 24%, respectively, of all the neoplasms. All the leukemias in the AT population were lymphoid in origin; in the general childhood population, in contrast, about a fifth of leukemias are myeloid. (This pattern also contrasts with the distribution of types of leukemia in FA (Table III) in which lymphoid leukemia is unknown). Also emphasizing the unusual sites and types of neoplasia in AT is the occurrence of a subtype of Hodgkin's disease that is

TABLE II. Frequency and Ages of Onset of Various Types of Malignant Neoplasia in AT (1981 Extract From the Immunodeficiency-Cancer Registry [129])

Type of neoplasm	Mean age of onset (range)	Number	
Lymphoma, non-Hodgkin's disease	9.3 (2.7–22.0)	48	
Histiocytic			14
Lymphosarcoma			12
Malignant			10
Other*			12
Hodgkin's disease	10.3 (4.0–18.8)	12	
Leukemia	11.0 (2.9–27.9, plus one aged 45.4)	26	
Lymphoid			23
Myeloid			0
"Acute"			3
Other	18.5 (2.3–30.0)	22	
Stomach			8
Brain, ovary, skin (3 each)			9
Liver			2
Parotid, larynx, breast			3
Total		108	

*Lymphoblastic, 4; undifferentiated, 4; Burkitt's, 3; immunocytoma.

TABLE III. Neoplasia in Persons with Fanconi's Anemia (FA) Reported to Have Occurred Before 1981 (For Those Five Entries Italicized, Uncertainty or Doubt Exists With Respect to the Diagnosis of FA, Neoplasia, or Both)*

Year patient was first reported [ref.]	Age of onset of hematological disorder	Age (and year) when neoplasia was diagnosed	Firmness† of diagnosis of		Neoplasm	Anabolic steroid administered before onset of neoplasia (years)	Comment
			FA	Neoplasia			
1952 [126]	*8*	*11 (1949)*	*A*	*D*	*Leukemoid reaction*	—	
1955 [29]	Never	27 (1952)	B	A	Acute leukemia, probably granulocytic	—	FA never diagnosed
1964 [116–120,125,126]	24	35 (1980)	A	A	Carcinoma, bronchus	—	Asymptomatic primary and metastases disclosed at autopsy.
1965 [*102*]; 1972 [67]	14	25 (c. 1962)	A	A	Hepatocellular carcinoma	+(4–6)	Cirrhosis; case reported twice. Testosterone begun in 1958.
1966 [34]	Never	25 (1961)	B	A	Epidermoid carcinoma, esophagus	—	Minimal hematological disturbance.
1966 [34]	*18*	*17 (1957)*	*C*	*E*	*Acute "aleukemic" leukemia, granulocytic*	—	*Ambiguous report; leukemia appears highly improbable*
1966 [3,15,16]	7	8 (pre-1965)	A	A	Acute monocytic leukemia	—	
1966 [3,15,16]	7	7½ (pre-1965)	A	A	Acute monocytic leukemia	+	Very short period of of steroid therapy before leukemia was diagnosed.

Year [ref]					Tumor		Notes
1966 [134]	30	31 (1963)	C	A	Squamous cell carcinoma, anal skin	—	2 primary lesions in 1 person.
		32 (1964)	C	A	Carcinoma-in-situ (Bowen's disease), vulva	—	
1968 [54]	8	11 (1964)	A	A	Acute monomyelocytic leukemia	+ (3)	
1970 [32]	2	7 (1968)	A	A	Acute myelomonocytic leukemia	+ (3)	
1970 [78]; 1971 [135]	Never	21 (1968)	C	A	Squamous cell carcinoma, gingiva	—	No hematological abnormalities; case reported twice. 2 primary lesions in 1 person.
1971 [134, 135]	35	38 (1968)	C	A	Squamous cell carcinoma, anus	+ (3)	
		39 (1969)	C	A	Carcinoma-in-situ, labium minus	+ (3)	
1971 [13]	20	21	A	A	Hepatoma, well-differentiated	+ (1)	
1971 [97]	<20	21	E	A	Hepatoma, well-differentiated	+ (3/4)	
1972 [67]	<14	21 (1970)	B(E)	A	Hepatocellular carcinoma, well-differentiated	+ (7)	
1973 [60]	≤14	≤38	A	A	Squamous cell carcinoma, tongue	+ 24	2 primary lesions in 1 patient. Androgen therapy only 4 mo.
	9½	38	A	A	Hepatocellular carcinoma	+ (24)	
1974 [58,59]	10 (1972)	B	A	Acute stemcell leukemia	+ (1/4)	
1974 [76]	4	10 (1973)	A	A	Acute leukemia ("monocytoid")	+ (4)	

(Continued on next page)

Year patient was first reported [ref.]	Age of onset of hematological disorder	Age (and year) when neoplasia was diagnosed	Firmness† of diagnosis of		Neoplasm	Anabolic steroid administered before onset of neoplasia (years)	Comment
			FA	Neoplasia			
1974 [12,24,25]	8	23 (1972)	A	A	Hepato-cholangioma	—	Cirrhosis. Extra-hepatic metastases.
1975 [38]	5	12 (1974)	B(E)	B(E)	Hepatic adenoma	+ (½)	
1975 [82]			A	A	Hepatic adenoma, well differentiated	+ (3)	
1975 [110]	7	21 (1971)	A	A	Squamous cell carcinoma, gingiva	+ (11)	2 primary lesions in 1 person.
		>21 (1971 or later)	A	A	Squamous cell carcinoma, tongue	+ (11)	
1975 [110]	10	21 (1973)	A	A	Acute leukemia, probably myelo- or myelomonocytic	+ (10)	2 primaries in 1 patient. Hepatoma asymptomatic.
		21 (1973)	A	A	Hepatoma	+ (10)	
1975 [28]	9	12 (1972)	A	A	Adenoma, liver multi-nodular, benign	+ (4)	
1975 [66]	5	10 (1973)	B	A	Hepatoma, well-differentiated	+ (4)	
1976 [148]	3	11 (1973)	A	A	Acute myeloid leukemia	+ (4)	
1976 [117,120]	24	29 (1974)	B	B(E)	Acute leukemia	+ (5)	Androgen therapy of unknown duration.
1976 [117,120]	10	24 (1974)	A	B(E)	Acute leukemia	+	
1976 [68]	12	34 (1974)	A	A	Hepatocellular cancer, well-differentiated	+ (12)	
1976 [143]	5	8-9	A	A	Benign hepatoma	+ (3-4)	
1977 [19]	3	5	A	A	Acute myelomonocytic leukemia	+ (2)	

					+ (4)	2 lesions in 1 person.
	4	12 13 (1974)		B C	+ (4) + (9)	"One grande pan- myelose (B.O.M.)" Hepatoma, benign Multiple hepato- cellular neoplasms (well- differentiated hepatocarcinomata)
1977 [81]	≤6	6	A	A	+ (1/6)	Hepatocellular carcinoma
1977 [6,80]	5	15 (1974)	A	A	+ (8)	Acute erythro- leukemia	Case reported twice.
1978 [127]	9	18 (1976)	A	A	+ (9)	Acute leukemia, myelomonocytic
1978 [99,121]	20	23 (1975)	A	B		Multiple skin cancers	Multiple primary lesions in 1 person.
		26 (1978)	A	A		Squamous cell carcinoma, lip
1978 [127]	6	16	A	A	+ (6)	Acute leukemia, myeloblastic
1979 [98]	5	15	A	A	+ (10)	Acute erythro- leukemia
1980 [87]	3	13 (1975)	A	A	+ (8)	Acute leukemia
1980 [10]	6	14 (1979)	A	A	+(1/4)	Erythroleukemia
1980 [75,130]	6	11	A	B	+ (3)	Acute myelomono- cytic leukemia
1981 [5,145]	5	5 (1980)	A	A	–	Acute myeloid leukemia

*Some data in this table are unpublished and were obtained through review of hospital records and, or, through personal communication with authors. Certain cases that have been reported were purged from the table after further information was so obtained, because of my doubt of a diagnosis or because of lack of adequate evidence to support such. Five entries appear in the table — those italicized — only because of their historical importance, i.e., citation repeatedly in papers on this subject. (I am grateful to Dr. B.P. Alter for helpful discussions during the preparation of this table, but the final decisions concerning the data to be included were mine.)

Blank = information lacking; + = yes; – = no.

†A, firm; B, probable; C, possible; D, probably not; E, inadequate information reported for a decision.

extremely rare in children under 10: the lymphocyte-depletion type of Hodgkin's disease occurred in four of the children with AT, between ages 4 and 18.

Thus, more instances of cancer have been documented in AT than in any other GDID. Although most persons with AT die from infections, about 1 in 10 apparently develops at least one neoplasm. The distribution of types of cancers is distinctive, somewhat different from that associated with other GDIDs and very different from that associated with immunodeficiency induced medically to prevent the rejection of engrafted organs.

The role, if any, of chromosome instability in the cancer proneness of AT is unknown [101]. Clones of cells with translocations affecting chromosome Nos. 7 and 14 are found much more often in AT T-lymphocytes than in cells from normal persons, in cells from persons with other conditions characterized by chromosome instability, or in cells exposed to clastogens; in this regard, it also is noteworthy that chromosome No. 14 is translocated preferentially in lymphomata generally. As yet, only preliminary understanding exists of the significance of the interrelationships in AT of chromosome instability, immunodeficiency, and the emergence of neoplastic clones, particularly in tissues of the immune system itself. The same may be said of BS, in which chromosome instability and immunodeficiency also figure prominently, along with neoplasia that often originates in lymphoid tissue. The situations in AT and BS are reminiscent of the extraordinarily complex interrelationships currently being elucidated in the case of Epstein-Barr virus infection, immunodeficiency both genetic and induced, under- and over-proliferation of various lymphoid cell sub-populations, and the eventual emergence of a monoclonal neoplasm that often is characterized by a marker chromosome rearrangement (the matters addressed in [72]). In AT as in BS, seemingly too many, rather than too few, possible explanations for cancer proneness present themselves. That a person with either BS or AT is at such an impressively greater-than-normal risk of malignant neoplasia conceivably indicates that more than a single factor is responsible in persons homozygous for these rare genes. The hypothesis, then, would be that in both BS and AT an increased occurrence of cellular events capable of initiating or, and, promoting neoplasia is coupled with a defective system of immune surveillance of either virus-infected cells or neoplastic clones. It would be assumed that, by different mechanisms, the primary defect in each disorder would result in both a characteristic genetic instability and a characteristically defective immune function.

FANCONI'S ANEMIA: WHY SO MANY CANCERS LATELY?

FA was described in 1927 [36], and reports of affected persons have been published regularly since. Reviews of the literature have been made from time to time. Thus, in 1952, C.O. Carter and his associates [105] found 21 bona fide cases; by 1958, 69 cases of FA could be assembled [12]; by 1964, 129 cases [53]; and, by 1973, 217 cases [8]. Alter [2] states that the literature now contains reports of more than 300 patients. Obviously, the number of affected persons being submitted to careful clinical observation has increased considerably in recent decades.

Papers associating FA and malignant neoplasia began to appear in the 1950s. Leukemia was reported in relatives of persons with FA: in a brother in 1955 [29] (but, as will be mentioned [ii, below], the brother probably had FA) and in cousins in 1951 [7] and 1952 [35,105]. With respect to this association, two well written and adequate case reports require comment [29,126]: (i) In 1952, Silver et al. [126] described a girl with features typical of FA who at age 7 presented symptoms of marrow aplasia. During the last several years of her life, the girl had major infections of the face, cervical nodes, and lungs accompanied by a striking leukocytosis, with blood-leukocyte counts as high as 59,000. Although such high leukocyte counts are unexpected in the face of chronic marrow aplasia, the child was not considered to have leukemia, either by the doctors who treated her or by the pathologists who performed a postmortem examination and concluded that she had a leukemoid reaction—"leukocytosis under the stimulus of multiple staphylococcus infection." If she actually had leukemia, this would become the first recorded instance of cancer in FA; unfortunately, the question cannot be settled from the information available to me. (ii) In 1955, Cowdell et al. [29] reported acute leukemia in a 27-year-old man with small stature, hyperpigmentation, and several anatomical defects characteristic of FA; however, the man had not been diagnosed as having FA and had been healthy and free of hematological symptoms up to the time of leukemia onset; the leukemic man's brother, with similar features of FA, developed marrow aplasia at age 22 and, except for the late onset, would seem to have had typical FA. It seems reasonable to conclude that both brothers had the genetic constitution for FA.

In 1959 Garriga and Crosby published an often-quoted paper [42] in which they, after reviewing the FA literature, raised the possibility that leukemia occurs with an increased frequency in members of families ascertained through persons with FA. As examples to support their hypothesis, they listed the leukemic brother just mentioned [29], the leukemic cousins [7,105], and a leukemic uncle (no reference to the uncle being provided, however) of FA patients. It is surprising that Garriga and Crosby, in referring to the child whom Silver et al. [126] had reported (summarized above, i), wrote simply and directly, "The patient died of leukemia," offering no explanation for having altered the diagnosis of the original authors. That particular "leukemia" was, in fact, the only neoplasm they tabulated in the 66 persons they considered to have had FA up to the time of their review. (Although both Reinhold et al. [105] and Cowdell et al. [29] in earlier papers had addressed themselves to the question of leukemia in relation to FA, neither group had viewed the patient of Silver et al. [126] as leukemic; also, Alter in her review of bone marrow-failure syndromes [3] rejects that patient as being leukemic.) Garriga and Crosby wrote [42], "The rate of leukemia in Fanconi *families* seems to be astonishingly high," and also, "Of special interest is the demonstration of a high incidence of leukemia *in the families* of patients with hereditary hypoplasia of the bone marrow" (emphasis added). My main point here is that the Garriga and Crosby paper does not show that persons with FA are predisposed to cancer; in fact, their tabulations, as presented, justify no conclusion concerning the incidence of neoplasia in families ascertained through persons with FA.

The first indisputable report of neoplasia in a person with FA appeared in 1965 [102] (Table III). Some may prefer to classify the leukemic brother of the man with FA described in 1955 by Cowdell et al. [29] (mentioned above, ii) as pre-aplastic Fanconi's anemia, or perhaps Fanconi's anemia *sans* anemia, and he is accepted as a bona fide case of FA in Table III and Figure 4. Thus, during the first 37 years of FA's existence as a recognized clinical entity, not a single indisputable report appeared of a malignant neoplasm in a person with FA. Dr. Fanconi himself did not emphasize cancer proneness as a feature of the syndrome when he reviewed FA in 1967 [37], four decades after having described the condition [36]; he did cite Garriga and Crosby [42] to support his belief that "In the families with F.A., leukemia and other myelopathies are more frequent than in normal families." In view of the paucity of FA cancer data before 1967, how can it be that FA nowadays is widely considered to be a disorder with a significant cancer predisposition?

The following relevant ideas and reports were published about the time this concept was becoming widespread and thus contributed to its acceptance. (i) The chromosome instability characteristic of cultured FA cells [101,116] and the striking hypersensitivity of FA cells to certain DNA-damaging agents [111,112] have been considered by many, including myself, to be of central importance to the origin of cancer in FA.

However, in retrospect, thinking about FA, cancer, and chromosome breakage can be seen to have been complicated and clouded by several contemporary trends and observations. At the same time that chromosome instability in FA was first being reported—1964-1966 [15,16,45,116]—reports of cancer in FA also were beginning to appear (Fig. 4). When we [45] discovered increased chromosome breakage in a child with FA, independently of Schroeder [116] and of G.E. Bloom, Warner, Gerald, and Diamond [15,16], we were unaware of any association between FA and cancer; Schroeder's report suggests that she also was unaware of it [116]. In contrast, G.E. Bloom et al. [15,16] apparently were aware of a link, but—anticipating comments to be made below—it could be significant that their report emanated from the center at which anabolic steroid therapy for aplastic anemia had been originated by Shahidi and Diamond [122,123]. BS was, at just the same time (1964, 1965), being reported to feature chromosome instability [44,45], and evidence was mounting [45,113] that BS is a genetically determined condition which predisposes strikingly to cancer. Quite naturally, BS and FA were lumped together [46] (and in some respects this has been valuable).

(ii) In 1967 hypertransformability of FA cells by the oncogenic virus SV40 was reported by Todaro et al. [143], fitting nicely with the newly emerging view that FA patients are predisposed to malignant transformation in vivo. (It was considerably later that hypersensitivy of FA cells to certain DNA-damaging agents was to be reported [111].)

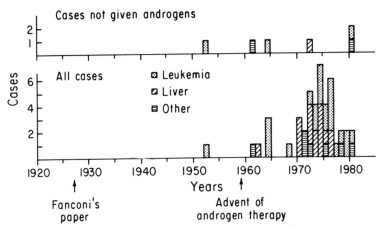

Fig. 4. Neoplasms in Fanconi's anemia, by year of occurrence.

(*iii*) In the mid-1960s, papers began appearing stating outright that FA is a cancer-predisposing condition. However, an impressive—and instructive—circularity of citations supporting such statements can be found in this literature, examples of which follow. Garriga and Crosby's 1959 paper [42] was quoted repeatedly—incorrectly—as evidence that persons with FA get cancer more often than expected [e.g., 16,24,54,61, 125,135,137,138], not that members of families ascertained through persons with FA may do so, which actually is what Garriga and Crosby suggested (*vide supra*). The other reference cited most regularly to support an author's statement that cancer occurs with increased frequency in FA itself was the paper by G.E. Bloom, Warner, Gerald, and Diamond [16], mentioned above. (Later, papers by Swift and his associates [134,135] and Dosik et al. [32] were cited also, but Dosik et al. had cited Swift and Hirschhorn [135] and Garriga and Crosby [42] as evidence that persons with FA "appear to have a significantly increased susceptibility to the development of acute leukemia," and Swift and Hirschhorn [134] had cited G.E. Bloom et al. [16] and Garriga and Crosby [42] as evidence that "leukemias and solid tumors have been remarkably frequent in patients with Fanconi's anemia and in members of their families." Swift et al. [135] had cited Garriga and Crosby [42] and G.E. Bloom et al. [16] as evidence that "individuals with Fanconi's anemia often die with acute leukemia," although he and his associates did report several unusual cancers in some older persons they considered to have FA (Table III) [134,135].) But, for the reasons already given, the two papers [16,42] cited as the main evidence that FA is a cancer-prone condition cannot be taken as supportive of this idea. Nevertheless, G.E. Bloom et al. [16] wrote, "Garriga and Crosby *described* the increased incidence of leukemia found in Fanconi's aplastic anemia" (emphasis added). G.E. Bloom et al. [16] did mention in the Discussion section of their paper, without presenting details, the observation that two FA patients known to them personally had died during childhood of acute monocytic leukemia. In 1966, Hoefnagel et al. [63] wrote, "The high incidence of leukemia in patients with Fanconi's panmyelopathy and members of the family has also been documented," citing as support for the statement two papers, the one by Garriga and Crosby and the one by G.E. Bloom et al. [16]. Similarly, Schuler et al. [125] in 1969 referred to "the surprisingly high frequency of leukemia in Fanconi's anemia," citing the same two references as their evidence. By 1968 the opinion had become widespread that persons with FA were at great risk of developing leukemia [46].

(*iv*) In 1971 Swift [136,137], because of FA's "striking predisposition to neoplastic disease" [136], re-examined the original proposal of Garriga and Crosby [42] that more relatives of persons with FA had died of cancer than expected. After comparing the number of cancers he had learned about in eight families ascertained through a person he believed had FA with the number he expected in the general population, he presented a convincing argument suggesting that they indeed had a striking predisposition to cancer [137]. (Only in 1979 was this claim retracted, after an appropriate number of FA families finally has been studied [23,139].)

In this way, FA became known widely as a condition with a striking predisposition to cancer. It also was widely believed, solely on the basis of Swift's prematurely reported and unconfirmed work [136,137], that relatives of persons with FA are cancer-prone.

It was when I first had occasion to tabulate the cancers known to have occurred in each of the several chromosome-breakage syndromes and to compare the distributions of types of cancers which occur in them [50] (already mentioned) that I realized that something unusual and dramatic had begun to happen in FA in the mid-1960s; at that time many reports of malignant neoplasia began to appear. Table III and Figure 4 detail the 45 neoplasms known to me to have occurred by the end of 1980 in persons with what can, with reasonable certainty, be accepted as FA. Obviously, a dramatic association between FA and cancer does exist! Furthermore, it became apparent that the distribution of malignancy being reported in FA is strikingly different from that in the other chromosome-breakage syndromes, as well as from that in the general population of people in their early decades of life: 22 acute leukemias, not one of which is ALL; 16 primary tumors of the liver; 2 lingual carcinomata; 1 gingival carcinoma; multiple skin cancers and a labial squamous cell carcinoma in a young woman; 1 bronchial carcinoma; and, 1 esophageal carcinoma. Inevitably, the question arises as to why, in view of the fact that FA has been known as a clinical entity for more than half a century, essentially all of the reports of associated neoplasia have appeared only during the last two decades? The simplest explanation, although possibly not the correct or only one, is that careful observation of a large number of affected persons has enabled previously overlooked features of the syndrome to be revealed — cancer proneness being one of them. But, another question also arises: what is the explanation for the unusual distribution of types of neoplasia that is emerging in FA? The following review of the relevant literature attempts to answer these questions.

Prior to 1959, FA was not diagnosed in the absence of marrow failure. In addition, most of the reports published during the first several decades after 1927 described a disorder of childhood; typically, marrow failure in a child with hyperpigmentation and several of the characteristic anatomical defects would lead to death a few months or years after the onset of some hematological symptom (see the 1958 review of Bernard et al. [12]). A few early papers reported what may have been the same entity in adults [e.g., 106]. In recent years, reports have appeared with increasing frequency of what is generally believed to be the same clinical entity in persons in their second through fourth decades of life. Furthermore, "Fanconi's anemia" is occasionally diagnosed even in the absence of hematological abnormalities, and many affected persons have been identified who lack the characteristic anatomical malformations. The clinical spectrum is widening for FA, for not-quite-clear reasons. The availability of cytogenetic techniques to demonstrate the characteristic chromosome instability in FA doubtless is partially responsible for the increasing numbers of cases diagnosed when the clinical picture is atypical. "FA" conceivably encompasses multiple clinical entities, and our deficient understanding of just what is and what is not FA compounds the problem of determining whether untreated FA is or is not a cancer-prone condition. (In Table III, I have attempted to cope with this problem by grading the certainty of diagnosis.)

Partially responsible for the older ages now achieved by persons with FA is the therapeutic use of anabolic steroids (certain androgens) introduced in 1959 [122,123] (Fig. 4), a popular regimen for controlling marrow failure and thereby prolonging life of both children and young adults with aplastic marrows of various etiologies [2]. In many cases, corticosteroids are administered simultaneously with the androgen. When employed in FA, this treatment regimen typically is maintained for many months to several years, generally for periods longer than are required in idiopathic aplastic anemias. However, this therapy may have done more to alter the clinical presentation of FA than merely to increase the lifespans of affected persons; the striking sequence of events shown in Figure 4 — pre-anabolic steroid therapy era, few reported cancers; anabolic steroid therapy era, many cancers — suggests that the therapy itself may be responsible in some way for the cancers in FA, or at least for the relatively recent recognition of such, as was suspected earlier by several authors [e.g., 8,67,124,133].

Because anabolic steroid therapy, using, for example, oxymetholone, testosterone esters, or nortestosterone decanoate, is employed in few conditions other than FA, its long-term side effects are not well known. But, leukemia has been reported in individuals so treated who had non-FA aplastic anemia [31,41,71], as have hepatocellular neoplasms [38,67,131,132]. The first cancer reported in a person with FA (referred to above) was a hepatocellular carcinoma found in 1964 at autopsy of a 27-year-old man who had advanced cirrhosis [67,102]; he had been treated with androgen since 1958, apparently for azospermia. (It was not until 1959 that androgen was reported to be useful in the treatment of aplastic anemia [*122*]!) Twenty-six cases of anabolic steroid-related neoplasms and neoplasm-like lesions of the liver have been reported [*132*], including those in persons with FA. Because of the rarity of such tumors in Europe and America, many of the authors reporting liver cancers in FA have suggested that anabolic steroid therapy is responsible to some degree [e.g., *38,82,124,133*]. The combination of ANLL and liver cancer constitutes an unusual pattern of types of neoplasia in FA; that pathological changes occur in the liver in persons with conditions other than FA who are given anabolic steroids suggests, at least for the liver neoplasia in FA, that the medical management, rather than the genome of the treated person, may be of prime importance. (Incidentally, the hepatocellular carcinoma of FA treated with anabolic steroids is clinically atypical [*124*]. It has histological features similar to that cancer which kills quickly after onset of symptoms in parts of the world in which it is a common disease, such as Africa, whereas in both FA and non-FA aplastic anemia its course usually is more benign, as seemingly is the case with many, but not all, cases of anabolic steroid-related hepatocellular neoplasms. Other pathologic changes in the liver including the condition referred to as peliosis hepatis [*83*] also occur in a proportion of persons given such treatment, including non-FA, and sometimes these changes are associated with hepatocellular neoplasms of various types, some benign in appearance (called adenomata) and others malignant (carcinomata) [*124*]. The question often arises as to whether the liver in which the neoplasm appears, which itself often shows extensive non-neoplastic change (fibrosis, peliosis hepatis, hemosiderosis, or frank cirrhosis), has been damaged by the multiple transfusions that usually are requried, and possibly by transfusion-borne hepatitis.)

As early as 1973, Beard et al. [*8*] pointed out that when leukemia does occur in FA, "it does not necessarily represent the natural course of FA as is often assumed," and suggested that bone-marrow aplasia per se and long-term androgen therapy were possible etiological factors in the emergence of leukemia. How might androgen therapy be responsible? It could be an indirect effect, simply by prolonging the life of some of those with FA and permitting clinical cancer the time it requires to emerge in a truly cancer-prone disorder. This is perhaps the most popularly held view today. Another possibility is that androgen therapy enhances an underlying cancer predisposition in FA. However, one could depart radically from present attitudes and ask whether such therapy, alone, might be responsible for FA's cancer propensity; such a view would account for the existence of a single report of cancer in FA [*29*] during the first 37 years of its known existence.

It should not be overlooked, however, that at least six cancers have occurred in individuals diagnosed as having FA who, as far as I can learn, had received no anabolic steroid treatment (Table III, column 7; Fig. 4, upper portion). Even six cancers in a young population of such small size probably would qualify FA as a cancer-prone disorder. In the future, detailed information concerning such cases can help to clarify this interesting and important matter. Because of the popularity of anabolic steroid therapy, few persons with FA now go without such treatment, which for the present makes the problem difficult to resolve. The clinical course of those who in the future will be treated by bone marrow transplantation rather than androgen therapy will be followed with interest. Of value, also, will be both the careful documentation and reporting of therapy duration in relation to the time of onset of malignant neoplasia in persons with non-FA aplastic states of the marrow and similar documentation of any cancer that may occur in non-hematological conditions for which anabolic steroid is given for protracted periods.

It cannot be concluded yet that the genetic constitution responsible for the phenotype referred to as FA predisposes per se to cancer. Also, it has not been determined what, if any, are the roles in the etiology of the cancers that emerge in FA of the chromosome instability, the purported hypertransformability by an oncogenic virus [143], the hypersensitivity to certain DNA-damaging agents [111], and the administration of anabolic steroids.

XERODERMA PIGMENTOSUM: WHY SO FEW CANCERS IN TISSUES UNEXPOSED TO SUNLIGHT?

Since the report made by Cleaver in 1968 [26] demonstrating that UV-irradiated cells of most persons with XP undergo an abnormally small amount of "repair replication" of damaged DNA, much attention has focused on this genetic disorder [74]. XP cells have been studied especially with respect to DNA synthesis, DNA-damage repair, and mutability, but the clinical condition also has been submitted to new and careful observation. The XP phenotype is produced by at least eight different genetic constitutions, a remarkable degree of genetic heterogeneity for what itself is a rare clinical entity. (In Egypt and certain other parts of North Africa, XP, although still rare, is relatively more common than in Western Europe and North America [27]; the frequency of the XP gene(s) also is known to be relatively high in Japan [86].)

After it was realized that XP cells in vitro are hypersensitive to 254-nm wavelength UV irradiation, additional related studies of various types were made. It was learned that XP cells are hypermutable and that they are hypersensitive to DNA-damaging agents other than UV, specifically to certain chemicals known to be mutagens and carcinogens. The most obvious clinical feature of XP is the drastic change resulting from damage sustained by sunlight-exposed tissues of the body. The freckling, corneal damage, cell death, and emergence of myriad neoplasms of diverse types of the skin and eye, sometimes even of the tip of the tongue, all are attributable to the DNA-damage repair deficiency in cells whose nuclei inevitably are irradiated by the sun. However, DNA-damaging agents, very common in our 20th-century environment, also must by various routes reach internal cells of the person with XP – by inhalation, ingestion, or absorption through the skin. Because XP cells are known to be hypersensitive to such chemicals, it might be expected that persons with XP would manifest an abnormally great number of the ordinary neoplasms that originate in internal tissues (breast, enteric, bronchial, and genitourinary epithelium; lymphoid and hematogenous cells; connective tissue cells), not just in the sun-exposed tissues. Therefore, when I first tabulated them in 1979 [50], it was a surprise to find that the medical literature contains very few examples of neoplasms other than those of the skin, eye, and tongue. How might this be explained?

One possibility that has been considered is the abbreviated life expectancy in XP. However, although the life expectancy in XP is shorter than normal [74], mainly because of premature death from skin cancer, many affected persons do live well past the age of 20. In this respect, a comparison of XP and BS is informative. In BS, the life expectancy certainly is curtailed greatly: only approximately a third of the persons known to have been affected have lived past 20 (Fig. 2). Because many internal cancers have been recognized in Bloom's syndrome, including carcinomata of the gastrointestinal tract (Table I), a shortened life expectancy per se is an inadequate explanation for their paucity in XP. Whatever the true explanation for the scarcity of "internal" cancers in XP may be, I again am able (Table IV) to find reports of only a very few such cancers (see also [73]); hence, the question posed in the title of this portion of my paper.

Analysis of the data being collected by the recently established Xeroderma Pigmentosum Registry [4], a primary objective of which is to identify and collect information concerning all persons with XP in the United States, in due time will determine whether an excessive number of internal neoplasms does or does not occur, in comparison with their occurrence in the general population. For the present, however, the available information (Table IV) affirms that the person with XP is predisposed to cancer only in tissues exposed to sunlight. Although various interesting theories have been advanced to explain the absence of an obvious increase in non-integumentary neoplasia in XP [20,22], I myself have no new interpretation of the situation and maintain the possibly naive optimism I expressed in 1979 [50]: the observation "suggests that many of the oncogenic chemicals identified in recent years, substances hopelessly ubiquitous in our modern environment, may not regularly reach body tissues in significant amounts."

TABLE IV. Neoplasms Reported in Xeroderma Pigmentosum, Exclusive of Those Affecting the Skin, Eye, and Tip of the Tongue

Type of neoplasm	Age	Complementation group	Reference
Leukemia			
Acute lymphocytic*	3	A	[103]
"Myelogenous"	32		[11]
Epithelial			
Buccal mucosa (gum†), squamous cell carcinoma	9		[104]
Palate, squamous cell carcinoma	18		[142]
Thyroid, adenoma‡	22		[104]
Breast, "infiltrating carcinoma"	38		[144]
Brain			
Medulloblastoma, cerebellar	14	C	[52,91,96]
Glioblastoma multiforme with sarcomatous component§	15	C	[55,56]

*A cell line from this patient is identified as CRL 1203 in the American Type Culture Collection (ATCC). Fluorouracil in propylene glycol was applied daily to the skin of this child as a therapeutic measure; the report [103] fails to state when this was given in relation to the onset of the leukemia.

† "... at the gum margin over the left upper cuspids...."

‡ A "small adenoma of the thyroid" was diagnosed in 1931 but could not be detected by physical examination in 1939.

§ A cell line from this person, XP15BE, is banked by the ATCC. The tumor, confirmed histologically at the time of brain surgery, developed after Goldstein and Hay-Roe published their report [55,56].

Note: A "bronchial carcinoma" in a 62-year-old man with "pigmented xerodermoid" was mentioned by Hofman et al. [65]. The patient's sister's fibroblasts have been identified as "variant" XP [39]. Dr. E.G. Jung has communicated to me the following information, by which I choose to exclude the patient from the above table: a carcinoma of a central bronchus was "suspected" in 1979; metastases were detected in the lungs. "Sputum contained melanoma cells." Biopsy was never made, nor was there an autopsy when he died in 1980.

NEOPLASIA IN WERNER'S SYNDROME

WS was first described in 1904 [146]; an eponymic designation for it has been in use since 1934 [88]. Reports of probably fewer than 200 affected persons in Western Europe and North America have been made; many are known in the Japanese literature [57,84,85]. The main reason for investigating WS is that here can be observed the premature emergence of many of the disorders—including neoplasia—that occur in the normal person as he ages.

The first report of neoplasia in a person with WS appeared in 1939 [1]. In their definitive 1966 paper on this disorder, Epstein et al. [33] recognized 125 instances of WS, and among these they found 17 instances of some form of "significant" benign or malignant neoplasia. They noted that the neoplasms were predominantly of connective tissue or mesenchyme derivation and that they constituted a principal cause of death in those with the syndrome. Scattered reports of neoplasia in WS have continued to appear since 1966, but no concerted effort has been made by any group to determine the true incidence of this complication. In view of the infrequency with which WS is diagnosed—its frequency has not been estimated, but certainly it is rare—the expectancy of neoplasia can be concluded to be considerably greater than in the general population. Table V lists the neoplasms about which I have been able to learn and which I consider reasonably reliably reported.

Table V shows that the distribution of types of neoplasms in WS is unusual, clearly different from that in the other rare disorders under consideration here. Furthermore, it also is very different from that in the general population. In fact, some of the commonest human cancers do not appear in the list, notably lung, colon, and stomach carcinomata and those of lymphoid tissue. This suggests caution in using WS as a model for studying natural human aging, at least with respect to neoplasia.

The excessive number of tumors affecting connective tissue in WS takes on special interest in light of the striking increase in occurrence of chromosome translocations that are observed in fibroblast cultures derived from WS skin biopsies [64,108,109,114,115], a matter discussed more extensively elsewhere [114,115].

CONCLUSIONS

The patterns of neoplasia, i.e., the frequency distributions of types of neoplasms, that characterize each of five rare genetic syndromes have been presented in the five sections of this paper. For each syndrome, unusual or distinctive aspects of the pattern were commented on and some

TABLE V. Neoplasia Reported in Werner's Syndrome

Type of neoplasm*	Age	Reference
Connective tissue or mesenchyme derivation, malignant		
Myosarcoma, uterus	36	[33]
Sarcoma, nerve sheath	37	[33]
Fibrosarcoma (or fibroliposarcoma), radius	38	[1,33,89]
Spindle-cell sarcoma	39	[33]
Fibroxanthoma	40	[14]
Hemangiolipoma with occasional mitoses	42	[33]
Melanotic sarcoma	42	[33]
Osteogenic sarcoma	49	[33]
Osteogenic sarcoma	50	[107]
Epithelial, malignant		
Carcinoma, gastric	38	[100]
Adenocarcinoma, papillary, thyroid	39	[33]
Carcinoma, basal cell	40	[100]
Carcinoma, basal cell	40	[100]
Carcinoma, breast	41	[33]
Cystadenocarcinoma, papillary, ovary	41	[14]
Carcinoma, liver	42	[33]
Carcinoma, liver (or adenocarcinoma, hepatic duct)	42	[33,89]
Carcinoma, undifferentiated, prostate	47	[33,79]
Carcinoma, basal cell, nose	52	[147]
Carcinoma, squamous cell, scalp	56	[147]
Carcinoma, squamous cell, forearm	56	[147]
Leukemia		
Myeloid, acute	24	[141]
Acute	32	[14]
Benign tumors		
"Uterine tumors"	22	[33]
Meningioma	34	[33]
Paraganglioma, bladder (submucosal) and adrenal (adjacent to)	37	[33]
Meningioma	41	[85]
Meningioma	42	[33]
Myoma, uterus	42	[33]
Meningioma and fibrosarcoma	44	[85]
Meningioma	51	[33]
Adenoma (several cases), pituitary, thyroid, adrenal		[33,85,89]

*Terminology of the authors in most cases.
Note: Goto et al. [57] listed the following malignant tumors in 11 of the 195 Japanese patients with Werner's syndrome (ages in parentheses), but with neither documentation nor statement of their method of ascertaining patients: fibrosarcoma on the left thigh (31), acute myelocytic leukemia (36), pseudomyxoma peritonei (ovarian origin) (37), leiomyosarcoma (40), melanoma (40), anaplastic carcinoma of the thyroid, giant cell type (42), breast cancer (histologically not defined) (46), cholangioma (49), carcinoma of the bladder (53), laryngeal carcinoma (60), and fibrosarcoma (61). They also listed the following benign tumors: 6 thyroidal adenomata, uterine myoma, lymphangioma, meningioma, and fibroadenoma of the breast.

questions were raised. Table VI summarizes the types of neoplasms in a comparative way, along with those of three other genetic disorders of immune function not known to be associated with chromosome instability.

The broad groupings of neoplasms in Table VI obscure the subtle but striking diffrences that exist among the syndromes, differences that can be revealed by more detailed analysis and comparison of the patterns than are presented here. The table does show that lymphoid neoplasia is more common in the GDIDs than in the other syndromes (also than in the general population). In addition, however, differences exist among the several GDIDs themselves. For example, 68% of the neoplasms in AT can be classified as lymphoid (78% if Hodgkin's disease is included as such), whereas in BS these comprise only 44% (or 48%). In WAS, none of the 8 leukemias recorded has been lymphoid, whereas in AT, 22 of the 26 (information unavailable on the other 4) have been [70]. Lymphoma, Hodgkin's disease, and lymphoid leukemia are unreported in FA, XP, and WS. Several other distinctive aspects of the patterns were mentioned earlier, such as the unusual types of carcinomata and the lack of lymphoid leukemia in FA, the unexpected deficiency of non-integumentary neoplasia in XP, and the excess of connective tissue-derived tumors in WS.

Some of the differences may have a trivial explanation. Thus, emergence of epithelial neoplasia may depend on longevity; the mean ages at the time of onset of neoplasia follow this order: BS > AT > WAS > SCID (Table VI), and the relative frequencies of neoplasms that originated in epithelium follow exactly the same order: 28%, 17%, 2%, and 0%, respectively (Table I in [70]). Most persons with SCID and WAS die at a very young age as a consequence of their immune defect, whereas those with AT and BS have a longer life expectancy, being relatively less severely crippled immunologically. Most patients with AT do die of infection; in BS, however, 15 of the 21 who have died experienced (as a group) severe and repeated infections that in another age would have been lethal, but survived the infections and went on to die—still prematurely—of cancer.

Comparisons made using Table VI are limited somewhat in their usefulness not only because the number of neoplasms for most of the entities is small but also because the data were collected in different ways, as mentioned at the outset: (*i*) the data for WAS, SCID, Bruton's agammaglobulinemia, and AT were collected retrospectively through the ICR; (*ii*) those for BS were collected prospectively through the Bloom's Syndrome Registry; and, (*iii*) those for FA, XP, and WS consist of what I could find in the medical literature.

The incidence estimates given in the last row of Table VI are provisional. The populations of persons affected with the various syndromes

TABLE VI. Comparison of the Distributions of Types of Malignant Neoplasms Reported in Certain Genetically Determined Immunodeficiencies and in the Chromosome-Breakage Syndromes*

	Genetically determined immunodeficiencies				Chromosome-breakage syndromes			
	WAS†	SCID†	Bruton's agamma†	AT†	BS‡	FA	XP§	WS
Median age (range) in years at diagnosis	6.0(1.5–22)	0.9 (0.1–2.25)	8.0 (1.1–20)	9.0 (2.0–31)	19 (4–44)	20 (5–38)	16 (3–38)	40 (24–56)
Neoplasm								
Lymphoma, Non-Hodgkin's disease	37	10	4	50	7	0	0	0
Hodgkin's disease	4	1	1	11	1	0	0	0
Leukemia	8	4	7	26	9	22	2	2
Lymphoid	0	1	3	22	4	0	0	0
Myeloid	4	3	2	0	0	5	1	1
Other/unspecified	4	0		4	5	17	1	
Other neoplasms	4		2	19	8	20	5	21
Total neoplasms	53	15	15	106	25	42	7	23
Risk of occurrence of neoplasia	15.4%¶	1.5%¶	0.7%¶	11.7%¶	25.2%¶			

*Abbreviations defined in text. Units in table indicate numbers of cases known. Blanks indicate lack of information.
†Data from ICR, May, 1979 [70]. Several distinct entities are combined as "severe combined-system immunodeficiency."
‡Data from Bloom's Syndrome Registry.
§Excluding neoplasia of the skin, eye, and tip of the tongue.
¶Estimate based on incidence of malignant neoplasia in the clinical experience of contributors to the ICR [128].

from whom those with cancer emerged are of unknown size. With the exception of BS, no registries have been developed of all those affected in a given geographical region or population. The estimates given for the GDIDs other than BS are based on the total numbers of patients with the various disorders being followed in the clinics that report the occurrence of cancers to the ICR, and possibly do not reflect the true incidences [128]. The estimate for BS, although reasonably accurate for the moment, is an underestimate, because as the registered BS population ages, more and more cancers are being reported to the Registry (Fig. 3). The mean age of those with BS who are alive is only 17 years, as few affected persons survived infancy before the antibiotic era. The same must be true for the other constitutional immunodeficiencies, and it is true also for FA, but for other reasons: many FA-affected persons have survived longer during the past two decades because of a reasonably efficacious treatment for the panmyelopathy. For FA, XP, and WS, I have made no estimates of the incidence of neoplasia because the only denominators available are the numbers of cases that have been reported in the medical literature.

I would like to think that tabulations of cancers in selected populations – such as those first made by Gatti and Good [43] for the constitutional immunodeficiencies, and then the periodic reports from the ICR [69,70,128,129], the DTTR [93,94], and the Bloom's Syndrome Registry [48, 49] – will in the future become available for many of the cancer-predisposing genetic disorders, as well as for populations exposed excessively to specific environmental carcinogens. From such population data, analyses and comparisons very well may reveal other unusual patterns of cancer. From these, in turn, will arise new questions, the laboratory investigations of which can then yield interesting basic information regarding the molecular basis for cancer itself.

LITERATURE CITED

1. AGATSTON, S.A., GARTNER, S.: Precocious cataracts and scleroderma (Rothmund's syndrome; Werner's syndrome). *Arch. Opthalmol.* 21:492–496, 1939.
2. ALTER, B.P., PARKMAN, R., RAPPAPORT, J.M.: Bone marrow failure syndromes. *In* Hematology of Infancy and Childhood, Second Edition (Nathan, D.G., Oski, F.A. eds.), Philadelphia: W.B. Saunders, pp. 168–249. 1981.
3. ALTER, B.P., POTTER, N.U.: Personal communication of unpublished data.
4. ANDREWS, A.D., GERMAN, J.L., III, KRAEMER, K.H., LAMBERT, W.C.: Xeroderma list. *Nature* 291:104, 292:490, 1981.
5. AUERBACH, A.D., WEINER, M.A., WARBURTON, D., YEBOA, K., LU, L., BROXMIRE, H.E.: Acute myeloid leukemia as the first hematologic manifestation of Fanconi anemia. *Am. J. Hematol.* 12:289–300, 1982.

6. BARGMAN, G.J., SHAHIDI, N.T., GILBERT, E.F., OPITZ, J.M.: Studies of malformation syndromes of man. XLVII: Disappearance of spermatogonia in the Fanconi anemia syndrome. *Eur. J. Pediatr.* 125:163-168, 1977.
7. BAUMANN, T.: Konstitutionelle Panmyelopathise mit multiplen Abartungen (Fanconi syndrome). *Ann. Paediatr.* 177:65-76, 142-174, 1951.
8. BEARD, M.E.J., YOUNG, D.E., BATEMAN, C.J.T., McCARTHY, G.T., SMITH, M.E., SINCLAIR, L., FRANKLIN, A.W., BODLEY SCOTT, R.: Fanconi's anaemia. *Q. J. Med., New Series* 42:403-422, 1973.
9. BERGER, R., BUSSEL, A., SCHENMETZLER, C.: Somatic segregation and Fanconi anemia. *Clin. Genet.* 11:409-415, 1977.
10. BERGER, R., BERNHEIM, A., LeCONIAT, M., VECCHIONE, D., SCHIASON, G.: Chromosomal studies of leukemic and preleukemic Fanconi's anemia patients. *Hum. Genet.* 56:59-62, 1980.
11. BERLIN, C., TAGER, A.: Xeroderma pigmentosum. Report of eight cases of mild to moderate type and course: A study of response to various radiations. *Dermatologica* 116:27-35, 1958.
12. BERNARD, J., MATHÉ, G., NAJEAN, Y.: Contribution a l'étude clinique et physiopathologique de la maladie de Fanconi. *Rev. Fr. Clin. Biol.* 3:599-612, 1958.
13. BERNSTEIN, M.S., HUNTER, R.L.,YACKNIN, S.: Hepatoma and peliosis hepatis developing in a patient with Fanconi's anemia. *N. Engl. J. Med.* 284:1135-1136, 1971.
14. BJÖRNBERG, A.: Werner's syndrome and malignancy. *Acta Derm Venereol.* (Stockh.) 56:149-150, 1976.
15. BLOOM, G.E., GERALD, P.S., WARNER, S., DIAMOND, L.K.: Chromosome aberrations in constitutional aplastic anemia and their possible relation to other hematopoietic disorders. *J. Pediat.* 67:924-925, 1965.
16. BLOOM, G.E., WARNER, S., GERALD, P.S., DIAMOND, L.K.: Chromosome abnormalities in constitutional aplastic anemia. *N. Engl. J. Med.* 274:8-14, 1966.
17. BODER, E., SEDGWICK, R.P.: Ataxia-telangiectasia. A familial syndrome of progressive cerebellar ataxia, oculo-cutaneous telangiectasia and frequent pulmonary infection. *U. So. Cal. Med. Bull.,* Spring issue:15-51, 1957.
18. BODER, E.: Discussion of paper by Peterson, R.D.A., Blaw, M., and Good, R.A. entitled "Ataxia-telangiectasia: A possible clinical counterpart of the animals rendered immunologically incompetent by thymectomy." *J. Pediatr.* 63:701-703, 1963.
19. BOURGEOIS, C.A., HILL, F.G.H.: Fanconi anemia leading to acute myelomonocytic leukemia. Cytogenetic studies. *Cancer* 39:1163-1167, 1977.
20. BRIDGES, B.: How important are somatic mutations and immune control in skin cancer? Reflections on xeroderma pigmentosum. *Carcinogenesis* 2:471-472, 1981.
21. BRUTON, O.C.: Agammaglobulinemia. *Pediatrics* 9:722-728, 1952.
22. CAIRNS, J.: The origin of human cancers. *Nature* 289:353-357, 1981.
23. CALDWELL, R., CHASE, C., SWIFT, M.: Cancer in Fanconi anemia families. *Am. J. Hum. Genet.* 31:132a, 1979.
24. CATTAN, D., VESIN, P., WAUTIER, J., KALIFAT, R., MEIGNAN, S.: Liver tumors and steroid hormones. *Lancet* 1:878, 1974.
25. CATTAN, D., KALIFAT, R., WAUTIER, J.-L., MEIGNAN, S., VESIN, P., PIET, R.: Maladie de Fanconi et cancer du foie. *Arch. Fr. Mal. App. Dig.* 63:41-48, 1974.
26. CLEAVER, J.E.: Defective repair replication of DNA in xeroderma pigmentosum. *Nature* 218:652-656, 1968.
27. CLEAVER, J.E., ZELLE, B., HASHEM, N., EL-HEFNAWI, M.H., GERMAN, J.: Xeroderma pigmentosum patients from Egypt: II. Preliminary correlations of epidemi

ology, clinical symptoms, and molecular biology. *J. Invest. Dermatol.* 77:96-101, 1981.
28. CORBERAND, J., PRIS, J., DUTAU, G., RUMEAU, J.-L., REGNIER, C.: Association d'une maladie de Fanconi et d'une tumeur hépatique. *Arch. Fr. Pediatr.* 33:275-283, 1975.
29. COWDELL, R.H., PHIZACKERLY, P.J.R., PYKE, D.A.: Constitutional anemia (Fanconi's syndrome) and leukemia in two brothers. *Blood* 10:788-801, 1955.
30. CUTLER, S.J., YOUNG, J.L.: Third National Cancer Survey: Incidence Data. National Cancer Institute Monograph No. 41, U.S. Government Printing Office, Washington, D.C., 1975.
31. DELAMORE, I.W., GEARY, C.G.: Aplastic anaemia, acute myeloblastic leukaemia, and oxymetholone. *Br. Med. J.* 2:743-745, 1971.
32. DOSIK, H., HSU, L.Y., TODARO, G.J., LEE, S.L., HIRSCHHORN, K., SELIRIO. E.S., ALTER, A.A.: Leukemia in Fanconi's anemia: Cytogenetic and tumor virus susceptibility studies. *Blood* 36:341-352, 1970.
33. EPSTEIN, C.J., MARTIN, G.M., SCHULTZ, A.L., MOTULSKY, A.: Werner's syndrome. A review of its symptomatology, natural history, pathologic features, genetics and relationship to the natural aging process. *Medicine* 45:177-221, 1966.
34. ESPARZA, A., THOMPSON, W.R.: Familial hypoplastic anemia with multiple congenital anomalies (Fanconi's syndrome) – Report of three cases. *R. I. Med. J.* 49:103-110, 1966.
35. FAIRBURN, E.A., BURGEN, A.S.V.: The skin lesions of monocytic leukaemia. *Br. J. Cancer* 1:352-362, 1947.
36. FANCONI, G.: Familiäre, infantile perniziosaartige Anämie (perniziöses Blutbild und Konstitution). *Jahrb. Kinderh.* 117:257-280, 1927.
37. FANCONI, G.: Familial constitutional panmyelopathy, Fanconi's anemia (F.A.). I. Clinical aspects. *Seminars in Hematology* 4:233-240, 1967.
38. FARRELL, G.C., JOSHUA, D.E., UREN, R.F., BAIRD, P.J., PERKINS, K.W., KRONENBERG, H.: Androgen-induced hepatoma. *Lancet* 1:430-432, 1975.
39. FISCHER, E., JUNG, E.G., CLEAVER, J.E.: Pigmented xerodermoid and XP-variants. *Arch. Dermatol. Res.* 269:329-330, 1980.
40. FUDENBERG, H., GOOD, R.A., GOODMAN, H.C., HITZIG, W., KUNKEL, H.G., ROITT, I.M., ROSEN, F.S., ROWE, D.S., SELIGMANN, M., SOOTHILL, J.R.: Primary immunodeficiencies. Report of a World Health Organization committee. *Pediatrics* 47:927-946, 1971.
41. GARDNER, F.H.: Androgen therapy of aplastic anaemia. *Clin. Haematol.* 7:571-585, 1978.
42. GARRIGA, S., CROSBY, W.H.: The incidence of leukemia in families of patients with hypoplasia of the marrow. *Blood* 24:1008-1014, 1959.
43. GATTI, R.A., GOOD, R.A.: Occurrence of malignancy in immunodeficiency diseases. A literature review. *Cancer* 28:89-98, 1971.
44. GERMAN, J.: Cytological evidence for crossing-over in vitro in human lymphoid cells. *Science* 144:298-301, 1964.
45. GERMAN, J., ARCHIBALD, R.M., BLOOM, D.: Chromosomal breakage in a rare and probably genetically determined syndrome of man. *Science* 148:506-507, 1965.
46. GERMAN, J.: Chromosomal breakage syndromes. *Birth Defects* 5(5):117-131, 1969.
47. GERMAN, J.: Bloom's syndrome. II. The prototype of human genetic disorders predisposing to chromosome instability and cancer. *In* Chromosomes and Cancer, (German, J. ed.), John Wiley and Sons, pp. 601-617, 1974.
48. GERMAN, J., BLOOM, D., PASSARGE, E.: Bloom's syndrome. V. Surveillance for

cancer in affected families. *Clin. Genet.* 12:162-168, 1977.
49. GERMAN, J., BLOOM, D., PASSARGE, E.: Bloom's syndrome. VII. Progress report for 1978. *Clin. Genet.* 15:361-367, 1979.
50. GERMAN, J.: The cancers in chromosome-breakage syndromes. *In* Radiation Research (Proc. 6th Internatl. Congress of Radiation Research, Tokyo, 1979) (Okada, S. et al., eds.) Publ. by Japan Ass. Radiation Res. Univ. of Tokyo, pp. 496-505, 1979.
51. GERMAN, J.: The chromosome-breakage syndromes: Different genes, different treatments, different cancers. *In* DNA Repair & Mutagenesis in Eukaryotes (Generoso, W.M., Shelby, M.D., de Serres, F.J., eds.) New York: Plenum Press, pp. 429-439, 1980.
52. GIANNELLI, F., AVERY, J., POLANI, P.E., TERRELL, C., GIAMMUSSO, V.: Xeroderma pigmentosum and medulloblastoma: Chromosome damage to lymphocytes during radiotherapy. *Radiat. Res.* 88:194-208, 1981.
53. GMYREK, D., SYLLM-RAPOPORT, I.: Zur Fanconi-Anämie (F.A.) Analyse von 129 beschriebenen Fällen. *Z. Kinderheilkd.* 91:297-337, 1964.
54. GMYREK, D., WITKOWSKI, R., SYLLM-RAPOPORT, I., JACOBASCH, G.: Chromosomal aberrations and abnormalities of red-cell metabolism in a case of Fanconi's anaemia before and after development of leukaemia. *German Med. Monthly* 13:105-111, 1968.
55. GOLDSTEIN, N., HAY-ROE, V.: Prevention of skin cancer with a PABA in alcohol sunscreen in xeroderma pigmentosum. *Cutis* 15:61-64, 1975.
56. GOLDSTEIN, N.: Personal communication of unpublished data.
57. GOTO, M., TANIMOTO, K., HORIUCHI, Y., SASAZUKI, T.: Family analysis of Werner's syndrome: A survey of 42 Japanese families with a review of the literature. *Clin. Genet.* 19:8-15, 1981.
58. GÖTZ, M., HANDEL, H., PICHLER, E., WEIPPL, G.: Fanconi-Anämie Übergang in Leukämie und Erythrozytenstoffwechselbefunde. *Kinderärztl. Prax.* 42:69-74, 1974.
59. GÖTZ, M.: Personal communication.
60. GUY, J.T., AUSLANDER, M.O.: Androgenic steroids and hepatocellular carcinoma. *Lancet* 1:148, 1973.
61. HARDISTY, R.M., WEATHERALL, D.J.: Blood and Its Disorders. Oxford: Blackwell Scientific Publ., p. 1131, 1974.
62. HEBRA, F., KAPOSI, M.: Lehrbuch der Hautkrankheiten. pp. 511, 515, 1870.
63. HOEFNAGEL, D., SULLIVAN, M., MCINTYRE, O.R., GRAY, J.A., STORRS, R.C.: Panmyelopathy with congenital anomalies (Fanconi) in two cousins. *Helv. Paediatr. Acta* 3:230-238, 1966.
64. HOEHN, H., BRYANT, E.M., AU, K., NORWOOD, T.H., BOMAN, H., MARTIN, G.M.: Variegated translocation mosaisism in human skin fibroblast cultures. *Cytogenet. Cell Genet.* 15:282-298, 1975.
65. HOFMAN, H., JUNG, E.G., SCHNYDER, U.W.: Pigmented xerodermoid: First report of a family. *Bull. Cancer* 65:347-350, 1978.
66. HOLDER, L.E., GNARRA, D.J., LAMPKIN, B.C., NISHIYAMA, H., PERKINS, P.: Hepatoma associated with anabolic steroid therapy. *Am. J. Roentgenol.* 124: 638-642, 1975.
67. JOHNSON, F.L., FEAGLER, J.R., LERNER, K.G., MAJERUS, P.W., SIEGEL, M., HARTMAN, J.R., THOMAS, E.D.: Association of androgenic-anabolic steroid therapy with development of hepatocellular carcinoma. *Lancet* 2:1273-1276, 1972.
68. KEW, M.C., COLLER, B. VAN, PROWSE, C.M., SKIKNE, B., WOLFSDORF, J.I., ISDALE,

J., KRAWITZ, S., ALTMAN, H., LEVIN, S.E., BOTHWELL, T.H.: The occurrence of primary hepatocellular cancer and peliosis hepatis after treatment with androgenic steroids. *S. Afr. Med. J.* 50:1233-1237, 1976.
69. KERSEY, J.H., SPECTOR, B.D., GOOD, R.A.: Primary immunodeficiency diseases and cancer: The Immunodeficiency-Cancer Registry. *Int. J. Cancer* 12:333-347, 1973.
70. KERSEY, J.H., FILIPOVICH, A.H., SPECTOR, B.D.: Immunodeficiency and malignancy. *In* Genetic and Environmental Factors in Experimental and Human Cancer, (Gelboin, H.V. et al., eds.), Tokyo: Japan Scientific Societies Press, pp. 111-126, 1980.
71. KING, J.B., BURNS, D.G.: Aplastic aneaemia, oxymetholone and acute myeloid leukaemia. *S. Afr. Med. J.* 46:1622-1623, 1972.
72. KLEIN, G., PURTILO, D.T. (EDS).: Symposium on Epstein-Barr virus induced lymphoproliferative diseases in immunodeficient patients. *Cancer Res.* 41:4209-4304, 1981.
73. KRAEMER, K.H.: Xeroderma pigmentosum. *In* Clinical Dermatology, Vol. 4 (Demis, D.J., Dobson, R.L., McGuire, J. eds.) Hagerstown: Harper and Row, Unit 19-7, pp. 1-33, 1980.
74. KRAEMER, K.H., LEE, M.M., SCOTTO, J.: Diseases of environmental-genetic interaction: Preliminary report of a retrospective study of neoplasia in 268 xeroderma pigmentosum patients. *In* Proc. Third Internatl. Conference on Envrionmental Mutagens and Carcinogens (Sugimura, T., Kondo, S., Takebe, H. eds.), Tokyo: Univ. of Tokyo Press, and New York: Alan R. Liss, Inc., pp. 603-611, 1982.
75. KUNZE, J.: Estren-Dameshek-Anämie mit myelomonocytärer Leukämie (Subtyp der Fanconi-Anämie?) *In* Klinische Genetik in der Pädiatrie (Spranger, J., Tolksdorf, M. eds.) Stuttgart: Georg Thieme Verlag, pp. 213-214, 1980.
76. LÉVY, J.M., STOLL, C., KORN, R.: Sur un cas de leucémie aiguë chez une fillette atteinte d'anémie de Fanconi. Revue de lat littérature. *Nouv. Rev. Fr. Hématol.* 14:713-720, 1974.
77. LOUIS-BAR, D.: Sur un syndrome progressif comprenant des télangiectasies capillaries cutanées et conjonctivales symétriques, à disposition naevoïde et des troubles cérébelleux. *Confin. Neurol.* 4:3242, 1941.
78. MCDONOUGH, E.R.: Fanconi anemia syndrome. *Arch. Otolaryngol.* 92:284-287, 1970.
79. MARTIN, G.M., SPRAGUE, C.A., EPSTEIN, C.J.: Replicative life-span of cultivated human cells. Effect of donor's age, tissue, and genotype. *Lab. Invest.* 23:86-92, 1970.
80. MEISNER, L.F., TAHER, A., SHAHIDI, N.T.: Chromosome changes and leukemic transformation in Fanconi's anemia. *In* Aplastic Anemia (Hibino, S., Takaku, F., Shahidi, N.T. eds.), Baltimore: University Park Press, pp. 253-271, 1978.
81. MOKROHISKY, S.T., AMBRUSO, D.R., HATHAWAY, W.E.: Fulminant hepatic neoplasia after androgen therapy. *N. Engl. J. Med.* 296:1411-1412, 1977.
82. MULVIHILL, J.J., RIDOLFI, R.L., SCHULTZ, F.R., BORZY, M.S., HAUGHTON, P.B.T.: Hepatic adenoma in Fanconi anemia treated with oxymetholone. *J. Pediatr.* 87:122-124, 1975.
83. NADELL, J., KOSEK, J.: Peliosis hepatis. *Arch. Pathol. Lab. Med.* 101:405-410, 1977.
84. NAKAO, Y., KISHIHARA, M., YOSHIMI, H., INOUE, Y., TANAKA, K., SAKAMOTO, N., MATSUKWA, S., IMURA, H., ICHIHASHI, M. FUJIWARA, Y.: Werner's syndrome. In vivo and in vitro characteristics as a model of aging. *Am. J. Med.* 65:919-932, 1978.
85. NAKAO, Y., HATTORI, T., TAKATSUKI, K., KURODA, Y., NAKAJI, T., FUJIWARA, Y.,

KISHIHARI, M., BABA, Y., FUJITA, T.: Immunological studies in Werner's syndrome. *Clin. Exp. Immunol.* 42:10-19, 1980.
86. NEEL, J.V., KODAI, M., BREWER, R., ANDERSON, R.C.: The incidence of consanguineous matings in Japan. *Am. J. Hum. Genet.* 1:156-178, 1949.
87. OBEID, D.A., HILL, F.G.H., HARNDEN, D., MANN, J.R., WOOD, B.S.B.: Oxymetholone hepatic tumors, and chromosome aberrations associated with leukemic transition. *Cancer* 46:1401-1404, 1980.
88. OPPENHEIMER, B.S., KUGEL, V.H.: Werner's syndrome – a heredo-familial disorder with scleroderma, bilateral juvenile cataract, precocious graying of the hair and endocrine stigmatization. *Trans. Assoc. Am. Physicians* 49:358-370, 1934.
89. OPPENHEIMER, B.S., KUGEL, V.H.: Werner's syndrome: Report of the first necropsy and of findings in a new case. *Am. J. Med. Sci.* 202:629-642, 1941.
90. PAGE, A.R., HANSEN, A.E., GOOD, R.A.: Occurrence of leukemia and lymphoma in patients with agammaglobulinemia. *Blood* 21:197-206, 1963.
91. PAWSEY, S.A., MAGNUS, I.A., RAMSAY, C.A., BENSON, P.F., GIANNELLI, F.: Clinical, genetic and DNA repair studies on a consecutive series of patients with xeroderma pigmentosum. *Q. J. Med.* 48:179-210, 1979.
92. PENN, I.: Some contributions of transplantation to our knowledge of cancer. *Transplant. Proc.* 12:676-680, 1980.
93. PENN, I.: Depressed immunity and the development of cancer: A review. *Clin. Exp. Immunol.* 46:459-474, 1981.
94. PENN, I.: Renal transplantation and cancer. *In* Proc. Eighth International Congress of Nephrology (Zurukzoglu, W. et al., eds.) Basel: S. Karger, pp. 527-538, 1981.
95. PERRIMOND, H., JUHAN-VAGUE, I., THÉVENIEAU, D., BAYLE, J., MURATORE, R., ORSINI, A.: Evolution médullaire et hépatique après androgénothérapie prolongée d'une anémie de Fanconi. *Nouv. Rev. Fr. Hématol.* 18:228, 1977.
96. POLANI, P.E.: DNA repair, defects and chromosome instability. *In* Human Genetics, Possibilities and Realities, Ciba Found. Sympos. 66 (New Series), Excerpta Medica, Amsterdam: p. 88, 1979.
97. PORT, R.P., PETASNICK, J.P., RANNINGER, K.: Angiographic demonstration of hepatoma in association with Fanconi's anemia. *Am. J. Roentgenol.* 113:82-83, 1971.
98. PRINDULL, G., JENTSCH, E., HANSMANN, I.: Fanconi's anaemia developing erythroleukaemia. *Scand. J. Haematol.* 23:59-63, 1979.
99. PULIGANDLA, B., STASS, S.A., SCHUMACHER, H.R., KENEKLIS, T.P., BOLLUM, F.J.: Terminal deoxynucleotidyl transferase in Fanconi's anemia. *Lancet* 2:1263, 1978.
100. RABBIOSI, G., BORRONI, G.: Werner's syndrome: Seven cases in one family. *Dermatalogica* 158:355-360, 1979.
101. RAY, J.H., GERMAN, J.: The chromosome changes in Bloom's syndrome, ataxia-telangiectasia, and Fanconi's anemia. *In* Genes, Chromosomes, and Neoplasia (Arrighi, F.E., Rao, P.N., Stubblefield, E., eds.), New York: Raven Press, pp. 351-378, 1981.
102. RECANT, L., LACY, P.: Fanconi's anemia and hepatic cirrhosis. *Am. J. Med.* 39:464-475, 1965.
103. REED, W.B., LANDING, B., SUGARMAN, G., CLEAVER, J.E., MELNYK, J.: Xeroderma pigmentosum. Clinical and laboratory investigation of its basic defect. *J.A.M.A.* 207:2073-2079, 1969.
104. REESE, A.B., AND WILBER, I.E.: The eye manifestations of xeroderma pigmentosum. *Am. J. Opthalmol.* 26:901-911, 1943.
105. REINHOLD, J.D.L., NEUMARK, E., LIGHTWOOD, R., CARTER, C.O.: Familial

hypoplastic anemia with congenital abnormalities (Fanconi's syndrome). *Blood* 7:915-926, 1952.
106. ROHR, K.: Familial panmyelophthisis: Fanconi syndrome in adults. *Blood* 4:103-141, 1949.
107. ROSEN, R.S., CIMINI, R., COBLENZ, D.: Werner's syndrome. *Br. J. Radiol.* 43:193-198, 1970.
108. SALK, D., AU, K., HOEHN, H., MARTIN, G.M.: Cytogenetics of Werner's syndrome cultivated skin fibroblasts: variegated translocation mosaicism. *Cytogenet. Cell Genet.* 30:92-107, 1981.
109. SALK, D., AU, K., STENCHEVER, M.R., MARTIN, G.M.: Evidence of clonal attenuation, clonal succession, and clonal expansion in mass cultures of aging Werner's syndrome skin fibroblasts. *Cytogenet. Cell Genet.* 30:108-117, 1981.
110. SARNA, G., TOMASULO, P., LOTZ, M.J., BUBINAK, J.F., SHULMAN, N.R.: Multiple neoplasms in two siblings with a variant form of Fanconi's anemia. *Cancer* 36:1029-1033, 1975.
111. SASAKI, M.S., TONOMURA, A.: A high susceptibility of Fanconi's anemia to chromosome breakage by DNA cross-linking agents. *Cancer Res.* 33:1829-1836, 1973.
112. SASAKI, M.S.: Fanconi's anemia. A condition possibly associated with a defective DNA repair. *In* DNA Repair Mechanisms (Hanawalt, P.C., Friedberg, E.C., Fox, C.F. eds.), New York: Academic Press pp. 675-684, 1978.
113. SAWITSKY, A., BLOOM, D., GERMAN, J.: Chromosomal breakage and acute leukemia in congenital telangiectatic erythema and stunted growth. *Ann. Intern. Med.* 65:487-495, 1966.
114. SCHONBERG, S., HENDERSON, E., NIERMEIJER, M.F., GERMAN, J.: Werner's syndrome: Preferential proliferation of clones with translocations. *Am. J. Hum. Genet.* 33:120a, 1981.
115. SCHONBERG, S., HENDERSON, E., NIERMEIJER, M.F., BOOTSMA, D., GERMAN, J.: Werner's syndrome: Preferential proliferation in vitro of clones with chromosome translocations. (In preparation).
116. SCHROEDER, T.M., ANSCHÜTZ, F., KNOPP, A.: Spontane Chromosomenaberrationen bei familiärer Panmyelopathie. *Humangenetik* 1:194-196, 1964.
117. SCHROEDER, T.M., TILGEN, D., KRÜGER, J., VOGEL, F.: Formal genetics of Fanconi's anemia. *Hum. Genet.* 32:257-288, 1976.
118. SCHROEDER, T.M., DRINGS, R., BEILNER, P., BUCHINGER, G.: Clinical and cytogenetic observations during a six-year period in an adult with Fanconi's anemia. *Blut* 34:119-132, 1976.
119. SCHROEDER, T.M., STAHL-MAUGÉ, C.: Mutagenic effects of isonicotinic acid hydracide in Fanconi's anemia. (Case 1). *Hum. Genet.* 52:309-321, 1979.
120. SCHROEDER, T.M.: Personal communication of unpublished information.
121. SCHUMACHER, H.R., GERMAN, J.: Unpublished information.
122. SHAHIDI, N.T., DIAMOND, L.K.: Testosterone-induced remission in aplastic anemia. *Am. J. Dis. Child.* 98:293-302, 1959.
123. SHAHIDI, N.T., DIAMOND, L.K.: Testosterone-induced remission in aplastic anemia of both acquired and congenital type. *N. Engl. J. Med.* 264:953-957, 1961.
124. SHAPIRO, P., IKEDA, R.M., RUEBNER, B.H., CONNORS, M.H., HALSTED, C.C., ABILDGAARD, C.F.: Multiple hepatic tumors and peliosis hepatis in Fanconi's anemia treated with androgens. *Am. J. Dis. Child.* 131:1104-1106, 1977.
125. SHULER, D., KISS, A., FÁBIÁN, F.: Chromosomal peculiarities and in vitro examinations in Fanconi's anaemia. *Humangenetik* 7:314-322, 1969.
126. SILVER, H.K., BLAIR, W.C., KEMPE, C.H.: Fanconi syndrome. Multiple congenital

anomalies with hypoplastic anemia. *Am. J. Dis. Child.* 83:14-25, 1952.
127. SKIKNE, B.S., LYNCH, S.R., BEZWODA, W.R., BOTHWELL, T.H., BERNSTEIN, R., KATZ, J., KRAMER, S., ZUCKER, M: Fanconi's anaemia, with special reference to erythrokinetic features. *S. Afr. Med. J.* 53:43-50, 1978.
128. SPECTOR, B.D., PERRY, G.S., III, KERSEY, J.H.: Genetically determined immunodeficiency disease (GDID) and malignancy: Report from the Immunodeficiency-Cancer Registry. *Clin. Immunol. Immunopathol.* 11:12-29, 1978.
129. SPECTOR, B.D., FILIPOVICH, A.H., PERRY, G.S., III, KERSEY, J.H.: Epidemiology of cancer in ataxia-telangiectasia. In Ataxia-telangiectasia – A Cellular and Molecular Link between Cancer, Neuropathology, and Immune Deficiency. (Bridges, B.A., Harnden, D.G. eds.), New York: John Wiley and Sons, pp. 103-138, 1982.
130. SPERLING, K., WEGNER, R.D., RIEHM, H., OBE, G.: Frequency and distribution of sister-chromatid exchanges in a case of Fanconi's anemia. *Humangenetik* 27:227-230, 1975.
131. STEINHERZ, P.G., CANALE, V.G., MILLER, D.R.: Hepatocellular carcinoma, transfusion-induced hemochromatosis, and congenital hypoplastic anemia (Blackfan-Diamond syndrome). *Am. J. Med.* 60:1032-1035, 1976.
132. STROMEYER, F.W., SMITH, D.H., ISHAK, K.A.: Anabolic steroid therapy and intrahepatic cholangiocarcinoma. *Cancer* 43:440-443, 1979.
133. SWEENEY, E.C., EVANS, D.J.: Hepatic lesions in patients treated with synthetic anabolic steroids. *J. Clin. Pathol.* 29:626-633, 1976.
134. SWIFT, M., HIRSCHHORN, K.: Fanconi's anemia. Inherited susceptibility to chromosome breakage in various tissues. *Ann. Intern. Med.* 65:496-503, 1966.
135. SWIFT, M., ZIMMERMAN, D., MCDONOUGH, E.R.: Squamous cell carcinomas in Fanconi's anemia. *J.A.M.A.* 216:325-326, 1971.
136. SWIFT, M.: Fanconi's anemia in the genetics of neoplasia. *Clin. Res.* 19:410, 1971.
137. SWIFT, M.: Fanconi's anemia in the genetics of neoplasia. *Nature* 230:370-373, 1971.
138. SWIFT, M.: Malignant disease in heterozygous carriers. *Birth Defects* 12(1):133-144, 1976.
139. SWIFT, M., CALDWELL, R.J., CHASE, C.: Reassessment of the cancer-predisposition of Fanconi's anemia heterozygotes. *J. Natl. Cancer Inst.* 65:863-867, 1980.
140. SYLLABA, L., HENNER, K.: Contribution a l'indépendance de l'athétose double idiopathique et congénitale. Atteinte familiale, syndrome dystrophique, signe du réseau vasculaire conjunctival, intégrite physique. *Rev. Neurol (Paris)* 1:541-562, 1926.
141. TAO, L.C., STECKER, E., GARDNER, H.A.: Werner's syndrome and acute myeloid leukemia. *Can. Med. Asso. J.* 105:951-954, 968, 1971.
142. THOMAS, J.S., PREMALATHA, S., YESUDIAN, P., THAMBIAH, A.S.: An analytical study of ten cases of xeroderma pigmentosum. *Indian Pediatr.* 16:327-329, 1979.
143. TODARO, G.J., GREEN, H., SWIFT, M.R.: Susceptibility of human diploid fibroblast strains to transformation by SV40 virus. *Science* 153:1232-1236, 1967.
144. VAN PATTER, H.T., DRUMMOND, J.A.: Malignant melanoma occurring in xeroderma pigmentosum. *Cancer* 6:942-947, 1953.
145. WEINER, M., YEBOA, K., WARBURTON, D., AUERBACH, A.: Acute myelomonocytic leukemia as the first hematologic manifestation in a patient with Fanconi anemia. *Ped. Res.* 15:590, 1981.
146. WERNER, C.W.O.: Über Katarakt in Verbindung mit Sklerodermie (Doctoral dissertation, Kiel University). Kiel: Schmidt and Klaunig, 1904.
147. ZALLA, J.A.: Werner'syndrome. *Cutis* 25:275-278, 1980.
148. ZAWARTKA, M., RESTORFF-LIBISZOWSKA, H., KOWALSKI, R., LITWINISZYN-KRZEWICKA, K., SCHEJBAL, J.: Bialaczka szpikowa w przebiegu anemii Fanconiego (AF). *Wiad. Lek.* 29:145-150, 1976.

THE CYTOGENETICS OF THE "CHROMOSOME-BREAKAGE SYNDROMES"

JAMES H. RAY AND JAMES GERMAN

INTRODUCTION

Chromosome instability and a predisposition to neoplasia are features common to the following rare, recessively transmitted disorders of man: Bloom's syndrome (BS), ataxia-telangiectasia (AT), Fanconi's anemia (FA), and Werner's syndrome (WS). The disorders sometimes are referred to collectively as the chromosome-breakage syndromes. Xeroderma pigmentosum (XP), another cancer-predisposing disorder, often is grouped with the chromosome-breakage syndromes even though excessive chromosome breakage is demonstrable only after exposure of the cells to certain DNA-damaging agents.

In each of these five disorders the clinical phenotype is distinctive, and the distribution of types of cancer that occurs is characteristic. Also, in each, the pattern of chromosome abnormalities is characteristic. The clinical features of the disorders and patterns of cancer that occur are presented elsewhere in this volume. Here, a concise description is made of the cytogenetic aberrations to be observed in untreated cells from each of the syndromes, i.e., spontaneously occurring aberrations, and in the case of XP, in cells exposed to DNA-damaging agents.

Hypersensitivity to DNA-damaging agents, manifested in several ways including cytogenetically, is a recurring feature encountered as persons with this class of disorders are examined experimentally. Our main objective here is to present the unusual cytogenetic features of such cells that appear to be constitutional (with the exception of XP, as just stated). Thus, other recent reviews should be consulted for complete descriptions of experimentally induced chromosome abnormalities and discussions of their possible bases [4,41,133].

BLOOM'S SYNDROME
Chromosome Aberrations

Chromosome instability was first reported to be a feature of BS in the mid-1960s [46,49]. Untreated BS cells in metaphase have a much greater than normal frequency of chromatid and isochromatid gaps and breaks, acentric fragments, polycentric chromosomes, sister-chromatid reunions, terminal association of telomeric regions, and chromatid-interchange configurations. Increased numbers of chromosomal aberrations have been detected in all types of BS cells that have been examined — blood T lymphocytes in short-term culture, uncultured and briefly cultured bone marrow aspirates, dermal fibroblasts in long-term culture, and Epstein-Barr virus (EBV)-transformed B lymphocytes in long-term culture. As a rule, 5–15% of BS metaphase figures have one or more aberrant chromosomes. A corresponding increase in the frequency of interphase cells with micronuclei is found in fibroblast and lymphoblastoid cells proliferating in vitro [50, J. German, unpublished results]; micronuclei also appear to be unusually frequent in direct bone marrow aspirates (Fig. 1), which suggests that increased chromosome instability occurs in vivo as well.

The most characteristic cytogenetic abnormality in BS cells is the quadriradial configuration (Qr) [51]. The Qr typical of BS, known as a Class I Qr (abbreviated Qr-I here), is one composed of homologous chromosomes; the centromeres are positioned in opposite arms as result of the symmetrical exchange of nonsister-chromatid segments [51,146,148]. The type of chromatid interchange that results in a Qr-I at metaphase does not produce a recognizable change in the chromosome complement, a fact that possibly explains why lymphocyte clones with rearranged chromosome complements, although they do occur in BS [27,124, S. Schonberg and J. German, unpublished results], have been observed relatively infrequently. When first reported, the Qr-I was interpreted as cytological evidence for the existence of somatic recombination in man [46]; confirmation of this interpretation came after the bromodeoxyuridine (BrdU) method for demonstrating sister-chromatid exchange (SCE) was developed [see discussion below and 92]. (Genetic evidence for so-

Fig. 1. Micronuclei (arrows) in nonmitotic cells of freshly aspirated bone marrow from a healthy, 9-year-old with Bloom's syndrome — 32(MiKo). The slides were stained with Giemsa after brief hyposmotic treatment of the specimen. Such cells, indicative of disturbed separation of a chromosome or chromosome fragments in the previous mitosis, are present in marrows from some normal persons but are found much more readily in Bloom's syndrome marrows. (Note: Acute lymphatic leukemia was diagnosed in 32(MiKo) 7 years after 3×10^9 marrow cells, a sample of which is shown here, were removed and stored in dimethyl sulfoxide (DMSO) in liquid nitrogen. Some of that marrow was infused into the patient after refractory postchemotherapy marrow aplasia became life-threatening.)

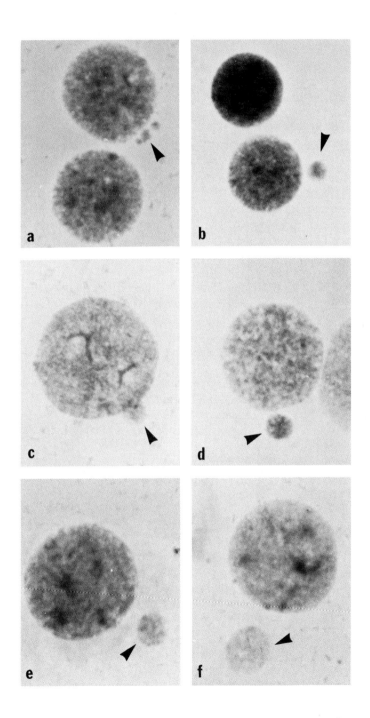

matic recombination in human cells has never been obtained, but see discussion below of cytogenetic evidence for its occurrence in vivo.)

The basis for the preponderance of Qr-Is in BS is not understood. One explanation considered but now discarded was that homologous chromosomes were paired more frequently than normal in BS cells. The absence of an increased incidence of Qr-Is following the x-irradiation of BS lymphocytes in G_2, a finding contrasting with the significantly increased incidences of Qr-Is to be observed in x-irradiated G_2 *Drosophila* chromosomes [1]—chromosomes normally are paired in *Drosophila* somatic cells—and Chinese hamster diplochromosomes [42], indicates absence of an abnormally increased degree of homologous chromosome pairing there [89]. In another study, the incidence of Qr-Is in endoreduplicated BS and normal fibroblasts was found to be greater than that in diploid BS and normal fibroblasts, with the Qr increase in endoreduplicated BS fibroblasts being much greater than that in endoreduplicated normal fibroblasts [88]. Taken together, the data suggest that excessive homologous chromosome pairing per se is not responsible for the increased incidence of Qr-Is in BS cells, but that, instead, an abnormally great tendency for chromatid interchange exists there. In fact, that homologous interchanges occur so frequently in BS can be interpreted as good evidence that homologous chromosome segments quite often associate normally during interphase (i.e., during S, when the homologous interchanges would occur).

Although cytological evidence for somatic recombination in BS cells can be demonstrated in vitro quite readily, the demonstration of Qr-Is in vivo, if they occur, has not been made [39,86,90,140,154,156]. (It should be noted, however, that the material available for study so far has been nothing more than a few bone marrow samples.) What may correspond to "twin-spotting" in the classical genetic sense, i.e., adjacent hypo- and hyperpigmented spots on affected persons' skin [39,48,164], has been suggested to represent evidence for the existence of somatic recombination in vivo in BS. Cytogenetic evidence for somatic recombination in vivo has been obtained [169]. Therman et al. studied the involvement in Qr formation of the satellites of the acrocentric autosomes (groups D(13–15) and G(21–22)) in BS and normal lymphocytes by taking advantage of the existence in man of inherited satellite polymorphisms with respect to size and brightness of fluorescence after quinacrine (Q) staining. Qr formations between acrocentric chromosomes occurred with a much greater frequency in BS lymphocytes than in normal lymphocytes (6.0/1,000 vs. 0.1/1,000). Furthermore, 31 different Q-bright satellite distribution patterns among the acrocentric chromosomes, ranging from a complete absence in some cells to a completely new distribution in others, were observed in 58 randomly selected blood T lymphocytes from a single individual with BS; homozygosity for Q-bright satellites was observed in

12 of the 58 cells, whereas constitutionally, the individual was heterozygous for the satellite polymorphisms. In each of 83 similarly selected lymphocytes from normal individuals, the distribution patterns of the Q-bright satellites on chromosomes were found to be unchanged. These cytogenetic data constitute impressive evidence that exchange of segments between chromosomes does occur surprisingly frequently in BS cells in vivo and rarely if at all in normal cells. The exchange of chromatid segments between homologous chromosomes, like that observed in BS, theoretically may have profound genetic consequences [45]. Depending on the segregation patterns of the affected chromatids in subsequent mitoses, traits that were suppressed in parental cells, because of their recessive nature, may be expressed. Conceivably, cells having such altered genetic constitutions are the progenitors of the malignant neoplasms that occur so commonly in BS, a suggestion already made by Passarge and Bartram [117].

Sister-Chromatid Exchange

Following the development of the BrdU method for studying SCE [91], a uniquely characteristic cytogenetic feature of BS cells was discovered. BS lymphocytes were shown to have an SCE frequency that was 10 to 14 times greater than that of normal (untreated) lymphocytes [7,23, 54,154,157]. A constitutively elevated SCE frequency now is recognized to be so characteristic of BS cells that it is used to confirm the clinical diagnosis [33,80,159]. Cells from other genetic disorders that have been studied, including the other chromosome-breakage syndromes, have a normal baseline SCE frequency; a few of many disorders that we have studied and found to have a normal frequency in untreated lymphocytes in short-term culture [our unpublished observations] are the following: unexplained severe growth deficiency both pre- and postnatal, dyskeratosis congenita, Thomson's syndrome, Rothmund's syndrome, Friedreich's ataxia, the Wiskott-Aldrich syndrome, X-linked agammaglobulinemia, the basal cell nevus syndrome, von Recklinghausen's disease, scleroderma, and lupus erythematosus. Dermal fibroblasts, bone marrow cells, and lymphoblastoid cells from BS also have strikingly elevated SCE frequencies [53] although their SCE frequencies tend to be somewhat lower than that of phytohemagglutinin (PHA)-stimulated BS lymphocytes.

An unexpected and as yet unexplained observation was made after the increased SCE frequency was discovered; sometimes two blood lymphocyte populations with respect to SCE frequency are present in persons with BS [54,80]. Of 21 affected persons studied in our laboratory [54], five had two lymphocyte populations. While the majority of the lymphocytes in each person exhibited the elevated baseline SCE frequency typical of BS, some lymphocytes, ranging from a few cells to as many as

47%, had a low SCE frequency like that found in cells from normal persons. A normal frequency of chromosome aberrations also is found in the BS lymphocytes that have low (normal) baseline SCE frequencies [unpublished results]. This dimorphism in terms of SCE frequency has not been detected in bone marrow and dermal fibroblasts derived from persons with BS, but lymphoblastoid cell lines (LCLs) derived from single B lymphocytes of circulating blood may have either an elevated or a normal SCE frequency [53,67].

The reason for the existence in BS of two lymphocyte populations with respect to chromosome instability and SCE frequency is unclear. In view of the recessive mode of inheritance of the disorder, all cells should be genetically identical at the BS locus. One possible explanation for the existence of two BS-lymphocyte populations is back-mutation, and the findings in two recent studies [58,177] indicating the spontaneous mutation frequency in BS fibroblasts to be approximately ten times greater than that in normal fibroblasts could be taken as support for this explanation. In our laboratory, evidence recently has been obtained from the study of PHA-stimulated lymphocytes which indicates that the spontaneous mutation rate in vivo in BS also is elevated markedly [Vijayalaxmi, H.J. Evans, J.H. Ray, and J. German, in preparation]. A spontaneous mutation, if it should occur at a BS locus, could restore normal gene function to one BS allele and give rise to a cell heterozygous at the BS locus, a cell that would display both the normal SCE frequency and the normal chromosome-aberration frequency characteristic of cells from BS heterozygotes. Back-mutation also might explain a somewhat similar phenotypic dimorphism observed in FA [5]. Not particularly favoring the mutational explanation, however, is an interesting and equally inexplicable phenotypic dimorphism that has been discovered in cells from some persons with XP [160] and mucolipidosis II [170]. The explanation for this recently detected phenomenon in persons that previously had been considered simply to be homozygous at some mutant locus probably will be interesting and important for understanding the genetics of the disorders.

The Basis for Chromosome Instability in BS

A number of reports have appeared in which the co-cultivation of BS and non-BS cells has been used experimentally to study the chromosome instability of BS. Tice et al. [171] reported that the SCE frequency of normal fibroblasts was increased following their co-cultivation with BS fibroblasts, as was the SCE frequency of PHA-stimulated normal human lymphocytes grown in medium "conditioned" by proliferating BS fibroblasts. Tice et al. concluded that chromosome instability in BS results from the over-production of some DNA-damaging cellular component.

However, most attempts by others to corroborate both the co-cultivation and conditioned medium data of Tice et al. have been unsuccessful [3,8, 112,127,141,155,172,181]. Emerit and Cerutti [36] did report, however, that greatly concentrated ultrafiltrates of medium conditioned by BS fibroblasts had both clastogenic and, to a much lesser extent, SCE-inducing properties when added to the growth medium of PHA-stimulated normal human lymphocytes.

Co-cultivation and conditioned medium experiments conducted in other laboratories have yielded data that contrast with those of Tice et al. Thus, the SCE frequency of BS fibroblasts was reported to be reduced significantly following co-cultivation with hamster (CHO) cells [172], CHO/BS cell hybrids [3], normal human fibroblasts [8,127], and XP and FA fibroblasts [9]. Furthermore, BS heterozygote fibroblasts were reported to have approximately 50% the SCE-reducing capability of normal human fibroblasts [9]. The normalizing effect that non-BS cells were reported to have on BS SCE frequency was dependent on the ratio of non-BS to BS cells, with higher ratios being the more effective [127,172]. Medium conditioned by 48 hr of normal fibroblast growth also reduced the SCE frequency of BS fibroblasts when it was used to support their growth, although the extent of the SCE reduction was less than that observed following co-cultivation with normal cells [127]. These experimental observations have been interpreted [8,9,127] to indicate that the cytogenetic disturbance in BS cells results from the absence or underproduction of some factor that is produced in normal cells and that is responsible for the low (normal) baseline SCE frequency there.

We have been unable to confirm the absence of what might be thought of as an SCE-correction factor in BS. In an extensive series of co-cultivation experiments we have conducted [unpublished results], fibroblasts and lymphoblasts derived from normal individuals had no effect on the SCE frequency of BS fibroblasts and lymphoblasts even when non-BS: BS cell ratios as great as 4:1 were used. Also, medium conditioned for 48 hr by proliferating normal fibroblasts, when it was used to support the growth of BS fibroblasts, was totally ineffective at reducing their elevated SCE frequency.

Schonberg and German [141] failed to observe a change in the SCE frequency of either Lesch-Nyhan syndrome (LN) or BS fibroblasts following their co-cultivation, using conditions that required the LN fibroblasts, if they were to survive, to be coupled metabolically to BS-fibroblasts. The absence of an effect of BS fibroblasts on LN fibroblast SCE frequency rules out the existence of a low molecular weight clastogenic factor in BS cells because such a compound, if it had been present, would have passed through gap junctions formed between the metabolically coupled LN and BS fibroblasts. Finally, the SCE frequencies of normal, BS heterozygote, and BS homozygote lymphocytes were found by

Shiraishi et al. [155] to be unchanged following the co-cultivation at 1:1 ratios for 24 hr of either normal and BS homozygote cells or BS heterozygote and BS homozygote cells.

Thus, the SCE frequency of BS cells has been reported to be increased, decreased, or unaffected following co-cultivation with normal cells. West et al. [181] have suggested that the seemingly conflicting results of the co-cultivation experiments might be explained on the basis of nothing more than cell-cycle perturbations brought about by the co-cultivation system itself. This suggestion is based on their finding that the SCE frequency of BS lymphocytes varies with the temperature at which the cells are incubated. (Schroeder and Stahl-Maugé [148] earlier had reported that the chromosome aberration frequency of BS lymphocytes was elevated markedly when the incubation temperature was raised from 37°C to 40°C, with a dramatic increase in the number of QrIs occurring at the higher temperature.) The SCE frequency of BS lymphocytes at 39°C was greater than that at 37.5°C, whereas at 35°C and 32°C it was decreased. The SCE frequency of normal human lymphocytes, on the other hand, was elevated above that at 37.5°C when the incubation temperature was raised to 39°C as well as when it was lowered to 35°C and 32°C. Comparison of the frequencies of first-, second-, and third-division cells at the various incubation temperatures demonstrated that the cell cycle of both BS and normal lymphocytes was prolonged at the two lower temperatures. In the view of West et al. [181], the decreased SCE frequency observed in BS fibroblasts co-cultivated with non-BS fibroblasts resulted from a lengthening of the cell cycle of the former because of growth suppression by the latter. Also according to West et al. [181], non-BS and BS lymphocyte co-cultivation affected neither the cell cycle nor, consequently, the SCE frequency because there was little interaction between the co-cultivated BS and non-BS lymphocytes. The decreased SCE frequency of BS cells observed following their growth in medium conditioned by normal cells [127] is difficult to explain on the basis of cell-cycle perturbation, however, inasmuch as conditioned medium is not known to affect adversely the cycling of cells. Therefore, further study of co-cultivated non-BS:BS cells is necessary before the basis for the results can be understood completely.

ATAXIA-TELANGIECTASIA
Chromosome Aberrations

Chromosome instability was reported to be a feature of AT in 1966 [63]. Spontaneous chromosome aberrations ordinarily are readily detectable in both blood T lymphocytes and dermal fibroblasts from AT patients, with the frequency in fibroblasts generally being greater [26-28,113], although exceptions clearly exist [123,138]. Also, T lymphocytes and dermal fibroblasts from AT heterozygotes may have elevated chromosome aberration frequencies [2,28,113]. Excessive chromosome instability has not been reported in AT LCLs, but clones of cells with stable chromosome rearrangements may be present in such cultures [27,81]. In only two of the seven AT bone marrows studied have elevated frequencies of chromosome aberrations been reported [2,64,65,93,95,136]; in one of the two, only 12 cells were analyzed [95], and in the other, the patient was leukemic [93]. The absence of widespread chromosome abnormalities in AT bone marrow aspirates and cultured lymphoblasts raises the possibility of tissue differences in the expression of chromosome instability.

The instability of AT is manifested by the presence in metaphase cells of chromatid and isochromatid gaps and breaks, acentric fragments, dicentric chromosomes, rearranged monocentric chromosomes, and triradial (Tr) and Qr chromatid-interchange configurations. The chromosomes involved in dicentric chromosome formation have been reported to be random [2,20,26,28,59,62,64,120,138,167]. Acentric chromosome fragments frequently are missing from cells with dicentric chromosomes, suggesting the interesting possibility that unbroken chromosomes can be involved in dicentric chromosome formation [62,167].

The most characteristic cytogenetic feature of AT is the existence in the circulating blood of populations of lymphocytes identifiable as members of mutant clones by the presence in their chromosome complements of marker rearrangements (e.g., Fig. 2). Clones of cells with rearranged chromosomes also may be present in cultures of dermal fibroblasts [26-28,113,179] and lymphoblastoid cells [27,81,83,84,175], and they have been observed in one leukemic bone marrow [93].

Certain chromosomes are involved preferentially in the rearrangements found in AT-lymphocyte clones, specifically, the Nos. 7 and 14 [78]. At least one group D(13-15) chromosome was involved in the formation of 62 of the 74 clones identified in the approximately 120 blood samples that have been studied in detail. The most frequently encountered chromosome rearrangements have been described as Dq+, t(Dq+;Dq−), and t(C;D). Dq+ chromosomes were present in 19 of the 74 clones reported [27,62,74,99,113,120]. The specific group D(13-15) chromosome involved in the Dq+ chromosome formation was not identifiable in most instances because many of the studies were made before G-banding came into widespread use; however, chromosome No. 14 was identified as the Dq+ chromosome when G-banding was applied successfully [62]. Clones of AT lymphocytes with two rearranged group D(13-15) chromosomes also are common−t(Dq+;Dq−)−and account

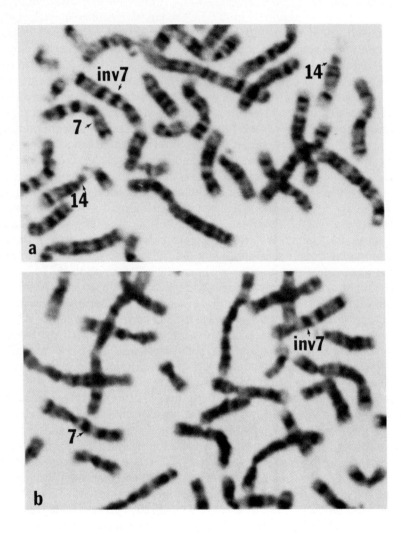

Fig. 2. Portions of two dividing, phytohemagglutinin (PHA)-stimulated lymphocytes from the blood of a girl with ataxia-telangiectasia. The same inversion in a No. 7 chromosome in each cell identifies them as members of a clone. A small population of cells with this mutated chromosome complement, along with other clones with different rearrangements affecting the Nos. 7, 14, or both, has been found each time the girl's blood has been sampled during recent years. Nevertheless, she has shown clinical evidence of neither malignant neoplasia nor immunodeficiency (unpublished observations).

for 17 of the 74 clones [2,20,27,59,93,99,106,123,158,167,175]. G-banding has revealed the two group D(13-15) chromosomes involved in the t(Dq+;Dq−) chromosome formation to be the Nos. 14. The distribution of breakpoints that leads to the translocation is highly nonrandom, almost always occurring at band 14q12 on one chromosome and band 14q31-34 on the other. The long arm of one chromosome No. 14 (14q12 to 14qter) then is translocated onto the distal end of the long arm of the other, leading to the production of one aberrant No. 14 with a tandem duplication of essentially the entire long arm − 14q+ − and another aberrant No. 14 with a deletion of essentially the entire long arm − 14q−.

Translocations between group C(6-X-12) and group D(13-15) chromosomes account for 17 of the 74 AT-lymphocyte clones reported [2,6,20, 59,99,167]. The group D(13-15) chromosome involved in the t(C;D) chromosome rearrangement is the No. 14; the group C(6-X-12) chromosome can be either an X or a No. 7. The breakpoint on No. 14 again is at band 14q12. The breakpoint on the group C(6-X-12) chromosome can be on either chromosome arm. Additional types of group D(13-15) chromosome abnormalities, including a Dq− chromosome [20,167], a ring No. 14 [99], a t(?;14) [6], an inversion No. 14 [6,167], and monosomy 14 [15], have been identified in lymphocyte clones from other AT patients.

A number of lymphocyte clones containing rearranged chromosome Nos. 7 only have been detected in AT [6,137, B. Scott de Martinville and J. German, unpublished results] and in immunodeficiency states probably related to AT [79,180]. The most common rearrangement is pericentric inversion in one No. 7 following breakage at bands 7p14 and 7q35 (Fig. 2). Both Nos. 7 may be involved; thus, clones of cells with a t(7;7) chromosome rearrangement have been observed, the result of breaks at bands 7p14 and 7q35.

The long-term study of several individuals with AT has revealed that the frequency of lymphocytes in the blood that bears a particular chromosome rearrangement often changes with time [2,20,62,64,99,106,113, 138,167]. During an 8-year study of one affected individual [113,167] the frequency of a clone of lymphocytes bearing a t(14q+;14q−) chromosome rearrangement increased from 1% to 91%; subclones with dicentric chromosomes derived from members of chromosome groups D(13-15) and G(21-22), F(19-20) and G(21-22), and F(19-20) and D(13-15) appeared during the course of the study. An affected sister of the patient just mentioned was found to have a t(X;14)-bearing clone that accounted for from 53% to 80% of her dividing T lymphocytes [113,167]; subclones of cells with dicentrics were detected also. Hecht et al. [64] studied an AT patient with a t(14;14)-bearing lymphocyte clone that increased in frequency over a 52-month period from 1.7% to 57-68%.

The observations just mentioned make it apparent that some lymphocyte lineages that can be detected in AT blood outgrow others, even though clinical leukemia is not diagnosable. The basis for the apparent proliferative advantage of lymphocytes that have undergone abnormal chromosome rearrangement is unknown. We have suggested already [126] that such clones may be viewed as neoplasms, in that they exhibit a degree of autonomy with respect to proliferation and increase in number. A functional significance—i.e., a clinical effect on immune function—might be expected of the preferential overgrowth of a single lineage of T lymphocytes, but, if there is one, it has not been discovered. What effect might this have? One of the major clinical features of AT is immunodeficiency, although not every patient exhibits it, at least at all stages of the disease. Excessive representation of the progeny of a single cell, as is regularly observed in AT blood, should disrupt the balanced interplay that occurs normally between different lymphocyte populations. Unless immunological diversity is generated continuously in such a clone, its excessive proliferation should compromise the diversity known to characterize cell populations of the immune system, here the diversity of the circulating T-lymphocyte population. We postulate that the inordinate proliferation of a clone of T lymphocytes as result of a chromosome mutation that gave the clone a degree of autonomy of proliferation is responsible for the immunodeficiency that often supervenes in AT.

The question arises also as to what role, if any, clones of lymphocytes that bear chromosome rearrangements play in the predisposition to lymphoid cancer that is so prominent a clinical feature of AT. Extensive cytogenetic study has been made of four AT patients with T-cell leukemia [93,99,136,175]. The leukemic cells (PHA-stimulated lymphocytes) of three of the four consisted of clones of cells that bore a reciprocal translocation between the two No. 14 chromosomes (t(14q+;14q−)) [93, 99,136]. The lymphocytes with the translocation accounted for 100% of the PHA-stimulated lymphocytes in two of the patients [99,136] and 98% of the PHA-stimulated lymphocytes in the other [93]. These same chromosome rearrangements had been detected in clones circulating in the blood of all three patients prior to the clinical diagnosis of leukemia [56, 93,99]. The breakpoints leading to the t(14;14) chromosome rearrangement were at bands 14q12 and 14q31–34, precisely the breakpoints detected in the many clones of nonleukemic lymphocytes already mentioned. Additional chromosome changes were present in the leukemic cells; however, the only one found consistently was the loss of the smaller of the two abnormal Nos. 14, the 14q− chromosome [93,99,158]. In the fourth AT patient with T-cell leukemia, 15.4% of the PHA-stimulated lymphocytes contained a t(14;14) chromosome rearrangement [175]; however, the translocation was not present in seven bone marrow samples taken from the patient during the course of leukemia but was observed in EBV-transformed B lymphocytes, which suggested that cells bearing the t(14;14) translocation were not members of the leukemic clone of cells.

One AT patient with Burkitt's lymphoma has been studied [83,84]. B-lymphocytes from the patient had the t(8;14) chromosome rearrangement typical of African Burkitt's lymphoma. Finally, in an AT patient in whom Hodgkin's disease subsequently was diagnosed, Bernstein et al. [15] found a clone of blood T lymphocytes that was monosomic for chromosome No. 14, along with a subclone in which a chromosome No. 6 also had undergone deletion.

That a chromosome No. 14 has been found to be abnormal in cells of all but one of the cytogenetically well-studied AT neoplasms is highly noteworthy. The long-term study of more individuals with AT before the onset of cancer and during its progression probably will yield valuable information regarding the role of chromosome change in lymphoid neoplasia.

The Basis for Chromosome Instability in AT

The molecular basis for chromosome instability in AT is unknown. Harnden and Brown [61] suggested that the diverse features of AT, including the increased spontaneous chromosome instability, might be explained on the basis of a molecular defect involving some cofactor necessary for the proper functioning of multiple enzymes. They stated that metals such as zinc, manganese, and copper that are necessary for the proper functioning of metalloenzymes were good candidates for being deficient in AT but offered no data derived from the study of AT to support their suggestion.

AT cells are hypersensitive to the chromosome-breaking action of x-rays [166] and bleomycin [168]. Recent studies have demonstrated that semiconservative DNA replication in AT cells following x-irradiation differs from that in irradiated normal cells [32,35,75,114]. Replication normally is suppressed following x-irradiation, and this presumably allows time for damaged DNA segments to be repaired before they replicate and possibly produce a permanent alteration of the base sequence of the damaged DNA segment. In AT cells, DNA synthesis following x-irradiation is suppressed only slightly, so that damaged segments of DNA are permitted to replicate; one of the presumed consequences of this defective response is a dramatically increased incidence of microscopically detectable chromosome abnormalities. Huang and Sheridan [77] suggested that replication of damaged DNA segments may account also for the elevated incidence of chromosome aberrations that is detectable in untreated AT cells.

Recently, some experimental evidence has been interpreted to suggest that a clastogenic agent produced by AT cells is responsible for the chromosome instability observed. Shaham et al. [152] co-cultivated lymphocyte-enriched plasma from AT and normal individuals and found that the frequency of chromosome aberrations in the normal lymphocytes was increased significantly while the chromosome-aberration frequency of the AT lymphocytes remained unchanged. Medium that had been used to support AT-fibroblast growth also effectively increased the chromosome-aberration frequency of normal human lymphocytes [152]. The purported clastogenic factor of AT has been reported to be a peptide with a 500–1,000 molecular weight [151]. Co-cultivation and conditioned medium experiments failed to suggest that a clastogenic factor is produced by AT LCLs [29], an observation that may correspond to the failure of at least some AT LCLs to exhibit an increased incidence of spontaneous chromosome abnormalities [27].

FANCONI'S ANEMIA
Chromosome Aberrations

Chromosome instability was discovered independently by three laboratories to be a feature of FA, the reports appearing between 1964 and 1966 [18,19,49,144]. In the first study published, by Schroeder and her associates [144], one or more aberrant chromosomes were found in 25% of the PHA-stimulated lymphocytes from each of two affected brothers. Most cytogenetic studies made subsequently have corroborated the early reports of increased breakage, but reports of FA individuals lacking chromosome instability also have appeared [19,22,44,68,94,105,147,174]. The negative reports may reflect the variability in the frequency with which chromosome aberrations are found in FA lymphocytes. That the frequency of aberrations varies has been demonstrated most dramatically when multiple blood samples from a single patient have been studied serially. In one such long-term study [14] the frequency of aberrant metaphases in six blood samples taken from one patient during a 3-year period varied from 5 to 42%; in another [145], the aberration frequency in 19 blood samples taken from one patient during a 6-year period varied from 11 to 52%. The basis for this variability and the proportion of patients that show it are unknown.

Long-term dermal fibroblast cultures [50,100,103,139,163,173,184] and LCLs [27, and unpublished results in our laboratory] derived from persons with FA also exhibit increased incidences of spontaneous chromosome aberrations, but the aberration frequencies of such cultures generally are lower than that in blood T lymphocytes in short-term culture. The existence of chromosome instability in vivo, although not in doubt, has been difficult to demonstrate. Direct bone marrow aspirates from some patients have been reported to show elevated frequencies of chromosome abnormalities [13,14,16,44,60,96,103,112,122,153,163,184,187], whereas marrows from others have shown no increase [11,19,21,30,34,44, 71,94,139,145].

Included among the cytogenetic abnormalities commonly observed in FA cells in metaphase are chromatid and isochromatid gaps and breaks, acentric fragments, rearranged monocentric chromosomes, and Tr and Qr chromatid-interchange configurations. An increased frequency of cells with endoreduplicated chromosome complements also has been observed in FA [11,19,34,44,57,101,104,118,144,149,174,178].

The Qr when found in FA characteristically is one resulting from the interchange of chromatid segments between nonhomologous chromosomes [146]. The preferential involvement of nonhomologs contrasts sharply with the preferential involvement of homologs in BS; 95% of 536 FA Qrs analyzed in two studies [146,148] involved the exchange of genetic material between nonhomologous chromosomes, whereas 90% of BS Qrs have been shown to affect homologs [51]. Such nonhomologous exchanges may lead to microscopically observable genomic changes and give rise either to partially monosomic and, or, partially trisomic cells or to cells with balanced chromosome rearrangements (see Fig. 3 in [126]). This presumably explains the clones of cells with rearranged chromosome complements that are found in specimens of FA tissues. Such clones have been observed in lymphocytes in short-term culture [14,60, 145,188, J. German and D. Warburton, unpublished results], in uncultured or briefly cultured bone marrow cells [14,21,30,60,71,96,103,111], and in dermal fibroblasts in long-term culture [11]. Specific chromosomes are not involved in the formation of the rearranged chromosomes found in FA clones (in contrast to the situation in AT), but group A(1–3) and group C(6–X–12) chromosomes are reported [126] to be affected more frequently than those of other groups.

Cytogenetic studies have been made in a number of FA patients with acute leukemia in search of clones of cells with structurally rearranged chromosome complements [14,21,55,60,103,111,121]. In five of seven acute leukemias studied, structural rearrangements were detected. Berger et al. [14] found multiple clones of cells with rearrangements involving chromosome Nos. 1, 3, 7, and 12 in both bone marrow cells and peripheral lymphocytes; some of the rearrangements also had been observed prior to the clinical diagnosis of leukemia. Bone marrow cells and peripheral blood lymphocytes from another acute leukemia patient had a t(1;6) chromosome rearrangement [60,111]; in addition to the t(1;6) rearrangement, some cells had an abnormally long (i.e., a rearranged) chromosome No. 11, its extra material being unidentifiable. One patient with acute myelomonocytic leukemia had a clone of cells with a t(B;E) rearrangement that comprised 50% of his dividing marrow population [21]. Cells from two patients with erythroleukemia exhibited chromosome rearrangements [13,103]; in one, 100% of the dividing bone marrow cells had the karyotype 46,XY,1p+,3q+,7p− [13], and in the other, 100% of both dividing bone marrow cells and blood lymphocytes had the karyotype 45,XY,−C [103].

The Basis for Chromosome Instability in FA

The molecular basis for FA is unknown. Numerous suggestions have been made to explain the chromosome instability of FA. Löhr and his colleagues [97] reported hexokinase deficiency to be the metabolic defect in FA. Later, a DNA ligase deficiency was reported [69,70]. Recently, Wunder et al. [185] have reported different intracellular distributions of topoisomerases in FA and normal placentae and suggested that defective accumulation of topoisomerases in FA nuclei is responsible for the cytogenetic disturbance observed; more than 95% of the topoisomerase activity in normal placentae was localized to the nuclei, whereas in FA most of the activity (approximately 87%) was localized to the cytoplasm.

A possible disturbance in oxygen metabolism in FA has been reported. Nordenson [108] compared the spontaneous chromosome-aberration frequencies of FA lymphocytes cultured in the presence and absence of superoxide dismutase (SOD) and catalase, enzymes that use superoxide radicals and hydrogen peroxide, respectively, as their substrates, and found the aberration frequencies to be reduced significantly in their presence. Along these same lines, Joenje et al. [82] compared the spontaneous chromosome-aberration frequencies of FA and normal PHA-stimulated lymphocytes cultured in oxygen concentrations ranging from 5 to 45%. The aberration frequencies of normal lymphocytes were unchanged at any oxygen concentration, whereas the aberration frequencies of lymphocytes from each of four persons with FA increased significantly with increasing oxygen concentrations.

Several types of experiments utilizing FA serum, co-cultivation, and cell fusion have been carried out in attempting to learn more about the molecular basis for FA: (1) In two early studies, lymphocytes from normal individuals were cultured in medium containing FA serum, and vice versa [58,118]; in neither study was evidence obtained for the existence of a clastogenic factor in FA. (2) Studies employing co-cultivation of FA and non-FA cells have yielded conflicting data as to the nature of the defect in FA. Germain and Requin [43] co-cultivated lymphocytes from each of four FA patients with lymphocytes from opposite-sexed, normal individuals and observed no effect on the chromosome-aberration frequency of either cell type. Nordenson et al. [109], on the other hand, conducted similar experiments and found the aberration frequencies of lymphocytes from two FA patients to be reduced significantly following co-cultivation. Zakrzewski and Sperling [188] co-cultivated FA fibroblasts and CHO cells at initial ratios of 1:2 and observed only a slight reduction in the spontaneous chromosome-aberration frequency of the FA fibroblasts. However, they found that when mitomycin C (MMC) was present during the final 24 hr of co-cultivation, the MMC-induced chromosome-aberration frequency of the FA fibroblasts was reduced significantly in comparison to that of similarly treated FA fibroblasts cultured alone. Furthermore, the MMC-treated CHO cells from the co-cultivation experiment had a significantly greater aberration frequency than similarly treated CHO cells cultured alone. Just how CHO cells can suppress the chromosome-aberration-inducing ability of MMC in FA cells while FA cells simultaneously enhance the chromosome-aberration-inducing ability of MMC in CHO cells is unclear. (3) The FA phenotype behaves as a recessive trait in non-FA/FA-fibroblast hybrids [186]. The aberration frequency of chromosomes in such hybrid cells did not differ from that of normal cells. Such non-FA/FA hybrids also behaved normally in their response to chromosome-aberration induction by MMC.

Results from the cytogenetic study of T lymphocytes from an FA patient into whom normal bone marrow cells were transplanted bear on the question of the production of a clastogenic factor in FA cells [12]. Berger et al. [12] transplanted bone marrow into a girl with FA, using marrow cells from her healthy brother. The chromosome complement of PHA-stimulated lymphocytes was analyzed on six occasions during the 3-year period following the transplantation; on each occasion the lymphocyte karyotype was male (46,XY), and no evidence was found of chromosome instability in the (male) cells. If chromosome instability in FA should be the result of a factor present in and released from the cells, then the male lymphocytes circulating in the blood of the girl with FA might have been expected to have exhibited some degree of chromosome instability similar to that observed in the FA cells themselves.

XERODERMA PIGMENTOSUM
Chromosome Aberrations

As stated at the outset, XP cells differ from those of the three syndromes just described by failing to exhibit spontaneously an elevated frequency of chromosome abnormalities [25]. Probably because of the lack of spontaneous chromosome breakage in XP, few reports have been published of the cytogenetic study of untreated XP patients. Nevertheless, for certain purposes, XP has been considered one of the chromosome-breakage syndromes since it first was so-categorized a decade ago [47]–thus, its inclusion in this review.

Five reports are known to us in which cytogenetic abnormalities have been described in untreated cells from XP patients. An elevated frequency of unstable chromosome aberrations was reported on two occasions when dermal fibroblasts were studied. Huang et al. [78] found the frequency of unstable chromosome aberrations in four late-passage XP cell lines to be significantly greater than that in comparable-passage normal cell lines. Bloch-Shtacher et al. [17] also reported the aberration frequency of some XP-fibroblast cell lines to be elevated, depending on the biopsy site of the tissue used for their development; when non-sun-exposed areas of the body were the source of the fibroblasts, 87% of the cells had no breaks, whereas only 57% of the cells in fibroblast cell lines derived from sun-exposed areas had none.

Stable chromosome rearrangements have been observed in untreated XP cells on three occasions. In one inadequately detailed report, blood T lymphocytes from an XP patient were reported to have the karyotype 45,XX,t(D;D), whereas lymphocytes from both her sib, who also had XP, and her parents lacked the translocation [176]. Hokkanen et al. [73], as reported in Kraemer [87], found an abnormal group D(13–15) chromosome in each of three fibroblast cell lines. German et al. [52] detected two clones of cells with abnormal chromosome complements in the fourth passage of a dermal fibroblast cell line derived from a noncancerous but sun-exposed area of the skin of a person with XP; no fibroblast cell lines derived from unaffected persons were studied for comparison, however, so the significance of the observation is unknown.

Even though excessive chromosome instability is not observed routinely in untreated XP cells, it manifests itself following the exposure of cells from at least some XP patients to certain DNA-damaging agents. Thus, Parrington et al. [116] compared the incidences of chromosome aberrations induced by ultraviolet (uv)-light (254 nm) in three patients (complementation groups unknown) and in normal individuals and found the induced chromosome-aberration frequency to be significantly greater in the XP cells. Hypersensitivity with respect to chromosome-aberration induction by uv-light has been confirmed using cells from XP patients belonging to complementation group A [98,133,150]; however, cells belonging to complementation groups B through G have not, to our knowledge, been studied for such hypersensitivity. Cells from XP "variants" (i.e., excision-repair-proficient cells) exhibit a normal amount of chromosome breakage in response to uv-light.

XP cells of at least some of the known complementation groups also are hypersensitive to the chromosome-breaking action of 4-nitroquinoline-1-oxide (4-NQO) [133,134,162], N-acetoxy-2-acetylaminofluorene [132], nitrogen mustard (HN2) [135], decarbamyl mitomycin C (DCMMC) [133,135], and the activated mycotoxins aflatoxin B_1 and sterigmatocystin [161]. Excessive chromosome breakage by these agents is accompanied in each instance by a decreased amount of DNA-repair synthesis. Thus, XP cells are unable to remove the damaged DNA segments that are produced by these agents, and the excessive amount of unrepaired damage appears to manifest itself in the form of an increased incidence of chromosome abnormalities.

Sister-Chromatid Exchange

Untreated cells from XP patients exhibit a normal baseline SCE frequency [7,24,31,85,143,182]. The sensitivity of XP cells to SCE induction following exposure to DNA-damaging agents generally is the same as their sensitivity to chromosome breakage by these agents [92 and 125 for detailed reviews]. Thus, following exposure to uv-light, the SCE frequencies of cells from most complementation groups tested have been greater than those of similarly treated control cells [7,24,31,119,143]. Cells from mentally deficient XP patients (the De Sanctis-Cacchione syndrome) [143] and from XP variants [24,31] respond normally to uv-light induction of SCEs. Several chemicals have been tested for their SCE-inducing abilities in XP T lymphocytes. 4-NQO [119], HN2 [119], and DCMMC [133] — the DNA damage of which is defectively repaired in XP — induced a greater-than-normal SCE response, whereas ethyl methanesulfonate (EMS) [119] and MMC [115] — the DNA damage of which is proficiently repaired — induced excessive and normal responses, respectively.

An SV40-transformed XP fibroblast cell line, XP12R0, was reported to be hypersensitive to all agents tested — methyl methanesulfonate, N-methyl-N'-nitro-N-nitrosoguanidine, ethylnitrosourea, dimethyl sulfate, EMS, 4-NQO, and MMC — whether or not the damage they produce is defectively repaired [183]; however, Heddle and Arlett [66] found several untransformed XP-fibroblast cell lines to exhibit a normal SCE induction by EMS, leading them to suggest that the cell that had been transformed by SV40 to give rise to the XP12R0 cell line was not a typical XP cell.

WERNER'S SYNDROME
Chromosome Aberrations

Few detailed cytogenetic studies of WS cells have been reported.

A normal frequency of chromosome aberrations in PHA-stimulated WS lymphocytes has been reported consistently [10,37,40,72,102,142,165,

S. Schonberg et al., in preparation], with a single exception [107]. Nordenson [107] reported finding an elevated frequency of unstable chromosome aberrations in blood cultures from four individuals with WS. WS dermal fibroblast cultures [72,128, S. Schonberg et al., in preparation] and LCLs [142, S. Schonberg et al., in preparation] also exhibit normal numbers of unstable chromosome aberrations. However, the available data demonstrate quite clearly that an unusual cytological difference, a quantitative one, exists between WS cells in long-term culture and those from normal individuals, viz., a strikingly increased frequency of cells that are members of clones with mutated chromosome complements.

The existence of rearranged chromosomes in WS dermal fibroblasts was reported first by Hoehn et al. in 1975 (although those authors stated that it had been detected earlier, but not reported, by W.W. Nichols) [72]. They observed at least one clone of cells with a rearranged chromosome complement in 13 of the 15 fibroblast cultures that had been developed from multiple samples of skin removed postmortem from a woman with WS. The term "variegated translocation mosaicism" (VTM) was coined [72] to describe the existence of multiple clones in cell lines derived from a single individual. Some concern was expressed by Hoehn et al. [72] that a terminal bacterial infection that their patient had might have contributed to the VTM observed in the multiple cultures they had developed from her skin. In retrospect, however, the infection may be considered to have been insignificant because dermal fibroblast cell lines developed subsequently from six other WS individuals, five of whom presumably were infection-free at the time skin samples were removed, also have been found to exhibit VTM [110,128,142, S. Schonberg et al., in preparation]. Also, three EBV-transformed LCLs we have developed from the B lymphocytes of one individual with WS also exhibit stable but unique chromosome rearrangements [142, S. Schonberg et al., in preparation].

Do the rearrangements so readily detected as clones in long-term cultures of WS cells occur in vivo or in vitro? In support of an in vivo origin for at least some of the rearranged chromosomes is the finding by Salk et al. [129,130] of identically rearranged chromosome complements in different WS-fibroblast cell lines. On one occasion the two cell lines had been derived from separate specimens taken postmortem from a WS patient; on the other, the two cell lines had been derived from different explants of one postmortem specimen. (The specimens of skin had been frozen at the time of autopsy – thus, the possibility for separate experiments.) At the same time as the evidence just cited for in vivo chromosome instability in WS exists, the appearance in dermal fibroblast cultures and LCLs derived from single WS individuals of many clones of cells with different chromosome rearrangements also suggests that rearranged chromosomes may arise in vitro.

The reason that long-term WS cell lines are comprised predominantly of clones of cells bearing rearranged chromosome complements is unknown. The poor growth potential of WS cells in vitro may be a contributing factor. Abnormally small numbers of fibroblasts migrate from tissue explants during initiation of WS-fibroblast cell lines [S. Schonberg et al., in preparation], and the cell lines that do develop, therefore, must be derived from just a few cells [131]. In some WS-fibroblast cultures, many clones with marker-chromosome rearrangements are found in early passages when the cultures exhibit their best growth. With increasing age of the cultures, these clones of cells become less numerous and eventually are replaced by one or at most a few clones [129,130]. In contrast, other WS cell cultures initially have only one or two clones of cells but then develop numerous clones of cells during later stages of cell growth. Schonberg et al. [142, S. Schonberg et al., in preparation] have suggested that any of the many chromosome rearrangements observed in WS cells may release the cells from some restriction on growth in vitro that is not a feature of fibroblasts from normal persons. This hypothesis would explain both the paucity of cells that emerge from the primary explants of WS skin and the increased incidence of clones of cells with unique chromosome rearrangements that characterize WS cell lines. By this hypothesis, the short in vitro life-span that is characteristic of WS cell lines would result from the fact that, much more often than when the explants are from normal persons, a single or at most a few cells give rise to the cultures, thus necessitating the occurrence of many more cell divisions in the progeny of those few cells in the primary culture flask before enough cells accumulate to permit the first subculture. A different interpretation is favored by Salk et al. [130], who suggest that the formation, evolution, and replacement of clones of cells with rearranged chromosomes in WS dermal fibroblast cell lines do not differ from the normal process often referred to as "aging" or "senescing" that occurs in cell lines derived from normal individuals, though the clones of cells in cell lines from normal individuals for the most part are not cytogenetically marked. According to Salk et al. [130], the seemingly early appearance in WS cell cultures of clones of cells having rearranged chromosome complement is the result merely of the poor growth potential of WS cells in vitro.

The Basis for Chromosome Instability in WS

Nordenson [107] suggested defective oxygen metabolism as the basis for the increased chromosome instability she observed in PHA-stimulated WS lymphocytes. This suggestion was based on her finding that the frequency of spontaneous chromosome abnormalities in WS lymphocytes was decreased when the cells were cultured in the presence of SOD and catalase. To determine whether defective oxygen metabolism might play a role in the appearance of clones of cells with rearranged chromosome complements in WS dermal fibroblasts, Salk et al. [129] compared the incidences of cytogenetic abnormalities in cell lines derived in the presence and in the absence of SOD and catalase as well as in cell lines established at ambient (18%) and reduced (1%) oxygen concentrations. In neither instance was an effect on the appearance of rearranged chromosome complements noted. Therefore, at the very least, defective oxygen metabolism does not explain all of the cytogenetic disturbances that are observed.

CONCLUDING REMARKS

For many decades the study of induced chromosome breakage and rearrangement has constituted a major aspect of the study of both plant and animal chromosomes. When techniques were developed in the 1950s that permitted the serious study of human chromosomes, it was natural that the microscopically visible response made by the human genome to insult by environmental agents known to produce mutations and cause cancer also would be examined, and many valuable studies of this nature have been conducted. In contrast, the study of genetically determined chromosome breakage and rearrangement — microscopic evidence of constitutional genomic instability — has consisted, with two major exceptions, predominantly of scattered reports of single examples in various species. One exception, in a sense, is the study of heritable transposable genetic elements in maize; the other is the subject of this review paper. In man, relatively more attention has been focused on genetic factors that can affect chromosome stability.

The emphasis on the study of constitutional human chromosome instability developed in large measure because, with a few exceptions, it was detected in clinical conditions that predispose to cancer. (It is interesting to note in this regard, however, that the discovery of the tendency to chromosome breakage and rearrangement in BS and FA preceded the realization that those disorders do predispose to cancer.) During the era of extremely active investigation of human chromosomes in relation to disease, i.e., post-1956, the study of constitutional chromosome instability moved at a steady pace, but it always was overshadowed by the much larger number of studies that in a general way can be referred to as karyotype-phenotype correlations, those concerning both developmental defects and cancers. Nevertheless, the study of the cytogenetics of the chromosome-breakage syndromes by relatively few laboratories has turned out to be a useful adjunct to the field of cancer cytogenetics, a field in which an extremely large and impressive amount of data has been accumulated by numerous laboratories and clinics detailing the changes in the chromosome complements that are found in association with various types of cancer.

Thus, a relatively small number of careful cytogenetic studies — summarized in this paper — utilizing cells from five rare genetic conditions, the chromosome-breakage syndromes, have documented an impressive genomic instability in each. The elucidation of the changes that occur in cells from persons with these rare conditions, in each case the consequence of a different, genetically determined cellular defect, has contributed to the general realization that, along with its remarkable stability, the eukaryotic genome is capable of a remarkable degree and variety of instability. That an excessive amount of genomic instability detectable at the cytological level in each of the chromosome-breakage syndromes is correlated with an increased frequency of malignant neoplasia seems particularly significant in view of the enormous amount of instability currently being demonstrated at the molecular level in eukaryotic DNA, i.e., transposable genetic elements, "oncogenes," tumor viruses, and transfectable "cancer genes."

Although the continued, careful accumulation of cytogenetic data in the chromosome-breakage syndromes is very much in order, cells homozygous for the genes responsible for these rare conditions also lend themselves to a new type of study. As we have summarized here, increased genomic instability in these cells has been carefully documented in the past by investigators at the cytological level, i.e., using the compound microscope. In the future, a look at genomic instability at a new level — the molecular level — should provide additional understanding and insight into the changes that occur and either occasionally are, or at least predispose to, the crucial change(s) by which neoplastic transformation originates and by which a transformed cell progresses toward what will constitute clinically significant (i.e., malignant) neoplasia.

Acknowledgments. This work was supported partially by research grants from NIH (HD 04134) and the American Cancer Society.

LITERATURE CITED

1. ABBADESSA R., BURDICH A.B.: The effect of X-irradiation on somatic crossing over in *Drosophila melanogaster. Genetics* 48:1345–1356, 1963.
2. AL SAADI A., PALUTKE M., KRISHNA KUMAR G.: Evolution of chromosomal abnormalities in sequential cytogenetic studies of ataxia telangiectasia. *Hum. Genet.* 55: 23–29, 1980.
3. ALHADEFF B., VELIVASAKIS M., PAGAN-CHARRY I., WRIGHT W.C., SINISCALCO M.: High rate of sister chromatid exchanges of Bloom's syndrome chromosomes is corrected in rodent human somatic cell hybrids. *Cytogenet. Cell Genet.* 27:8–23, 1980.

4. ARLETT C.F., LEHMANN A.R.: Human disorders showing increased sensitivity to the induction of genetic damage. *Annu. Rev. Genet.* 12:95–115, 1978.
5. AUERBACH A.D., ADLER B., CHAGANTI R.S.K.: Prenatal and postnatal diagnosis and carrier detection of Fanconi anemia by a cytogenetic method. *Pediatrics* 67: 128–135, 1981.
6. AURIAS A., DUTRILLAUX B., BURIOT D., LEJEUNE J.: High frequencies of inversions and translocations of chromosomes 7 and 14 in ataxia telangiectasia. *Mutat. Res.* 69:369–374, 1980.
7. BARTRAM C.R., KOSKE-WESTPHAL T., PASSARGE E.: Chromatid exchanges in ataxia telangiectasia, Bloom's syndrome, Werner syndrome, and xeroderma pigmentosum. *Ann. Hum. Genet.* 40:79–86, 1976.
8. BARTRAM C.R., RUDIGER H.W., PASSARGE E.: Frequency of sister chromatid exchanges in Bloom syndrome fibroblasts reduced by cocultivation with normal cells. *Hum. Genet.* 46:331–334, 1979.
9. BARTRAM C.R., RUDIGER H.W., SCHMIDT-PREUSS U., PASSARGE E.: Functional deficiency of fibroblasts heterozygous for Bloom syndrome as specific manifestation of the primary defect. *Am. J. Hum. Genet.* 33:928–934, 1981.
10. BEADLE G.F., MACKAY I.R., WHITTINGHAM S., TAGGART G., HARRIS A.W., HARRISON L.C.: Werner's syndrome, a model of premature aging? *J. Med.* 9:377–403, 1978.
11. BEARD M.E.J., YOUNG D.E., BATEMAN C.J.T., MCCARTHY G.T., SMITH M.E., SINCLAIR L., FRANKLIN A.W., SCOTT R.B.: Fanconi's anaemia. *Q. J. Med.* 42:403–422, 1973.
12. BERGER R., BERNHEIM A., LE CONIAT M., VECCHIONE D., DEVERGIE A., GLUCKMAN E.: Bone marrow graft of a Fanconi's anemia patient. Cytogenetic study. *Cancer Genet. Cytogenet.* 2:127–130, 1980.
13. BERGER R., BERNHEIM A., LE CONIAT M., VECCHIONE D., SCHAISON G.: Chromosomal studies of leukemic and preleukemic Fanconi's anemia patients. Examples of acquired 'chromosomal amplification'. *Hum. Genet.* 56:59–62, 1980.
14. BERGER R., BUSSEL A., SCHENMETZLER C.: Somatic segregation in Fanconi anemia. *Clin. Genet.* 11:409–415, 1977.
15. BERNSTEIN R., PINTO M., JENKINS T.: Ataxia telangiectasia with evolution of monosomy 14 and emergence of Hodgkin's disease. *Cancer Genet. Cytogenet.* 4: 31–37, 1981.
16. BERSI M., GASPARINI C.: Anomalie cromosomiche in un caso di anemia di Fanconi. *Minerva Med.* 64:1633–1637, 1973.
17. BLOCH-SHTACHER N., SLOR H., GOODMAN R.H.: Xeroderma pigmentosum: A model for studying chromosomal changes in neoplasia. *In* Wahrman J., Lewis K.R. (eds.): "Chromosomes Today." Jerusalem: Israel Universities Press, 4:427, 1973.
18. BLOOM G.E., GERALD P.S., WARNER S., DIAMOND L.K.: Chromosome aberrations in constitutional aplastic anemia and their possible relation to other hematopoietic disorders. *J. Pediatr.* 67:924–925, 1965.
19. BLOOM G.E., WARNER S., GERALD P.S., DIAMOND L.K.: Chromosome abnormalities in constitutional aplastic anemia. *N. Eng. J. Med.* 274:8–14, 1966.
20. BOCHKOV N.P., LOPUKHIN Y.M., KULESHOV N.P., KOVALCHUK L.V.: Cytogenetic study of patients with ataxia-telangiectasia. *Humangenetik* 24:115–128, 1974.
21. BOURGEOIS C.A., HILL F.G.H.: Fanconi anemia leading to acute myelomonocytic leukemia: Cytogenetic studies. *Cancer* 39:1163–1167, 1977.
22. BUSHKELL L.L., KERSEY J.H., CERVENKA J.: Chromosomal breaks in T and B lymphocytes in Fanconi's anemia. *Clin. Genet.* 9:583–587, 1976.

23. CHAGANTI R.S.K., SCHONBERG S.A., GERMAN J.: A manyfold increase in sister chromatid exchanges in Bloom's syndrome lymphocytes. *Proc. Natl. Acad. Sci. USA* 71:4508-4512, 1974.
24. CHENG W.-S., TARONE R.E., ANDREWS A.D. WHANG-PENG J.S., ROBBINS J.H.: Ultraviolet light-induced sister chromatid exchanges in xeroderma pigmentosum and in Cockayne's syndrome lymphocyte cell lines. *Cancer Res.* 38:1601-1609, 1978.
25. CLEAVER J.E., BOOTSMA D.: Xeroderma pigmentosum: Biochemical and genetic characteristics. *Annu. Rev. Genet.* 9:19-38, 1975.
26. COHEN M.M., KOHN G., DAGAN J.: Chromosomes in ataxia-telangiectasia. *Lancet* ii: 1500, 1973.
27. COHEN M.M., SAGI M., BEN-ZUR Z., SCHAAP T., VOSS R., KOHN G., BEN-BASSAT H.: Ataxia telangiectasia: Chromosomal stability in continuous lymphoblastoid cell lines. *Cytogenet. Cell Genet.* 23:44-52, 1979.
28. COHEN M.M., SHAHAM M., DAGAN J., SHMUELI E., KOHN G.: Cytogenetic investigations in families with ataxia-telangiectasia. *Cytogenet. Cell Genet.* 15:338-356, 1975.
29. COHEN M.M., SIMPSON S.J.: Absence of a clastogenic factor in ataxia telangiectasia lymphoblastoid cells. *Cancer Genet. Cytogenet.* 2:327-334, 1980.
30. CROSSEN P.E., MELLOR J.E.L., ADAMS A.C., GUNZ F.W.: Chromosome studies in Fanconi's anaemia before and after treatment with oxymetholone. *Pathology* 4: 27-33, 1972.
31. DE WEERD-KASTELEIN E.A., KEIJZER W., RAINALDI G., BOOTSMA D.: Induction of sister chromatid exchanges in xeroderma pigmentosum cells after exposure to ultraviolet light. *Mutat. Res.* 45:253-261, 1977.
32. DE WIT J., JASPERS N.G.J., BOOTSMA D.: The rate of DNA synthesis in normal human and ataxia telangiectasia cells after exposure to x-irradiation. *Mutat. Res.* 80:221-226, 1981.
33. DICKEN C.H., DEWALD G., GORDON H.: Sister chromatid exchanges in Bloom's syndrome. *Arch. Dermatol.* 114:755-760, 1978.
34. DOSIK H., HSU L.Y., TODARO G.J., LEE S.L., HIRSCHHORN K., SELIRIO E.S., ALTER A.A.: Leukemia in Fanconi's anemia: Cytogenetic and tumor virus susceptibility studies. *Blood* 36:341-352, 1970.
35. EDWARDS M.J., TAYLOR A.M.R.: Unusual levels of (ADP-ribose) and DNA synthesis in ataxia telangiectasia cells following gamma-ray irradiation. *Nature* 287: 745-747, 1980.
36. EMERIT I., CERUTTI P.: Clastogenic activity from Bloom syndrome fibroblast cultures. *Proc. Natl. Acad. Sci. USA* 78:1868-1872, 1981.
37. EPSTEIN C.J., MARTIN G.M., SCHULTZ A.L., MOTULSKY A.G.: Werner's syndrome: A review of its symptomatology, natural history, pathologic features, genetics and relationship to the natural aging process. *Medicine* 45:177-221, 1966.
38. FERAK V., BENKO J., CAJKOVA E.: Genetik der Ataxie-Teleangiectasie. *Acta Genet. Med. Gemellol.* (Roma) 14:57-72, 1965.
39. FESTA R.S., MEADOWS A.T., BOSHES R.A.: Leukemia in a black child with Bloom's syndrome: Somatic recombination as a possible mechanism for neoplasia. *Cancer* 44:1507-1510, 1979.
40. FRACCARO M., BOTT M.G., CALVERT H.T.: Chromosomes in Werner's syndrome. *Lancet* i:536, 1962.
41. FRIEDBERG E.C., EHMANN U.K., WILLIAMS J.I.: Human diseases associated with defective DNA repair. *Adv. Radiat. Biol.* 8:85-174, 1979.

42. GATTI M., RIZZONI M., PALITTI F., OLIVIERI G.: Studies on induced aberrations in diplochromosomes of Chinese hamster cells. *Mutat. Res.* 20:87-99, 1973.
43. GERMAIN D., REQUIN C.: Anomalies chromosomiques dans les cytopénies constitutionelles. *Nouv. Rev. Fr. Hematol.* 10:107-117, 1970.
44. GERMAIN D., REQUIN C., ROBERT J., VIALA J.J., FREYCON F.: Les anomalies chromosomiques dans l'anémie de Fanconi (A propos de le observations personnelles). *Pediatrie* 23:153-167, 1968.
45. GERMAN J.: Biological role for chromatid exchange in mammalian somatic cells? *In* Schimke R.T. (ed.): "Gene Amplication." Cold Spring Harbor: Cold Spring Harbor Laboratory, pp. 307-312, 1982.
46. GERMAN J.: Cytological evidence for crossing-over *in vitro* in human lymphoid cells. *Science* 144:298-301, 1964.
47. GERMAN J.: Genes which increase chromosomal instability in somatic cells and predispose to cancer. *Prog. Med. Génét.* 8:61-101, 1972.
48. GERMAN J.: Genetic disorders associated with chromosomal instability and cancer. *J. Invest. Dermatol.* 60:427-434, 1973.
49. GERMAN J., ARCHIBALD R., BLOOM D.: Chromosomal breakage in a rare and probably genetically determined syndrome of man. *Science* 148:506-507, 1965.
50. GERMAN J., CRIPPA L.P.: Chromosomal breakage in diploid cell lines from Bloom's syndrome and Fanconi's anemia. *Ann Génét.* (Paris) 9:143-154, 1966.
51. GERMAN J., CRIPPA L.P., BLOOM D.: Bloom's syndrome. III. Analysis of the chromosome aberration characteristic of this disorder. *Chromosoma* 48:361-366, 1974.
52. GERMAN J., GILLERAN T.G., SETLOW R.B., REGAN J.D.: Mutant kaytotypes in a culture of cells from a man with xeroderma pigmentosum. *Ann. Genet.* (Paris) 16:23-27, 1973.
53. GERMAN J., SCHONBERG S.: Bloom's syndrome. IX. Review of cytological and biochemical aspects. *In* Gelboin H.V., MacMahon B., Matsushima T., Sugimura T., Takayama S., Takebe H. (eds.): "Genetic and Environmental Factors in Experimental and Human Cancer." Tokyo: Japan Scientific Societies Press, pp. 175-186, 1980.
54. GERMAN J., SCHONBERG S., LOUIE E., CHAGANTI R.S.K.: Bloom's syndrome. IV. Sister-chromatid exchanges in lymphocytes. *Am. J. Hum. Genet.* 29:248-255, 1977.
55. GMYREK D., WITKOWSKI R., SYLLM-RAPOPORT I., JACOBASCH G.: Chromosomal aberrations and abnormalities of red-cell metabolism in a case of Fanconi's anaemia before and after development of leukaemia. *Germ. Med. Monthly* 13:105-111, 1968.
56. GOODMAN W.N., COOPER W.C., KESSLER G.B., FISCHER M.S., GARDNER M.B.: Ataxia-telangiectasia. A report of two cases in siblings presenting a picture of progressive spinal muscular atrophy. *Bull. Los Angeles Neurol. Soc.* 34:1-22, 1969.
57. GUANTI G., PETRINELLI P., SCHETTINI F.: Cytogenetical and clinical investigations in aplastic anaemia (Fanconi's type). *Humangenetik* 13:222-233, 1971.
58. GUPTA R.S., GOLDSTEIN S.: Diptheria toxin resistance in human fibroblast cell strains from normal and cancer-prone individuals. *Mutat. Res.* 73:331-338, 1980.
59. HARNDEN D.G.: Ataxia telangiectasia syndrome: Cytogenetic and cancer aspects. *In* German J. (ed.): "Chromosomes and Cancer." New York: John Wiley and Sons, pp. 619-636, 1974.
60. HARNDEN D.G.: The relationships between induced chromosome aberrations and chromosome abnormality in tumour cells. *In* Armendares S., Lisker R. (eds.): "Human Genetics." Amsterdam: Excerpta Medica, pp. 355-366, 1977.

61. HARNDEN D.G., BROWN K.W.: Genes, chromosomes and cancer. *In* Belyaev D.K. (ed.): "Problems in General Genetics." Moscow: MIR Publishers, pp. 263-273, 1981.
62. HAYASHI K., SCHMID W.: Tandem duplication q14 and dicentric formation by end-to-end chromosome fusions in ataxia telangiectasia (AT): Clinical and cytogenetic findings in 5 patients. *Humangenetik* 30:135-141, 1975.
63. HECHT F., KOLER R.D., RIGAS D.A., DAHNKE G.S., CASE M.P., TISDALE V., MILLER R.W.: Leukemia and lymphocytes in ataxia-telangiectasia. *Lancet* ii:1193, 1966.
64. HECHT F., MCCAW B.K., KOLER R.D.: Ataxia-telangiectasia – Clonal growth of translocation lymphocytes. *N. Engl. J. Med.* 289:286-291, 1973.
65. HECHT F., MCCAW B.K., PEAKMAN D., ROBINSON A.: Non-random occurrence of 7-14 translocations in human lymphocyte cultures. *Nature* 255:243-244, 1975.
66. HEDDLE J.A., ARLETT C.F.: Untransformed xeroderma pigmentosum cells are not hypersensitive to sister-chromatid exchange production by ethyl methane sulfonate – implications for the use of transformed cell lines and for the mechanism by which SCE arise. *Mutat. Res.* 72:119-126, 1980.
67. HENDERSON E., GERMAN J.: Development and characterization of lymphoblastoid cell lines (LCLs) from "chromosome breakage syndromes" and related genetic disorders. *J. Supramol. Struct. (Suppl.)* 2:83, 1978.
68. HIGURASHI M., CONEN P.E.: In vitro chromosomal radiosensitivity in "chromosomal breakage syndromes." *Cancer* 32:380-383, 1973.
69. HIRSCH-KAUFMANN M., SCHWEIGER M., WAGNER E.F., SPERLING K.: Deficiency of DNA ligase activity in Fanconi's anemia. *Hum. Genet.* 45:25-32, 1978.
70. HIRSCH-KAUFMANN M., SCHWEIGER M., WAGNER E.F., SPERLING K.: DNA-ligase is deficient in Fanconi's anaemia. *In* Altmann H., Riklis E., Slor H. (eds.): "DNA Repair and Late Effects." Proc. International Symposium of the "IGEGM". Negev: Nuclear Research Center, pp. 263-270, 1980.
71. HIRSCHMAN R.J., SHULMAN N.R., ABUELO J.G., WHANG-PENG J.: Chromosomal aberrations in two cases of inherited aplastic anemia with unusual clinical features. *Ann. Intern. Med.* 71:107-117, 1969.
72. HOEHN H., BRYANT E.M., AU K., NORWOOD T.H., BOMAN H., MARTIN G.M.: Variegated translocation mosaicism in human skin fibroblast cultures. *Cytogenet. Cell Genet.* 15:282-298, 1975.
73. HOKKANEN E., TIVANIANEN M., WALTIMO D.: Zu den neurologischen Manifestationen des Xeroderma pigmentosum. *Dtsch. Z. Nervenheilkd.* 196:206-216, 1969.
74. HOOK E.B., HATCHER N.H., CALKA O.J.: Apparent "in situ" clone of cytogenetically marked ataxia-telangiectasia lymphocytes. *Humangenetik* 30:251-257, 1975.
75. HOULDSWORTH J., LAVIN M.E.: Effect of ionizing radiation on DNA synthesis in ataxia telangiectasia cells. *Nucleic Acids Res.* 8:3709-3720, 1980.
76. HUANG C.C., BANERJEE A., HOU Y.: Chromosomal instability in cell lines derived from patients with xeroderma pigmentosum. *Proc. Soc. Exp. Biol. Med.* 148:1244-1248, 1975.
77. HUANG P.C., SHERIDAN R.B.: Genetic and biochemical studies with ataxia telangiectasia. A review. *Hum. Genet.* 59:1-9, 1981.
78. Human Gene Mapping 6. *Cytogenet. Cell Genet.* Vol. 32:205-207, 1982.
79. HUSTINX T.W.J., SCHERES J.M.J.C., WEEMAES C.M.R., TER HAAR B.G.A., JANSSEN A.H.: Karyotype instability with multiple 7/14 and 7/7 rearrangements. *Hum. Genet.* 49:199-208, 1979.
80. HUSTINX T.W.J., TER HAAR B.G.A., SCHERES J.M.J.C. RUTTEN F.J., WEEMAES C.M.R., HOPPE R.L.E., JANSSEN A.H.: Bloom's syndrome in two Dutch families. *Clin. Genet.* 12:85-96, 1977.

81. JEAN P., RICHER C.L., MURER-ORLANDO M., LUU D.H., JONCAS J.H.: Translocation 8;14 in an ataxia telangiectasia-derived cell line. *Nature* 277:56-58, 1979.
82. JOENJE H., ARWERT F., ERIKSSON A.W., DE KONING H., OOSTRA A.B.: Oxygen-dependence of chromosomal aberrations in Fanconi's anaemia. *Nature* 290:142-143, 1981.
83. KAISER-MCCAW B., EPSTEIN A.L., OVERTON K.M., KAPLAN H.S., HECHT F.: The cytogenetics of human lymphomas: Chromosome 14 in Burkitt's, diffuse histiocytic and related neoplasms. *In* De La Chapelle A., Sorsa M. (eds.): "Chromosomes Today." Amsterdam: Elsevier/North-Holland Biomedical Press, 6:383-390, 1977.
84. KAISER-MCCAW B., HECHT F.: Ataxia-telangiectasia; chromosomes and cancer. *In* Bridges B.A., Harnden D.G. (eds.): "Ataxia Telangiectasia – A Cellular and Molecular Link Between Cancer, Neuropathology, and Immune Deficiency." Chichester: John Wiley and Sons, pp. 243-257, 1982.
85. KATO H., STICH H.F.: Sister chromatid exchanges in ageing and repair-deficient human fibroblasts. *Nature* 260:447-448, 1976.
86. KIOSSOGLOU K., MOSCHOS A., MANTALENAKI-LAMBROU K., HAIDAS S.: Acute lymphoblastic leukemia in Bloom's syndrome. *Hippocrates* 3:29-35, 1979.
87. KRAEMER K.H.: Xeroderma pigmentosum. *In* Demis D.J., Dobson R.L., McGuire J. (eds.): "Clinical Dermatology." Hagerstown: Harper and Row, pp. 1-33, 1980.
88. KUHN E.M.: A high incidence of mitotic chiasmata in endoreduplicated Bloom's syndrome cells. *Hum. Genet.* 58:417-421, 1981.
89. KUHN E.M.: Effects of X-irradiation in G_1 and G_2 on Bloom's syndrome and normal chromosomes. *Hum. Genet.* 54:335-341, 1980.
90. LANDAU J.W., SASAKI M.S., NEWCOMER V.D., NORMAN A.: Bloom's syndrome: The syndrome of telangiectatic erythema and growth retardation. *Arch. Dermatol.* 94:687-694, 1966.
91. LATT S.A.: Microfluorometric detection of deoxyribonucleic acid replication in human metaphase chromosomes. *Proc. Natl. Acad. Sci. USA* 70:3395-3399, 1973.
92. LATT S.A., SCHRECK R.R., DOUGHERTY C.P., GUSTASHAW K.M., JUERGENS L.A., KAISER T.N.: Sister-chromatid exchange – The phenomenon and its relationship to chromosome-fragility diseases. This volume.
93. LEVITT R., PIERRE R.V., WHITE W.L., SIEKERT R.G.: Atypical lymphoid leukemia in ataxia telangiectasia. *Blood* 52:1003-1011, 1978.
94. LIEBER E., HSU L., SPITLER L., FUDENBERG H.H.: Cytogenetic findings in a parent of a patient with Fanconi's anemia. *Clin. Genet.* 3:357-363, 1972.
95. LISKER R., COBO A.: Chromosome breakage in ataxia-telangiectasia. *Lancet* i:618, 1970.
96. LISKER R., DE GUTIERREZ A.C.: Cytogenetic studies in Fanconi's anemia. Description of a case with bone marrow clonal evolution. *Clin. Genet.* 5:72-76, 1974.
97. LOHR G.W., WALLER H.D., ANSCHUTZ F., KNOPP A.: Biochemische Defekte in den Blutzellen bei familiärer Panmyelopathie (Typ Fanconi). *Humangenetik* 1:383-387, 1965.
98. MARSHALL R.R., SCOTT D.: The relationship between chromosome damage and cell killing in uv-irradiated normal and xeroderma pigmentosum cells. *Mutat. Res.* 36:397-400, 1976.
99. MCCAW B.K., HECHT F., HARNDEN D.G., TEPLITZ R.L.: Somatic rearrangement of chromosome 14 in human lymphocytes. *Proc. Natl. Acad. Sci. USA* 72:2071-2075, 1975.
100. MCDOUGALL J.K.: Spontaneous and adenovirus type 12-induced chromosome aberrations in Fanconi's anaemia fibroblasts. *Int. J. Cancer* 7:526-534, 1971.
101. MCINTOSH S., BERG W.R., LUBINIECKI A.S.: Fanconi's anemia: The preanemic

phase. Am. J. Pediatr. Hematol. Oncol. 1:107-110, 1979.
102. McKusik V.A.: Medical genetics 1962. J. Chronic Dis. 16:599-601, 1963.
103. Meisner L.F., Taher A., Shahidi N.T.: Chromosome changes and leukemic transformation in Fanconi's anemia. In Hibino S., Takaku F., Shahidi N.T. (eds.): "Aplastic Anemia." Baltimore: University Park Press, pp. 253-271, 1978.
104. Meme J.S., Gripenberg U., Kahkonen M.: Fanconi's anaemia; chromosome breakage in a large African family. Hereditas 93:255-260, 1980.
105. Nathanson S.D., van Biljon S.M., Kallmeyer J.: Constitutional aplastic anaemia (Fanconi type): Case presentation and review of the literature. S. Afr. Med. J. 42:1159-1161, 1968.
106. Nelson M.M., Blom A., Arens L.: Chromosomes in ataxia telangiectasia. Lancet i:518-519, 1975.
107. Nordenson I.: Chromosome breaks in Werner's syndrome and their prevention in vitro by radical-scavenging enzymes. Hereditas 87:151-154, 1977.
108. Nordenson I.: Effect of superoxide dismutase and catalase on spontaneously occurring chromosome breaks in patients with Fanconi's anemia. Hereditas 86:147-150, 1977.
109. Nordenson I., Bjorksten B., Lundh B.: Prevention of chromosomal breakage in Fanconi's anemia by cocultivation with normal cells. Hum. Genet. 56:169-171, 1980.
110. Norwood T.H., Hoehn H., Salk D., Martin G.M.: Cellular aging in Werner's syndrome: A unique phenotype? J. Invest. Dermatol. 73:92-96, 1979.
111. Obeid D.A., Hill F.G.H., Harnden D., Mann J.R., Wood B.S.B.: Fanconi anemia: Oxymetholone, hepatic tumors, and chromosome aberrations associated with leukemic transition. Cancer 46:1401-1406, 1980.
112. Ockey C.H.: Quantitative replicon analysis of DNA synthesis in cancer-prone conditions and the defects in Bloom's syndrome. J. Cell Sci. 40:125-144, 1979.
113. Oxford J.M., Harnden D.G., Parrington J.M., Delhanty J.D.A.: Specific chromosome aberrations in ataxia telangiectasia. J. Med. Genet. 12:251-262, 1975.
114. Painter R.B., Young B.R.: Radiosensitivity in ataxia-telangiectasia: A new explanation. Proc. Natl. Acad. Sci. USA 77:7315-7317, 1980.
115. Pant G.S., Kamada N.: Sister chromatid exchanges and their relevance to defective DNA repair in xeroderma pigmentosum cells. Indian J. Exp. Biol. 16:1194-1196, 1978.
116. Parrington J.M., Delhanty J.D.A., Baden H.P.: Unscheduled DNA synthesis, U.V.-induced chromosome aberrations and SV_{40} transformation in cultured cells from xeroderma pigmentosum. Ann. Hum. Genet. 35:149-160, 1971.
117. Passarge E., Bartram C.R.: Somatic recombination as possible prelude to malignant transformation. In Bergsma D. (ed.): "Cancer and Genetics." New York: Alan R. Liss, pp. 177-180, 1976.
118. Perkins J., Timson J., Emery A.E.H.: Clinical and chromosome studies in Fanconi's aplastic aneamia. J. Med. Genet. 6:28-33, 1969.
119. Perry P.E., Jager M., Evans H.J.: Mutagen-induced sister chromatid exchanges in xeroderma pigmentosum and normal lymphocytes. In Evans H.J., Lloyd D.C. (eds.): "Mutagen-Induced Chromosome Damage in Man." New Haven: Yale University Press, pp. 201-207, 1978.
120. Pfeiffer R.A.: Chromosomal abnormalities in ataxia-telangiectasia (Louis Bar's syndrome). Humangenetik 8:302-306, 1970.
121. Prindull G., Jentsch E., Hansmann I.: Fanconi's anaemia developing erythroleukaemia. Scand. J. Haematol. 23:59-63, 1979.
122. Puliglandla B., Stass S.A., Schumacher H.R., Keneklis T.P., Bollum F.J.: Ter-

minal deoxynucleotidyl transferase in Fanconi's anaemia. *Lancet* ii:1263, 1978.
123. RARY J.M., BENDER M.A., KELLY T.E.: A 14/14 marker chromosome lymphocyte clone in ataxia telangiectasia. *J. Hered.* 66:33-35, 1975.
124. RAUH J.L., SOUKUP S.W.: Bloom's syndrome. *Am. J. Dis. Child.* 116:409-413, 1968.
125. RAY J.H., GERMAN J.: Sister chromatid exchange in the chromosome breakage syndromes. *In* Sandberg A.A. (ed.): "Sister Chromatid Exchange." New York: Alan R. Liss, pp. 553-577, 1982.
126. RAY J.H., GERMAN J.: The chromosome changes in Bloom's syndrome, ataxia-telangiectasia, and Fanconi's anemia. *In* Arrighi F.E., Rao P.N., Stubblefield E. (eds.): "Genes, Chromosomes, and Neoplasia." New York: Raven Press, pp. 351-378, 1981.
127. RUDIGER H.W., BARTRAM C.R., HARDER W., PASSARGE E.: Rate of sister chromatid exchanges in Bloom syndrome fibroblasts reduced by co-cultivation with normal fibroblasts. *Am. J. Hum. Genet.* 32:150-157, 1980.
128. SALK D., AU K., HOEHN H., MARTIN G.M.: Cytogenetics of Werner's syndrome cultured skin fibroblasts: Variegated translocation mosaicism. *Cytogenet. Cell Genet.* 30:92-107, 1981.
129. SALK D., AU K., HOEHN H., MARTIN G.M.: Effects of radical-scavenging enzymes and reduced oxygen exposure on growth and chromosome abnormalities of Werner syndrome cultured skin fibroblasts. *Hum. Genet.* 57:269-275, 1981.
130. SALK D., AU K., HOEHN H., STENCHEVER M.R., MARTIN G.M.: Evidence of clonal attenuation, clonal succession, and clonal expansion in mass cultures of aging Werner's syndrome skin fibroblasts. *Cytogenet. Cell Genet.* 30:108-117, 1981.
131. SALK D., BRYANT E., AU K., HOEHN H., MARTIN G.M.: Systematic growth studies, cocultivation, and cell hybridization studies of Werner syndrome cultured skin fibroblasts. *Hum. Genet.* 58:310-316, 1981.
132. SAN R.H.C., STICH W., STICH H.F.: Differential sensitivity of xeroderma pigmentosum cells of different repair capacities towards the chromosome breaking action of carcinogens and mutagens. *Int. J. Cancer* 20:181-187, 1977.
133. SASAKI M.S.: Chromosome aberration formation and sister chromatid exchange in relation to DNA repair in human cells. *In* Generoso W.M., Shelby M.D., de Serres F.J. (eds.): "DNA Repair and Mutagenesis in Eukaryotes." New York: Plenum Press, pp. 285-313, 1980.
134. SASAKI M.S.: DNA repair capacity and susceptibility to chromosome breakage in xeroderma pigmentosum cells. *Mutat. Res.* 20:291-293, 1973.
135. SASAKI M.S., TODA K., OZAWA A.: Role of DNA repair in the susceptibility to chromosome breakage and cell killing in cultured human fibroblasts. *In* Seiji M., Bernstein I.A. (eds.): "Biochemistry of Cutaneous Epidermal Differentiation." Baltimore: University Park Press, pp. 167-179, 1977.
136. SAXON A., STEVENS R.H., GOLDE D.W.: Helper and suppressor T-lymphocyte leukemia in ataxia telangiectasia. *N. Engl. J. Med.* 300:700-704, 1979.
137. SCHERES J.M.J.C., HUSTINX T.W.J., WEEMAES C.M.R.: Chromosome 7 in ataxia telangiectasia. *J. Pediatr.* 97:440-441, 1980.
138. SCHMID W., JERUSALEM F.: Cytogenetic findings in two brothers with ataxia telangiectasia (Louis Bar's syndrome). *Arch. Genet. (Zur.)* 45:49-52, 1972.
139. SCHMID W., SCHARER K., BAUMANN T., FANCONI G.: Chromosomenbrüchigkeit bei der familiären Panmyelopathie (Typus Fanconi). *Schweiz. Med. Wochenschr.* 95:1461-1464, 1965.
140. SCHOEN E.J., SHEARN M.A.: Immunoglobulin deficiency in Bloom's syndrome. *Am. J. Dis. Child.* 113:594-596, 1967.

141. SCHONBERG S., GERMAN J.: Sister chromatid exchange in cells metabolically coupled to Bloom's syndrome cells. *Nature* 284:72-74, 1980.
142. SCHONBERG S.A., HENDERSON E., NIERMEIJER M.F., GERMAN J.: Werner's syndrome: Preferential proliferation of clones with translocations. *Am. J. Hum. Genet.* 33:120A, 1981.
143. SCHONWALD A.D., PASSARGE E.: UV-light induced sister chromatid exchanges in xeroderma pigmentosum lymphocytes. *Hum. Genet.* 36:213-218, 1977.
144. SCHROEDER T.M., ANSCHUTZ F., KNOPP A.: Spontane Chromosomenaberrationen bei familiärer Panmyelopathie. *Humangenetik* 1:194-196, 1964.
145. SCHROEDER T.M., DRINGS P., BEILNER P., BUCHINGER G.: Clinical and cytogenetic observations during a six-year period in an adult with Fanconi's anaemia. *Blut* 34:119-132, 1976.
146. SCHROEDER T.M., GERMAN J.: Bloom's syndrome and Fanconi's anemia: Demonstration of two distinctive patterns of chromosome disruption and rearrangement. *Humangenetik* 25:299-306, 1974.
147. SCHROEDER T.M., POHLER E., HUFNAGI H.D., STAHL-MAUGE C.: Fanconi's anemia: Terminal leukemia and "Forme fruste" in one family. *Clin Genet.* 16:260-268, 1979.
148. SCHROEDER T.M., STAHL-MAUGE C.: Spontaneous chromosome instability, chromosome reparation and recombination in Fanconi's anemia and Bloom's syndrome. *In* Altmann H. (ed.): "DNA Repair and Late Effects." International Symposium of the "IGEGM". Eisenstadt, Instut für Biologie, Forschungszentrum Seibersdorf/Wien, pp. 35-50, 1976.
149. SCHULER D., KISS A., FABIAN F.: Chromosomal peculiarities and "in vitro" examinations in Fanconi's anaemia. *Humangenetik* 7:314-322, 1969.
150. SCOTT D., MARSHALL R.R.: An investigation of uv-induced chromosome changes in relation to lethality in normal and uv-sensitive human cells. *In* Evans H.J., Lloyd D.C. (eds.): "Mutagen-Induced Chromosome Damage in Man." Edinburgh: Edinburgh University Press, pp. 129-141, 1978.
151. SHAHAM M., BECKER Y.: The ataxia telangiectasia clastogenic factor is a low molecular weight peptide. *Hum. Genet.* 58:422-424, 1981.
152. SHAHAM M., BECKER Y., COHEN M.M.: A diffusable clastogenic factor in ataxia telangiectasia. *Cytogenet. Cell Genet.* 27:155-161, 1980.
153. SHAHID M.J., KHOURI E.P., BALLAS S.K.: Fanconi's anaemia: Report of a patient with significant chromosomal abnormalities in bone marrow cells. *J. Med. Genet.* 9:474-478, 1972.
154. SHIRAISHI Y., FREEMAN A.I., SANDBERG A.A.: Increased sister chromatid exchange in bone marrow and blood cells from Bloom's syndrome. *Cytogenet. Cell Genet.* 17:162-173, 1976.
155. SHIRAISHI Y., MATSUI S.I., SANDBERG A.A.: Normalization by cell fusion of sister chromatid exchange in Bloom syndrome lymphocytes. *Science* 212:820-822, 1981.
156. SHIRAISHI Y., SANDBERG A.A.: The relationship between sister chromatid exchanges and chromosome aberrations in Bloom's syndrome. *Cytogenet. Cell Genet.* 18:13-23, 1977.
157. SHIRAISHI S., SANDBERG A.A.: Evaluation of sister chromatid exchanges in Bloom's syndrome. *Cytobios* 21:175-184, 1978.
158. SPARKES R.S., COMO R., GOLDE D.W.: Cytogenetic abnormalities in ataxia telangiectasia with T-cell chronic lymphocytic leukemia. *Cancer Genet. Cytogenet.* 1:329-336, 1980.
159. SPERLING K., GOLL U., KUNZE J., LUDTKE E.K., TOLKSDORF M., OBE G.: Cytogenetic investigations in a new case of Bloom's syndrome. *Hum. Genet.* 31:47-52, 1976.

160. STEFANINI M., KEIJZER W., DALPRA L., ELLI R., PORRO M.N., NICOLETTI B., NUZZO F.: Differences in the levels of uv repair and in clinical symptoms in two sibs affected by xeroderma pigmentosum. *Hum. Genet.* 54:177-182, 1980.
161. STICH H.F., LAISHES B.A.: The response of xeroderma pigmentosum cells and controls to the activated mycotoxins, aflatoxins and sterigmatocystin. *Int. J. Cancer* 16:266-274, 1975.
162. STICH H.F., STICH W., SAN R.H.C.: Chromosome aberrations in xeroderma pigmentosum cells exposed to the carcinogens, 4-nitroquinoline-1-oxide and N-methyl-N'-nitro-nitrosoguanidine. *Proc. Soc. Exp. Biol. Med.* 142:1141-1144, 1973.
163. SWIFT M.R., HIRSCHHORN K.: Fanconi's anemia: Inherited susceptibility to chromosome breakage in various tissues. *Ann. Intern. Med.* 65:496-503, 1966.
164. SZALAY G.C., WEINSTEIN E.D.: Questionable Bloom's syndrome in a Negro girl. *Am. J. Dis. Child.* 124:245-248, 1972.
165. TAO L.C., STECKER E., GARDNER H.A.: Werner's syndrome and acute myeloid leukemia. *Can. Med. Assoc. J.* 105:952-954, 1971.
166. TAYLOR A.M.R., METCALFE J.A., OXFORD J.M., HARNDEN D.G.: Is chromatid-type damage in ataxia telangiectasia after irradiation at G_0 a consequence of defective repair? *Nature* 260:441-443, 1976.
167. TAYLOR A.M.R., OXFORD J.M., METCALFE J.A.: Spontaneous cytogenetic abnormalities in lymphocytes from thirteen patients with ataxia telangiectasia. *Int. J. Cancer* 27:311-319, 1981.
168. TAYLOR A.M.R., ROSNEY C.M., CAMPBELL J.B.: Unusual sensitivity of ataxia telangiectasia cells to bleomycin. *Cancer Res.* 39:1046-1050, 1979.
169. THERMAN E., OTTO P.G., SHAHIDI N.T.: Mitotic recombination and segregation of satellites in Bloom's syndrome. *Chromosoma* 82:627-636, 1981.
170. THOMAS G.H., MILLER C.S., TOOMEY K.E., REYNOLDS L.W., REITMAN M.L., VARKI A., VANNIER A., ROSENBAUM K.N., BIAS W.B., SCHOFIELD B.H.: Two clonal cell populations (mosaicism) in a 46,XY male with mucolipidosis II (I-cell disease) – An autosomal recessive disorder. *Am. J. Hum. Genet.* 34:611-622, 1982.
171. TICE R., WINDLER G., RARY J.M.: Effect of cocultivation on sister chromatid exchange frequencies in Bloom's syndrome and normal fibroblast cells. *Nature* 273:538-540, 1978.
172. VAN BUUL P.P.W., NATARAJAN A.T., VERDEGAAL-IMMERZEEL E.A.M.: Suppression of the frequencies of sister chromatid exchanges in Bloom's syndrome fibroblasts by co-cultivation with Chinese hamster cells. *Hum. Genet.* 44:187-189, 1978.
173. VARELA M.A., STERNBERG W.H.: Preanaemic state in Fanconi's anaemia. *Lancet* ii: 566-567, 1967.
174. VON KOSKULL H., AULA P.: Distribution of chromosome breaks in measles, Fanconi's anemia, and controls. *Hereditas* 87:1-10, 1977.
175. WAKE N., MINOWADA J., PARK B., SANDBERG A.A.: Chromosomes and causation of human cancer and leukemia. XLVIII. T-cell acute leukemia in ataxia telangiectasia. *Cancer Genet. Cytogenet.* 6:345-357, 1982.
176. WALTIMO O., IVANIANEN M., HOKKANEN E.: Xeroderma pigmentosum with neurological manifestations. Family studies of two affected sisters, one of them with a chromosome abnormality, and report of one separate case. *Acta Neurol. Scand.* 43 [Suppl.] 31:66-67, 1967.
177. WARREN S.T., SCHULTZ R.A., CHANG C.C., WADE M.H., TROSKO J.E.: Elevated spontaneous mutation rate in Bloom syndrome fibroblasts. *Proc. Natl. Acad. Sci. USA* 78:3133-3137, 1981.

178. WASSERMANN H.P., FRY R., COHN H.J.: Fanconi's anaemia: Cytogenetic studies in a family. *S. Afr. Med. J.* 42:1162-1165, 1968.
179. WEBB T., HARDING M.: Chromosome complement and SV40 transformation of cells from patients susceptible to malignant disease. *Br. J. Cancer* 36:583-591, 1977.
180. WEEMAES C.M.R., HUSTINX T.W.J., SCHERES J.M.J.C., VAN MUNSTER P.J.J., BAKKEREN J.A.J.M., TAALMAN R.D.F.M.: A new chromosomal instability disorder: The Nijmegen breakage syndrome. *Acta Paediatr. Scand.* 70:557-564, 1981.
181. WEST J., LYTTLETON M.J., GIANNELLI F.: Effect of incubation temperature on the frequency of sister chromatid exchanges in Bloom's syndrome lymphocytes. *Hum. Genet.* 59:204-207, 1981.
182. WOLFF S., BODYCOTE J., THOMAS G.H., CLEAVER J.E.: Sister chromatid exchange in xeroderma pigmentosum cells that are defective in DNA excision repair or postreplication repair. *Genetics* 81:349-355, 1975.
183. WOLFF S., RODIN B., CLEAVER J.E.: Sister chromosome exchanges induced by mutagenic carcinogens in normal and xeroderma pigmentosum cells. *Nature* 265: 347-349, 1977.
184. WOLMAN S.R., SWIFT M.: Bone marrow chromosomes in Fanconi's anaemia. *J. Med. Genet.* 9:473-474, 1972.
185. WUNDER E., BURGHARDT U., LANG B., HAMILTON L.: Fanconi's anemia: Anomaly of enzyme passage through the nuclear membrane? *Hum Genet.* 58:149-155, 1981.
186. YOSHIDA M.C.: Suppression of spontaneous and mitomycin C-induced chromosome aberrations in Fanconi's anemia by cell fusion with normal human fibroblasts. *Hum. Genet.* 55:223-226, 1980.
187. ZAIZOV R., MATOTH Y., MAMON Z.: Long-term observations in children with Fanconi's anemia. *In* Hibino S., Takaku F., Shahidi N.T. (eds.): "Aplastic Anemia." Baltimore: University Park Press, pp. 243-251, 1978.
188. ZAKRZEWSKI S., SPERLING K.: Antagonistic effect of cocultivation on mitomycin C-induced aberration rate in cells of a patient with Fanconi's anemia and in Chinese hamster ovary cells. *Hum. Genet.* 56:85-88, 1980.

SISTER-CHROMATID EXCHANGE – THE PHENOMENON AND ITS RELATIONSHIP TO CHROMOSOME-FRAGILITY DISEASES

SAMUEL A. LATT, RHONA R. SCHRECK, CHARLOTTE P. DOUGHERTY, KAREN M. GUSTASHAW, LOIS A. JUERGENS, AND TIM N. KAISER

INTRODUCTION

Sister-chromatid exchanges are reciprocal interchanges between DNA-replication products. These exchanges, referred to as SCEs, typically are scored in metaphase chromosomes. Except in the case of ring chromosomes, SCE detection requires some means of sister-chromatid differentiation. This was initially accomplished by Taylor et al. [123], by use of tritiated thymidine (^{3}HdT). More recently, BrdU-dye techniques [56,64,67,95] have greatly simplified detection of SCEs (Figs. 1 and 2) and have stimulated research employing SCE analysis.

Interest in SCEs has centered on their use in the detection of mutagenic-carcinogen effects on cells [e.g., 57,72,73,93,138] and in the characterization of chromosome-fragility diseases. However, although SCEs are easy to measure, they have proved difficult to understand at a molecular level. Applications of SCE analysis thus have been largely empirical, whereas mechanistic studies have focused primarily on correlations between SCEs and other events. The present chapter will summarize data that characterize the SCE phenomenon. Particular attention will be given to new understanding derived from analysis of SCE formation in cells from patients with chromosome-fragility diseases.

BASIC FEATURES OF THE SCE PHENOMENON

The fundamental observation made possible by sister-chromatid differentiation is that newly replicated DNA is distributed between sister

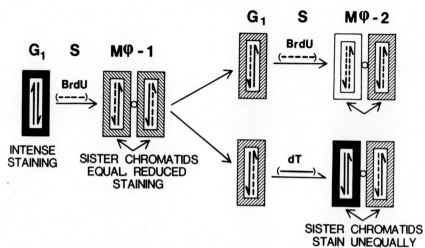

Fig. 1. Sister-chromatid differentiation by BrdU-dye techniques. Cells are allowed to incorporate BrdU (---) for one cycle, followed by a second cycle of replication in which the presence of BrdU is optional. Sister chromatids in metaphase chromosomes from such second division cells will exhibit unequal fluorescence, if stained, e.g., with 33258 Hoechst, or unequal intensity following Giemsa staining, reflecting different numbers of BrdU-substituted polynucleotide chains. Solid, hatched, and open areas surrounding each rectangle represent intense, intermediate, and pale staining, respectively.

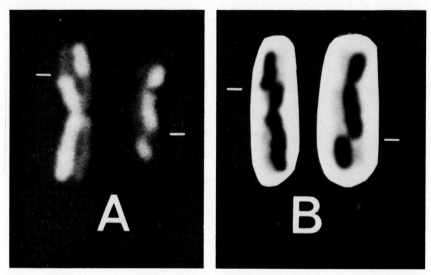

Fig. 2. Sister-chromatid exchanges (SCEs). The chromosomes in this figure are from human lymphocytes which replicated twice in medium containing 10^{-5} M BrdU, 6×10^{-6} M U, and 4×10^{-7} M FdU. Those in panel A were stained with 33258 Hoechst and photographed under conditions described for fluorescence microscopy. Those in panel B were exposed to fluorescent light while mounted in buffer containing 10^{-4} M 33258 Hoechst, incubated in $2 \times$ SSC, and stained with Giemsa [95]. SCEs are indicated by white lines.

TABLE I. Factors That Can Influence SCE Induction

Type and amount of DNA damage
Metabolism of agent tested
SCE test system itself, e.g., BrdU
DNA replication and repair in the target system
Cell-cycle position of the target cells

chromatids in a semiconservative manner [123]. This cytological characterization of DNA replication was soon corroborated and extended at a molecular level by the elegant experiments of Meselson and Stahl [86]. Taylor [122] also demonstrated that SCE formation obeyed constraints expected for an interchange between DNA molecules that conserved polynucleotide-chain polarity. This result was subsequently confirmed by experiments employing BrdU-dye methodology [125,137].

Recent studies have illustrated conditions that can alter the SCE frequency and have provided information about the spatial and temporal distribution of SCEs (Table I). Most, if not all, of the SCEs observed in chromosomes labeled with ^{3}HdT [42] or BrdU [36] are induced by the agent employed for sister-chromatid differentiation. The baseline SCE frequency in human tissue usually falls in the range of 10 ± 5 per cell; values in other material are comparable, especially when normalized for cellular DNA content. BrdU induction of SCEs has been shown to involve at least two effects, one due to substitution of BrdU into DNA [22, 65,85] and another, of perhaps greater magnitude, due to high concentrations of BrdU, which presumably perturb nucleotide metabolism [22].

INDUCTION OF SCEs BY EXOGENOUS AGENTS

Much of the interest in SCEs derives from the ability of a large number of agents, primarily mutagenic carcinogens [94], to increase SCE frequencies to levels appreciably above the BrdU-induced baseline (Table II). Alkylating agents are especially effective. SCE frequencies of more than 100 per cell are easily produced and scored. Beyond this amount, toxic effects due to the inducing agent become evident, and resolution of closely spaced exchanges becomes difficult. The existence of an upper limit to SCE formation, which would reflect a fundamental participating

TABLE II. Types of Mutagenic-Carcinogens Exhibiting Strong SCE Induction

Type	Example
Alkylating agents	EMS, MMC, 4-NQO
DNA-binding dyes	Daunorubicin
Base analogues	BrdU
Irradiation	BrdU + light, UV, x-ray, ^{3}H (intrinsic)
Viruses	SV40
Miscellaneous agents	Acetaldehyde

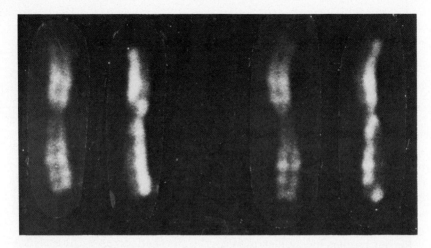

Fig. 3. SCEs in human chromosome 1. Chromosomes were prepared from peripheral leukocytes obtained from normal human subjects and grown 70-72 hours in medium containing 0.01 or 0.02 mM BrdU. Chromosomes were stained with quinacrine (left-hand member of each pair), then destained and restained with 33258 Hoechst (right-hand member of each pair) [66]. SCEs are evident as abrupt reciprocal alternations in 33258 Hoechst fluorescence along chromatids.

unit, e.g., a cluster of replicons or even a region corresponding to a metaphase chromosome band, has not yet been excluded. Suspicion of such a structural constraint on SCE formation is reinforced by observations [20, 44,66] that SCEs occur preferentially between quinacrine-bright bands, perhaps at the junction of quinacrine-bright and quinacrine-dull bands (Fig. 3). The basis for such a preferential chromosomal location of SCEs is unknown. One plausible hypothesis is that interband junctions are the sites of special DNA sequences that mediate SCE formation.

Most of the agents that are effective at inducing SCEs interact in some manner with DNA. More than 150 agents have been tested for SCE induction, with positive results observed for more than half [68]. Analysis of the correlation between SCE induction and carcinogenesis, though of obvious interest, has been limited primarily by the paucity of carcinogenicity data. However, results thus far [68] suggest that such a correlation may prove to be high. Other factors that can alter SCE frequencies include an elevated BrdU concentration (see above), certain steroid hormones [11], and the presence of inhibitors of poly(ADP-ribose) synthesis [127]. However, in contrast to results of early studies [59], the tumor promotor tetradecanoyl-phorbol-acetate (TPA) does not appear to be highly effective at SCE induction [32,80,88,97,124]. It is possible that SCE induction may be a common endpoint for a variety of processes that affect cells.

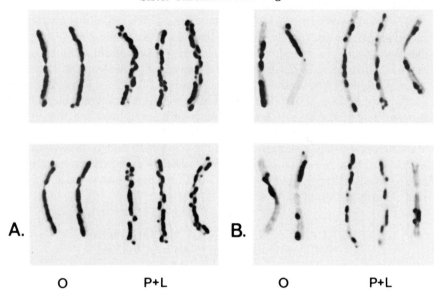

Fig. 4. Induction of SCE in second- and third-division metaphases of hamster cells. CHO cells synchronized at the G1-S boundary were treated with 8-methoxypsoralen (6×10^{-6} M) plus light (1.1×10^4 ergs/mm^2, mainly 365 nm) and were then allowed to replicate in BrdU (2.5×10^{-5} M) for (A.) 32 or (B.) 51 hours. Each set of five chromosomes consists of two controls (at the left) and three from cells treated before transfer to medium containing BrdU (at the right) [71].

Exposure of cells to 8-methoxypsoralen plus near ultraviolet light, a combination highly effective at SCE induction (Fig. 4), can be used to characterize the SCE response [70]. For example, cells treated in this manner can exhibit SCE elevations for at least three cycles after reaction with DNA (Fig. 4; see also [52]), which suggests that SCEs may serve to reflect persistent DNA damage. By use of tritiated 8-methoxypsoralen, it was possible to determine that as many as 200 8-methoxypsoralen-DNA monoadducts and crosslinks were produced for each SCE detected [15]. The ratio of DNA damage by ultraviolet light [100] or mitomycin C [110] to SCE formation is even greater. Thus, although SCEs may in some instances be more frequent than chromosome aberrations, they still do not account for all of the DNA damage produced in cells. Rather, the frequencies of SCEs and chromosome breaks each can vary within a range of several orders of magnitude, with the number of lesions in the DNA serving as a likely upper boundary for both. DNA alkylation tends to foster SCE formation, whereas agents such as bleomycin that produce both single- and double-strand DNA breaks [21] are more likely to cause chromosome aberrations. The ratio of SCEs to structural aberrations can thus vary widely, depending on the specific conditions to which cells are exposed.

The relative efficiency of SCE induction by DNA crosslinks and DNA monoadducts is not totally clear. Both mitomycin C and 8-methoxypsoralen-plus-light can crosslink DNA. However, the monofunctional derivative of mitomycin C, decarbamoyl-mitomycin C, is at least as effective at inducing SCEs as is the parent compound [14]. Also, the results of a study employing controlled production of 8-methoxypsoralen-DNA monoadducts [103] by a single brief pulse of laser light [53] indicate that these monoadducts can account for most of the SCEs produced by 8-methoxypsoralen, although SCE induction by DNA crosslinks was not excluded. Heddle and associates (this volume) have performed experiments implicating 8-methoxypsoralen-DNA crosslinks in SCE induction. This question is still under investigation.

RELATIONSHIP OF SCE FORMATION TO THE DNA-REPLICATION FORK

SCE induction by a variety of agents appears to occur at or near the DNA-replication fork. Wolff et al. [136] demonstrated that exposure of rodent cells to ultraviolet light led to elevated SCE frequencies only if DNA replication intervened between irradiation and metaphase-chromosome harvest. More precise timing of SCE induction, within S phase, was possible by use of synchronized Chinese hamster ovary (CHO) cells and 8-methoxypsoralen plus carefully timed pulses of near ultraviolet light [70]. The conclusion of this study was that there is a regional requirement for DNA replication after DNA damage in order for SCE induction to occur. More recent experiments with synchronized CHO cells, which employed BrdU incorporation plus near ultraviolet light, support this conclusion [74]. Painter [91] recently has hypothesized that SCE formation occurs at junctions between clusters of replicons. It should now prove rewarding to attempt to examine SCE formation in terms of details of events at the DNA-replication fork (see [25] for a review).

CORRELATES OF SCE FORMATION

SCE formation, whatever its molecular details, can be correlated with a number of events of biological significance. For example, SCE induction by different agents has been shown to be proportional to mutagenesis [13,14,22,105]. However, the ratios of SCEs to mutation at one locus (that for hypoxanthine-guanosine phosphoribosyl transferase) can vary over at least a 100-fold range, depending on the agent employed, and wider variations in this ratio might be expected under different conditions. Strict proportionality between SCE induction and mutagenesis may not be universal, however (see, e.g., [105]). Data presented thus far do not differentiate between two extremes: one in which SCE formation leads to mutation, and the other in which SCEs and mutations reflect divergent consequences of an earlier event, such as DNA damage. This latter model also may account for the correlation between SCE induction and the induction of viruses from transformed cells [55,141], even though the details of these two processes probably differ greatly. The large ratio between the numbers and sites of DNA damage and SCE formation provides ample opportunity for many events to correlate with SCEs without their being connected to SCEs by a causative relationship.

SCE FORMATION IN CHROMOSOME-FRAGILITY DISEASES

Additional insight about the SCE process can be derived from studies of cells from patients with diseases that are characterized by chromosome fragility and, or, a defect in DNA repair. These diseases include Bloom's syndrome, xeroderma pigmentosum, ataxia-telangiectasia, and Fanconi's anemia [38,115]. Some type of relationship between SCEs and these diseases was anticipated [64], because both involve DNA breakage. Alterations in SCE formation have been observed in some of these conditions, although the underlying mechanisms are unclear.

Bloom's Syndrome

The cytogenetic hallmark of Bloom's syndrome (BS) is increased chromosome exchange. German [37] described an increase in quadriradial figures (QRs) which involved homologous chromosomes in cells from patients with BS and interpreted this as a reflection of increased interhomolog somatic exchange. Ten years later, German and associates [16] utilized BrdU-dye techniques both to detect a tenfold increase in SCE formation in lymphocytes from patients with BS and to demonstrate directly the interchange of homologous chromatids during the course of QR formation. QRs formed as a response of cells to mitomycin C exposure also were shown to involve interchromosome exchange in somatic cells [51].

Studies with BS cells have indicated similarities between interhomolog and intrahomolog (sister) chromatid exchange. Both SCEs and interhomolog QRs are increased in BS fibroblasts and in most BS lymphocytes. The absence of an SCE elevation in a subset of lymphocytes from some BS patients [40] and in at least one lymphoblast line derived from a patient with Bloom's syndrome (Ray et al., this volume) is accompanied by a normal QR frequency. Santesson et al. [106] have suggested that the cells with fewer SCEs may be B-type lymphocytes. This interesting hypothesis has not yet been fully tested. Kuhn [62,63] observed that the location of points of exchange in QRs in BS cells is either within quinacrine-dull bands or at the junctions of quinacrine-dull and quinacrine-bright chromosome bands. This is precisely the localization previously observed for SCEs. Such a coincidence prompts the speculation that specific DNA sequences, located at such chromosomal positions, might facilitate or mediate the formation of both QRs and SCEs.

The basis for the elevated SCE frequency in BS remains unknown. Hand and German [45,46] have observed a retardation in replication fork progression in this condition, ranging from 20 to 40% of normal, in fibroblasts and lymphocytes, respectively, and it is tempting to link this observation with current ideas that place the site of SCE formation at the replication fork. Ockey [90] has confirmed the existence of retarded replication fork movement at low cell densities, even though at cell densities higher than those employed by Hand and German for autoradiographic work or those used for SCE analysis, replication-fork progression did not differ significantly from normal. Whatever its underlying basis, the elevated SCE frequency in BS cells appears to be associated with an elevated mutation frequency [132] and so attests to the correlation between SCEs and mutational events.

Inasmuch as most if not all baseline SCEs in normal cells are induced by BrdU, it is tempting to suspect that the elevated baseline SCE frequency in BS reflects a hypersensitivity to BrdU. However, this point has not been tested adequately. It is known that stimulation of SCE formation above baseline levels is significantly greater in BS than in normal cells. This has been most thoroughly demonstrated using the agent ethylmethane sulfonate (EMS) and either lymphocytes [60] or fibroblasts [61], and it has been observed to a lesser extent in cells treated with mitomycin C [74,117]. Giannelli et al. [41] described an accentuated reduction in survival in BS cells after ultraviolet irradiation, although they did not report SCE data. It is worth noting that treatment of cells with either alkylating agents or ultraviolet light can retard DNA replication-fork progression [18,114], and it is at least plausible to suspect that superposition of this effect upon the already retarded replication-fork progression in BS might somehow relate to the exaggerated SCE response.

Perhaps the most promising hint about the mechanism of SCE elevation in BS has been obtained from studies in which BS cells have somehow been mixed with normal cells. Bryant et al. [10] used cell fusion to effect this admixture and observed a normalization of SCE frequency in the BS chromosomes without an elevation of SCEs in control chromosomes. Alhadeff et al. [1] obtained similar results upon fusing BS cells with CHO cells. The simplest interpretation of these results is that there is a *trans*-acting agent, found in non-BS cells, that normalizes the SCE frequency in BS cells. Co-cultivation studies, either limited to human material [7,102] or mixing human and rodent cells [128], generally support this view, although the extent of correction is less than that observed after cell fusion. As further evidence in favor of *trans*-acting correction, Rüdiger has observed that BS heterozygote cells are less effective than are normal cells in correcting SCE frequencies during co-cultivation.

In all of the above studies, the "correcting" cell line itself did not experience any cytogenetic alterations. Even when non-BS cells were metabolically coupled to the BS cells via gap junctions, by using correction of HGPRT deficiency as a selective force, no clastogenic effect that emanated from the BS cells was detected [111]. However, Tice et al. [126] did observe an increase in SCEs in normal cells that were exposed to medium used to grow BS fibroblasts, whereas Emerit and Cerutti [29] observed a similar effect, which was reversible by free radical scavengers, on chromosome aberrations (though not SCEs). These latter studies might have detected agents, capable of damaging DNA, which are inactivated in normal cells. In this case, *trans*-correction might be an event, not necessarily specific to BS, that is associated with neutralization of these agents. Alternatively, there might be a macromolecule of which a deficiency in BS leads to enhanced SCE formation. By use of SCE frequency as an assay, it should be possible to purify the correcting molecule and answer these questions.

Xeroderma Pigmentosum

Xeroderma pigmentosum (XP), like Bloom's syndrome, is associated with a predisposition to neoplasia [19]. Chromosome aberrations are not increased in XP cells, even though a defect in the repair of DNA damage by ultraviolet light is thought to underlie this disease. SCE studies in XP have indicated the effectiveness with which unremoved DNA damage, as opposed to the process of damage removal, can induce SCEs. Ultraviolet (UV) irradiation of lymphocytes [6,112], lymphoblasts [17], or fibroblasts [26] from the most prevalent complementation groups of XP leads to a greater induction of SCEs than does irradiation of normal cells. A notable exception to this exaggerated SCE response is the "postreplicational repair variant" [76] of XP. In contrast to SCE induction, induction of mutations by UV light has been described as elevated in both classic and variant XP complementation groups [82–84]. Results with the variant cells constitute a departure from the correlation between SCEs and mutations. Alkylating agents such as ethylnitrosourea lead to an exaggerated SCE response in one SV40-transformed line of group A, XP cells: XP12RO [138]. Removal of O^6 alkyl groups in this same line of XP cells is reduced [43]; this is consistent with the idea that it is unremoved DNA damage, such as O^6 alkylation, that leads to SCE induction. This observation complements that of a reduced SCE response of cells that have been pretreated with multiple low doses of N-methyl-N'-netro-N-nitrosoguanidine (MNNG) to subsequent alkylation [104], presumably because of induction of increased removal of O^6 alkylation. Heddle and Arlett [48] have presented data suggesting that the SCE induction by alkylating agents in the XP12RO cells might reflect primarily the effects of SV40 transformation and not the defects of the underlying disease. However, this qualification does not remove the correlation between the persistence of DNA damage and the induction of SCEs. In fact, Day et al. [23] recently have shown that a reduction in the ability to excise O^6 methyl guanine is a property of many cells (designated "mer-" cells) [24], some (but not all) of which are SV40-transformed. Reduced ability to support the growth of MNNG-treated adenovirus 5, and greater sensitivity to killing by, and greater sensitivity to SCE induction by MNNG, are common features of such cells.

Ataxia-Telangiectasia

Ataxia-telangiectasia (AT) is not associated with any characteristic change in SCE formation. Both the baseline SCE frequency [35,47] and the induction of SCEs by x-rays, adriamycin, or mitomycin C [34] are the same as the rates in normal cells. Information about the underlying defect in this disease thus serves to identify reaction pathways which probably do not play an important role in the SCE process.

Cells from patients with AT are unusually sensitive to x-rays [120] and to bleomycin [78,121], agents that are relatively more effective at inducing chromosome breaks than SCEs. X-irradiation under anoxic conditions produces gamma endonuclease-sensitive damage that is removed poorly by AT cells [92]. Bleomycin is known to be capable of producing double-strand DNA breaks [21], as can x-irradiation [77], and it is possible that repair of this damage also may be defective in AT.

Chemical evidence for decreased DNA-break repair has not been obtained in AT [119,131], although, as suggested by Taylor [119], only a small fraction of unrepaired breaks could account for the biological effects. X-ray-induced mutagenesis is reduced in AT [2], as is unscheduled DNA synthesis induced by γ-irradiation [130], and some impairment of the repair of DNA damage due to high-energy radiation is probable. Reports that AT cells are hypersensitive to MNNG [113] or 4-nitroquinoline oxide (4NQO) [118] suggest that the DNA-repair defect in AT may be more generalized. Edwards and Taylor [28] recently have reported that the usual reduction in DNA synthesis after DNA damage is not observed in AT lymphoblasts, and poly(ADP-ribose) synthesis is not stimulated. It will be important to determine whether the effect on poly(ADP-ribose) synthesis is primary or secondary to a defect in repair at the DNA level. One report, by Shaham et al. [116], describes a diffusible clastogenic factor produced by AT cells. At present, the primary conclusion regarding SCEs that can be drawn from studies of

TABLE III. Effect of Mitomycin C (MMC) on SCE Formation in Fibroblasts From Patients With Fanconi's Anemia*

	0 MMC	10 ng/ml MMC
Patients (8)†	9.5 (8.0–11.0)	23.4 (10.0–34.9)
Parents (4)	8.2 (6.5–10.2)	32.8 (27.6–40.1)
Controls (5)‡	8.0 (7.0–09.9)	33.8 (26.4–41.5)

*Cells were cultured in medium containing (usually 2.5×10^{-5} M) BrdU and dC (10^{-4} M); mitomycin C, when added, was present during an interval corresponding approximately to the second replication cycle of cells scored. Data given are the average SCE frequency for all samples examined, together with the range of the averages for individual samples.
†Includes data on three cell lines obtained from cell repositories.
‡Material from one control was examined on two separate occasions and the data from each entered as separate values.

AT cells is that lesions that lead to DNA breakage need not also stimulate SCE.

Fanconi's Anemia

The fourth major autosomal recessive disease associated with cytogenetic abnormalities and an increased risk for neoplasia is Fanconi's anemia (FA). The hallmark of FA is hypersensitivity to agents capable of inducing DNA crosslinks. This is evident as reduced cell survival [107], increased chromosome breakage [5,8,75,109], or micronucleus formation [49] after exposure to potentially bifunctional alkylating agents. Chemical evidence for reduced DNA crosslink removal in FA has not been obtained consistently [31,33,58,108]. This may reflect a vast excess of total DNA crosslinks compared with those resulting in cell death or chromosome aberrations, and it might also be due to interindividual genetic heterogeneity within the clinical entity FA. It is not clear to what extent the isolated reports of increased sensitivity of FA cells to high doses of UV light [96] or high-energy radiation [98] can be related to the increased sensitivity of such cells to DNA crosslinks.

Similarly, data on SCE formation in FA are mixed. Baseline SCE frequencies are normal [16,75]; but mitomycin C- or nitrogen mustard-induced SCE formation in lymphocytes has been reported to be depressed [75], normal [89], or elevated [9]. Suppression of SCE induction by mitomycin C was appreciable in lymphocytes but minimal, compared with intersample variation, in fibroblasts [75] (Table III). In contrast, mitomycin-C treatment led to significantly greater numbers of chromosome aberrations in both lymphocytes and fibroblasts (Tables IV,V) from patients with FA, as compared with those cells from controls (Tables IV,V). Exaggerated chromosome breakage in response to another crosslinking agent, diepoxybutane, has been used by Auerbach et al. [3,4] as a method

TABLE IV. Effect of Mitomycin C (MMC) on Chromosomal Aberrations in Peripheral Lymphocytes From Patients With Fanconi's Anemia*

	0 MMC	30 ng/ml MMC
Patients (10)	0.22 (0.02–0.70)†	1.34 (0.85–2.34)
Parents (6)	0.02 (0.00–0.05)	0.01 (0.00–0.02)
Controls (10)	0.01 (0.00–0.03)	0.04 (0.00–0.13)

*Cells were cultured as described in the legend to Figure 3. MMC, when added, was present for the third and final day of culture.
†Mostly chromatid breaks; also quadriradials and other aberrations. Table entries are averages per cell, with the range in different patients given in parentheses.

TABLE V. Effect of Mitomycin C (MMC) on Chromosome Aberration Formation in Fibroblasts From Patients With Fanconi's Anemia*

	0 MMC	10 ng/ml MMC
Patients (8)	0.10 (0.00–0.21)	2.49 (0.82–8.15)
Parents (4)	0.04 (0.00–0.07)	0.10 (0.00–0.21)
Controls (5)	0.01 (0.00–0.04)	0.08 (0.00–0.17)

*Material and data presentation as described in the footnotes to Table III.

that is both simpler and more reliable than SCE analysis for the diagnosis of FA.

Although the yield of metaphases in FA blood lymphocyte cultures is low [39,134], the SCE results that were reported initially with just four patients appear to be reproducible, for they were found in six additional patients (Fig. 5) in which the diagnosis of FA was supported by clinical data (Table VI). The results are at least empirically consistent with the studies of Finkelberg et al. [30], which showed that mutagenesis of FA cells treated with mitomycin C was less than that of normal cells. In view of the data of Carrano et al. [14], which suggest that SCEs induced by mitomycin C might actually be due to DNA monoadducts rather than to crosslinks, it is entirely possible that the formation of chromosome aberrations in FA lymphocytes after exposure to mitomycin C may be due to lesions (crosslinks) different from those that produce SCEs.[1] If this is true, then the results observed might be very sensitive to such procedural details as the duration of cell culture and harvest, and perhaps [9], to additional agents (e.g., FUdR) which were used in one study [75] but not in the others.

---

[1] It is important to note that more than one-third of the chromatid breaks formed in response to mitomycin C occur at sites of SCE (Fig. 6), termed incomplete SCEs [75] – as do many of the breaks formed in FA chromosomes that are exposed only to BrdU [27]. Thus, in this condition the events of chromatid exchange and breakage may not be entirely independent.

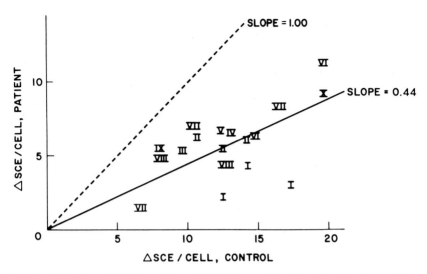

Fig. 5. Induction of SCEs by mitomycin C in human blood lymphocytes from patients with FA. Lymphocytes from patients and from control individuals were cultured for 3 days in the presence of 1.2×10^{-5} M BrdU, 10^{-4} M dC, 4×10^{-7} M FdU, and 6×10^{-6} M U. Exposure to mitomycin C (0.03 μg/ml) occurred during the last day of culture. Plotted on the Y axis are SCE increments of patient's cells, compared with cultures not exposed to mitomycin C. The X axis for each point is the SCE increment due to 0.03 μg/ml mitomycin C in a parallel (and similarly treated) culture of control lymphocytes. Roman numerals, positioned at these X, Y coordinates, refer to individual patients. For some patients, multiple data points are shown. The least squares, best-fit, straight line through the data and the origin has a slope of 0.44. For comparison, a line with a slope of 1.0 also is shown.

TABLE VI. Clinical Features in Patients With Fanconi's Anemia*

Patient	Sex	Pancytopenia	Radial anomalies	Renal anomalies	Hearing deficit	Skin hyperpig- mentation	Short stature
I†	M	+	+		+	+	+
II†	M	+	−	+		+	+
III	M	+	−	−		+	+
IV‡	M	+	+	+	+	−	+
V‡	M	+	+	−	+	+	+
VI	F	+	−	−	−	+	−
VII	M	+	−	+	−	−	+
VIII§	M	+	+	−	+	−	+
IX	M	+				+	+
X‡	F	+	+	+			+

*+, present; −, absent; blank, no information.
†Sibs.
‡One sib affected.
§Parents are first cousins.

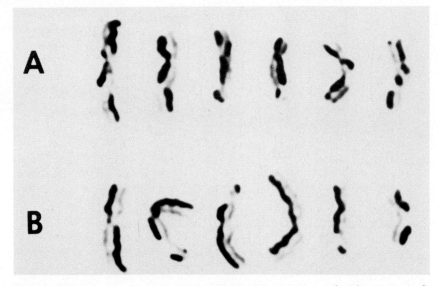

Fig. 6. Aberrations in chromosomes from FA lymphocytes treated with mitomycin C. These chromosomes were selected from chromosomes observed in cells grown in medium containing 10^{-5} M BrdU, 10^{-4} M dC, 4×10^{-7} FdU, and 6×10^{-6} M U. Mitomycin C (0.03 µg/ml) was added for the third (and final) day of culture. Slides were stained first with 33258 Hoechst, subjected to intense illumination required for fluorescence photomicroscopy, washed, stained with Giemsa, and rephotographed. Chromosomes in row A exhibit chromatid breaks at sites of incomplete SCE while those in row B exhibit breaks away from such sites [69].

Interpretation of the differences between the results in various studies of SCE formation in FA is incomplete. The report [89] that described a normal SCE response in FA was unusual in that increased chromosomal breaks were not found except at very high mitomycin C doses. The most recent study, which described SCE hyperreactivity by FA lymphocytes to nitrogen mustard [9], was based on SCE increments that were of the same magnitude as the statistical errors in their measurement. Also, the absolute SCE increment described for normal cells treated with nitrogen mustard seemed low. Subsequent research has indicated that resolution of these questions may require studies of interphase, rather than metaphase, cells.

Kaiser et al. [54] have employed flow cytometry to detect and characterize the sensitivity of FA cells to mitomycin C. FA fibroblasts, 24 hours after treatment with mitomycin C, exhibit DNA histograms with an exaggerated accumulation of cells in the G2 + M peak. Very few of the cells sorted from this peak were in metaphase, which suggests that mitomycin C caused an accumulation of cells in G2, or perhaps in very late S.

In contrast to the response to mitomycin C, the G2+M peak accumulation of FA fibroblasts that were exposed to ethylmethane sulfonate, a monofunctional alkylating agent, did not differ significantly from normal.

The basis for accumulation of FA cells in G2 is unknown. If mitomycin C merely blocked DNA synthesis, cells might be expected to accumulate throughout S phase. G2 accumulation of plant cells has been found to be due to the presence of DNA breaks between clusters of replicons [129], but analogous data on FA cells are not yet available. It also is not known whether the isolated observation of decreased DNA-ligase activity [50] in cells from one patient with FA can be related to the cell-cycle data described above. Whatever the underlying mechanism, it is the exaggerated block of FA cells treated with mitomycin C just before mitosis that accounts for the difficulty in characterizing the response of these cells to this agent by the analysis of metaphase chromosomes. Such studies sample a very small subset of the cell population; i.e., the cellular response to be examined actually thwarts the cytological method commonly used for its detection. This is not the case for DNA-flow histogram analysis, which might well find a role in detecting and characterizing FA cells. Analogous flow cytometric studies of other conditions such as AT might be possible, through the use of different agents (e.g., x-rays, bleomycin) as selective cell perturbants.

An exciting recent development in the analysis of FA is the cross-correction of mitomycin C-treated cells either by fusion with normal fibroblasts [139] or co-cultivation. The co-cultivation studies [140] employed CHO cells; the CHO cells underwent an increase in chromosome aberrations, but they caused a decrease in aberrations in the FA cells. These results can be interpreted as indicative of a clastogenic factor produced by mitomycin C-treated FA cells, but as yet they do not permit conclusions to be drawn about the relationship of this clastogenic activity to the presumptive defect in response to DNA crosslinks in this disease.

MOLECULAR ANALYSIS OF SCE FORMATION

The accumulating data about cells from patients with chromosome-breakage diseases help to characterize the SCE process, but definitive analysis probably will require information at a molecular level. Biochemical studies of these conditions might prove especially useful in characterizing the relationship between the number of sites of DNA damage and the considerably smaller numbers of cytological events. A chemical assay for SCE formation would be invaluable for such work. However, DNA in CHO and human cells with a density somewhat greater than that of mainband DNA [79,81] has thus far confounded attempts [87,99,

101,133] to detect DNA sequences at the site of SCE. Perhaps recombinant DNA technology can provide a means to amplify these latter sequences so that they can be studied. Also, DNA-cloning studies might prove useful for examining possible relationships between SCE formation and DNA-sequence transposition (e.g., [12]). The enzymology of SCE formation thus far has been obscure, but the existence of *trans*-acting factors that correct the cytological behavior of diseased cells ultimately might provide a way to examine the sequence of steps resulting in SCE formation.

BIOLOGICAL SIGNIFICANCE OF SCE FORMATION

Finally, there remains the question of the biological significance of SCE formation. The SCE process is widespread, observed in numerous animal and plant species, and it seems reasonable to link its evolutionary conservation to some function of biological importance. Alternatively, SCE formation might utilize enzymatic steps that are preserved for other purposes, e.g., meiotic interchange. Resolution of this question will be necessary in order to know whether SCEs have intrinsic importance to cells or whether they serve instead as an empirically useful but indirect signature of one type of cellular response to DNA damage.

Acknowledgments. This chapter includes illustrative and tabular material from previous, referenced publications. The research is supported by grants from the National Institutes of Health (GM 21121), the American Cancer Society (CD-36G), and the National Foundation – March of Dimes (1-353). We are indebted to Drs. Murray Feingold, Joel Rappaport, Samuel Lux, Blanche Alter, Morris Medalie, and William A. Robinson for their help in patient referral.

LITERATURE CITED

1. ALHADEFF, B., VELIVASAKIS, M., PAGAN-CHARRY, I., WRIGHT, W.C., SINISCALCO, M.: High rate of sister chromatid exchanges of Bloom's syndrome chromosomes is corrected in rodent human somatic cell hybrids. *Cytogenet. Cell Genet.* 27:8–23, 1980.
2. ARLETT, C.F., HARCOURT, S.A.: Cell killing and mutagenesis in repair-defective human cells. *In* DNA Repair Mechanisms (Hanawalt, P.C., Friedberg, E.C., Fox, C.F., eds.), New York: Academic Press, 633–636, 1978.
3. AUERBACH, A.D., ADLER, B., CHAGANTI, R.S.K.: Pre- and postnatal diagnosis and carrier detection of Fanconi's anemia by a cytogenetic method. *Pediatrics* 67:128–135, 1981.
4. AUERBACH, A.D., WARBURTON, D., BLOOM, A.D., CHAGANTI, R.S.K.: Prenatal diagnosis of the Fanconi anemia gene by cytogenetic methods. *Am. J. Hum. Genet.* 31: 77–81, 1979.
5. AUERBACH, A.D., WOLMAN, S.R.: Susceptibility of Fanconi's anemia fibroblasts to chromosome damage by carcinogens. *Nature* 261:494–496, 1976.
6. BARTRAM, C.R., KOSKE-WESTPHAL, T., PASSARGE, E.: Chromatid exchange in ataxia telangiectasia, Bloom's syndrome, Werner's syndrome, and xeroderma pigmentosum. *Ann. Hum. Genet.* 40:79–86, 1976.

7. BARTRAM, C.R., RUDIGER, H.W., PASSARGE, E.: Frequency of sister chromatid exchanges in Bloom's syndrome fibroblasts reduced by cocultivation with normal cells. *Hum. Genet.* 46:331-334, 1979.
8. BERGER, R., BERNHEIM, A., LE CONIAT, M., VECCHIONE, D., SCHAISON, G.: Nitrogen mustard-induced chromosome breakage: A tool for Fanconi's anemia diagnosis. *Cancer Genet. Cytogenet.* 2:269-274, 1980.
9. BERGER, R., BERNHEIM, A., LE CONIAT, M., VECCHIONE, D., SCHAISON, G.: Sister chromatid exchanges induced by nitrogen mustard in Fanconi's anemia. Application to the detection of heterozygotes and interpretation of the results. *Cancer Genet. Cytogenet.* 2:259-267, 1980.
10. BRYANT, E.M., HOEHN, H., MARTIN, G.M.: Normalization of sister chromatid exchange frequencies in Bloom's syndrome by euploid cell hybridization. *Nature* 279: 795-796, 1979.
11. BYNUM, G., KRAM, D., DEAN, R., HADLEY, E., MONTICONE, R., BICKINGS, C., SCHNEIDER, E.: Steroid modulation of sister chromatid exchange induction by mitomycin C and UV light. *Environ. Mutagen.* 2:247, 1980.
12. CALOS, M.P., MILLER, J.H.: Transposable elements. *Cell* 20:579-595, 1980.
13. CARRANO, A.V., THOMPSON, L.H., LINDL, P.A., MINKLER, J.L.: Sister chromatid exchange as an indicator of mutagenesis. *Nature* 271:551-553, 1978.
14. CARRANO, A.V., THOMPSON, L.H., STETKA, D.G., MINKLER, J.L., MAZRIMAS, J.A., FONG, S.: DNA crosslinking, sister chromatid exchange, and specific locus mutations. *Mutat. Res.* 63:175-188, 1979.
15. CASSEL, D.M., LATT, S.A.: Relationship between DNA adduct formation and sister chromatid exchange induction by ^{3}H-8-methoxypsoralen in Chinese hamster ovary cells. *Exp. Cell Res.* 128:15-22, 1980.
16. CHAGANTI, R.S.K., SCHONBERG, S., GERMAN, J.: A manyfold increase in sister chromatid exchanges in Bloom's syndrome lymphocytes. *Proc. Natl. Acad. Sci. USA* 71:4508-4512, 1974.
17. CHENG, W.S., TARONE, R.F., ANDREWS, A.D., WHANG-PENG, J.S., ROBBINS, J.H.: Ultraviolet light-induced sister chromatid exchanges in xeroderma pigmentosum and in Cockayne's syndrome lymphocyte cell lines. *Cancer Res.* 38:1601-1609, 1978.
18. CLEAVER, J.E.: DNA repair and its coupling to DNA replication in eukaryotic cells. *Biochim. Biophys. Acta* 516:489-516, 1978.
19. CLEAVER, J.E.: DNA damage, repair systems and human hypersensitive diseases. *J. Environ. Pathol. Toxicol.* 3:53-68, 1980.
20. CROSSEN, P.E., DRETS, M.E., ARRIGHI, F.E., JOHNSTON, D.A.: Analysis of the frequency and distribution of sister chromatid exchanges in cultured human lymphocytes. *Hum. Genet.* 35:345-352, 1977.
21. D'ANDREA, A.D., HASELTINE, W.A.: Specific cleavage of DNA by the antitumor antibiotics neocarzinostatin and bleomycin. *Proc. Natl. Acad. Sci. USA* 75: 3608-3612, 1978.
22. DAVIDSON, R.L., KAUFMAN, E.R., DOUGHERTY, C.P., OUELLETTE, A.M., DIFOLCO, C., LATT, S.A.: Induction of sister chromatid exchanges by bromodeoxyuridine in DNA. *Nature* 284:74-76, 1980.
23. DAY, R.S., ZIOLKOWSKI, C.H.J., SCUDIERO, D.A., MEYER, S.A., LUBINIECKI, A.S., GIRARDI, A.J., GALLOWAY, S.M., BYNUM, G.D.: Defective repair of alkylated DNA by human tumor and SV40-transformed human cell strains. *Nature* 288:724-727, 1980.
24. DAY, R.S., ZIOLKOWSKI, C.H.J., SCUDIERO, D.A., MEYER, S.A., MATTERN, M.R.: Human tumor cell strains defective in the repair of alkylation damage. *Carcinogenesis* 1:21-32, 1980.

25. DePamphilis, M.L., Wassarman, P.M.: Replication of eukaryotic chromosomes: A close-up of the replication fork. *Annu. Rev. Biochem.* 49:627-666, 1980.
26. De Weerd-Kastelein, E.A., Keijzer, W., Rainaldi, G., Bootsma, D.: Induction of sister chromatid exchanges in xeroderma pigmentosum cells after exposure to ultraviolet light. *Mutat. Res.* 45:253-261, 1977.
27. Dutrillaux, B., Couturier, J., Viegas-Pequignot, E., Schaison, G.: Localization of chromatid breaks in Fanconi's anemia, using three consecutive stains. *Hum. Genet.* 37:65-71, 1977.
28. Edwards, M.J., Taylor, A.M.R.: Unusual levels of (ADP-ribose)$_n$ and DNA synthesis in ataxia telangiectasia cells following gamma-ray irradiation. *Nature* 287:745-747, 1980.
29. Emerit, I., Cerutti, P.: Clastogenic activity from Bloom's syndrome fibroblast cultures. *Proc. Natl. Acad. Sci. USA* 78:1868-1872, 1981.
30. Finkelberg, R., Buchwald, M., Siminovich, L.: Decreased mutagenesis in cells from patients with Fanconi's anemia. *Am. J. Hum. Genet.* 29:42A, 1977.
31. Fornace, A.J., Jr., Little, J.B., Weichselbaum, R.R.: DNA repair in a Fanconi's anemia fibroblast cell strain. *Biochim. Biophys. Acta* 561:99-109, 1979.
32. Fujiwara, Y., Kano, Y., Tatsumi, M., Paul, P.: Effects of a tumor promotor and an anti-promotor on spontaneous and UV-induced 6-thioguanine-resistant mutations and sister chromatid exchanges in V79 Chinese hamster cells. *Mutat. Res.* 71: 243-251, 1980.
33. Fujiwara, Y., Tatsumi, M., Sasaki, M.S.: Crosslink repair in human cells and its possible defect in Fanconi's anemia cells. *J. Mol. Biol.* 113:634-649, 1977.
34. Galloway, S.M.: Ataxia telangiectasia: The effects of chemical mutagens and X-rays on sister chromatid exchanges in blood lymphocytes. *Mutat. Res.* 45: 343-349, 1977.
35. Galloway, S.M., Evans, H.J.: Sister chromatid exchange in human chromosomes from normal individuals and patients with ataxia telangiectasia. *Cytogenet. Cell Genet.* 15:17-29, 1975.
36. Gatti, M., Pimpinelli, S., Santini, G., Olivieri, G.: Lack of spontaneous sister chromatid exchange (SCE) in somatic cells of *Drosophila melanogaster*. *Genetics* 91:255-274, 1979.
37. German, J.: Cytological evidence for crossing-over in vitro in human lymphoid cells. *Science* 144:298-301, 1964.
38. German, J.: Genes which increase chromosomal instability in somatic cells and predispose to cancer. *Prog. Med. Genet.* 8:61-101, 1972.
39. German, J., Caskie, S., Schonberg, S.: A simple cytogenetic test for increased mutagen-sensitivity. *J. Supramol. Struct. Suppl.* 2:89, 1978.
40. German, J., Schonberg, S., Louie, E., Chaganti, R.S.K.: Bloom's syndrome. IV. Sister chromatid exchanges in lymphocytes. *Am. J. Hum. Genet.* 29:248-255, 1977.
41. Giannelli, F., Benson, P.F., Pawsey, S.A., Polani, P.E.: Ultraviolet light sensitivity and delayed DNA-chain maturation in Bloom's syndrome fibroblasts. *Nature* 265:466-469, 1977.
42. Gibson, D.A., Prescott, D.M.: Induction of sister chromatid exchanges in chromosomes of rat kangaroo cells by tritium incorporated into DNA. *Exp. Cell Res.* 74:397-402, 1972.
43. Goth-Goldstein, R.: Repair of DNA damage by alkylating carcinogens is defective in xeroderma pigmentosum-derived fibroblasts. *Nature* 267:81-92, 1977.
44. Haglund, U., Zech, L.: Simultaneous staining of sister chromatid exchanges and Q-bands in human chromosomes after treatment with methyl methanesulfonate, quinacrine mustard, and quinacrine. *Hum. Genet.* 49:307-317, 1979.

45. HAND, R., GERMAN, J.: A retarded rate of DNA chain growth in Bloom's syndrome. *Proc. Natl. Acad. Sci. USA* 72:758-762, 1975.
46. HAND, R., GERMAN, J.: Bloom's syndrome: DNA replication in cultured fibroblasts and lymphocytes. *Hum. Genet.* 38:297-306, 1977.
47. HATCHER, N.H., BRINSON, P.S., HOOK, E.B.: Sister chromatid exchanges in ataxia telangiectasia. *Mutat. Res.* 35:333-336, 1976.
48. HEDDLE, J.A., ARLETT, C.F.: Untransformed xeroderma pigmentosum cells are not hypersensitive to sister chromatid exchange production by ethylmethane sulfonate – implications for the use of transformed cell lines and for the mechanism by which SCE arise. *Mutat. Res.* 72:119-125, 1980.
49. HEDDLE, J.A., LUE, C.B., SAUNDERS, E.F., BENZ, R.D.: Sensitivity to five mutagens in Fanconi's anemia as measured by the micronucleus method. *Cancer Res.* 38: 2983-2988, 1978.
50. HIRSCH-KAUFFMANN, M., SCHWEIGER, M., WAGNER, E.F., SPERLING, K.: Deficiency of DNA ligase activity in Fanconi's anemia. *Hum. Genet.* 45:25-32, 1978.
51. HUTTNER, K.M., RUDDLE, F.H.: Study of mitomycin C-induced chromosomal exchange. *Chromosoma* 56:1-13, 1975.
52. ISHII, Y., BENDER, M.: Factors influencing the frequency of mitomycin C-induced sister chromatid exchanges in 5-bromodeoxyuridine substituted human lymphocytes in culture. *Mutat. Res.* 51:411-418, 1978.
53. JOHNSTON, B.H., JOHNSON, M.A., MOORE, C.B., HEARST, J.E.: Psoralen-DNA photoreaction: Controlled production of mono- and diadducts with nanosecond ultraviolet laser pulses. *Science* 197:906-908, 1977.
54. KAISER, T.N., DOUGHERTY, C.P., LOJEWSKI, A.J., LATT, S.A.: Flow cytometric analysis of the effect of mitomycin C in Fanconi's anemia. *Am. J. Hum. Genet.* 31:75A, 1980.
55. KAPLAN, J.C., ZAMANSKY, G.B., BLACK, P.H., LATT, S.A.: Parallel induction of sister chromatid exchanges and infectious virus from SV40-transformed cells by alkylating agents. *Nature* 271:662-663, 1978.
56. KATO, H.: Spontaneous sister chromatid exchanges detected by BudR-labelling method. *Nature* 251:70-72, 1974.
57. KATO, H.: Spontaneous and induced sister chromatid exchanges as revealed by the BudR-labelling method. *Int. Rev. Cytol.* 49:55-97, 1977.
58. KAYE, J., SMITH, C.A., HANAWALT, P.C.: DNA repair in human cells containing photoadducts of 8-methoxypsoralen or angelicin. *Cancer Res.* 40:696-702, 1980.
59. KINSELLA, A.R., RADMAN, M.: Tumor promotor induces sister chromatid exchanges: Relevance to mechanisms of carcinogens. *Proc. Natl. Acad. Sci. USA* 75: 6149-6153, 1978.
60. KREPINSKY, A.B., HEDDLE, J.A., GERMAN, J.: Sensitivity of Bloom's syndrome lymphocytes to ethylmethanesulfonate. *Hum. Genet.* 50:151-156, 1979.
61. KREPINSKY, A.B., RAINBOW, A.J., HEDDLE, J.A.: Studies on the ultraviolet light sensitivity of Bloom's syndrome fibroblasts. *Mutat. Res.* 69:357-368, 1980.
62. KUHN, E.M.: Localization by Q-banding of mitotic chiasmata in cases of Bloom's syndrome. *Chromosoma* 57:1-11, 1976.
63. KUHN, E.M.: Mitotic chiasmata and other quadriradials in mitomycin C-treated Bloom's syndrome lymphocytes. *Chromosoma* 66:287-297, 1978.
64. LATT, S.A.: Microfluorometric detection of deoxyribonucleic acid replication in human metaphase chromosomes. *Proc. Natl. Acad. Sci. USA* 70:3395-3399, 1973.
65. LATT, S.A.: Sister chromatid exchanges, indices of human chromosome damage and repair: Detection by fluorescence and induction by mitomycin C. *Proc. Natl. Acad. Sci. USA* 71:3162-3166, 1974.
66. LATT, S.A.: Localization of sister chromatid exchanges in human chromosomes. *Science* 185:74-76, 1974.

67. LATT, S.A.: Sister chromatid exchanges: Newer methods for their detection. In Sister Chromatid Exchanges, (Wolff, S., ed.), New York: J. Wiley & Son, 17-40, 1982.
68. LATT, S.A., ALLEN, J.W., BLOOM, S.E., CARRANO, A.V., FALKE, E., KRAM, D., SCHNEIDER, E., SCHRECK, R.R., TICE, R., WHITFIELD, B., WOLFF, S.: Sister chromatid exchanges. Mutat. Res. 87:17-62, 1981.
69. LATT, S.A., JUERGENS, L.A.: Determinants of sister chromatid exchange frequencies in human chromosomes. In Population Cytogenetics, Studies in Humans (Hook, E.B., Porter, I.H., eds), New York: Academic Press, 217-236, 1976.
70. LATT, S.A., LOVEDAY, K.S.: Characterization of sister chromatid exchange induction by 8-methoxypsoralen plus near UV light. Cytogenet. Cell Genet. 21:184-200, 1978.
71. LATT, S.A., MUNROE, S.H., DISTECHE, C.M., ROGERS, W.E., CASSELL, D.M.: Uses of fluorescent dyes to study chromosome structure and replication. In Chromosomes Today (de la Chapelle, A., Sorsa, M., eds.), Amsterdam: Elsevier/North Holland Biomedical Press, 27-36, 1977.
72. LATT, S.A., SCHRECK, R.R.: Sister chromatid exchange analysis. Am. J. Hum. Genet. 32:297-313, 1980.
73. LATT, S.A., SCHRECK, R.R., LOVEDAY, K.S., DOUGHERTY, C.P., SHULER, C.F.: Sister chromatid exchanges. In Advances in Human Genetics, Vol. 10 (Harris, H., Hirschhorn, K., eds.), New York: Plenum Publishing, pp. 267-331, 1980.
74. LATT, S.A., SCHRECK, R.R., POWERS, M.M., JUERGENS, L.A., SHULER, C.F., LOVEDAY, K.S., PAIKA, I.J.: Induction of sister chromatid exchanges by mutagens: Modulating factors. In Biological and Population Aspects of Human Mutation (Hook, E., Porter, I., eds.) New York: Academic Press, 191-206, 1981.
75. LATT, S.A., STETTEN, G., JUERGENS, L.A., BUCHANAN, G.R., GERALD, P.S.: Induction by alkylating agents of sister chromatid exchanges and chromatid breaks in Fanconi's anemia. Proc. Natl. Acad. Sci. USA 72:4066, 1975.
76. LEHMANN, A.R., KIRK-BELL, S., ARLETT, C.F., PATERSON, M.C., LOHMAN, P.H.M., DE WEERD-KASTELEIN, E.A., BOOTSMA, D.: Xeroderma pigmentosum cells with normal levels of excision repair have a defect in DNA synthesis after UV-irradiation. Proc. Natl. Acad. Sci. USA 72:219-223, 1975.
77. LEHMANN, A.R., STEVENS, S.: The production and repair of double strand breaks in cells from normal humans and from patients with ataxia telangiectasia. Biochim. Biophys. Acta 474:49-60, 1977.
78. LEHMANN, A.R., STEVENS, S.: The response of ataxia telangiectasia cells to bleomycin. Nucleic Acids Res. 6:1953-1960, 1979.
79. LOVEDAY, K.S., LATT, S.A.: Search for DNA interchange corresponding to sister chromatid exchanges in Chinese hamster ovary cells. Nucleic Acid Res. 5: 4087-4104, 1978.
80. LOVEDAY, K.S., LATT, S.A.: The effect of a tumor promotor, 12-0-tetradecanoylphorbol-13-acetate (TPA), on sister chromatid exchange formation in cultured Chinese hamster cells. Mutat. Res. 67:343-348, 1979.
81. LOVEDAY, K.S., LATT, S.A.: Is there biochemical evidence for sister chromatid exchange formation? Am. J. Hum. Genet. 31:103A, 1979.
82. MAHER, V.M., DORNEY, D.J., HEFLICH, R.H., LEVINSON, J.W., MENDRALA, A.L., MCCORMICK, J.J.: Biological and biochemical evidence that DNA repair processes in normal human cells act to reduce the lethal and mutagenic effects of exposure to carcinogens. In DNA Repair Mechanisms (Hanawalt, P.C., Friedberg, E.C., Fox, C.F., eds.) ICN-UCLA Symposium on Molecular and Cellular Biology, vol. IX, New York: Academic, 717-722, 1978.
83. MAHER, V.M., OUELLETTE, L.M., CURREN, R.D., MCCORMICK, J.J.: Caffeine enhancement of the cytotoxic and mutagenic effect of ultraviolet irradiation in a

xeroderma pigmentosum variant strain of human cells. *Biochem. Biophys. Res. Commun.* 71:228-234, 1976.
84. MAHER, V.M., OUELLETTE, L.M., CURREN, R.D., McCORMICK, J.J.: Frequency of ultraviolet light-induced mutations is higher in xeroderma pigmentosum variant cells than in normal cells. *Nature* 261:593-595, 1976.
85. MAZRIMAS, J.A., STETKA, D.G.: Direct evidence for the role of incorporated BudR in the induction of sister chromatid exchanges. *Exp. Cell Res.* 117:23-30, 1978.
86. MESELSON, M., STAHL, F.: The replication of DNA in E coli. *Proc. Natl. Acad. Sci. USA* 44:671-682, 1958.
87. MOORE, P.D., HOLLIDAY, R.: Evidence for the formation of hybrid DNA during mitotic recombination in Chinese hamster cells. *Cell* 8:573-579, 1976.
88. NAGASAWA, H., LITTLE, J.B.: Effect of tumor promotors, protease inhibitors, and repair processes on X ray-induced sister chromatid exchanges in mouse cells. *Proc. Natl. Acad. Sci. USA* 76:1943-1947, 1979.
89. NOVOTNA, B., GOETZ, P., SURKOVA, N.I.: Effects of alkylating agents on lymphocytes from controls and from patients with Fanconi's anemia. *Hum. Genet.* 49:41-50, 1979.
90. OCKEY, C.H.: Quantitative replicon analysis of DNA synthesis in cancer-prone conditions and the defects in Bloom's syndrome. *J. Cell Sci.* 40:125-144, 1979.
91. PAINTER, R.B.: A replication model for sister chromatid exchange. *Mutat. Res.* 70:337-341, 1980.
92. PATERSON, M.C., SMITH, B.P., LOHMAN, P.H.M., ANDERSON, A.K., FISHMAN, L.: Defective excision repair of gamma-ray-damaged DNA in human (ataxia telangiectasia) fibroblasts. *Nature* 260:444-447, 1976.
93. PERRY, P.: Chemical mutagens and sister chromatid exchange. In Chemical Mutagens, Vol. 6, (Hollaender, A., DeSerres, F. ed.), New York: Plenum Press, 1-39, 1980.
94. PERRY, P., EVANS, H.J.: Cytological detection of mutagen-carcinogen exposure by sister chromatid exchange. *Nature* 258:121-125, 1975.
95. PERRY, P., WOLFF, S.: New Giemsa method for differential staining of sister chromatids. *Nature* 261:156-158, 1974.
96. POON, P.K., O'BRIEN, R.L., PARKER, J.W.: Defective DNA repair in Fanconi's anemia. *Nature* 250:223-225, 1974.
97. POPESCU, N.C., AMSBAUGH, S.C., DIPAOLO, J.A.: Enhancement of N-methyl-N'-nitro-N-nitrosoguanidine transformation of Syrian hamster cells by a phorbol diester is independent of sister chromatid exchanges and chromosome aberrations. *Proc. Natl. Acad. Sci. USA* 77:7282-7286, 1980.
98. REMSEN, J.F., CERUTTI, P.A.: Deficiency of gamma-ray excision repairs in skin fibroblasts from patients with Fanconi's anemia. *Proc. Natl. Acad. Sci. USA* 73:2419-2423, 1976.
99. RESNICK, M.A., MOORE, P.D.: Molecular recombination and the repair of DNA double-stranded breaks in CHO cells. *Nucleic Acids Res.* 6:3145-3160, 1979.
100. REYNOLDS, R.J., NATARAJAN, A.T., LOHMAN, P.H.M.: *Micrococcus luteus* UV-endonuclease sensitive sites and sister chromatid exchanges in Chinese hamster ovary cells. *Mutat. Res.* 64:353-356, 1979.
101. ROMMELAERE, J., MILLER-FAURES, A.: Detection by density equilibrium centrifugation of recombinant-like DNA molecules in somatic mammalian cells. *J. Mol. Biol.* 98:195-218, 1975.
102. RUDIGER, H.W., BARTRAM, C.R., HARDER, W.: Rate of sister chromatid exchanges in Bloom syndrome fibroblasts reduced by cocultivation with normal fibroblasts. *Am. J. Hum. Genet.* 32:150-157, 1980.
103. SAHAR, E., KITTREL, C., FULGHUM, S., FELD, M., LATT, S.A.: Sister chromatid exchange induction in Chinese hamster ovary cells by 8-methoxypsoralen and short

pulses of laser light. Assessment of the relative importance of 8-methoxy-psoralen-DNA monoadducts and crosslinks. *Mutat. Res.* 83:91–105, 1981.
104. SAMSON, L., SCHWARTZ, J.L.: Evidence for an adaptive DNA repair pathway in CHO and human skin fibroblast cell lines. *Nature* 287:861–863, 1980.
105. SAN SEBASTIAN, J.R., O'NEILL, J.P., HSIE, A.W.: Induction of chromosome aberrations, sister chromatid exchanges, and specific locus mutations in Chinese hamster ovary cells by 5-bromodeoxyuridine. *Cytogenet. Cell Genet.* 28:47–54, 1980.
106. SANTESSON, B., LINDAHL-KIESSLING, K., MATTSSON, A.: SCE in B and T lymphocytes. Possible implications for Bloom's syndrome. *Clin. Genet.* 16:133–135, 1979.
107. SASAKI, M.S.: Is Fanconi's anemia defective in a process essential to the repair of DNA crosslinks? *Nature* 257:501–503, 1975.
108. SASAKI, M.S.: Fanconi's anemia, a condition possibly associated with a defective DNA repair. *In* DNA Repair Mechanisms (Hanawalt, P.C., Friedberg, E.C., Fox, C.F., eds.), ICN-UCLA Symposium on Molecular and Cellular Biology, vol. IX, New York: Academic, 675–684, 1978.
109. SASAKI, M.S., TONOMURA, A.: A high susceptibility of Fanconi's anemia to chromosome breakage by DNA crosslink agents. *Cancer Res.* 33:1829–1835, 1973.
110. SCHNEIDER, R.L., MONTICONE, R.E.: Cellular aging and sister chromatid exchange. II. Effect of in vitro passage of human fetal lung fibroblasts on baseline and mutagen induced sister chromatid exchange frequency level. *Exp. Cell Res.* 115:269–276, 1978.
111. SCHONBERG, S., GERMAN, J.: Sister chromatid exchange in cells metabolically coupled to Bloom's syndrome cells. *Nature* 284:72–74, 1980.
112. SCHONWALD, A.D., PASSARGE, E.: UV-light induced sister chromatid exchanges in xeroderma pigmentosum lymphocytes. *Hum. Genet.* 36:213–218, 1977.
113. SCUDIERO, D.A.: Decreased DNA repair synthesis and defective colony-forming ability of ataxia telangiectasia fibroblast cell strains treated with N-methyl-N'-nitro-N-nitrosoguanidine. *Cancer Res.* 40:984–990, 1980.
114. SCUDIERO, D.A., STRAUSS, B.: Accumulation of single-stranded regions in DNA and the block to replication in a human cell line alkylated with methyl methanesulfonate. *J. Mol. Biol.* 83:17–34, 1974.
115. SETLOW, R.B.: Repair deficient human disorders and cancer. Review article. *Nature* 271:713–717, 1978.
116. SHAHAM, M., BECKER, Y., COHEN, M.M.: A diffusable clastogenic factor in ataxia telangiectasia. *Cytogenet. Cell Genet.* 27:155–161, 1980.
117. SHIRIASHI, Y., SANDBERG, A.A.: Effects of mitomycin C on normal and Bloom's syndrome cells. *Mutat. Res.* 49:239–248, 1978.
118. SMITH, P.J., PATERSON, M.C.: Defective DNA repair and increased lethality in ataxia telangiectasia cells exposed to 4-nitroquinoline-1-oxide. *Nature* 287:747–749, 1980.
119. TAYLOR, A.M.R.: Unrepaired DNA strand breaks in irradiated ataxia telangiectasia lymphocytes suggested from cytogenetic observations. *Mutat. Res.* 50:407–418, 1978.
120. TAYLOR, A.M.R., HARNDEN, D.G., ARLETT, C.F., HARCOURT, S.A., LEHMANN, A.R., STEVENS, S., BRIDGES, B.A.: Ataxia telangiectasia: A human mutation with abnormal radiation sensitivity. *Nature* 258:427–429, 1975.
121. TAYLOR, A.M.R., ROSNEY, C.M., CAMPBELL, J.B.: Unusual sensitivity of ataxia telangiectasia cells to bleomycin. *Cancer Res.* 39:1046–1050, 1979.
122. TAYLOR, J.H.: Sister chromatid exchanges in tritium-labeled chromosomes. *Genetics* 43:515–529, 1958.
123. TAYLOR, J.H., WOODS, P.S., HUGHES, W.L.: The organization and duplication of chromosomes as revealed by autoradiographic studies using tritium-labeled thymidine. *Proc. Natl. Acad. Sci. USA* 43:122–128, 1957.

124. THOMPSON, L.H., BAKER, R.M., CARRANO, A.V., BROOKMAN, K.W.: Failure of the phorbol ester 12-0-tetradecanoylphorbol-13-acetate to enhance sister chromatid exchange, mitotic segregation, or expression of mutations in Chinese hamster cells. *Cancer Res.* 40:3245-3251, 1980.
125. TICE, R., CHAILLET, J., SCHNEIDER, E.L.: Evidence derived from sister chromatid exchanges of restricted rejoining of chromatid sub-units. *Nature* 256:642-644, 1975.
126. TICE, R., WINDLER, G., RARY, J.M.: Effect of cocultivation on sister chromatid exchange frequencies in Bloom's syndrome and normal fibroblast cells. *Nature* 273: 538-540, 1978.
127. UTAKOJI, T., HOSODA, K., UMEZAWA, K., SAWAMURA, M., MATSUSHIMA, T., MIWA, M., SUGIMURA, T.: Induction of sister chromatid exchanges by nicotinamide in Chinese hamster lung fibroblasts and human lymphoblastoid cells. *Biochem. Biophys. Res. Commun.* 90:1147-1152, 1979.
128. VAN BUUL, P.P.W., NATARAJAN, A.T., VERDEGAAL-IMMERZEEL, A.M.: Suppression of the frequencies of sister chromatid exchanges in Bloom's syndrome fibroblasts by cocultivation with Chinese hamster cells. *Hum. Genet.* 44:187-189, 1978.
129. VAN'T HOF, J.: Pea (*Pisum sativum*) cells arrested in G2 have nascent DNA with breaks between replicons and replication clusters. *Exp. Cell Res.* 129:231-237, 1980.
130. VINCENT, R.A. JR., FINK, A.J., HUANG, P.C.: Unscheduled DNA synthesis in cultured ataxia telangiectasia fibroblast-like cells. *Mutat. Res.* 72:245-249, 1980.
131. VINCENT, R.A., JR., SHERIDAN, R.B., HUANG, P.C.: DNA strand breakage repair in ataxia telangiectasia fibroblast-like cells. *Mutat. Res.* 33:357-366, 1975.
132. WARREN, S.T., SCHULTZ, R.A., CHANG, C.C., TROSKO, J.E.: Elevated spontaneous mutation rate in Bloom syndrome fibroblasts. *Am. J. Hum. Genet.* 32:161A, 1980.
133. WATERS, R., REGAN, J.D., GERMAN, J.: Increased amounts of hybrid (heavy/heavy) DNA in Bloom's syndrome fibroblasts. *Biochem. Biophys. Res. Commun.* 83: 536-541, 1978.
134. WEKSBERG, R., BUCHWALD, M., SARGENT, P., THOMPSON, M.W., SIMINOVITCH, L.: Specific cellular defects in patients with Fanconi's anemia. *J. Cell. Physiol.* 101: 311-324, 1979.
135. WOLFF, S.: Sister chromatid exchanges. *Annu. Rev. Genet.* 11:183-201, 1977.
136. WOLFF, S., BODYCOTE, J., PAINTER, R.B.: Sister chromatid exchanges induced in Chinese hamster cells by UV irradiation of different stages of the cell cycle: The necessity for cells to pass through S. *Mutat. Res.* 25:73-81, 1974.
137. WOLFF, S., PERRY, P.: Insights of chromatid structure from sister chromatid exchange ratios and the lack of both isolabelling and heterolabelling as determined by the FPG technique. *Exp. Cell Res.* 93:23-30, 1975.
138. WOLFF, S., RODIN, B., CLEAVER, J.E.: Sister chromatid exchanges induced by mutagenic carcinogens in normal and xeroderma pigmentosum cells. *Nature* 265: 345-347, 1977.
139. YOSHIDA, M.C.: Suppression of spontaneous and mitomycin C-induced chromosome aberrations in Fanconi's anemia by cell fusion with normal human fibroblasts. *Hum. Genet.* 55:223-226, 1980.
140. ZAKRZEWSKI, S., SPERLING, K.: Antagonistic effect of cocultivation on mitomycin C-induced aberration rate in cells of a patient with Fanconi's anemia and in Chinese hamster ovary cells. *Hum. Genet.* 56:85-88, 1980.
141. ZAMANSKY, G.B., LATT, S.A., KAPLAN, J.C., KLEINMAN, L.F., DOUGHERTY, C.P., BLACK, P.H.: The co-induction of sister chromatid exchanges and virus synthesis in mammalian cells. *Exp. Cell Res.* 126:473-476, 1980.

THE INTERRELATIONSHIPS IN ATAXIA-TELANGIECTASIA OF IMMUNE DEFICIENCY, CHROMOSOME INSTABILITY, AND CANCER

BARBARA KAISER-McCAW AND FREDERICK HECHT

The chromosome instability syndromes provide models for the study of the interrelationships between immunodeficiency, chromosome instability, and cancer. Although in ataxia-telangiectasia (AT) all three of these features may coexist in a given patient, no one of them is found consistently in every patient. Immunodeficiency may be manifested by a deficiency or absence of one or more of the immunoglobulins, particularly IgA, and, or a diminished response of lymphocytes to phytohemagglutinin (PHA). However, not all patients with AT are demonstrably immunodeficient. Chromosome breakage and rearrangements are generally present. However, not all patients exhibit increased chromosome instability. To complicate the situation further, an AT patient may have immunodeficiency and, or chromosome instability at one stage of the disease but not at another stage. As a group, patients with AT have an increased risk of developing a malignancy, particularly of the lymphoreticular system. However, not all patients develop it. Even if a patient develops a malignancy, death may occur from some other cause, and cancer accounts for only 10% of the deaths.

AT PATIENTS WITH CANCER STUDIED CYTOGENETICALLY

Today, we are just beginning to piece together the observations made in AT by clinicians and laboratory investigators. In order to begin to understand the interrelationships between immunodeficiency, chromosome instability, and cancer, we have chosen here to consider only the few AT patients with cancer on whom detailed serial studies have been made.

Case 1

This patient (J.Z.) has been the subject of several reports [5, 21, 15, 14, 2]. She presented the classical clinical picture of AT, with striking neurological findings, oculocutaneous telangiectases, and sinopulmonary infections. Serum IgA was undetectable at age 23, the age when immunoglobulin determinations were first made.

At age 29, her blood leukocyte count gradually rose from normal to 26,900/mm^3, and abnormal circulating lymphocytes were seen. Cell-surface-marker studies of the blood lymphocytes by E-rosetting revealed 81% T-cells, and the diagnosis of chronic lymphocytic leukemia (CLL) was made. The patient was asymptomatic until age 31 when her blood leukocyte count reached 121,900/mm^3. She failed to respond to chemotherapy, developed pleural effusion and pulmonary infiltration, and died at age 32. In addition to CLL, the patient also had multiple neoplasms of other types, including a leiomyoma of the ileum, adenomatous polyps of the rectum, leiomyomata and a leiomyosarcoma of the uterus, and a benign cystadenofibroma of the ovary.

In 1967, when J.Z. was 24 years old and before the clinical onset of leukemia, at a time when IgA reportedly was absent from her serum, studies of PHA-stimulated blood lymphocyte chromosomes disclosed the presence of a population of lymphocytes with an apparently balanced reciprocal translocation involving two group D(13-15) chromosomes, subsequently identified as t(14;14)(q12;q32) (Fig. 1). At age 31, after the onset of CLL, the same translocation was observed in 100% of her PHA-responsive lymphocytes; however, the smaller of the two rearranged chromosomes (the 14q−) had been lost from the cells. Chromosome "instability" was represented only by chromosome translocations. The spontaneous chromosome breakage affecting various chromosome regions at random that earlier had been observed no longer was present. IgA now had become detectable in her serum.

Case 2

This patient (M.P.) also has been the subject of numerous reports [9,14 (Case A), 18,23,19]. She walked at 14 months of age with an unsteady gait and had frequent upper respiratory infections during childhood. When she was examined at 38 years of age, she had telangiectases and neurological findings consistent with AT. Serum IgA was undetectable on two separate occasions. At age 48, her symptoms again were consistent with the diagnosis of AT. Quantitative serum IgA studies revealed

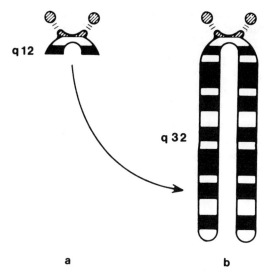

Fig. 1. Diagrammatic representation in metaphase chromosomes of a translocation between the two chromosomes 14 often reported in AT lymphocytes. a. 14q−, due to a break at band 14q12 and deletion distal to the break. b. 14q+, due to a break and tandem translocation at band 14q32. The translocation is described t(14;14)(q12;q32).

less than 2 μm/ml, and immunoelectrophoresis revealed no precipitin arc for IgA.

An elevated blood leukocyte count was first noted in the patient at age 46. When a detailed hematologic study was made some 14 months later, the leukocyte count was 66,000/mm^3, 69% of which were lymphocytes; 73% of the peripheral blood mononuclear cells formed rosettes with sheep erythrocytes, and 6% of them had Fc receptors. It was at that time that T-cell CLL was diagnosed.

When the patient had been studied at age 38, her lymphocyte response to PHA had been described as impaired; however, after the onset of leukemia, her lymphocytes responded normally to PHA. One possible explanation for the improved lymphocyte response to PHA is that the neoplastic T-cells had the capacity to respond to PHA.

The patient was given chemotherapy, but her blood leukocyte count rose to 364,000/mm^3, 99% of which were T-lymphocytes. She developed polyclonal hyperimmunoglobulinemia, with elevated concentrations of IgG and IgM but still without detectable IgA. She had increasing symptoms of leukemic pleural infiltration and pleural effusion and died 15 months after leukemia was diagnosed.

Eleven years before the diagnosis of leukemia in this patient, at age 38, all PHA-stimulated lymphocytes had the chromosome complement 45,XX,−D,+t(D;D). At the time of the diagnosis of leukemia and twice thereafter, banded chromosomes from 72-hour PHA-stimulated lymphocyte cultures showed the 14q+ marker chromosome to be t(14;14) (q12;q32). The modal chromosome number was 44: a No. 20 and the deleted chromosome 14, the (14q−), were absent, and the chromosome 14 with the tandem translocation (the 14q+) remained [23]. At that time, all metaphase cells in cultures of E-rosetted T-lymphocytes bearing both IgM and IgG Fc receptors had the 14q+ marker chromosome but lacked the 14q−. The retention of the 14q+ translocation chromosome and the loss of the 14q− chromosome in this case are similar to the cytological events in Case 1.

Cases 3 and 4

Two sisters with AT were studied by Levitt et al. [13]. The younger (Case 3) was diagnosed as having AT at age 3. When seen again at age 21, this sister had typical clinical symptoms of AT, with low serum concentration of IgA. Chromosome studies of PHA-stimulated lymphocytes at age 23 disclosed cells with a modal number of 45 as a result of the absence of one chromosome 14. A 14q+ chromosome which was present in all metaphases appeared to be the same translocation detected in Cases 1 and 2 (above) − t(14;14) (q12;q32). Although her PHA-stimulated lymphocytes had this clone, she was entirely normal from a hematologic viewpoint, and at age 24 (when last examined), she presented no sign of malignancy.

The older sister, Case 4, was accorded the diagnosis of AT at 15 months of age. Although she had a normal peripheral blood profile at 19 years of age (complete count and smear), she was diagnosed as having leukemia at age 25. At that time her serum concentration of IgA was low. The leukocyte count was increased to 48,400/mm^3, of which 73% were lymphocytes. Many of the lymphocytes were unusually large, had prominent nucleoli, and appeared to be atypical and immature. Cell-surface-marker studies showed that 58% of peripheral blood lymphocytes were T-cells and 11% were B-cells. A diagnosis of atypical, subacute T-cell lymphocytic leukemia was made.

The leukemia was not treated initially. Twelve months later the patient experienced increasing fatigue and was found to have a mild anemia and an increased blood leukocyte count of 613,000/mm^3, 95% of which were large, atypical lymphocytes. Chemotherapy was begun, but after 8 weeks the patient developed pneumonia and died from respiratory failure.

In the initial study of uncultured bone marrow from Case 4 no cytogenetic abnormality was detected. In the initial study of PHA-stimulated blood lymphocytes a modal chromosome number of 43 and an abnormal karyotype that included a 14q+ chromosome were found. The 14q+ chromosome resulted from a (14q;14q) tandem translocation. The mitotic response of the leukemic cells to PHA suggested that the leukemic cells were T-cells. Two subsequent lymphocyte cultures (also using PHA) disclosed a modal chromosome number of 43 and an abnormal karyotype that included a 14q+ chromosome. Uncultured bone marrow examined a year after the diagnosis of leukemia had been made again showed cells with normal karyotypes only, as did peripheral blood cells that entered metaphase without PHA stimulation.

Presumably the primary cytogenetic change was the t(14;14), creating the smaller 14q− and the larger 14q+ chromosomes. The cytogenetic origin and evolution of this clone can only be conjectured, because the patient's chromosomes were studied only after the onset of leukemia. One key finding was the absence of the deleted 14 (the 14q−) chromosome, as in Cases 1 and 2. The 14q+ chromosome still was present in the population of neoplastic cells.

CONSISTENT FEATURES IN INFORMATIVE PATIENTS WITH LEUKEMIA

Reports of several cases of AT with chromosome instability, clonal proliferation, and cancer now may be found in the literature. However, the three just described were selected for consideration here because they were studied over a number of years and because the dynamic aspects of their cells and immune function were documented carefully. The following similarities were observed in them.

AT was associated in each case with a diminished concentration of serum IgA, a decreased response of blood lymphocytes to PHA, and hypoplastic lymphoid tissue. Initial chromosome studies disclosed not only random breakage of chromosomes but also one or more cells marked by a translocation that involved chromosome 14. A clone of lymphocytes marked by a t(14;14)(q12;q32) proliferated in the blood of each patient. The number of cells responsive to PHA appeared to increase correspondingly to the number of cells in the clone. The smaller proximal part of chromosome 14 (the 14q− marker chromosome) was lost, and only the larger tandem translocation (the 14+ marker) remained. The clone of cells marked by the 14q+ translocation chromosome became malignant, and the diagnosis of chronic or subacute T-cell lymphocytic leukemia was made.

STAGES IN AT

Based on the observations in these and a few additional AT cases [for example, *14, 3, 20, 25*], we have proposed the existence of a series of stages in the chromosome changes observed in AT lymphocytes [*12*], as follows:

Stage I: Origination

Cells with increased chromosome breakage and rearrangements are observed. The chromosomes are involved randomly.

Stage II: Competition

Cells with specific rearrangements are seen, usually balanced translocations or inversions. Although no one rearrangement predominates, chromosomes 7 and 14 are involved nonrandomly.

Stage III: Proliferation

One cell line with a specific rearrangement has emerged from the competition to be predominant. That rearrangement is present in from 2 to 3% up to almost 100% of PHA-responsive blood cells. The rearrangement is usually a translocation involving either the short or the long arm of chromosome 7 and, or one or both of two sites on the long arm of chromosome 14. Several minor clones may be present as well.

Stage IV: Fixation

This stage has been reached when 100% of the PHA-stimulated lymphocytes contain the same chromosome rearrangement. To date, only 14;14 translocations have been observed in 100% of the cells.

Stage V: Conversion

An overtly malignant clone of cells now is identifiable. In many AT patients, however, the clone of cells marked by the 14;14 translocation never becomes malignant, and so does not progress (or has not progressed). The patient's death may be due to an infection or a neoplasm affecting some other cell type.

PREFERENTIAL INVOLVEMENT OF CHROMOSOME 14

No explanation is available yet for the preferential involvement of chromosome 14 in premalignant and malignant clones in AT. The gene for nucleoside phosphorylase (NP) has been mapped to chromosome 14 in the region of bands 14q11-21, and persons with NP deficiency are immunodeficient. However, NP deficiency has never been reported in AT patients. Enzyme assays of NP in lymphocytes and bone marrow red cells in Case 2 showed normal NP activity [23]. Other genes mapped to chromosome 14 include those for tryptophanyl-tRNA synthetase, creatine kinase (brain type), protease inhibitor, an Epstein-Barr virus (EBV)-integration site, and those of the immunoglobulin heavy-chain gene family [10]. Regional assignments of the EBV integration site and, or, the immunoglobulin heavy-chain family on chromosome 14 may show them to be of significance in relation to the clonal evolution that occurs in lymphocytes of AT patients.

The immunologic studies in Case 2 were extensive and showed that the leukemic T-cells provided normal helper and suppressor activity [18]. While some of the leukemic cells bore a receptor for the Fc portion of IgM, others bore a comparable receptor for IgG. Both groups of cells contained the 14q+ marker chromosome. Malignant transformation in this patient apparently occurred in an uncommitted T-lymphocyte. The development of polyclonal hyperimmunoglobulinemia with rising IgG and IgM concentrations may simply have reflected B-cell stimulation. Another possibility is that malignant transformation involved an uncommitted precursor lymphocyte that had not undergone T- or B-cell differentiation.

Comprehensive immunological studies of AT patients in the future may provide valuable insights not only into the interactions of various components of the lymphoproliferative system but also into the connection between the immune system and the cytogenetic changes.

PROLIFERATIVE ADVANTAGE OF A CLONE MARKED BY A CHROMOSOME 14 REARRANGEMENT

Why does the number of lymphocytes marked by rearrangements of chromosome 14 increase in some patients? Are cells with a rearrangement of chromosome 14 more responsive to mitogenic stimulus? Are the chromosomes in cells with a rearrangement of chromosome 14 more resistant to random breakage? Any explanations for the perpetuation of lymphocytes marked by rearrangements of chromosome 14 must remain as speculations at this time.

At some point in the proliferation of the t(14;14) clone in the cases described above, the proximal part of one chromosome 14 was lost (i.e., the part in the 14q− chromosome), leaving only the 14q+ component of the translocation. Concurrently, leukemia was diagnosed. Therefore, the loss of the segment 14pter-14q12 may be the significant event in the development of malignancy. Another clue may be the case of an AT patient diagnosed as having Hodgkin's disease [4]. Before the diagnosis of Hodgkin's disease was made, 33% of peripheral blood lymphocyte had the complement 45,XX,−14. Although a direct connection between monosomy 14 and Hodgkin's disease cannot be made, additional cases ultimately may confirm that the loss of part, or all, of chromosome 14 is a significant step in malignant transformation.

This discussion has been confined to chromosome rearrangements, particularly those involving chromosome 14, and lymphocyte populations in AT patients. However, it must be recognized that rearrangments of chromosome 14 also have been observed in fibroblasts [17,6,7], lymphoblastoid cell lines [11,7], bone marrow cells [13], and amniotic fluid cells [22] derived from a fetus with AT. However, the dynamics of chromosome 14 rearrangement and clonal evolution in cell types other than lymphocytes are far from clear. In fact, the chromosome 14 changes in the other cell types may have arisen in vitro.

CHROMOSOMES AND CANCER IN AT PATIENTS

The clinical association between AT and malignancy is well established [16]. The Immunodeficiency-Cancer Registry lists 108 patients with AT who developed malignancies of various types: 48 (44%) non-Hodgkin's lymphomas, 26 (24%) leukemias, 22 (20%) carcinomas, and 12 (11%) Hodgkin's disease [24]. Although lymphoid malignancies are the most frequent (70%), a direct relationship between chromosome breakage with rearrangements of chromosome 14 in lymphocytes, clonal proliferation, and cancer cannot be postulated as yet. Wake et al [26] have observed one AT patient who had a chromosome 14-translocation clone in blood lymphocytes but who developed an atypical lymphocytic leukemia that apparently was derived from cells that did not bear the t(14;14) chromosome. Another person with AT and a t(14;14) clone in 87% of lymphocytes analyzed developed a hepatocellular carcinoma [1]; no direct relationship can be proposed between the clone of lymphocytes and the hepatocellular carcinoma.

The interrelationships between immunodeficiency, chromosome instability, and predisposition to cancer in AT probably all are tied to a defect, or defects, at the cellular level. It long has been known that cells from AT patients are hypersensitive to x-ray-irradiation. It now appears that a defect in the DNA-damage-recognition system exists in AT cells. Because x-ray-induced DNA damage is not recognized immediately by AT cells, the cycling cells will progress unchecked through the S phase of the cell-division cycle and not be permitted to complete necessary DNA repair, so that DNA replication occurs on damaged templates [8]. (Cessation of S-phase synthesis in normal cells after x-irradiation would permit DNA repair to occur.) Increased frequencies of chromosome aberrations such as chromatid breaks are the direct consequence.

The whole story is far from complete, but important steps are being made toward understanding the chromosome-instability syndromes.

Acknowledgments. This work is supported in part by a research grant from the National Cancer Institute on the "Cytogenetics of Clonal Neoplasias," No. CA-25055.

LITERATURE CITED

1. AL-SAADI A., PALUTKE M., KUMAR G.K.: Evolution of chromosomal abnormalities in sequential cytogenetic studies of ataxia telangiectasia. *Hum. Genet.* 55:23-29, 1980.
2. AMROMIN G.D., BODER E., TEPLITZ R.: Ataxia-telangiectasia with a 32 year survival: A clinicopathological report. *J. Neuropathol. Exp. Neurol.* 38:621-643, 1979.
3. AURIAS A., DUTRILLAUX B., BURIOT D., LEJEUNE J.: High frequencies of inversions and translocations of chromosomes 7 and 14 in ataxia telangiectasia. *Mutat. Res.* 69:369-374, 1980.
4. BERNSTEIN R., PINTO M., JENKINS T.: Ataxia-telangiectasia with evolution of monosomy 14 and emergence of Hodgkin's disease. *Cancer Genet. Cytogenet.* 4:31-38,1981.
5. BODER E., SEDGWICK R.P.: Ataxia-telangiectasia. A familial syndrome of progressive cerebellar ataxia, oculocutaneous telangiectasia and frequent pulmonary infection. Pediatrics 21:526-544, 1958.
6. COHEN M.M., KOHN G., DAGAN, J.: Chromosomes in ataxia-telangiectasia. *Lancet* 2:1500, 1973.
7. COHEN M.M., SAGI M., BEN-ZUR Z., SCHAAP T., VOSS R., KOHN G., BEN-BASSAT H.: Ataxia-telangiectasia: Chromosomal stability in continuous lymphoblastoid cell lines. *Cytogenet. Cell Genet.* 23:44-52, 1979.
8. CRAMER P., PAINTER R.B.: Bleomycin-resistant DNA synthesis in ataxia telangiectasia cells. *Nature* 291:671-672, 1981.
9. GOODMAN W.N., COOPER W.C., KESSLER G.B., RISCHER M.D., GARDNER M.B.: Ataxia-telangiectasia: A report of two cases in siblings presenting a picture of progressive spinal muscular atrophy. *Bull. Los Angeles Neurol. Soc.* 34:1-22, 1969.
10. Human Gene Mapping: Sixth International Workshop. *Cytogenet. Cell Genet.* 32, 1982.
11. JEAN P., RICHER C.L., MUER-ORLANDO M., LUV D.H., JONCAS J.H.: Translocation 8;14 in an ataxia-telangiectasia-derived cell line. *Nature* 277:56-58, 1979.
12. KAISER-MCCAW B.K., HECHT F.: Ataxia-telangiectasia: Chromosomes and cancer. In Harnden D.G., Bridges B.A. (eds.): "Ataxia-Telangiectasia." New York: John Wiley and Sons, Ltd., 1981.
13. LEVITT R., PIERRE R.V., WHITE W.L., SIEKERT R.G.: Atypical lymphoid leukemia in ataxia telangiectasia. *Blood* 52:1003-1011, 1978.

14. McCaw B.K., Hecht F., Harnden D.G., Teplitz R.L.: Somatic rearrangement of chromosome 14 in human lymphocytes. *Proc. Natl. Acad. Sci.* U.S.A. 72: 2071-2075, 1975.
15. McFarlin D.E., Strober W., Waldmann T.A.: Ataxia-telangiectasia. *Medicine* 51:281-314, 1972.
16. Miller R.W.: Highlights in clinical discoveries relating to ataxia-telangiectasia. In Harnden D.G., Bridges B.A. (eds.): "Ataxia-Telangiectasia." New York: John Wiley and Sons, Ltd., 1981.
17. Oxford J.M., Harnden D.G., Parrington J.M., Delhanty J.D.A.: Specific chromosome aberrations in ataxia telangiectasia. *J. Med. Genet.* 12:251-262, 1975.
18. Saxon A., Stevens R.H., Golde D.W.: Helper and suppressor T-lymphocyte leukemia in ataxia telangiectasia. *New England J. Med.* 300:700-704, 1979.
19. Saxon A., Stevens R.H., Golde D.W.: T-cell leukemia in ataxia telangiectasia. *New England J. Med.* 301:945, 1980.
20. Scheres J.M.J.C., Hustinex T.W.J., Weemaes C.M.R.: Chromosome 7 in ataxia-telangiectasia. *J. Pediatr.* 97:440-441, 1980.
21. Sedgwick R.P., Boder E.: Ataxia-telangiectasia. In Winken P.J., Bruyn G.W. (eds.): "Handbook of Clinical Neurology." Amsterdam: North-Holland Publ. Co., 14:267-339, 1972.
22. Shaham M., Voss R., Becker Y., Yarkoni S., Ornoy A., Kohn G.: Prenatal diagnosis of ataxia telangiectasia. *J. Pediatr.* 100:134-137, 1982.
23. Sparkes R.S., Como R., Golde D.W.: Cytogenetic abnormalities in ataxia telangiectasia with T-cell chronic lymphocytic leukemia. *Cancer Genet. Cytogenet.* 1:329-338, 1980.
24. Spector B.D., Flilipovich A.H., Perry G.S., Kersey J.H.: Epidemiology of cancer in ataxia-telangiectasia. In Harnden D.G., Bridges B.A. (eds.): "Ataxia-Telangiectasia." New York: John Wiley and Sons, Ltd., 1981.
25. Taylor A.M.R., Oxford J.M., Metcalfe J.A.: Spontaneous cytogenetic abnormalities in lymphocytes from thirteen patients with ataxia-telangiectasia. *Int. J. Cancer* 27:311-319, 1981.
26. Wake N., Minowada J., Park B., Sandberg A.A.: Chromosomes and causation of human cancer and leukemia. XLVIII. T-cell acute leukemia in ataxia telangiectasia. *Cancer Genet. Cytogenet.* 6:345-357, 1982.

CELLULAR SENSITIVITY TO MUTAGENS AND CARCINOGENS IN THE CHROMOSOME-BREAKAGE AND OTHER CANCER-PRONE SYNDROMES

JOHN A. HEDDLE, ALENA B. KREPINSKY, AND RICHARD R. MARSHALL

The term "chromosome-breakage syndromes" was coined by James German [59] to draw attention to the common features of three rare, genetically determined human syndromes: ataxia telangiectasia, Bloom's syndrome, and Fanconi's anemia. In all three disorders, untreated lymphocyte cultures show an abnormally high frequency of structural chromosomal aberrations. Interest in these syndromes has been enhanced by the discovery of their high rates of cancer [58,60,64,142,179]. More recently, it has been found that all three are characterized by specific, but different, mutagen hypersensitivities in vitro. They thus belong to a larger category, the mutagen-hypersensitive syndromes, which includes xeroderma pigmentosum and several other cancer-prone genetic diseases. Although mutagen hypersensitivity is only one aspect of these syndromes, it can be exploited experimentally to achieve further understanding of the mechanisms of DNA repair, chromosomal breakage, mutagenesis, and carcinogenesis in human cells. In this article we have summarized the data on mutagen sensitivity in cancer-prone syndromes of genetic origin and have discussed their reliability, completeness, and implications.

It was difficult to decide which syndromes to include in this review. Indeed, the next section deals with the adequacy of the identification of the chromosome-breakage syndromes and justifies, in part, our inclusion of many cancer-prone conditions not known to have abnormally high levels of spontaneous chromosomal aberrations. The classification of syndromes and the summaries of their mutagen sensitivity follow. As will be seen, the data on mutagen sensitivity are often fragmentary and inconsistent. Possible reasons for such inconsistencies, particularly the extent of human genetic variability, are outlined in a separate section. Although xeroderma pigmentosum (XP) is not a chromosome-breakage syndrome in the usual sense (in that the spontaneous rate of chromosomal breakage in lymphocyte cultures is normal), it is the best-studied of the mutagen-hypersensitive syndromes and is a model for many studies of the chromosome-breakage syndromes. As such it is discussed first. XP arises from a repair deficiency, but alternate models for the origin of mutagen hypersensitivity are also discussed, inasmuch as not all mutagen hypersensitivities may arise in this way. Finally, general implications of the data obtained on mutagen sensitivity are considered, particularly in terms of hypothetical mechanisms of carcinogenesis.

UNDISCOVERED CHROMOSOME-BREAKAGE SYNDROMES

Intercomparisons of the features of cancer-prone syndromes are useful in drawing general conclusions about the mechanism of carcinogenesis. If the syndromes have a common characteristic, as do the chromosome-breakage syndromes, then the similarities and differences are particularly significant. For example, the spontaneous mutation rate in the chromosome-breakage syndromes is of great interest with respect to the somatic mutation theory of cancer. It is obvious that the more the mutagen-hypersensitive and chromosome-breakage syndromes are compared, the greater will be the likelihood that incidental factors will be eliminated and the important factors revealed. It must be admitted, however, that it is by no means certain that all or even most of the chromosome-breakage syndromes have been discovered. There are two reasons for this. The first is that, in the typical analysis of the karyotype performed on human lymphocyte cultures, too few cells are analyzed to reveal anything less than an extraordinary increase in chromosomal breakage. For this reason, some syndromes, in which the karyotypes are normal, but in which the spontaneous levels of chromosomal aberrations are elevated, may have been missed. The second reason is that the only cells that normally are available for analysis are lymphocytes and fibroblasts. Since both chemical carcinogens and genetically determined cancers are usually tissue specific, it seems likely that tissue-specific chromosome-breakage syndromes can exist. Although none is known, this could well be due to the difficulty in obtaining and analyzing other human tissue.

Any one of the three characteristics mentioned so far – spontaneous chromosomal aberrations, cancer proneness, and mutagen hypersensitivity – could by itself define a group of syndromes. These groups may overlap, but they are not synonymous. It is clear, for example, that there are cancer-prone syndromes in which the spontaneous rate of chromosomal breakage in lymphocyte cultures is normal, e.g., familial retinoblastoma. It is also clear that there can be mutagen hypersensitivity in the absence of a high spontaneous rate of chromosomal breakage, for example, in XP and Down's syndrome. In these syndromes only the *induced* rate of chromosomal breakage is abnormally high [131,154]. German [61] has called one such syndrome (XP) a "conditional chromosome-breakage syndrome." We are not aware of a situation in which there is an abnormally high spontaneous rate of chromosomal breakage but no cancer proneness. A simple scheme that is consistent with the information available is that all chromosome-breakage syndromes will be mutagen hypersensitive, and that all mutagen hypersensitive syndromes will be cancer prone, although in neither case will the converse hold. Nevertheless, the lack of information on chromosomal breakage in most tissues has led us to review all cancer-prone conditions in which there is evidence of mutagen hypersensitivity.

THE SYNDROMES

Those cancer-prone syndromes in which mutagen sensitivity has been studied have been subdivided into several groups below. This classification is based on the chromosomal and genetic characteristics:

I. chromosome-breakage syndromes with autosomal recessive mode of inheritance;

II. syndromes with unstable karyotypes and other modes of inheritance;

III. autosomal recessive syndromes with some characteristics similar to the classic chromosome-breakage syndromes;

IV. autosomal dominant disorders without known karyotypic instability;

V. disorders with a specific chromosomal abnormality.

We have not attempted to give a complete list of disorders for each of the groups. It is to be expected that some of these syndromes will be reclassified when more information becomes available (e.g., after further cytogenetic analysis); indeed, we hope to stimulate such research.

I. Autosomal Recessive Chromosome-Breakage Syndromes

This group is represented by the three classic chromosome-breakage syndromes – ataxia telangiectasia, Bloom's syndrome, and Fanconi's anemia. The clinical symptoms of these disorders have been described in detail elsewhere [51,60,139]. Recently, another autosomal recessive disorder, Werner's syndrome [45,105], known for its predisposition to cancer, has been confirmed as a chromosome-breakage syndrome and included in this group (see chapter by W.T. Brown, this volume). All four syndromes show typical pigmentation or lesions on specific skin areas and have been classified as genodermatoses [122]. The unifying characteristics of these conditions are chromosomal instability, cancer, mode of inheritance, and skin involvement. The cellular characteristics of these disorders are summarized in Table II and their mutagen sensitivities in Tables III-V.

II. Syndromes With Unstable Karyotypes and Various Modes of Inheritance

Several other syndromes have been reported to have various chromosomal abnormalities and to be cancer prone (Table VII). The chromosomal instability in these disorders is not nearly as well established as in group I [23,70]. In some cases putative instability involves changes in ploidy or endoreduplication rather than chromosomal breakage [161]. Further study is needed to discriminate between genetic heterogeneity and other reasons for the interlaboratory differences. Mutagen sensitivity has been studied in two members of this group:

Dyskeratosis congenita. This is a very rare genodermatosis, apparently usually transmitted as an X-linked recessive [171]. In this condition there is, typically, a thickening and whitening of the mucous membranes, particularly the oral mucosa. These lesions are premalignant; the patients usually die from squamous cell carcinoma [171,105].

Basal cell nevus syndrome. This is an autosomal dominant condition [121] with numerous skin lesions and a large spectrum of abnormalities in various other organs [33,71,105].

No data on mutagen sensitivity have been reported on the other two disorders of this group—incontinentia pigmenti [25,93] and scleroderma [79,96].

III. Autosomal Recessive Syndromes With Some Characteristics Similar to the Classic Chromosome-Breakage Syndromes

The classic example of this group is xeroderma pigmentosum, which is discussed in detail later. Its spontaneous chromosomal aberration frequency is normal but reaches abnormally high levels after treatment with certain mutagens. The other syndromes that we have assigned to this group are the following (Table VI):

The Chediak-Higashi syndrome. This syndrome is characterized by large granules in leukocytes; diluted pigmentation of skin, eyes, and hair; sun sensitivity, and increased susceptibility to bacterial infection. Death usually occurs before the age of 10 years from either an infection or cancer [182].

The Rothmund-Thomson syndrome. Those affected are normal at birth and up to 2 years of age; affected individuals then develop typical persistent reticulate bluish lesions with whitish centers, mainly on face and limbs. Various degrees of sun sensitivity, as well as hypo- and hyperpigmentation, proportionate dwarfism, juvenile cataracts, and skeletal abnormalities are symptoms often associated with the disorder [105,150].

Cockayne's syndrome. The disorder is characterized by an arrest of development starting usually in the second year of life, subsequent truncal dwarfism, atrophy of skeletal muscles, deafness, mental deficiency, sun sensitivity, and pigmentary degeneration of the retina [123]. In one case, a trisomy of chromosome 20 has been reported [68]; however, cytogenetic analysis in another patient revealed a normal karyotype [200].

Progeria (the Hutchinson-Gilford syndrome). The most striking features are stunted growth and premature senility. Accompanying characteristics are skin atrophy, loss of hair, loss of subcutaneous fat, hypogonadism, and arteriosclerosis, the last of which is usually the cause of death before the age of 10 years [41].

The cellular characteristics of the syndromes are summarized in Tables I and VI. In the last two disorders, progeria and Cockayne's syndrome, an increased risk of malignancy has not been reported. However, it may be masked by early deaths from other causes and the very rare occurrence of both syndromes.

IV. Autosomal Dominant Disorders Without Known Karyotypic Instability

This group has many members, among them familial retinoblastoma, Gardner's syndrome, Huntington's chorea, porokeratosis of Mibelli, Cowden's disease (multiple hamartoma syndrome), neurofibromatosis (von Recklinghausen's disease), Pringle's disease (tuberous sclerosis), tylosis, cutaneous leiomyomata, hereditary lipomatoses, blue-rubber-bleb-nevus syndrome, and multiple chemodectoma. Most of these disorders are genodermatoses [122]. Little is known about their spontaneous rates of mutation and chromosomal breakage characteristics, but it seems that chromosomal instability is not typical of fibroblasts or lymphocytes of these conditons. In porokeratosis of Mibelli, an increase in

TABLE I. MUTAGEN SENSITIVITY IN EXCISION-DEFICIENT XERODERMA PIGMENTOSUM*

	Nontransformed				Transformed			References
	Chromosomal aberrations a	SCE b	Cell killing c	Mutation d	SCE e	Cell killing f	Mutation g	
None uv 254 nm	N ↑	N ↑	↑	N ↑	N ↑			a[131,137]; b[202]; d[6,124]; e[203] a[131,137]; b[157,195]; c[2,31,118,127]; d[6,69,124,134]; e[29,203]
4-Nitroquinoline-1-oxide (4NQO)	↑	↑	↑		↑			a[151,152]; b[144]; c[155,177,178,181]; e[203]
4NQO derivatives			↑					c[177]
N-acetoxy-2-acetylaminofluorene (NAAF)	↑		↑					a[151]; c[126]
N-hydroxy-2-acetylaminofluorene (HAAF)			↑					c[116]
Various polycyclic hydrocarbons			↑	↑				c[83,129]; d[129]
Activated aflatoxin B₁	↑		↑					a[176]; c[176]
Activated sterigmatocystin	↑		↑					a[176]; c[176]

Cellular Sensitivity and Cancer Proneness

Agent				References
Decarbamoyl MMC	↑			a[155]; c[53,155]
Mitomycin C (MMC)	↑;N			a[75, 155]; c[53,116,155]; e[203]
Nitrogen mustard (HN2)		↑		b[144];c[155]
8-MOP+UVA	↑			c[155]
Busulphan	↑			c[155]
Ionizing radiation	N;N$_v$			a[44]; c[8,155,199; 9]
N-methyl-N′-nitro-N-nitrosoguanidine (MNNG)	N		↑	a[151,178]; c[83,177]; e[203]; g[12]
N-methyl-N-nitrosourea (MNU)	N			c[125]
N-ethyl-N-nitrosourea (ENU)			↑	e[203]
Methyl methanesulfonate (MMS)	N		↑	a[152]; c[116,125,155]; e[203]
Ethyl methanesulfonate (EMS)	↑;N		↑	b[144,81]; c[155,190; 81,116]; e[81,203]; f[81]; g[12]
Dimethyl sulfate			↑	e[203]
Daunomycin	N			a[151]
Formaldehyde		↑		c[34]
Antipain		↑		c[94]

*↑, Hypersensitivity; N, normal sensitivity; v, variable response observed among patients.

TABLE II. Spontaneous Characteristics of the Chromosome-Breakage Syndromes*

Syndrome	Chromosomal aberrations a	Sister-chromatid exchanges b	Mutation c	Immunity d	Cancer e	References
Ataxia telangiectasia	↑	N	N	↓	↑	a[74,80]; b[16, 56,76]; c[6]; d[143]; e[142]
Bloom's syndrome	↑	↑	↑	↓	↑	a[62,63]; b[28]; c[72,193]; d[92]; e[60,64]
Fanconi's anemia	↑	N	N; ↓		↑	a[159]; b[78, 115,175]; c[72;47]; e[58,179]
Werner's syndrome	↑	N		↓	↑	a[45,208]; b[16]; d[43]; e[65]

*Refer to Table I for explanation of symbols.

chromosomal aberrations has been found only in the cells originating from the skin lesions [185]. Only familial retinoblastoma, Gardner's syndrome, and Huntington's chorea have been used in sensitivity studies (Table VII).

V. Disorders With a Specific Chromosomal Abnormality and Predisposition to Cancer

Positive correlation between high risk of cancer and a specific structural or numerical change in the karyotype has been observed in various human conditions, particularly in trisomy 21 (Down's syndrome), trisomy 18, trisomy 13, and those retinoblastomas characterized by a deletion in chromosome 13. Thus the incidence of leukemia in Down's syndrome is reported to be about 20 times greater than in normal individuals [145]. In deletion-13 retinoblastoma, the chromosomal abnormality occurs prezygotically, is present in all or most somatic cells, and consistently predisposes to cancer. The tumors, whether unilateral or bilateral, do not differ histologically from those in autosomal dominant familial retinoblastoma. The cellular characteristics of these disorders are given in Table VII.

HUMAN DIVERSITY

It is appropriate at this point to emphasize the need for care in the interpretation of results obtained on individual cell lines or strains. It is evident from the tables that the characteristics of a strain (or line) may not be representative of the syndrome to which it belongs. Broad diversity is a characteristic feature of the hypersensitivity syndromes and the spectra of responses can be very wide. The range of ultraviolet (uv) sensitivity [2] and excision-repair characteristics [51] among XP cell strains provides a good example of this. Furthermore, people are highly diverse and often have unique characteristics which are unrelated to the condi-

TABLE III. Mutagen Sensitivity in Ataxia Telangiectasia*

	Chromosomal aberrations a	SCE b	Cell killing c	Mutation d	References
uv 254 nm			N	N	c[7,164,172]; d[7]
uv 313 nm			↑V		c[172]
4-Nitroquinoline-1-oxide			↑V		c[14,173]
N-hydroxy-2-acetylamino-fluorene			N		c[7]
Mitomycin C		N	↑V		b[55]; c[86]
Diepoxybutane	↑				a[11]
Ionizing radiation	↑	N	↑	↓	a[85,135,146,187]; b[55]; c[139]; d[6,7]
Bleomycin	↑		↑		a[186]; c[186]
N-methyl-N′-nitro-N-nitrosoguanidine			↑V; N		c[141,163,164; 14]
N-ethyl-N-nitrosourea			↑		c[141]
Methyl methanesulfonate			↑V; N		c[14,86,144,164; 7]
Ethyl methanesulfonate		N	↑; N		b[55]; c[14,116]
Actinomycin D			↑		c[86]
Adriamycin		N			b[55]
6-Mercaptopurine			N		c[49]
6-Azauridine			N		c[49]
Azaserine			N		c[49]
8-Azaadenine			N		c[49]
5-Fluorouracil			N		c[49]

*Refer to Table I for explanation of symbols.

TABLE IV. Mutagen Sensitivity in Bloom's syndrome†

	Chromosomal aberrations a	SCE b	Cell killing c	Mutation d	References
uv 254 nm	N_V	N_V	↑; N; N_V		a[109]; b[109]; c[66; 5,7,109,165,166; 95]
uv 313 nm			N_V		c[207]
4-Nitroquinoline-l-oxide	N	↑			a[170]; b[170]
Decarbanoyl MMC	N	↑			a[170]; b[170]
Mitomycin C	N	↑; N	↑		a[170]; b[169,170; 108]; c[85]
Diepoxybutane	N				a[11]
Ionizing radiation	↑; N; ↓		N		a[85,113; 44; 189]; c[5,7]
N-ethyl-N'-nitro-N-nitrosoguanidine		↑	N*		b[110]
N-methyl-N'-nitro-N-nitrosoguanidine				N	d[72]
N-ethyl-N-nitrosourea		↑			b[110]
Methyl methanesulfonate		↑	N		b[108]; c[5,7]
Ethyl methanesulfonate	↑*	↑	↑		b[107,108]; c[5,7]
Caffeine	N	N			a[170]; b[170]
5-Bromo-2'-deoxyuridine	N*				

†Refer to Table I for explanation of symbols.
*Unpublished observations [Heddle, J.A., Krepinsky, A.B., Gingerich, J., German, J.].

tion being studied. For example, an individual with XP was recently recognized whose cells are not only uv sensitive but also sensitive to ionizing radiation [9]. The x-ray sensitivity may be the result not of the XP gene, but of another completely unrelated gene which he has also inherited. It may also be, of course, that this is another XP locus or a new allele in a complementation group already identified. For these reasons, it is not surprising that the pattern of sensitivity is sometimes unclear (see Tables), and the classification of syndromes as hypersensitive or not can be a considerable oversimplification.

This problem is further complicated by the definition of the normal response. With respect to radiosensitivity, for example, the normal range of D_o was defined as 97–180 rads by Arlett and Harcourt [8] and as 128–164 rads by Weichselbaum et al. [199]. A number of strains not considered hypersensitive under the criteria of the former survey would have been considered so under those of the latter. Both reports noted the problems that the inclusion of strains in the "control" category from clinically normal individuals heterozygous for radiosensitive genes might create. In view of this, the biological significance of "slight" hyper-

TABLE V. MUTAGEN SENSITIVITY IN FANCONI'S ANEMIA*

	Chromosomal aberrations a	SCE b	Cell killing c	Mutation d	References
uv 254 nm	↑		N; ↓		a[153]; c[48,53; 155]
uv 355 nm	N				a[153]
4NQO derivatives	N		↑		a[153]; c[155]
Decarbamoyl MMC	N		N		a[153]; c[53,54,155]
Mitomycin C	↑	N; ↓	↑		a[82,115,153]; b[115,136;115]; c[46,48,52-54,155,196]; d[47]
Nitrogen mustard	↑		↑		a[18,19,153]; b[18,20]; c[155]
8-Methoxypsoralen (8-MOP)	N	↑			a[153]
8-MOP + UVA	↑		↑		a[153]; c[155,196]
Busulphan			↑		c[155]
Diepoxybutane	↑				a[10,11]
Ionizing radiation	↑; N		↑; N		a[82,84,85,90; 153]; c[102; 46,48,155]
N-methyl-N'-nitro-N-nitrosoguanidine	N		N		a[153]; c[196]
Methyl methanesulfonate	N		N		a[82,153]; c[155]
Ethyl methanesulfonate	↑; N	N	↑; N	↑	a[10; 82,115]; b[115; 46,196]; c[155; 46,196]; d[47]
Triethylene phosphoramide		N			b[136]
Actinomycin D	N		N		a[153]; c[196]
Chloramphenicol	↑				a[153]
Caffeine	↑				a[153]
6-Mercaptopurine			↑		c[49]
6-Thioguanine			↑		c[49]
6-Azauridine			N		c[49]
Azaserine			N		c[49]
8-Azaadenine			N		c[49]
5-Bromo-2'-deoxyuridine	N				a[82]
Cyclophosphamide	N				a[17]
Cyclophosphamide metabolites	↑				a[17]
Elevated temperature	N				a[158]

*Refer to Table I for explanation of symbols.

TABLE VI. CELLULAR CHARACTERISTICS OF THE AUTOSOMAL RECESSIVE SYNDROMES: CLASSIFICATION GROUP III*‡

Syndrome	Spontaneous characteristics					Mutagen sensitivity	References
	Chromosomal aberrations a	SCE b	Mutation c	Immunity d	Cancer e	f	
Chediak-Higashi syndrome				↓	↑	Cell killing:↑ (uv, 4N QO)†; N (MNNG)†	d[204,]; e[42,182]; f[183]
Rothmund-Thomson syndrome					↑	Cell killing: ↑$_V$ (γ-rays)	e[147]; f[174]
Cockayne's syndrome	N	N	N			SCE: ↑(uv); N (uv) Cell killing:↑ (uv, 4NQO, MMC, NAAF, EMS); N (x-, γ-rays, EMS, MMS, ICR 170, HAAF) Mutation: ↑ (uv)	a[200]; b[29†,132] c[7]; f[3,5–8,29†, 87, 132,156,192]
Progeria						Cell killing: ↑ (γ-rays)	f[7,8]

*For cellular characteristics of XP see Table I.
†Experiments involving transformed cell lines.
‡Refer to Table I for explanation of symbols.

sensitivity when only a limited number of controls have been used is doubtful. In our opinion, *an absolute minimum requirement* is measurement of the response in cells of two unaffected and two affected persons. Both the controls and the affected persons should be genetically unrelated to one another.

The technical difficulties sometimes inherent in experiments involving cell strains make the use of transformed cell lines a temptation. Whether transformed material can always be considered to have a representative phenotype has, however, been questioned. Heddle and Arlett [81] have reported that, although an SV40-transformed XP-cell line from group A (XP12RO-SV40) was hypersensitive to both sister-chromatid exchange (SCE) [203] and cell killing by ethyl methanesulfonate, this was not so in three nontransformed cell strains from the same complementation group. Furthermore, increased lethality after ethyl methanesulfonate treatment was not observed in XP12RO cells in the absence of transformation. In accord with the data of Heddle and Arlett [81] are those of Ishizaki et al. [94] who observed an increased sensitivity in several XP cell strains to the protease inhibitor antipain. In an SV40-transformed derivative of one of these strains (XP2OS), this sensitivity was lost, although excision repair remained the same. A difference in response between transformed and nontransformed cells has also been observed in terms of uv-induced SCE in Cockayne's syndrome cells [29,132] and ionizing radiation-induced aberrations in Down's syndrome cells [91,154].

XERODERMA PIGMENTOSUM AS A MODEL

The main features of XP are of great interest as a model for other mutagen hypersensitive syndromes because so much is known about it. Not only was XP the first to be discovered but also its etiology was apparent from the direct relationship between exposure to sunlight and the predisposition to develop cancer. The extensive information on uv damage and repair in microorganisms and in cultured mammalian cells provided a framework not available for other mutagen hypersensitivities. Thus, in contrast to other syndromes, the discovery of a repair defect [30] was followed quickly by the demonstration of a cellular hypersensitivity [31]. There is, nevertheless, much to be learned even about XP, as no specific enzyme or protein defect has yet been identified as causal in XP, nor has the mechanistic relationship between the cellular sensitivity and the cancers or other clinical symptoms been established clearly.

Like the chromosome-breakage syndromes, XP is inherited as an autosomal recessive trait. Of the clinical features of XP, which have been reviewed elsewhere [21,149], only two will be mentioned here. The most revealing feature is the development of erythema, pigmentation, and cancer in those parts of the skin exposed to light, but not elsewhere. The most intriguing feature is the involvement of neural disorders in some forms of XP, but not all. The clinical indications that there is more than one form of the condition have been confirmed in genetic and biochemical studies of fibroblast cultures [100]. The most important information to emerge from the study of XP in vitro, however, is that cells of all affected persons are defective in their ability to handle uv-induced DNA damage [50] and are hypersensitive to the cytotoxic effects of uv light, in consequence. In most XP cell strains this manifests itself as a reduced capacity to excise pyrimidine dimers [30,140,148,168,184], the major photoproduct produced by 254-nm light. It is noteworthy that the wavelength of uv used for most studies of XP is not a major component of sunlight. This is an important point because it is likely that most of the known mutagen

TABLE VII. Cellular Characteristics of Syndromes Belonging to Groups II, IV, and V†

Syndrome (group)	Chromosomal aberrations a	SCE b	Mutation c	Immunity d	Cancer e	Mutagen sensitivity f	References
Dyskeratosis congenita (II)	↑; N	↑;N		↓	↑	SCE: ↑ (trimethylpsoralen + UVA)	a[67,161; 23]; b[23; 26,27];d[67, 161,171]; e[171];f[26,27]
Basal cell nevus syndrome (II)	↑; N	N	N		↑	Cell killing: N (γ-rays)	a[73,88; 70]; b[70];c[6]; e[71]; f[187]
Familial retinoblastoma (IV)		↑	N		↑	Chromosomal aberrations ↑(x-rays) SCE: ↑ (x-rays) Cell killing: ↑V (x-rays, γ-rays)	b[162]; c[6]; e[103]; f[6–8,162,197, 198]
Gardner's syndrome (IV)	↑V	N			↑	SCE: N (MMC)	a[37–40; 188]; b[188];

Huntington's chorea (IV)	N		Cell killing: ↑(MNU); ↑$_V$ (x-, γ-rays) Cell killing: ↑$_V$ (γ-rays)	e[57]; f[99,120] a[138]; b[138]; f[6]
Del[13] retinoblastoma (V)		↑	Cell killing: ↑$_V$ (x-, γ-rays)	b[191]; e[104]; f[8,197,198]
Down's syndrome and certain other trisomics (V)	N	N	Chromosomal aberrations: ↑ (x-, γ-rays, MNU, trenimon, "zitostop"); N (γ-rays, MMC, DEB) SCE: ↑ (x-rays, trenimon); N (x-rays, MMC) Cell killing: N (uv, x-rays) Mutation: N (x-rays)	a[112]; b[119,206]; c[205]; e[111]; f[1,10,13*, 35,36, 91*,97,112, 114,154, 160,205]

†Refer to Table I for explanation of symbols.
*Experiments involving transformed cell lines.

hypersensitivities are, similarly, experimental substitutes for those agents that are important clinically.

The defect in excision repair has made it possible, by means of cell-fusion techniques, to subdivide XP cell strains into seven complementation groups, A–G [4,98,106,194]. In addition, excision-proficient XP cells, XP "variants" [32], have been identified and shown to have a defect in postreplication repair of uv damage [117,118]. uv-induced lethality is less pronounced in variant strains [2,32,127]; no complementation groups have yet been identified [22]. The cellular studies thus show that mutations at any one of at least eight loci can cause clinical XP. One might expect that, in excision-defective cells at least, each complementation group would represent a genetic defect at a different step of a biochemical pathway. This expectation has not been realized, however. The weight of evidence would indicate that although the residual level of repair varies from less than 5% in group A to 60% in group E, the defect in each case lies at the same point, i.e., the incision step of excision repair [77,168,184]. In spite of this, no specific enzyme defect has been identified in XP. Furthermore, the results of Mortelmans et al. [133] have shown that extracts of XP cells *can* excise pyrimidine dimers from purified DNA and that extracts from all except group A cells can excise dimers from chromatin. These authors propose that XP cells may be defective in nonenzymic factors associated with making DNA lesions accessible to repair. In contrast to other hypersensitive syndromes, the ease with which the repair defect can be measured in XP has made it possible to explore not only genetic variability but also the quantitative relationship between the repair defect and the clinical severity of the disease. Agreement regarding this point has not been reached. Bootsma et al. [21] and Takebe [180] have provided data which suggest that an inverse relationship exists. Although this has not been confirmed by Robbins et al. [149], such a correlation may be obscured by the influence of environmental factors on the expression of clinical symptoms.

Two other features of XP are of particular interest because of their implications for the mechanisms of DNA repair in human cells: the response made by cells to mutagens other than uv, and the variety of cellular endpoints affected. XP cells for example, are hypersensitive to 4-nitroquinoline-1-oxide, a uv-mimetic agent in bacteria. A number of other cross-sensitivities have also been reported and an attempt has been made to review these data in Table I. Cross-sensitivity to several mutagens is a general feature of mutagen-hypersensitive syndromes, but the hypersensitivity need not be reflected by all endpoints. Ultimately the aim of cellular studies is to elucidate the mechanisms responsible for the particular disease. A significant observation in view of the increased cancer susceptibility in XP is that uv-induced mutability is elevated in vitro. This has been shown in both excision-deficient and excision-proficient cell strains with a variety of selective systems [6,69,128, 130,134]. At equal cytotoxicity (but different doses), similar numbers of mutants are induced in normal strains and excision-defective XP strains, but more are induced in variants [128,134]. It follows that postreplication repair plays an important role as a mutagenic pathway, although it is not important for cell survival.

XP cells also show increased mutability after treatment with various carcinogenic hydrocarbons [129] (Table I). In view of this and the somatic mutation theory of cancer, one might expect to see an increased susceptibility of XP individuals to internal tumors. No such susceptibility has been found [24]. This may be because mutagens to which XP is hypersensitive do not occur in the environment in amounts sufficient to cause detectable levels of cancer. Alternately, the scarcity of autopsy records and the early death of most XP patients may have rendered any susceptibility to internal tumors undetectable [51].

DEFICIENCIES IN DNA REPAIR AND ALTERNATE MODELS

The somatic mutation theory of cancer has received support from the finding that the majority of chemical carcinogens are either mutagenic or are metabolized to mutagens. Furthermore, of the three classes of carcinogens recognized—viruses, radiations, and chemicals—only chemicals seem likely to vary from place to place sufficiently to account for the differences in cancer rates in different populations, differences that have been shown to be environmental rather than intrinsic [101]. XP, although intrinsically hypersensitive, fits the theory because the induction of somatic mutations is elevated in XP cells exposed to ultraviolet light [124]. It is tempting, therefore, to assume that the other mutagen-hypersensitive syndromes have similar etiologies. The situation is more complex, however, because there is no obvious involvement of environmental agents. Furthermore, the process of transformation of a cell into a malignant tumor by a chemical has a number of different steps. The steps that are potentially involved under the somatic mutation theory can be classified as follows: (1) events that affect the extent of DNA damage (which could arise from changes in the uptake, metabolism, transport, or excretion of the chemical or its metabolites); (2) events that affect the proportions of different DNA lesions (which, again, could be produced by metabolic differences); (3) the repair or misrepair of lesions in DNA; and

(4) events unrelated to dose that must occur after exposure to a carcinogen before a cell expresses its malignant potential. Presumably, any of these processes could be altered genetically in a manner that would increase the frequency of malignant tumors. It is necessary, therefore, to consider four etiological models, corresponding to the four classes listed above, any of which could explain increased susceptibility to cancer.

The existence of mutagen hypersensitivity or of spontaneous chromosomal damage in cells cultured from cancer-prone persons eliminates any model in which the primary biochemical defect is in the process of carcinogenesis *following* DNA repair (i.e., eliminates class 4 above). Such defects would not be expected to lead to either mutagen hypersensitivity or spontaneous aberrations. (We do not rule out the possibility that chromosomal aberrations arising from a primary biochemical defect can have secondary effects on the later stages of carcinogenesis.) Three models remain, corresponding to (1) quantitative alterations in DNA damage, (2) qualitative alterations in DNA damage, and (3) changes in the repair of DNA damage. All could cause the mutagen hypersensitivity and would be consistent with an increased spontaneous frequency of aberrations under certain circumstances [167]. Obviously, measurements of all DNA lesions produced would distinguish among the three models. Unfortunately, measurements of DNA damage are often difficult and expensive, and rarely, if ever, are all lesions detectable. There is, however, an indirect approach.

If cells are hypersensitive because of a genetic defect that alters, quantitatively, the damage done by a given exposure to a mutagen, then all cellular endpoints ought to be affected to the same extent. If, for example, the DNA damage were doubled in the hypersensitive cells as compared to normal cells under the same conditions of exposure, then any endpoint measured in the hypersensitive cells ought to be that of the normal cells at twice this exposure. Hence, all cellular endpoints should be modified by the same multiplier of dose under model 1. In contrast, this would not necessarily occur as a result of a defect in DNA repair (model 3), nor as a result of qualitative changes in damage (model 2), because different lesions may be repaired by different pathways which have different efficiencies at producing mutations, aberrations, SCE, and other effects. Models 2 and 3 are not eliminated by a failure to detect different responses at different endpoints. Nevertheless, the relative sensitivity of cells, as judged by different endpoints, can provide important information concerning the underlying metabolic defect. This, in turn, can reveal the role of normal processes, such as DNA repair, in mutagenesis and carcinogenesis.

IMPLICATIONS OF MUTAGEN HYPERSENSITIVITY

The first cancer-prone syndrome to be studied systematically for a variety of mutagens was Fanconi's anemia (FA) (Table V). In a now-classical study, Sasaki and Tonomura [153] showed that in this syndrome the pattern of sensitivity was consistent with a defect in the repair of DNA cross-links, but more or less normal repair of other lesions in DNA. Studies of repair in cells of FA support this hypothesis [52]. It is particularly interesting that the hypersensitivity is much more pronounced for chromosomal aberrations and cell killing than for SCE and mutations, which are either unaffected or reduced in frequency [47,115]. This result is one of the clearest demonstrations of a difference in origin of aberrations and SCE. It is also an example of the usefulness of measurements of cellular sensitivity to mutagenic carcinogens in distinguishing among possible mechanisms. The spontaneous mutation rate in FA has been reported to be normal or subnormal [47].

In Bloom's syndrome (BS) (Table IV), no specific biochemical defect is known, but studies of mutagen sensitivity are consistent with either a defect in DNA repair or a qualitative alteration in DNA damage. The only hypersensitivity found in cells of all patients studied so far is to ethylating agents, as measured by SCE induction [107,108,110]. Cell survival does not reveal this hypersensitivity. In studies of sensitivity to uv light, some cell strains have been found to be hypersensitive and others not by measures of cell survival and SCE induction. Such variability also has been reported for SCE induction by mitomycin C. Whether the reason for the differing results is genetic diversity or something else, is not known. The spontaneous mutation rate is elevated in BS [72,193], as is the spontaneous rate of SCE formation and of chromosomal breakage. One cell strain has been studied for methylnitrosoguanidine mutagenesis and found to be normal [72].

An increased sensitivity to ionizing radiation and certain "radiomimetic" chemicals has been demonstrated in cells from patients with ataxia telangiectasia (AT) in terms of both chromosomal aberrations and cell killing (Table III). This has been correlated with a reduced level of DNA-repair synthesis in some strains, among which two complementation groups have been defined [139]. The defect in DNA repair may, however, be a secondary consequence of a different primary defect, inasmuch as in *all* cell strains of AT studied to date DNA synthesis is depressed to a lesser extent by ionizing radiations than in normal cells [89,201]. The spontaneous mutation rate appears to be normal.

It is evident that there is some relationship between DNA-repair defects, mutagen hypersensitivity, mutagenesis, and cancer in several syndromes. The most intellectually satisfying theory, and one that fits well with the finding that most chemical carcinogens are mutagens, is that specific locus mutations are the initial events in carcinogenesis and that defects in DNA repair result in enhanced mutagenesis. In practice, however, the data are not easily reconciled with this hypothesis. Only in BS is the spontaneous mutation rate elevated. Furthermore, although the mutation yield is abnormally elevated in XP by uv light, it may be subnormal in FA after mitomycin C treatment, and seems to be nonexistent in AT after treatment with ionizing radiations. Although it is not possible to exclude technical artefacts in measurements of mutation frequencies entirely, it is noteworthy that uv-induced mutations arise at normal rates in AT [7]. It may be, therefore, that chromosomal rearrangements play a more significant role in carcinogenesis than specific locus mutations. Certainly such rearrangements are known to be capable of altering gene expression. Furthermore, most, if not all, chemical mutagens are clastogenic. Even cell-lethal chromosomal aberrations may be important in carcinogenesis, because each dead cell must be replaced; this ultimately may lead to an increased probability of a quiescent but potentially tumorigenic cell being called upon to divide. The recent demonstration that diethylstilbestrol (a known clastogen) did not cause an increase in specific locus mutations in Syrian hamster cells, which are easily transformed by it, adds support to the concept that chromosomal mutations may play a more important role in carcinogenesis than specific locus mutation [15].

CONCLUSIONS

A review of the literature on mutagen sensitivity of the chromosome-breakage and other cancer-prone syndromes has revealed that for many syndromes the data are few and, sometimes, contradictory. We have speculated that some of the contradictions may arise from genetic diversity within the syndromes, or simply from unrelated genes present in the few individuals tested. It is important that sufficient cell donors be studied to define the response adequately in both the controls and the syndromes under investigation.

Measurements of mutagen sensitivity have proved useful in studies of the etiology of the syndromes, particularly when a hypersensitivity has been found. Such a hypersensitivity provides an experimental tool with which to investigate the mechanisms involved. Measurements of induced and spontaneous mutation rates, while consistent with the somatic mutation theory of cancer for XP, have provided data for the other syndromes which are not easily accommodated. It may be, therefore, that chromosomal aberrations, which occur in close parallel to the cancers, play an important role.

LITERATURE CITED

1. ALDENHOFF, P., WEGNER, R.-D., SPERLING, K.: Differential sensitivity of diploid and trisomic cells from patients with Down syndrome mosaic after treatment with the trifunctional alkylating agent trenimon. *Hum. Genet.* 56:123-125, 1980.
2. ANDREWS, A.D., BARRETT, S.F., ROBBINS, J.H.: Xeroderma pigmentosum neurological abormalities correlate with colony-forming ability after ultraviolet radiation. *Proc. Natl. Acad. Sci. USA* 75:1984-1988, 1978.
3. ANDREWS, A.D., BARRETT, S.F., YODER, F.W., ROBBINS, J.H.: Cockayne's syndrome fibroblasts have increased sensitivity to ultraviolet light but normal rates of unscheduled DNA synthesis. *J. Invest. Dermatol.* 70:237-239, 1978.
4. ARASE, S., KOZUKA, T., TANAKA, K., IKENAGA, M., TAKEBE, H.: A sixth complementation group in xeroderma pigmentosum. *Mutat. Res.* 59:143-146, 1979.
5. ARLETT, C.F., LEHMANN, A.R.: Human disorders showing increased sensitivity to the induction of genetic damage. *Annu. Rev. Genet.* 12:95-115, 1978.
6. ARLETT, C.F.: Survival and mutation in gamma-irradiated human cell strains from normal or cancer-prone individuals. *In* Proc. 6th Int. Congr. Radiat. Res. (Okada, S., Imamura, M., Terashima, T., Yamaguchi, H., eds.), Tokyo: Publ. Jpn. Assoc. Radiat. Res., pp. 596-602, 1979.
7. ARLETT, C.F., HARCOURT, S.A.: Cell killing and mutagenesis in repair-defective human cells. *In* DNA Repair Mechanisms (Hanawalt, P.C., Friedberg, E.C., Fox, C.F., eds.), New York: Academic Press, pp. 633-636, 1978.
8. ARLETT, C.F., HARCOURT, S.A.: Survey of radiosensitivity in a variety of human cell strains. *Cancer Res.* 40:926-932, 1980.
9. ARLETT, C.F., HARCOURT, S.A., LEHMANN, A.R., STEVENS, S., FERGUSON-SMITH, M.A., MORLEY, W.N.: Studies on a new case of xeroderma pigmentosum (XP3BR) from complementation group G with cellular sensitivity to ionizing radiation. *Carcinogenesis* 1:745-751, 1980.
10. AUERBACH, A.D., WOLMAN, S.R.: Susceptibility of Fanconi's anaemia fibroblasts to chromosome damage by carcinogens. *Nature* 261:494-496, 1976.
11. AUERBACH, A.D., WOLMAN, S.R.: Carcinogen-induced chromosome breakage in chromosome instability syndromes. *Cancer Genet. Cytogenet.* 1:21-28, 1979.
12. BAKER, R.M., ZUERNDORFER, G., MANDEL, G.: Enhanced susceptibility of a xeroderma pigmentosum cell line to mutagenesis by MNNG and EMS. *Environ. Mutagen.* 2:269-270, 1980.
13. BANERJEE, A., JUNG, O., HUANG, C.C.: Response of hematopoietic cell lines derived from patients with Down's syndrome and from normal individuals to mitomycin C and caffeine. *J. Natl. Cancer Inst.* 59:37-39, 1977.
14. BARFKNECHT, T.R., LITTLE, J.B.: Ataxia telangiectasia fibroblasts are abnormally sensitive to some DNA alkylating agents. *Proc. Am. Assoc. Cancer Res.* 21:44, 1980.
15. BARRETT, J.C.: Cell transformation, mutation and cancer. *In* The Use of Mammalian Cells for Detection of Environmental Carcinogens. Mechanisms and Application (Heidelberger, C., Inui, N., Kuroki, T., Yamada, M., eds.). Gann Monograph on Cancer Research (in press).

16. BARTRAM, C.R., KOSKE-WESTPHAL, T., PASSARGE, E.: Chromatid exchanges in ataxia telangiectasia, Bloom syndrome, Werner syndrome and xeroderma pigmentosum. *Ann. Hum. Genet.* 40:79-86, 1976.
17. BERGER, R., BERNHEIM, A., GLUCKMAN, E., GISSELBRECHT, C.: In vitro effect of cyclophosphamide metabolites on chromosomes of Fanconi anaemia patient. *Br. J. Haematol.* 45:565-568, 1980.
18. BERGER, R., BERNHEIM, A., LE CONIAT, M., VECCHIONE, D., SCHAISON, G.: Effect du chlorhydrate de chlorméthine sur les chromosomes dans l'anémie de Fanconi: Application au diagnostic et a la détection des hétérozygotes. *C. R. Acad. Sci. Ser. D* 290:457-459, 1980.
19. BERGER, R., BERNHEIM, A., LE CONIAT, M., VECCHIONE, D., SCHAISON, G.: Nitrogen mustard-induced chromosome breakage: A tool for Fanconi's anemia diagnosis. *Cancer Genet. Cytogenet.* 2:269-274, 1980.
20. BERGER, R., BERNHEIM, A., LE CONIAT, M., VECCHIONE, D., SCHAISON, G.: Sister chromatid exchanges induced by nitrogen mustard in Fanconi's anemia. Application to the detection of heterozygotes and interpretation of the results. *Cancer Genet. Cytogenet.* 2:259-267, 1980.
21. BOOTSMA, D.: Defective DNA repair and cancer. *In* Research in Photobiology (Castellani, A., ed.), New York: Plenum Press, pp. 455-468, 1976.
22. BOOTSMA, D.: Xeroderma pigmentosum. *In* DNA Repair Mechanisms (Hanawalt, P.C., Friedberg, E.C., Fox, C.F., eds.), New York: Academic Press, pp. 589-601, 1978.
23. BURGDORF, W., KURVINK, K., CERVENKA, J.: Sister chromatid exchange in dyskeratosis congenita lymphocytes. *J. Med. Genet.* 14:256-257, 1977.
24. CAIRNS, J.: The origin of human cancers. *Nature* 289:353-357, 1981.
25. CARNEY, R.G.: Incontinentia pigmenti. A world statistical analysis. *Arch. Dermatol.* 112:535-542, 1976.
26. CARTER, D.M., GAYNOR, A., MCGUIRE, J.: Sister chromatid exchange in Dyskeratosis congenita after exposure to trimethyl psoralen and UV light. *In* DNA Repair Mechanisms (Hanawalt, P.C., Friedberg, E.C., Fox, C.F., eds.), New York: Academic Press, pp. 671-674, 1978.
27. CARTER, D.M., PAN, M., GAYNOR, A., MCGUIRE, J.S., SIBRACK, L.: Psoralen-DNA cross-linking photoproducts in Dyskeratosis congenita: Delay in excision and promotion of sister chromatid exchange. *J. Invest. Dermatol.* 73:97-101, 1979.
28. CHAGANTI, R.S.K., SCHONBERG, S., GERMAN, J.: A manyfold increase in sister chromatid exchanges in Bloom's syndrome lymphocytes. *Proc. Natl. Acad. Sci. USA* 71:4508-4512, 1974.
29. CHENG, W.S., TARONE, R.E., ANDREWS, A.D., WANG-PENG, J.S., ROBBINS, J.H.: Ultraviolet light-induced sister chromatid exchange in xeroderma pigmentosum and Cockayne's syndrome lymphocyte cell lines. *Cancer Res.* 38:1601-1609, 1978.
30. CLEAVER, J.E.: Defective repair replication of DNA in xeroderma pigmentosum. *Nature* 218:652-656, 1968.
31. CLEAVER, J.E.: DNA repair and radiation sensitivity in human (xeroderma pigmentosum) cells. *Int. J. Radiat. Biol.* 18:557-565, 1970.
32. CLEAVER, J.E.: Xeroderma pigmentosum: Variants with normal DNA repair and normal sensitivity to ultraviolet light. *J. Invest. Dermatol.* 58:124-128, 1972.
33. CLENDENNING, W.E., BLOCK, J.B., RADDE, G.: Basal cell nevus syndrome. *Arch. Dermatol.* 90:38-53, 1964.
34. COPPEY, J., NOCENTINI, S.: Survival and herpes virus production of normal and

xeroderma pigmentosum fibroblasts after treatment with formaldehyde. *Mutat. Res.* 62:355-361, 1979.
35. COUNTRYMAN, P.I., HEDDLE, J.A., CRAWFORD, E.: The repair of X-ray-induced chromosomal damage in trisomy-21 and normal diploid lymphocytes. *Cancer Res.* 37:52-58, 1977.
36. CROSSEN, P.E., MORGAN, W.F.: Sensitivity of Down's syndrome lymphocytes to mitomycin C and X-irradiation measured by sister chromatid exchange frequency. *Cancer Genet. Cytogenet.* 2:281-285, 1980.
37. DANES, B.S.: The Gardner syndrome–A study in cell culture. *Cancer* 36:2327-2333, 1975.
38. DANES, B.S.: Increased tetraploidy: Cell specific for the Gardner gene in the cultured cell. *Cancer* 38:1983-1988, 1976.
39. DANES, B.S.: The Gardner syndrome: Increased tetraploidy in cultured skin fibroblasts. *J. Med. Genet.* 13:52-56, 1976.
40. DANES, B.S., KRUSH, A.J.: The Gardner syndrome: A family study in cell culture. *J. Natl. Cancer Inst.* 58:771-775, 1977.
41. DEBUSK, F.L.: Hutchinson-Gilford progeria syndrome. *J. Pediatr.* 80:697-724, 1972.
42. DENT, P.B., FISH, L.A., WHITE, J.G., GOOD, R.A.: Chediak-Higashi syndrome: Observations on the nature of the associated malignancy. *Lab. Invest.* 15:1634-1642, 1966.
43. DJAWARI, D., LUKASCHEK, E., JECHT, E.: Altered cellular immunity in Werner's syndrome. *Dermatologica* 161:233-237, 1980.
44. EVANS, H.J., ADAMS, A.C., CLARKSON, J.M., GERMAN, J.: Chromosome aberrations and unscheduled DNA synthesis in X- and UV-irradiated lymphocytes from a boy with Bloom's syndrome and a man with xeroderma pigmentosum. *Cytogenet. Cell Genet.* 20:124-140, 1978.
45. EPSTEIN, C.J., MARTIN, G.M., SCHULTZ, A.L., MOTULSKY, A.G.: Werner's syndrome: A review of its symptomatology, natural history, pathologic features, genetics and relationship to the natural aging process. *Medicine (Baltimore)* 45:77-221, 1966.
46. FINKELBERG, R., THOMPSON, M., SIMINOVITCH, L.: Survival after treatment with EMS, γ-rays, and mitomycin C of skin fibroblasts from patients with Fanconi's anemia. *Am. J. Hum. Genet.* 26:A30, 1974.
47. FINKELBERG, R., BUCHWALD, M., SIMINOVITCH, L.: Decreased mutagenesis in cells from patients with Fanconi's anaemia. *Am. J. Hum. Genet.* 29:42A, 1977.
48. FORNACE, A.J., JR., LITTLE, J.B., WEICHSELBAUM, R.R.: DNA repair in a Fanconi's anemia fibroblast cell strain. *Biochim. Biophys. Acta* 561:99-109, 1979.
49. FRAZELLE, J.H., HARRIS, J.S., SWIFT, M.: Response of Fanconi anemia fibroblasts to adenine and purine analogues. *Mutat. Res.* 80:373-380, 1981.
50. FRIEDBERG, E.C.: Xeroderma pigmentosum: Recent studies on the DNA repair defects. *Arch. Pathol. Lab. Med.* 102:3-7, 1978.
51. FRIEDBERG, E.C., EHMANN, U.K., WILLIAMS, J.I.: Human diseases associated with defective DNA repair. *Adv. Radiat. Biol.* 8:85-174, 1979.
52. FUJIWARA, Y., TATSUMI, M.: Repair of mitomycin C damage to DNA in mammalian cells and its impairment in Fanconi's anemia cells. *Biochem. Biophys. Res. Commun.* 66:592-598, 1975.
53. FUJIWARA, Y., TATSUMI, M.: Cross-link repair in human cells and its possible defect in Fanconi's anemia cells. *J. Mol. Biol.* 113:635-649, 1977.

54. FUJIWARA, Y., SASAKI, M.S.: Cross-link repair in mammalian cells and its impairment in Fanconi's anemia cells. *Mutat. Res.* 46:120, 1977.
55. GALLOWAY, S.M.: Ataxia telangiectasia: The effects of chemical mutagens and X-rays on sister chromatid exchanges in blood lymphocytes. *Mutat. Res.* 45:343-349, 1977.
56. GALLOWAY, S.M., EVANS, H.J.: Sister chromatid exchange in human chromosomes from normal individuals and patients with ataxia telangiectasia. *Cytogenet. Cell Genet.* 15:17-29, 1975.
57. GARDNER, E.J.: Discovery of the Gardner syndrome. *Birth Defects* 8(13):48-51, 1972.
58. GARRIGA, S., CROSBY, W.: The incidence of leukemia in families of patients with hypoplasia of the marrow. *Blood* 14:1008-1014, 1959.
59. GERMAN, J.: Chromosomal breakage syndromes. *Birth Defects* 5(5): 117-131, 1969.
60. GERMAN, J.: Bloom's syndrome. II. The prototype of human genetic disorders predisposing to chromosome instability and cancer. *In* Chromosomes and Cancer (German, J., ed.), New York: Wiley and Sons, pp. 601-617, 1974.
61. GERMAN, J.: The association of chromosome instability, defective DNA repair, and cancer in some rare human genetic diseases. *In* Human Genetics (Armendares, S., Lisker, R., eds.), Amsterdam: Excerpta Medica, pp. 64-68, 1977.
62. GERMAN, J., CRIPPA, L.P.: Chromosomal breakage in diploid cell lines from Bloom's syndrome and Fanconi's anemia. *Ann. Génét.* 9:143-154, 1966.
63. GERMAN, J., ARCHIBALD, R., BLOOM, D.: Chromosomal breakage in a rare and probably genetically determined syndrome of man. *Science* 148:506-507, 1965.
64. GERMAN, J., BLOOM, D., PASSARGE, E.: Bloom's syndrome. V. Surveillance for cancer in affected families. *Clin. Genet.* 12:162-168, 1977.
65. GARTLER, H.: Karzinombildung beim Werner Syndrom. *Dermatol. Wochenschr.* 49:606-616, 1964.
66. GIANNELLI, F., BENSON, P.F., PAWSEY, S.A., POLANI, P.E.: Ultraviolet light sensitivity and delayed DNA-chain maturation in Bloom's syndrome fibroblasts. *Nature* 265:466-469, 1977.
67. GIANNETTI, A., SEIDENARI, S.: Deficit of cell-mediated immunity, chromosomal alterations and defective DNA repair in a case of dyskeratosis congenita. *Dermatologica* 160:113-117, 1980.
68. GIVANTOS, F.: Human chromosomal abnormalities. *Bull. Tulane Med. Fac.* 20:241-253, 1961.
69. GLOVER, T.W., CHANG, C.C., TROSKO, J.E., LI, S.S.L.: Ultraviolet light induction of diptheria toxin-resistant mutants in normal and xeroderma pigmentosum fibroblasts. *Proc. Natl. Acad. Sci. USA* 76:3982-3986, 1979.
70. GORLIN, R.J.: Monogenic disorders associated with neoplasia. *In* Genetics of Human Cancer (Mulvihill, J.J., Miller, R.W., Fraumeni, J.F., Jr., eds.), New York: Raven Press, pp. 169-178, 1977.
71. GORLIN, R.J., SEDANO, H.O.: The multiple nevoid basal cell carcinoma syndrome revisited. *Birth Defects* 7(8):140-148, 1972.
72. GUPTA, R.S., GOLDSTEIN, S.: Diphtheria toxin resistance in human fibroblast cell strains from normal and cancer-prone individuals. *Mutat. Res.* 73:331-338, 1980.
73. HAPPLE, R., HOEHN, H.: Cytogenetic studies on cultured fibroblast-like cells derived from basal cell carcinoma tissue. *Clin. Genet.* 4:17-24, 1973.
74. HARNDEN, D.G.: Ataxia telangiectasia syndrome: Cytogenetic and cancer aspects. *In* Chromosomes and Cancer (German, J., ed.), New York: Wiley and Sons, pp. 619-636, 1974.

75. HARTLEY-ASP, B.: The influence of caffeine on the mitomycin-C induced chromosome aberration frequency in normal human and xeroderma pigmentosum cells. *Mutat. Res.* 49:117-126, 1977.
76. HATCHER, N.H., BRINSON, P.S., HOOK, E.B.: Sister chromatid exchanges in Ataxia telangiectasia. *Mutat. Res.* 35:333-336, 1976.
77. HAYAKAWA, H., ISHIZAKI, K., INOUE, M., YAGI, T., SEKIGUCHI, M., TAKEBE, H.: Repair of ultraviolet radiation damage in xeroderma pigmentosum cells belonging to complementation group F. *Mutat. Res.* 80:381-388, 1981.
78. HAYASHI, K., SCHMID, W.: The rate of sister chromatid exchanges parallel to spontaneous chromosome breakage in Fanconi's anemia and to trenimon-induced aberrations in human lymphocytes and fibroblasts. *Humangenetik* 29:201-206, 1975.
79. HECHT, F., KAISER McCAW, B.: Chromosome instability syndromes. *In* Genetics of Human Cancer (Mulvihill, J.J., Miller, R.W., Fraumeni, J.F., Jr., eds.), New York: Raven Press, pp. 105-123, 1977.
80. HECHT, F., KOLER, R.D., RIGAS, D.A., DAHNKE, G.S., CASE, M.P., TISDALE, V., MILLER, R.W.: Leukaemia and lymphocytes in ataxia-telangiectasia. *Lancet* ii:1193, 1966.
81. HEDDLE, J.A., ARLETT, C.F.: Untransformed xeroderma pigmentosum cells are not hypersensitive to sister chromatid exchange production by EMS. Implication for the use of transformed cell lines and the mechanism by which SCE arise. *Mutat. Res.* 72:119-125, 1980.
82. HEDDLE, J.A., LUE, C.B., SAUNDERS, F., BENZ, R.D.: Sensitivity to five mutagens in Fanconi's anemia as measured by the micronucleus method. *Cancer Res.* 38:2983-2988, 1978.
83. HEFLICH, R.H., DORNEY, D.J., MAHER, V.M., McCORMICK, J.J.: Reactive derivatives of benzo(a)pyrene and 7,12-dimethylbenz(a) anthracene cause S_1 nuclease sensitive sites in DNA and "UV-like" repair. *Biochem. Biophys. Res. Commun.* 77:634-641, 1977.
84. HIGURASHI, M., CONEN, P.E.: In vitro chromosomal radiosensitivity in Fanconi's anemia. *Blood* 38:336-342, 1971.
85. HIGURASHI, M., CONEN, P.E.: In vitro chromosomal radiosensitivity in chromosomal breakage syndromes. *Cancer* 32:380-383, 1973.
86. HOAR, D.I., SARGENT, P.: Chemical mutagen hypersensitivity in ataxia telangiectasia. *Nature* 261:590-592, 1976.
87. HOAR, D.I., WAGHORNE, C.: DNA repair in Cockayne syndrome. *Am. J. Hum. Genet.* 30:590-601, 1978.
88. HORLAND, A.A., WOLMAN, S.R., COX, R.P.: Cytogenetic studies in patients with the basal cell nevus syndrome and their relatives. *Am. J. Hum. Genet.* 27:47A, 1975.
89. HOULDSWORTH, J., LAVIN, M.F.: Effect of ionizing radiation on DNA synthesis in ataxia telangiectasia cells. *Nucleic Acid Res.* 8:3709-3720, 1980.
90. HUANG, C.C.: Sensitivity to radiation and chemicals of cell cultures derived from patients with Down syndrome (DS), Fanconi anemia (FA) and normal persons. *Proc. Am. Assoc. Cancer Res.* 17:20, 1976.
91. HUANG, C.C., BANERJEE, A., TAN, J.C., HOU, Y.: Comparison of radiosensitivity between human hematopoietic cell lines derived from patients with Down's syndrome and from normal persons. *J. Natl. Cancer Inst.* 59:33-36, 1977.
92. HUTTEROTH, T.H., LITWIN, S.D., GERMAN, J.: Abnormal immune response of Bloom's syndrome lymphocytes in vitro. *J. Clin. Invest.* 56:1-7, 1975.
93. IANCU, T., KOMLOS, L., SHABTAY, F., ELIAN E., HALBRECHT, I., BOOK, J.A.: Incontinentia pigmenti. *Clin. Genet.* 7:103-106, 1975.

94. ISHIZAKI, K., YAGI, T., TAKEBE, H.: Cytotoxic effects of protease inhibitors on human cells. I. High sensitivity of xeroderma pigmentosum cells to antipain. *Cancer Lett.* 10:199–205, 1980.
95. ISHIZAKI, K., YAGI, T., INOUE, M., NIKAIDO, O., TAKEBE, H.: DNA repair in Bloom's syndrome fibroblasts after UV irradiation or treatment with mitomycin C. *Mutat. Res.* 80:213–219, 1981.
96. JARETT, A., SPEARMAN, R.I.C., RILEY, P.A.: Dermatology. A Functional Introduction. London: English Universities Press Ltd., 246 pp., 1966.
97. KAINA, B., WALLER, H., RIEGER, R.: The action of N-methyl-N-nitrosourea on nonestablished human cell lines in vitro. I. Cell cycle inhibition and aberration induction in diploid and Down's fibroblasts. *Mutat. Res.* 43:387–400, 1977.
98. KEIJZER, W., JASPERS, N.G.J., ABRAHAMS, P.J., TAYLOR, A.M.R., ARLETT, C.F., ZELLE, B., TAKEBE, H., KINMONT, P.D.S., BOOTSMA, D.: A seventh complementation group in excision deficient xeroderma pigmentosum. *Mutat. Res.* 62:183–190, 1979.
99. KINSELLA, T.J., LITTLE, J.B., NOVE, J., WEICHSELBAUM, R.R.: In vitro study of skin fibroblasts from two families with Gardner's syndrome to the lethal effects of X-ray, ultraviolet light and mitomycin-C. *28th. Annu. Meet. Radiat. Soc.:*110, 1980.
100. KLEIJER, W.J., DE WEERD-KASTELEIN, E.A., SLUYTER, M.L., KEIJZER, W., DE WIT, J., BOOTSMA, D.: UV-induced DNA repair synthesis in cells of patients with different forms of xeroderma pigmentosum and of heterozygotes. *Mutat. Res.* 20:417–428, 1973.
101. KMET, J.: The role of migrant population in studies of selected cancer sites: A review. *J. Chronic. Dis.* 23:305–335, 1970.
102. KNOX, S.J., SHIFRINE, M., WILSON, F.D., GREENBERG, B., ROSENBLATT, L.S.: Increased radiosensitivity of T-lymphocyte progenitor cells from patients with Fanconi's anemia and other immunologic and hematologic disorders. *29th. Annu. Meet. Radiat. Res. Soc.:*111, 1981.
103. KNUDSON, A.G., JR.: Mutation and cancer: Statistical study of retinoblastoma. *Proc. Natl. Acad. Sci. USA* 68:820–823, 1971.
104. KNUDSON, A.G., MEADOWS, A.T., NICHOLS, W.W., HILL, R.: Chromosomal deletion and retinoblastoma. *N. Engl. J. Med.* 295:1120–1123, 1976.
105. KOBLENZER, P.J.: Genetic disorders involving various systems. Skin. *In* Genetic Disorders of Man (Goodman, R.M., ed.), Boston: Little, Brown and Co., pp. 265–327, 1970.
106. KRAEMER, K.H., DE WEERD-KASTELEIN, E.A., ROBBINS, J.H., KEIJZER, W., BARRETT, S.F., PETINGA, R.A., BOOTSMA, D.: Five complementation groups of xeroderma pigmentosum. *Mutat. Res.* 33:327–340, 1976.
107. KREPINSKY, A.B., HEDDLE, J.A., GERMAN, J.: Sensitivity of Bloom's syndrome lymphocytes to ethyl methanesulfonate. *Hum. Genet.* 50:151–156, 1979.
108. KREPINSKY, A.B., HEDDLE, J.A., RAINBOW, A.J., KWOK, A.: Sensitivity of Bloom's syndrome cells to specific mutagens. *Environ. Mutagen.* 1:188–189, 1979.
109. KREPINSKY, A.B., RAINBOW, A.J., HEDDLE, J.A.: Studies on the ultraviolet light sensitivity of Bloom's syndrome fibroblasts. *Mutat. Res.* 69:357–368, 1980.
110. KREPINSKY, A.B., GINGERICH, J., HEDDLE, J.A.: Further evidence for the hypersensitivity of Bloom syndrome cells to ethylating agents. *In* Chemical Mutagenesis, Human Population Monitoring and Genetic Risk Assessment (Bora, K.C., Douglas, G.R., Nestman, E.R., eds.), Amsterdam: Elsevier/North-Holland Biomedical Press, pp. 175–178, 1982.

111. KRIVIT, W., GOOD, R.A.: Simultaneous occurrence of mongolism and leukemia: Report of nation wide survey. *Am. J. Dis. Child.* 94:289-293, 1957.
112. KUCEROVA, M., POLIVKOVA, Z.: In vitro comparison of normal and trisomic cell sensitivity to physical and chemical mutagens. In Mutagen-Induced Chromosome Damage in Man (Evans, H.J., Lloyd, D.C., eds.), Edinburgh: University Press, pp. 185-190, 1978.
113. KUHN, E.M.: Effects of X-irradiation in G_1 and G_2 on Bloom's syndrome and normal chromosomes. *Hum. Genet.* 54:335-341, 1980.
114. LAMBERT, B., HANSSON, K., BUI, T.H., FUNES-CRAVIOTO, F., LINDSTEN, J., HOLMBERG, M., STRAUSMANIS, R.: DNA repair and frequency of X-ray and U.V.-light induced chromosome aberrations in leukocytes from patients with Down's syndrome. *Ann. Hum. Genet.* 39:293-303, 1976.
115. LATT, S.A., STETTEN, G., JUERGENS, L.A., BUCHANAN, G.R., GERALD, P.S.: Induction by alkylating agents of sister chromatid exchanges and chromatid breaks in Fanconi's anemia. *Proc. Natl. Acad. Sci. USA* 72:4066-4070, 1975.
116. LEHMANN, A.R., ARLETT, C.F.: Human genetic disorders with defects in repair of deoxyribonucleic acid. *Biochem. Soc. Trans.* 5:1199-1203, 1977.
117. LEHMANN, A.R., KIRK-BELL, S., ARLETT, C.F., PATERSON, M.C., LOHMAN, P.H.M., DE WEERD-KASTELEIN, E.A., BOOTSMA, D.: Xeroderma pigmentosum cells with normal levels of excision repair have a defect in DNA synthesis after ultraviolet light. *Proc. Natl. Acad. Sci. USA* 72:219-223, 1975.
118. LEHMANN, A.R., KIRK-BELL, S., ARLETT, C.F., HARCOURT, S.A., DE WEERD-KASTELEIN, E.A., KEIJZER, W., HALL-SMITH, P.: Repair of UV damage in a variety of human fibroblast cell strains. *Cancer Res.* 37:904-910, 1977.
119. LEZANA, E.A., BIANCHI, N.O., BIANCHI, M.S., ZABALA-SUAREZ, J.E.: Sister chromatid exchanges in Down syndromes and normal human beings. *Mutat. Res.* 45:85-90, 1977.
120. LITTLE, J.B., NOVE, J., WEICHSELBAUM, R.R.: Abnormal sensitivity of diploid skin fibroblasts from a family with Gardner's syndrome to the lethal effects of X-irradiation, ultraviolet light and mitomycin-C. *Mutat. Res.* 70:241-250, 1980.
121. LORENZ, R., FUHRMANN, W.: Familial basal cell nevus syndrome. *Hum. Genet.* 44:153-163, 1978.
122. LUTZNER, M.A.: Nosology among the neoplastic genodermatoses. In Genetics of Human Cancer (Mulvihill, J.J., Miller, R.W., Fraumeni, J.F. Jr., eds.), New York: Raven Press, pp. 145-167, 1977.
123. MACDONALD, W.B., FITCH, K.D., LEWIS, I.: Cockayne's syndrome. An heredofamilial disorder of growth and development. *Pediatrics* 25:997-1007, 1960.
124. MAHER, V.M., MCCORMICK, J.J.: Effect of DNA repair on the cytotoxicity and mutagenicity of UV-irradiation and of chemical carcinogens in normal and XP cells. In Biology of Radiation Carcinogenesis (Yuhas, J.M., Tennant, R.W., Regan, J.B., eds.), New York: Raven Press, pp. 129-145, 1976.
125. MAHER, V.M., MCCORMICK, J.J.: Comparison of mutagenic effect of UV radiation and chemicals in normal and DNA-repair deficient human cells in culture. In Chemical Mutagens Vol. 6 (de Serres, F.J., Hollaender, A., eds.), New York-London: Plenum Press, pp. 309-329, 1980.
126. MAHER, V.M., BIRCH, N., OTTO, J.R., MCCORMICK, J.J.: Cytotoxicity of carcinogenic aromatic amides in normal and xeroderma pigmentosum fibroblasts with different DNA repair capabilities. *J. Natl. Cancer Inst.* 54:1287-1294, 1975.
127. MAHER, V.M., CURREN, R.D., OUELLETTE, L.M., MCCORMICK, J.J.: Role of DNA repair in the cytotoxic and mutagenic action of physical and chemical carcinogens.

In In Vitro Metabolic Activation in Mutagenesis Testing (de Serres, F.J., Fouts, J.R., Bend, J.R., Philpot, R.M., eds.), Amsterdam: Elsevier/North-Holland Biomedical Press, pp. 313-335, 1976.
128. MAHER, V.M., OUELLETTE, L.M., CURREN, R.D., McCORMICK, J.J.: Frequency of ultraviolet light induced mutations is higher in xeroderma pigmentosum variant than in normal human cells. *Nature* 261:593-595, 1976.
129. MAHER, V.M., McCORMICK, J.J., GROVER, P.L., SIMS, P.: Effects of DNA repair on the cytotoxicity and mutagenicity of polycyclic hydrocarbon derivatives in normal and xeroderma pigmentosum fibroblasts. *Mutat. Res.* 43:117-138, 1977.
130. MAHER, V.M., DORNEY, D.J., MENDRALA, A.L., KONZE-THOMAS, B., McCORMICK, J.J.: DNA excision repair process in human cells can eliminate the cytotoxic and mutagenic consequences of UV-irradiation. *Mutat. Res.* 62:311-323, 1979.
131. MARSHALL, R.R., SCOTT, D.: The relationship between chromosome damage and cell killing in UV-irradiated normal and xeroderma pigmentosum cells. *Mutat. Res.* 36:397-400, 1976.
132. MARSHALL, R.R., ARLETT, C.F., HARCOURT, S.A., BROUGHTON, B.A.: Increased sensitivity of cell strains from Cockayne's syndrome to sister-chromatid-exchange induction and cell killing by UV light. *Mutat. Res.* 69:107-112, 1980.
133. MORTELMANS, K., FRIEDBERG, E.C., SLOR, H., THOMAS, G., CLEAVER, J.E.: Defective thymine dimer excision by cell free extracts of xeroderma pigmentosum cells. *Proc. Natl. Acad. Sci. USA* 73:2757-2761, 1976.
134. MYHR, B.C., TURNBULL, D., DiPAOLO, J.A.: UV-mutagenesis of normal and xeroderma pigmentosum variant human fibroblasts. *Mutat. Res.* 62:341-353, 1979.
135. NATARAJAN, A.T., MEYERS, M.: Chromosomal radiosensitivity of ataxia telangiectasia cells at different cell cycle stages. *Hum. Genet.* 52:127-132, 1979.
136. NOVOTNA, B., GOETZ, P., SURKOVA, N.I.: Effects of alkylating agents on lymphocytes from controls and from patients with Fanconi's anemia. Studies of sister chromatid exchanges, chromosomal aberrations, and kinetics of cell division. *Hum. Genet.* 49:41-50, 1979.
137. PARRINGTON, J.M., DELHANTY, J.D.A., BADEN, H.P.: Unscheduled DNA synthesis, U.V.-induced chromosome aberrations and SV40 transformation in cultured cells from xeroderma pigmentosum. *Ann. Hum. Genet.* 35:149-160, 1971.
138. PARRINGTON J.M., CASEY, G., WEST, L., DE VASCONCELOS MAIA, V.: Frequency of chromosome aberrations and chromatid exchange in cultured fibroblasts from patients with xeroderma pigmentosum, Huntington's chorea and normal controls. *Mutat. Res.* 46:146, 1977.
139. PATERSON, M.C., SMITH, P.J.: Ataxia telangiectasia: An inherited human disorder involving hypersensitivity to ionizing radiation and related DNA-damaging chemicals. *Annu. Rev. Genet.* 13:291-318, 1979.
140. PATERSON, M.C., LOHMAN, P.H.M., SLUYTER, M.L.: Use of a new endonuclease from micrococcus luteus to monitor the progress of DNA repair in UV-irradiated human cells. *Mutat. Res.* 19:245-256, 1973.
141. PATERSON, M.C., SMITH, P.J., MIDDLESTADT, M.V., SMITH, B.P., BECH-HANSEN, N.T.: Hypersensitivity to ethylnitrosourea in cultured fibroblast strains from patients with ataxia telangiectasia. *Environ. Mutagen.* 1:153-154, 1979.
142. PETERSON, R.D.A., KELLY, W.D., GOOD, R.A.: Ataxia-telangiectasia. Its association with a defective thymus, immunological-deficiency disease, and malignancy. *Lancet* 1:1189-1193, 1964.
143. PETERSON, R.D.A., COOPER, M.D., GOOD, R.A.: Lymphoid tissue abnormalities associated with ataxia-telangiectasia. *Am. J. Med.* 41:342-359, 1966.

144. PERRY, P.E., JAGER, M., EVANS, H.J.: Mutagen-induced sister chromatid exchanges in xeroderma pigmentosum and normal lymphocytes. *In* Mutagen-Induced Chromosome Damage in Man (Evans, H.J., Lloyd, D.C., eds.), Edinburgh University Press, pp. 201-207, 1978.
145. PORTER, I.H., PAUL, B.: Chromosomal anomalies and malignancy. *Birth Defects* 10:54-59, 1974.
146. RARY, J.M., BENDER, M.A., KELLY, T.E.: Cytogenetic studies on ataxia telangiectasia. *Am. J. Hum. Genet.* 26:70A, 1974.
147. REED, W.B., BODER, E., GARDNER, M.: Congenital and genetic skin disorders with tumor formation. *Birth Defects* 10:265-284, 1974.
148. REGAN, J.D., SETLOW, R.B., LEVY, R.D.: Normal and defective repair of damaged DNA in human cells: A sensitive assay utilizing the photolysis of bromodeoxyuridine. *Proc. Natl. Acad. Sci. USA* 68:708-712, 1971.
149. ROBBINS, J.H., KRAEMER, K.H., LUTZNER, M.A., FESTOFF, B.W., COON, H.G.: Xeroderma pigmentosum: An inherited disease with sun sensitivity, multiple cutaneous neoplasms and abnormal DNA repair. *Ann. Intern. Med.* 80:221-248, 1974.
150. ROOK, A., DAVIS, R., STEVANOVIC, D.: Poikiloderma congenitale Rothmund-Thomson's syndrome. *Acta. Dermatol. Venereol.* 34:392-420, 1959.
151. SAN, R.H.C., STICH, W., STICH, H.F.: Differential sensitivity of xeroderma pigmentosum cells of different repair capacities to the chromosome breaking action of carcinogens and mutagens. *Int. J. Cancer* 20:181-187, 1977.
152. SASAKI, M.S.: DNA repair capacity and susceptibility to chromosome breakage in xeroderma pigmentosum. *Mutat. Res.* 20:291-293, 1973.
153. SASAKI, M.S., TONOMURA, A.: A high susceptibility of Fanconi's anemia to chromosome breakage by DNA cross-linking agents. *Cancer Res.* 33:1829-1836, 1973.
154. SASAKI, M.S., TONOMURA, A., MATSUBARA, S.: Chromosomal constitution and its bearing on the chromosomal radiosensitivity in man. *Mutat. Res.* 10:617-633, 1970.
155. SASAKI, M.S., TODA, K., OZAWA, A.: Role of DNA repair in the susceptibility to chromosome breakage and cell killing in cultured human fibroblasts. *In* Biochemistry of Cutaneous Epidermal Differentiation (Seiji, M., Bernstein, A., eds.), Tokyo: University of Tokyo Press, pp. 167-179, 1977.
156. SCHMICKEL, R.D., CHU, E.H.Y., TROSKO, J.E., CHANG, C.C.: Cockayne syndrome: A cellular sensitivity to ultraviolet light. *Pediatrics* 60:135-139, 1977.
157. SCHONWALD, A.D., PASSARGE, E.: UV-light induced sister chromatid exchange in xeroderma pigmentosum lymphocytes. *Hum. Genet.* 36:213-218, 1977.
158. SCHROEDER, T.M., MAUGE, C.: Influence of higher incubation temperature on spontaneous and induced chromosome instability. *Mutat. Res.* 29:283, 1975.
159. SCHROEDER, T.M., ANSCHUTZ, F., KNOPP, A.: Spontane Chromosomenaberrationen bei familiärer Panmyelopathie. *Humangenetik* 1:194-196, 1964.
160. SCHULER, D., FEKETE, G., DOBOS, M.: Mit einem alkylierendem Agens (Zitostop) in vitro induzierbare Mutationen bei Malignomen und bei Syndromen, die zur Malignität disponieren. *Humangenetik* 16:329-336, 1972.
161. SCOGGINS, R.B., PRESCOTT, K.J., ASHER, G.H., BLAYLOCK, W.K., BRIGHT, R.W.: Dyskeratosis congenita with Fanconi-type anemia: Investigations of immunologic and other defects. *Clin. Res.* 19:409, 1971.
162. SCOTT, D., ZAMPETTI-BOSSELER, F., BENSON, W.J.: Radiation sensitivity of human ataxia telangiectasia and retinoblastoma cells: Various end-points. *Radiat. Environ. Biophys.* 17:347, 1980.
163. SCUDIERO, D.A.: Repair deficiency in N'-methyl-N'-nitro-N'-nitrosoguanidine

treated ataxia telangiectasia fibroblasts. *In* DNA Repair Mechanisms (Hanawalt, P.C., Friedberg, E.C., Fox, C.F., eds.), New York: Academic Press, pp. 655-658, 1978.
164. SCUDIERO, D.A.: Decreased DNA repair synthesis and defective colony-forming ability of Ataxia telangiectasia fibroblast cell strains treated with N-methyl-N'-nitro-N-nitrosoguanidine. *Cancer Res.* 40:984-990, 1980.
165. SELSKY, C., WEICHSELBAUM, R., LITTLE, J.B.: Defective host-cell reactivation of UV-irradiated herpes simplex virus by Bloom's syndrome skin fibroblasts. *In* DNA Repair Mechanisms (Hanawalt, P.C., Friedberg, E.C., Fox, C.F., eds.), New York: Academic Press, pp. 555-558, 1978.
166. SELSKY, C.A., HENSON, P., WEICHSELBAUM, R.R., LITTLE, J.B.: Defective reactivation of ultraviolet light-irradiated herpesvirus by a Bloom's syndrome fibroblast strain. *Cancer Res.* 39:3392-3396, 1979.
167. SETLOW, R.B.: Repair deficient human disorders and cancer. *Nature* 271:713-717, 1978.
168. SETLOW, R.B., REGAN, J.D., GERMAN, J., CARRIER, W.L.: Evidence that xeroderma pigmentosum cells do not perform the first step in the repair of ultraviolet damage to their DNA. *Proc. Natl. Acad. Sci. USA* 64:1035-1041, 1969.
169. SHIRAISHI, Y., SANDBERG, A.A.: Effects of mitomycin C on sister chromatid exchange in normal and Bloom's syndrome cells. *Mutat. Res.* 49:233-238, 1978.
170. SHIRAISHI, Y., SANDBERG, A.A.: Effects of chemicals on the frequency of sister chromatid exchanges and chromosome aberrations in normal and Bloom's syndrome lymphocytes. *Cytobios* 26:97-108, 1979.
171. SIRINAVIN, C., TROWBRIDGE, A.: Dyskeratosis congenita: Clinical and genetic aspects. *J. Med. Genet.* 12:339-354, 1975.
172. SMITH, P.J., PATERSON, M.C.: In vitro co-sensitivity to near-UV light and ionizing radiation in two human syndromes. *Environ. Mutagen.* 1:153-154, 1979.
173. SMITH, P.J., PATERSON, M.C.: Defective DNA repair and increased lethality in ataxia telangiectasia cells exposed to 4-nitroquinoline-1-oxide. *Nature* 287:747-749, 1980.
174. SMITH, P.J., PATERSON, M.C.: Rothmund Thomson syndrome: In vitro radiosensitivity and defective DNA repair in cultured fibroblasts. *28th Annu. Meet. Radiat. Res. Soc.*:110, 1980.
175. SPERLING, K., WEGNER, R.-D., RIEHM, H., OBE, G.: Frequency and distribution of sister-chromatid exchanges in a case of Fanconi's anemia. *Humangenetik* 27:227-230, 1975.
176. STICH, H.F., LAISHES, B.A.: The response of xeroderma pigmentosum cells and controls to the activated mycotoxins, aflatoxins and sterigmatocystin. *Int. J. Cancer* 16:266-274, 1975.
177. STICH, H.F., SAN, R.H.C., KAWAZOE, Y.: Increased sensitivity of xeroderma pigmentosum cells to some chemical carcinogens and mutagens. *Mutat. Res.* 17:127-137, 1973.
178. STICH, H.F., STICH, W., SAN, R.H.C.: Chromosome aberrations in xeroderma pigmentosum cells exposed to the carcinogens, 4NQO and MNNG. *Proc. Soc. Exp. Biol. Med.* 142:1141-1144, 1973.
179. SWIFT, M.R., HIRSCHHORN, K.: Fanconi's anemia. Inherited susceptibility to chromosome breakage in various tissues. *Ann. Intern. Med.* 65:496-503, 1966.
180. TAKEBE, H.: Xeroderma pigmentosum: DNA repair defects and skin cancer. *Gann* 24:103-117, 1979.
181. TAKEBE, H., FURUYAMA, J., MIKI, Y., KONDO, S.: High sensitivity of xeroderma pig-

mentosum cells to the carcinogen 4NQO. *Mutat. Res.* 15:98, 1972.
182. TAN, C., ETCUBANAS, E., LIEBERMAN, P., ISENBERG, H., KING, O., MURPHY, M.L.: Chediak-Higashi syndrome in a child with Hodgkin's disease. *Am. J. Dis. Child.* 121:135-139, 1971.
183. TANAKA, H., ORII, T.: High sensitivity but normal DNA-repair activity after UV irradiation in Epstein-Barr virus-transformed lymphoblastoid cell lines from Chediak-Higashi syndrome. *Mutat. Res.* 72:143-150, 1980.
184. TANAKA, K., HAYAKAWA, H., SEKIGUCHI, M., OKADA, Y.: Specific action of T4 endonuclease V on damaged DNA in xeroderma pigmentosum cells in vitro. *Proc. Natl. Acad. Sci. USA* 74:2958-2962, 1977.
185. TAYLOR, A.M.R., HARNDEN, D.G., FAIRBURN, E.A.: Chromosomal instability associated with susceptibility to malignant disease in patients with porokeratosis of Mibelli. *J. Natl. Cancer Inst.* 51:371-378, 1973.
186. TAYLOR, A.M.R., ROSNEY, C.M., CAMPBELL, J.B.: Unusual sensitivity of ataxia telangiectasia cells to bleomycin. *Cancer Res.* 39:1046-1050, 1979.
187. TAYLOR, A.M.R., HARNDEN, D.G., ARLETT, C.F., HARCOURT, S.A., LEHMANN, A.R., STEVENS, S., BRIDGES, B.A.: Ataxia telangiectasia: A human mutation with abnormal radiation sensitivity. *Nature* 258:427-429, 1975.
188. TICE, R., KRUSH, A., CHAILLET, J., SCHNEIDER, E.: Chromosomal studies on Gardner syndrome cells. Increased aneuploidy in cultured lymphocytes. *Am. J. Hum. Genet.* 27:89A, 1975.
189. TICE, R.R., RARY, J.M., BENDER, M.A.: An investigation of DNA repair in Bloom's syndrome. *In* DNA Repair Mechanisms (Hanawalt, P.C., Friedberg, E.C., Fox, C.F., eds.), New York: Academic Press, pp. 659-662, 1978.
190. THIELMANN, H.W., WITTE, I.: Correlation of the colony forming abilities of xeroderma pigmentosum fibroblasts with repair-specific DNA incision reactions catalyzed by cell free extracts. *Arch. Toxicol.* 44:197-207, 1980.
191. TURLEAU, C., CABANIS, M-O., DEGROUCHY, J.: Augmentation des échanges de chromatides dans les fibroblastes d'un enfant de del(13)-rétinoblastome. *Ann. Genet.* (Paris) 23:169-170, 1980.
192. WADE, M.H., CHU, E.H.Y.: Effects of DNA damaging agents on cultured fibroblasts derived from patients with Cockayne syndrome. *Mutat. Res.* 59:49-60, 1979.
193. WARREN, S.T., SCHULTZ, R.A., CHANG, C-C., WADE, M.H., TROSKO, J.E.: Elevated spontaneous mutation rate in Bloom syndrome fibroblasts. *Proc. Natl. Acad. Sci. USA* 78:3133-3137, 1981.
194. DE WEERD-KASTELEIN, E.A., KEIJZER, W., BOOTSMA, D.: Genetic heterogeneity of xeroderma pigmentosum demonstrated by somatic cell hybridization. *Nature* 238:80-83, 1972.
195. DE WEERD-KASTELEIN, E.A., KEIJZER, W., RAINALDI, G., BOOTSMA, D.: Induction of sister chromatid exchange in xeroderma pigmentosum cells after exposure to ultraviolet light. *Mutat. Res.* 45:253-261, 1977.
196. WEKSBERG, R., BUCHWALD, M., SARGENT, P., THOMPSON, M.W., SIMINOVITCH, L.: Specific cellular defects in patients with Fanconi anemia. *J. Cell. Physiol.* 101:311-324, 1979.
197. WEICHSELBAUM, R.R., NOVE, J., LITTLE, J.B.: Skin fibroblasts from a D-deletion type retinoblastoma patient are abnormally X-ray sensitive. *Nature* 266:726-727, 1977.
198. WEICHSELBAUM, R.R., NOVE, J., LITTLE, J.B.: X-ray sensitivity of diploid fibroblasts from patients with hereditary or sporadic retinoblastoma. *Proc. Natl. Acad.*

Sci. USA 75:3962-3964, 1978.
199. WEICHSELBAUM, R.R., NOVE, J., LITTLE, J.B.: X-ray sensitivity of 53 human diploid fibroblast cell strains from patients with characterized genetic disorders. Cancer Res. 40:920-925, 1980.
200. WINDMILLER, J., WHALLEY, P.J., FINK, C.W.: Cockayne's syndrome with chromosomal analysis. Am. J. Dis. Child. 105:118-122, 1963.
201. DE WIT, J., JASPERS, N.G.J., BOOTSMA, D.: The rate of DNA synthesis in normal and ataxia telangiectasia cells after exposure to X-irradiation. Mutat. Res. 80:221-226, 1981.
202. WOLFF, S., BODYCOTE, J., THOMAS, G.H., CLEAVER, J.E.: Sister chromatid exchange in xeroderma pigmentosum cells that are defective in DNA excision repair or postreplication repair. Genetics 81:349-355, 1975.
203. WOLFF, S., RODIN, B., CLEAVER, J.E.: Sister chromatid exchanges induced by mutagenic carcinogens in normal and xeroderma pigmentosum cells. Nature 265:347-349, 1977.
204. WOLFF, S.M., DALE, D.C., CLARK, R.A., ROOT, R.K., KIMBALL, H.R.: The Chediak-Higashi syndrome: Studies of host defenses. Ann. Intern. Med. 76:293-306, 1972.
205. YOTTI, L.P., GLOVER, T.W., TROSKO, J.E., SEGAL, D.J.: Comparative study of X-ray and UV induced cytotoxicity, DNA repair, and mutagenesis in Down's syndrome and normal fibroblasts. Pediatr. Res. 14:88-92, 1980.
206. YU, C.W., BORGAONKAR, D.S.: Normal rate of sister chromatid exchange in Down syndrome. Clin. Genet. 11:397-401, 1977.
207. ZBINDEN, I., CERUTTI, P.: Near-ultraviolet sensitivity of skin fibroblasts of patients with Bloom's syndrome. Biochem. Biophys. Res. Commun. 98:579-587, 1981.
208. NORDENSON, I.: Chromosome breaks in Werner's syndrome and their prevention in vitro by radical-scavenging enzymes. Hereditas 87:151-154, 1977.

REGULATION OF THE RESPONSES TO DNA DAMAGE IN THE HYPERSENSITIVITY DISEASES AND CHROMOSOME-BREAKAGE SYNDROMES

JAMES E. CLEAVER

INTRODUCTION

Many human genetic disorders, which affect a diverse set of organs, appear to exhibit an increased cellular sensitivity to DNA-damaging agents (Table I). Some diseases, notably xeroderma pigmentosum (XP), ataxia-telangiectasia (AT), and Fanconi's anemia (FA), show large increases in sensitivity (five- to tenfold) whereas others show slight increases of no more than a factor of two. Among the limited group that shows considerable hypersensitivity, only in XP is there a clear correlation between the hypersensitivity and the etiology of sunlight-induced skin cancers [12]. The hypersensitivity to x-rays in AT and to DNA cross-linking agents in FA does not have a clear correlation with the respective clinical features. AT and Bloom's syndrome also exhibit immunological disorders that may be more important than the hypersensitivity to DNA-damaging agents.

XP shows hypersensitivity not only to sunlight but also to most other DNA-damaging chemicals [12,14], and this may play a role in the etiology of the few internal cancers that have been reported in this disease (Table II). However, it is difficult to conclude that an increased frequency of internal cancers occurs in this disease, because the dosage of different environmental chemicals to which the patients may have been exposed is low in comparison to the relatively intense exposure of their skin to sunlight. Also, the reporting of internal cancers in XP may be incomplete, and in

TABLE I. Diseases Hypersensitive to DNA-Damaging Agents

Disease	DNA-damaging agent	D_0 ratio*	Reference
Xeroderma pigmentosum (Groups A,C, and D)	UV chemical carcinogens	5.0 –10	2
XP variant	UV	1.6	2
Ataxia-telangiectasia			
Homozygotes	x-rays	2.9 - 3.5	44,46
Heterozygotes	x-rays	0.9 - 1.2	44,46
Fanconi's anemia	Mitomycin-C	4.0 –15	22
	x-rays	1.0 - 2	56
Cockayne's syndrome			
Homozygotes	UV	4.6	53
Heterozygotes	UV	1.8	53
Chediak-Higashi syndrome	UV	2.2	27
Retinoblastoma (hereditary)	x-rays	1.2 - 1.5	55
Huntington's chorea	x-rays	1.25- 2.0	3
Partial trisomy 13	x-rays	1.6 - 2.0	56
Progeria	x-rays	1.1 - 1.6	56
Werner's syndrome	x-rays	1.1 –16	56
Gardner's syndrome	x-rays	1.0 - 2	56
(hereditary polyposis coli)	UV	2.3	35

* D_0 is the dose to reduce survival to 37% (l/e) along the exponential portion of the survival curves. The ratio is that between normal cells and the disease in question.

view of the small number of XP patients worldwide, even a 100-fold increase in internal cancers over the rate in the general population may be difficult to recognize.

MUTATIONS AND SISTER-CHROMATID EXCHANGES IN XP AND BLOOM'S SYNDROME – INSIGHTS INTO MECHANISMS OF CARCINOGENESIS

Strikingly absent from the list of hypersensitive disorders is Bloom's syndrome. This disease exhibits high frequencies of lymphatic cancer, immunological deficiencies [23], spontaneous mutations [54], and spontaneous sister-chromatid exchanges (SCEs) [8], but hypersensitivity to DNA-damaging agents is not a consistent feature [30]. The only major cellular abnormality that is expressed consistently in Bloom's syndrome is the production of high frequencies of SCEs [30].

Bloom's syndrome and XP, therefore, appear to be at opposite extremes in the context we are considering. Bloom's syndrome is, *par excellence*, the spontaneous chromosome-breakage syndrome; all of the abnormalities associated with it point to a cellular defect in the normal regula-

TABLE II. Internal Cancers in Xeroderma Pigmentosum

Type of cancer	Number of cases reported*
Tongue†	10
Leukemias	2
Brain	2
Lung	2
Peritoneal	1
Testis	1
Thyroid	1
Breast	1

* Statistics from Kraemer [30] and Hofmann et al. [24].
† Particularly common in Egyptian cases; about 25% of those reported show tongue tumors [16].

tion of chromosome stability. By contrast, XP shows no abnormal chromosome breakage or SCE production, or even clinical abnormalities of the skin unless exposed to DNA-damaging agents. Consistent with these features, Bloom's syndrome shows normal repair of DNA damage [1,48], whereas XP shows reduced repair [9,12,14].

A comparison between Bloom's syndrome and the various forms of XP allows interesting correlations among carcinogenesis, mutagenesis, and chromosome breakage or rearrangement. Two major classes of XP are known: the excision-defective complementation groups A–G [14] and the XP variant [10,17]. The excision-defective groups fail to excise DNA damage at normal rates [16,50,59] and exhibit increased frequencies of ultraviolet (UV)-induced point mutations [36] and SCEs [19]. The XP variant can excise DNA damage but has abnormalities in the replication of damaged DNA [15,17,33,45]. In addition, the XP variant exhibits increased UV-induced mutagenesis [37] but normal frequencies of UV-induced SCEs [19] (Fig. 1). Therefore, a comparison of the responses to UV light of Bloom's syndrome, excision-defective XP, and XP-variant cells indicates a constant correlation between increased mutagenesis and cancer, but not between increased SCEs or chromosome breakage and cancer. If SCEs and chromosome breakage are indicators of increased frequencies of genetic rearrangements, distinct from the point mutations [36,37,54], then genetic rearrangement does not appear to correlate with carcinogenesis.

MECHANISM OF SCE PRODUCTION

These observations (Fig. 1) also provide clues to the mechanisms of SCE production. The presence of increased amounts of DNA damage during DNA replication can give rise to SCEs [57]. Several models for

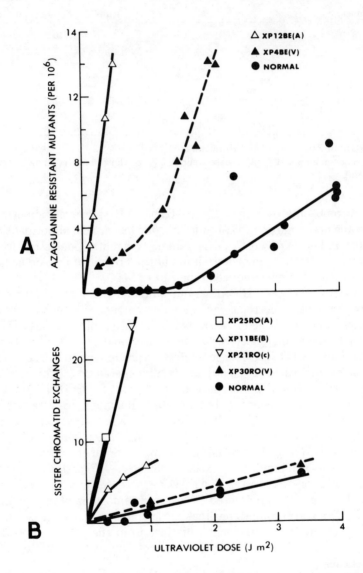

Fig. 1. Relative frequencies of UV-induced mutations to 8-azaguanine (20 μM) resistance or of SCEs in normal (●), excision-defective XP (○, ▽, △, □), and XP variant (▲) cells as a function of ultraviolet dose. A. Mutation frequencies redrawn from Maher et al. [36,37]. B. SCE frequencies redrawn from deWeerd-Kastelein et al. [19].

SCE production during replication of damaged DNA involve concepts of postreplication repair [26,39]. Both the normal frequencies of SCEs in UV-damaged XP variants, which have abnormalities in replication of damaged DNA [15,17,33], and the high frequencies of spontaneous SCEs in Bloom's syndrome which occur despite the absence of any major abnormality in repair or replication can be used to develop a new model for SCE production.

Because the XP variant characteristically replicates damaged DNA by stopping at almost every damaged site [15,17,33], it is not these blocked forks or single-strand gaps that stimulate increased SCE production (Fig. 1). Rather, it is the replication past damaged sites, without interruptions in newly replicated DNA, and the persistence of lesions behind replication forks on the parental strands that appear to stimulate production. Newly replicated DNA may be produced initially in an entwined form because of the higher-order structures found in the packaging of eukaryotic DNA [18] (Fig. 2). Recent discoveries of DNA topoisomerases that can unwind and unknot DNA by making and rejoining double-strand breaks may provide a mechanism that enables newly replicated DNA to be unraveled [4,25,32] (Fig. 2). The presence of unrepaired damage on recently replicated DNA could interfere with the function of topoisomerases so that the homologous chromatids are rejoined incorrectly, thus generating SCEs (Fig. 2). Recent discoveries that pyrimidine dimers [J.E. Cleaver, unpublished observations], bromodeoxyuridine [47], and alkylations [7] can have highly specific effects on the recognition and cleavage of double-stranded DNA by restriction enzymes illustrate how DNA damage may divert the action of topoisomerases.

Cells that replicate DNA from closely spaced origins (i.e., cells with small replicons) would probably require less unravelling of newly replicated DNA than cells with large replicons. Support for this idea comes from measurements of baseline SCE frequencies in various human and hamster cell lines grown in the same concentration of bromodeoxyuridine (Fig. 3). The SCE frequency is lowest in cells with small replicons and greatest in cells with large replicons.

Excision-defective XP cells presumably show high UV-induced SCE production because of the high frequency of damaged DNA that has replicated. Bloom's syndrome, however, may show high spontaneous frequencies of SCEs because of a defect in the topoisomerase mechanism itself, which is required for normal DNA replication. The increased mutagenesis in these diseases may be associated with both replication of damaged DNA and errors in SCE. Errors in SCE would be the major mechanism in Bloom's syndrome, but a less important mechanism in XP and XP variant cells.

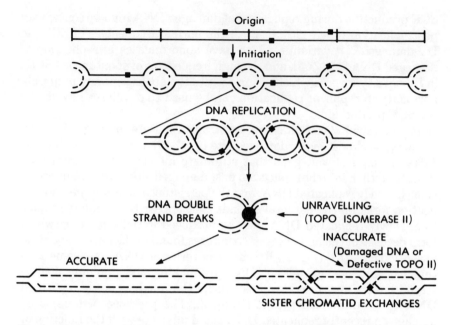

Fig. 2. Mechanism of SCE production. Replication that has gone past damaged sites leaves the damage in parental strands, in intact daughter double helices. Since replication will produce an entwined configuration of newly replicated DNA, this must be unraveled by double-strand breakage and reunion, probably involving the action of DNA topoisomerases. Errors introduced by DNA damage that causes faulty recognition and cutting of DNA by the topoisomerases may cause SCEs. Newly synthesized strands denoted by ----, damaged sites by ■. Site of each exchange and of damaged sites are shown coincident, but this has not been proven.

HYPERSENSITIVITY DISEASES AND REGULATORY SIGNALS

Replication of DNA before the cell has had the opportunity to repair damage is detrimental to cell survival and to the minimization of mutations and SCEs. Replication of damaged DNA may also be part of the chain of events that leads to transformation and carcinogenesis. DNA damage is always accompanied by an inhibition of DNA replication [38,40]. Until recently, this inhibition was assumed to be a passive process that was caused by the damage. However, recent observations indicate that this inhibition is an active part of the regulatory processes that are involved in recovery from damage. AT cells [45,46], and to a lesser extent retinoblastoma cells [55,56], are hypersensitive to killing by x-rays,

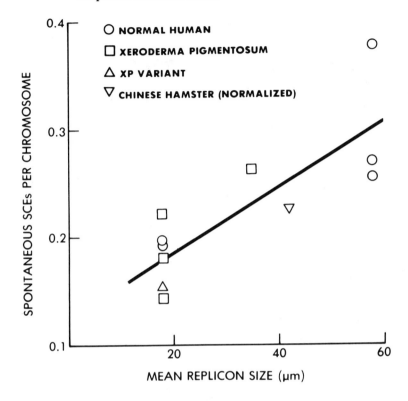

Fig. 3. Baseline (spontaneous) frequency of SCEs in various cell types as a function of average replicon size [28]. All cells were grown in 10 μM bromodeoxyuridine to determine baseline SCEs [52,58]. The data for Chinese hamster ovary cells were corrected by the ratio 20:46 to reduce the frequency to that for a hypothetical hamster cell with 46 chromosomes, as found in human cells. "Normal" human cells consisted of a variety of primary and transformed (SV40) fibroblasts, the latter having larger replicon sizes.

but DNA synthesis in these cells is more resistant than normal to inhibition by x-rays [20,42, J.E. Cleaver, N. Rand, and D.H. Char, unpublished observations] (Fig. 4). At first, this seems paradoxical. But if the inhibition of DNA synthesis in normal cells is an active process, then it may be part of a system that reduces the probability that cells will reach mitosis with unrepaired damage in DNA. When DNA synthesis fails to be inhibited in AT cells, replication of damaged regions of DNA will lead to cell death, mutagenesis, and transformation.

The regulatory signal that communicates DNA damage into an inhibition of DNA replication in AT cells may involve the synthesis of poly (ADP-ribose). The synthesis of this polymer usually is stimulated by the

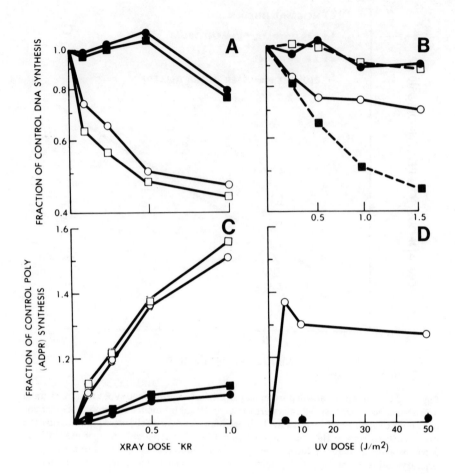

Fig. 4. Inhibition of DNA synthesis (A, B) and stimulation of poly (ADP-ribose) synthesis (C, D) by x-rays (A, C) or UV (B, D) in normal, AT, and XP cells. Data on AT cells measured 30 min after irradiation were redrawn from Edwards and Taylor [20]. Data on poly (ADP-ribose) synthesis in lymphoid XP cells were redrawn from Berger et al. [5]. Data on DNA synthesis in XP fibroblasts are from Cleaver [unpublished results]. The measurements of UV effects reflect general trends, but not necessarily precise quantitative relationships. A, C: ○, □ normal cells; ●, ■ AT cells. B, D: ○ normal cells measured 30 min (B) or 90 min after irradiation. D shows contrasting active and inactive poly(ADP-ribose) synthesis in normal and XP cells, respectively, at all times after irradiation until repair is complete in normal cells. □ normal cells measured 5 h after irradiation; ●, ■ XP group C cells measured 30 min or 5 h after irradiation, respectively. Solid lines connect the 5 h data, dashed lines connect the 30–90-min data.

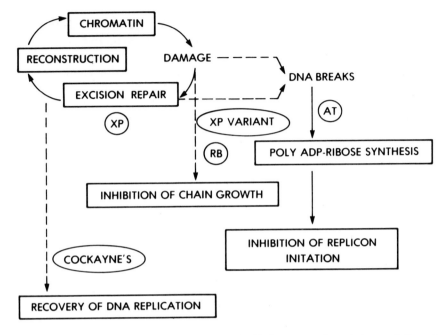

Fig. 5. Heuristic regulatory scheme relating DNA damage and repair to regulation of inhibition and recovery of DNA synthesis. The cycle shown in the top left represents the transient changes that occur when chromatin (i.e., DNA plus histones in nucleosomal structures) is damaged, then repaired, and the nucleosomes reconstructed. Damage and repair generate DNA breaks and the inhibition of chain growth occurs as a result of damaged sites blocking DNA polymerases. DNA breaks from direct damage or from repair stimulate poly (ADP-ribose) synthesis and subsequently inhibit replicon initiation. Solid lines connect the excision repair cycle and the poly (ADP-ribose) pathway; dashed lines set apart the repair cycle from events associated with DNA replication. The steps in the various pathways at which abnormalities have been seen in various human disorders are marked (XP, XP variant, etc.; RB = retinoblastoma).

presence of single-strand breaks in DNA. In AT cells, neither the x-ray-induced breaks nor the reduced repair breaks [44,46] seem to stimulate poly (ADP-ribose) synthesis as much as they do in irradiated normal cells, even though DNAse I-induced breaks do stimulate synthesis [20]. In addition, the coordination of replication in clusters of replicons seems to be lost in AT cells, and low doses of x-rays do not inhibit replicon initiation as they do in normal cells [41,42]. Poly (ADP-ribose) therefore may be part of a complex cellular regulatory process that translates DNA damage into an inhibition of DNA replication (Fig. 5).

The situation in XP cells irradiated with UV light is much more complex because the nature of the DNA damage and the inhibition of DNA replication are different. Base damages (pyrimidine dimers) are a major fraction of damage from UV light, whereas direct breaks are a major fraction of damage from x-rays. Single-strand breaks are generated to a significant extent only during excision repair of pyrimidine dimers soon after irradiation, and these breaks stimulate poly (ADP-ribose) synthesis [5]. Base damage from UV light blocks DNA-chain elongation directly, whereas single-strand breaks from x-rays and from excision of base damage inhibit the initiation of coordinated replicon synthesis [13,16,42, 45]. Consequently, after UV irradiation of XP (excision-defective) cells, the decreased frequency of excision-repair breaks results in decreased poly (ADP-ribose) synthesis [5]; this should result in less inhibition of DNA synthesis, but the unrepaired base damage independently inhibits chain elongation. Therefore, DNA replication is less sensitive in XP cells than in normal cells only shortly after irradiation (Fig. 4). It becomes more sensitive later, because the relative importance of excision-repair breaks and unrepaired base damage changes with time. Thus, the relationships among hypersensitivity, poly (ADP-ribose) synthesis, and inhibition of DNA replication exhibit characteristic alterations in AT and XP, according to the biochemical defects and the nature of the damage.

Once repair is complete and chromatin is reconstructed in the repaired region [6,11,51], DNA replication recovers. That this recovery is also an active process is illustrated by Cockayne's syndrome. This disorder is UV-sensitive (Table I) but exhibits no abnormalities in excision repair [1, 49,53]; DNA replication is inhibited by UV damage but fails to recover as it does in normal cells once extensive repair has occurred [34]. The size distributions of DNA synthesized throughout the period of declining DNA synthesis, as well as its assembly into high molecular weight DNA, appear to be normal [J.E. Cleaver, unpublished observations]. What seems to be aberrant in this disease is the lack of a regulatory signal that should switch replication on again during recovery. Although the nature of this signal is unknown, it may be something antagonistic to the synthesis of poly (ADP-ribose).

Fanconi's anemia is another disease in which there may be altered cellular regulation. The hypersensitivity of FA to DNA-crosslinking agents is not associated consistently with DNA-repair defects [21,22,29], but some abnormalities have been described in purine metabolism [M. Buchwald, unpublished observations]. FA, however, is less well understood than many other extreme hypersensitive diseases.

CONCLUSIONS

The discovery of hereditary human diseases, which often involve cancer, with characteristic cellular abnormalities in chromosome stability, SCE, and sensitivity to DNA-damaging agents has stimulated their investigation with respect to the regulation of DNA structure and function. The most hypersensitive diseases—XP, XP variant, Cockayne's syndrome and Fanconi's anemia—all have some abnormalities in DNA repair or replication, or the regulation of these processes. Bloom's syndrome, although not hypersensitive, may also have an abnormality involving replication and daughter-strand exchange. Many of the other diseases that show slight hypersensitivity may involve minor changes in these regulatory pathways or more distant alterations in cellular metabolism that indirectly influence regulation. The close association of defects in the regulation of DNA function, whether repair, replication, or chromosome stability, with clinical abnormalities that often lead to cancer highlights the importance of these phenomena and the necessity for their extensive investigation.

SUMMARY

Numerous human genetic diseases show various degrees of association between neoplasia and *in vitro* chromosome breakage and increased sensitivity to DNA damage. At one extreme, Bloom's syndrome shows major increases in spontaneous neoplasia, chromosome breakage, and SCE, but without major alterations in DNA repair or replication. At the other extreme, XP shows major increases in sunlight-induced skin cancer and possible increases in environmentally induced internal cancer as a result of defective DNA repair and replication. Other disorders, including AT, retinoblastoma, and Fanconi's anemia are intermediate between Bloom's syndrome and XP in terms of the association of neoplasia, chromosome breakage, and disturbed DNA metabolism. Bloom's syndrome may involve abnormal functioning of DNA topoisomerases that unravel newly replicated DNA. XP involves defective repair or replication that produces a constant correlation between mutagenesis and neoplasia but not between SCE (gene rearrangement?) and neoplasia. The other disorders may involve defects in the regulatory pathways that control the delays in DNA replication that normally provide time for completion of repair before replication.

Acknowledgments. This work was supported by the U.S. Department of Energy.

LITERATURE CITED

1. AHMED, F.E., SETLOW, R.B.: Excision repair in ataxia telangiectasia, Fanconi's anemia, Cockayne syndrome, and Bloom's syndrome after treatment with ultraviolet radiation and N-acetoxy-2-acetylaminofluorene. *Biochim. Biophys. Acta* 521: 805-817, 1978.
2. ANDREWS, A.D., BARRETT, S.F., ROBBINS, J.H.: Xeroderma pigmentosum neurological abnormalities correlate with colony-forming ability after ultraviolet radiation. *Proc. Natl. Acad. Sci. USA* 75:1984-1988, 1978.
3. ARLETT, C.F.: Survival and mutation in gamma-irradiated human cell strains from normal or cancer-prone individuals. *In* Radiation Research (Okada, S., Imamura, M., Terashima, T., Yamaguchi, H., eds.), Japanese Association for Radiation Research, Japan, 596-602, 1980.
4. BALDI, M.I., BENEDETTI, P., MATTOCCIA, E., TOCCHINI-VALENTINI, G.P.: In vitro catenation and decatenation of DNA and a novel eucaryotic ATP-dependent topoisomerase. *Cell* 20:461-467, 1980.
5. BERGER, N.A., SIKORSKI, G.W., PETZOLD, S.J., KUROHARA, K.K.: Defective poly (adenosine diphosphoribose) synthesis in xeroderma pigmentosum. *Biochemistry* 19:289-293, 1980.
6. BODELL, W.J., CLEAVER, J.E.: Transient conformation changes in chromatin during excision repair of ultraviolet damage to DNA. *Nucleic Acids Res.* 9:203-213, 1981.
7. BOEHM, T.L.J., DRAHOVSKY, D.: Impaired restriction endonuclease cleavage of DNA modified with N-methyl-N-nitrosourea. *Carcinogenesis* 1:729-731, 1980.
8. CHAGANTI, R.S.K., SCHONBERG, S., GERMAN, J.: A manyfold increase in sister chromatid exchanges in Bloom's syndrome lymphocytes. *Proc. Natl. Acad. Sci. USA* 71: 4508-4513, 1974.
9. CLEAVER, J.E.: Defective repair replication of DNA in xeroderma pigmentosum. *Nature* 218:652-656, 1968.
10. CLEAVER, J.E.: Xeroderma pigmentosum: Variants with normal DNA repair and normal sensitivity to ultraviolet light. *J. Invest. Dermatol.* 58:124-128, 1972.
11. CLEAVER, J.E.: Nucleosome structure controls rates of excision repair in DNA of human cells. *Nature* 270:451-453, 1977.
12. CLEAVER, J.E.: Xeroderma pigmentosum. *In* The Metabolic Basis of Inherited Disease (Stanbury, J.B., Wyngaarden, J.B., Frederickson, D.S., eds.), New York: McGraw-Hill Book Co., 1072-1095, 1978.
13. CLEAVER, J.E.: DNA repair and its coupling to DNA replication in eukaryotic cells. *Biochim. Biophys. Acta* 516:489-516, 1978.
14. CLEAVER, J.E., BOOTSMA, D.: Xeroderma pigmentosum: biochemical and genetic characteristics. *Annu. Rev. Genet.* 9:19-38, 1975.
15. CLEAVER, J.E., THOMAS, G.H., PARK, S.D.: Xeroderma pigmentosum variants have a slow recovery of DNA synthesis after irradiation with ultraviolet light. *Biochim. Biophys. Acta* 564:122-131, 1979.
16. CLEAVER, J.E., TROSKO, J.E.: Absence of excision of ultraviolet-induced cyclobutane dimers in xeroderma pigmentosum. *Photochem. Photobiol.* 11: 547-550, 1970.

17. CLEAVER, J.E., ZELLE, B., HASHEM, N., GERMAN, J.: Xeroderma pigmentosum patients from Egypt. II. Epidemiology, clinical symptoms and molecular biology. *J. Invest. Dermatol.* 77:96-101, 1981.
18. COOK, P.R., BRAZELL, I.A.: Detection and repair of single-strand breaks in nuclear DNA. *Nature* 263:679-682, 1976.
19. DE WEERD-KASTELEIN, E.A., KEIJZER, W., RAINALDI, G., BOOTSMA, D.: Induction of sister chromatid exchanges in xeroderma pigmentosum cells after exposure to ultraviolet light. *Mutat. Res.* 45:253-261, 1977.
20. EDWARDS, M.J., TAYLOR, A.M.R.: Unusual levels of (ADP-ribose)$_n$ and DNA synthesis in ataxia telangiectasia cells following γ-ray irradiation. *Nature* 287:745-747, 1980.
21. FORNACE, A.J., JR., LITTLE, J.B., WEICHSELBAUM, R.R.: DNA repair in a Fanconi's anemia fibroblast cell strain. *Biochim. Biophys. Acta* 561:99-109, 1979.
22. FUJIWARA, Y., TATSUMI, M., SASAKI, M.S.: Cross-link repair in human cells and its possible defect in Fanconi's anemia cells. *J. Mol. Biol.* 113:635-649, 1977.
23. GERMAN, J.: Genes which increase chromosomal instability in somatic cells and predispose to cancer. *Prog. Med. Genet.* 8:61-101, 1972.
24. HOFMANN, H., JUNG, E.G., SCHNYDER, U.W.: Pigmented xerodermoid: First report of a family. *Bull. Cancer (Paris)* 65:347-350, 1978.
25. HSIEH, T-S., BRUTLAG, D.: ATP-dependent DNA topoisomerase from D. melanogaster reversibly catenates duplex DNA rings. *Cell* 21:115-125, 1980.
26. ISHII, Y., BENDER, M.A.: Effects of inhibitors of DNA synthesis on spontaneous and ultraviolet light-induced sister-chromatid exchanges in Chinese hamster cells. *Mutat. Res.* 79:19-32, 1980.
27. KAPP, L.N., PARK, S.D., CLEAVER, J.E.: Replicon sizes in non-transformed and SV40-transformed cells, as estimated by a bromodeoxyuridine photolysis method. *Exp. Cell Res.* 123:375-377, 1979.
28. KAYE, J., SMITH, C.A., HANAWALT, P.C.: DNA repair in human cells containing photoadducts of 8-methoxypsoralen or angelicin. *Cancer Res* 40:696-702, 1980.
29. KRAEMER, K.H.: Xeroderma pigmentosum. *In* Clinical Dermatology (Demis, D.J., Dobson, R.L., McGuire, J., eds.), vol. 4, unit 19-7, 1-33, Hagerstown: Harper & Row, 1980.
30. KREPINSKY, A.B., RAINBOW, A.J., HEDDLE, J.A.: Studies on the ultraviolet light sensitivity of Bloom's syndrome fibroblasts. *Mutat. Res.* 69:357-368, 1980.
31. KREUZER, K.N., COZZARELLI, N.R.: Formation and resolution of DNA catenates by DNA gyrase. *Cell* 20:245-254, 1980.
32. LEHMANN, A.R., KIRK-BELL, S., ARLETT, C.F., PATERSON, M.C., LOHMAN, P.H.M., DEWEERD-KASTELEIN, E.A., BOOTSMA, D.: Xeroderma pigmentosum cells with normal levels of excision repair have a defect in DNA synthesis after UV-irradiation. *Proc. Natl. Acad. Sci. USA* 72:219-223, 1975.
33. LEHMANN, A.R., KIRK-BELL, S., MAYNE, L.: Abnormal kinetics of DNA synthesis in ultraviolet light irradiated cells from patients with Cockayne's syndrome. *Cancer Res.* 39:4237-4241, 1979.
34. LITTLE, J.B., NOVE, J., WEICHSELBAUM, R.R.: Abnormal sensitivity of diploid skin fibroblasts from a family with Gardner's syndrome to the lethal effects of x-irradiation, ultraviolet light and mitomycin-C. *Mutat. Res.* 70:241-250, 1980.
35. MAHER, V.M., DORNEY, D.J., MENDRALA, A.L., KONZE-THOMAS, B., MCCORMICK, J.J.: DNA excision-repair processes in human cells can eliminate the cytotoxic and mutagenic consequences of ultraviolet irradiation. *Mutat. Res.* 62:311-323, 1979.
36. MAHER, V.M., OUELLETTE, L.M., CURREN, R.D., MCCORMICK, J.J.: Frequency of

ultraviolet light-induced mutations is higher in xeroderma pigmentosum variant cells than in normal cells. *Nature* 261:593-595, 1976.
37. PAINTER, R.B.: Rapid test to detect agents that damage human DNA. *Nature* 265: 650-651, 1977.
38. PAINTER, R.B.: A replication model for sister-chromatid exchange. *Mutat. Res.* 70: 337-341, 1980.
39. PAINTER, R.B., HOWARD, R.: A comparison of the HeLa DNA-synthesis inhibition test and the Ames test for screening of mutagenic carcinogens. *Mutat. Res.* 54: 113-115, 1978.
40. PAINTER, R.B., YOUNG, B.R.: Formation of nascent DNA molecules during inhibition of replicon initiation in mammalian cells. *Biochim. Biophys. Acta* 418:146-153, 1976.
41. PAINTER, R.B., YOUNG, B.R.: Radiosensitivity in ataxia telangiectasia: A new explanation. *Proc. Natl. Acad. Sci. USA* 77:7315-7317, 1980.
42. PARK, S.D., CLEAVER, J.E.: Recovery of DNA synthesis after ultraviolet irradiation of xeroderma pigmentosum cells depends on excision repair and is blocked by caffeine. *Nucleic Acids Res.* 6:1151-1159, 1979.
43. PARK, S.D., CLEAVER, J.E.: Post-replication repair: Questions of its definition and possible alteration in xeroderma pigmentosum cell strains. *Proc. Natl. Acad. Sci. USA* 76:3927-3931, 1979.
44. PATERSON, M.C., ANDERSON, A.K., SMITH, B.P., SMITH, P.J.: Enhanced radiosensitivity of cultured fibroblasts from ataxia telangiectasia heterozygotes manifested by defective colony-forming ability and reduced DNA repair replication after hypoxic λ-irradiation. *Cancer Res.* 39:3725-3734, 1979.
45. PATERSON, M.C., SMITH, P.J.: Ataxia telangiectasia: An inherited human disorder involving hypersensitivity to ionizing radiation and related DNA-damaging chemicals. *Annu. Rev. Genet.* 13:291-318, 1979.
46. PETRUSKA, J., HORN, D.: Bromodeoxyuridine substitution in mammalian DNA can both stimulate and inhibit restriction cleavage. *Biochem. Biophys. Res. Commun.* 96:1317-1324, 1980.
47. REMSEN, J.F.: Repair of damage by N-acetoxy-2-acetylaminofluorene in Bloom's syndrome. *Mutat. Res.* 72:151-154, 1980.
48. SCHMICKEL, R.D., CHU, E.H.Y., TROSKO, J.E., CHANG, C.C.: Cockayne syndrome: A cellular sensitivity to ultraviolet light. *Pediatrics* 60:135-139, 1977.
49. SETLOW, R.B., REGAN, J.D., GERMAN, J., CARRIER, W.L.: Evidence that xeroderma pigmentosum cells do not perform the first step in the repair of ultraviolet damage to their DNA. *Proc. Natl. Acad. Sci. USA* 64:1035-1041, 1969.
50. SMERDON, M.J., TLSTY, T.D., LIEBERMAN, M.W.: Distribution of ultraviolet-induced DNA repair synthesis in nuclease sensitive and resistant regions of human chromatin. *Biochemistry* 17:2377-2386, 1978.
51. TAKEHISA, S., WOLFF, S.: The induction of sister-chromatid exchanges in Chinese hamster ovary cells by prolonged exposure to 2-acetylaminofluorene and S-9 Mix. *Mutat. Res.* 58:103-106, 1978.
52. TANAKA, H., ORII, T.: High sensitivity but normal DNA-repair activity after UV irradiation in Epstein-Barr virus-transformed lymphoblastoid cell lines from Chediak-Higashi syndrome. *Mutat. Res.* 72:143-150, 1980.
53. WADE, M.H., CHU, E.H.Y.: Effects of DNA damaging agents on cultured fibroblasts derived from patients with Cockayne syndrome. *Mutat. Res* 59:49-60, 1979.

54. WARREN, S.T., SCHULTZ, R.A., CHANG, C.C., TROSKO, J.E.: Elevated spontaneous mutation rate in Bloom's syndrome fibroblasts. Proceedings of the the 31st meeting of the American Society of Human Genetics, New York, 161A, 1980.
55. WEICHSELBAUM, R.R., NOVE, J., LITTLE, J.B.: X-ray sensitivity of diploid fibroblasts from patients with hereditary or sporadic retinoblastoma. *Proc. Natl. Acad. Sci. USA* 75:3962-3964, 1978.
56. WEICHSELBAUM, R.R., NOVE, J., LITTLE, J.B.: X-ray sensitivity of fifty-three human diploid fibroblast cell strains from patients with characterized genetic disorders. *Cancer Res.* 40:920-925, 1980.
57. WOLFF, S., BODYCOTE, J., PAINTER, R.B.: Sister-chromatid exchanges induced in Chinese hamster cells by UV irradiation of different stages of the cell cycle: The necessity for cells to pass through S. *Mutat. Res.* 25:73-81, 1974.
58. WOLFF, S., BODYCOTE, J., THOMAS, G.H., CLEAVER, J.E.: Sister chromatid exchange in xeroderma pigmentosum cells that are defective in DNA excision repair or postreplication repair. *Genetics* 81:349-355, 1975.
59. ZELLE, B., LOHMAN, P.H.M.: Repair of UV-endonuclease susceptible-sites in the 7 complementation groups of xeroderma pigmentosum A through G. *Mutat. Res.* 62: 363-368, 1979.

II. GENOMIC ALTERATIONS, THE MOLECULAR TO THE MICROSCOPIC: THEIR RELEVANCE IN NEOPLASIA

EFFECTS ON CHROMOSOMES OF CARCINOGENIC RAYS AND CHEMICALS

H.J. EVANS

INTRODUCTION

The recognition of an association between carcinogenesis and an abnormal chromosome constitution dates back to von Hansemann's work in 1890 [27], and particularly to Boveri's early studies (published in 1902) on abnormal chromosome content and abnormal development in sea urchin eggs [6] and to his classical treatise (published in 1914) on the origin of malignant tumors [7]. In the last work, Boveri expounded the notion that each malignant tumor developed from a single cell that had acquired some kind of abnormality in its chromosome constitution. Even before Boveri began to propound his hypothesis, the Hertwigs [29] had shown that exposure of cells to certain chemical agents could result in irregularities in mitotic cell division, and Koernicke [37] had demonstrated that exposure of plant chromosomes to roentgen or radium rays resulted in their breakage – a finding confirmed later by Boveri's own studies on the effect of x-rays on *Ascaris* embryos. In the late 1920s, Muller's [47] and Stadler's [61] classic experiments demonstrated the mutagenicity and the chromosome damaging effects of x-rays, and later of ultraviolet light [62] – physical agents that had already been shown to be carcinogens. Similarly, the work of Auerbach [5], Oehklers [53], and Darlington and Koller [11] during the Second World War showed that various chemical agents mimicked the effects of radiations in producing chromosome damage and mutations, and it turned out that these agents also are carcinogens. Thus, the association between the induction of chromosome damage and the induction of cancer is of long standing, although the debate on the importance of chromosome changes as causal factors in carcinogenesis continues.

In recent years, since the development of techniques to study human chromosomes, much attention has been given to the actions of both physical and chemical carcinogens on the somatic chromosomes of man. I shall review briefly some of the principal findings in this area and will draw heavily upon recent and current results obtained in my own laboratory.

THE NATURE OF CHROMOSOME CHANGES INDUCED BY PHYSICAL AND CHEMICAL AGENTS

If we expose human lymphocytes, or indeed any other kind of cell, to x-rays whilst they are in the interphase stage of the cycle, there will be almost immediate breakages of the single DNA duplex that is present in each chromatid. Many of these breaks will be repaired in minutes, or at most hours, after their formation, and some will be repaired abnormally. After this kind of DNA damage, especially damage giving double-strand breakage within the DNA duplex, broken or rearranged chromosomes are seen when the chromosomes appear at their first mitosis postirradiation [16]. Ionizing radiations also induce other kinds of damage in DNA that do not develop directly into DNA scissions, but there is evidence that some of this damage, if not repaired before the cell undergoes DNA replication, may result also in structural changes as a consequence of interference with the normal processes of replication.

The structural changes, or aberrations, that we observe embrace a variety of types that have been categorized elsewhere [13,14]. They include single breaks or terminal deletions, intercalary deletions, inversions, duplications, rings, and exchanges between chromosomes. In all cases the unit of breakage and exchange is the single chromatid; thus, if the chromosome is irradiated in the unreplicated state in G_1, then the aberration is normally formed at that time and is duplicated when the chromosome passes through S. Such a duplicated aberration is referred to as a chromosome-type aberration (Fig. 1). If the chromatid is broken during S or G_2 and appears at mitosis without having undergone a replication at the site of damage, then the aberration is of the chromatid type (Fig. 1). It follows that if a chromatid-type aberration is present at the first mitosis after exposure to x-rays, or indeed to any other agent, and the cell is allowed to proceed through yet another cell cycle posttreatment, then the chromatid aberration will itself be duplicated and appear as a derived chromosome-type change at the second posttreatment mitosis.

The distinction between the two fundamental types of aberrations on the basis of the time of development during the cell cycle is an important

Chromosome and chromatid-type aberrations and the cell cycle.

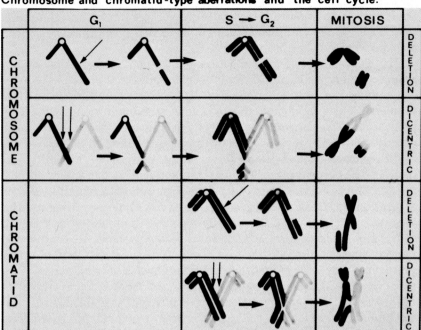

Fig. 1. Diagrammatic representation of some of the consequences of breakage and exchange of single chromatids in G_1 versus S or G_2 cells, and the distinction between chromosome and chromatid-type aberrations.

one: this is because all ionizing radiations, but very few chemical carcinogens, are capable of producing chromosome-type aberrations, that are seen at the first mitosis following treatment, in cells exposed in G_1, or chromatid-type aberrations in cells exposed in G_2. With most chemical mutagens, and also with nonionizing ultraviolet light, aberrations are observed at the first mitosis following treatment only if the cells are allowed to undergo DNA replication after exposure, and the aberrations observed always appear to be of the chromatid type. Very few chemical agents (e.g., bleomycin [65]) will produce direct breakages in the phosphodiester backbone of the DNA so that chromatid-type aberrations that develop as a consequence of either UV-light-induced thymine dimers in the DNA [26], the intercalation of agents between the bases, or the alkylation of N7 or O6 of guanine, etc., are believed to arise from misreplication at sites of unexcised lesions that are present in the DNA during replication. Thus, it is thought that if no replication occurs after the induction and failure of removal or repair of a DNA lesion that is not a double-strand DNA break induced by an "S-dependent" mutagen or carcinogen, then no aberration is formed.

The dependence upon replication for aberration formation by the majority of known chemical carcinogens implies that unexcised, chemically induced lesions could remain in DNA for very long periods, and lead to aberrations only if the damaged cell is stimulated to undergo DNA replication — weeks, months, or even years after exposure. In contrast, damage induced by x-rays would be expressed as aberrations in the interphase nucleus almost immediately after exposure. This difference between the chromosome-damaging properties of x-rays and of alkylating agents was evident in early studies on the induction of mutations and aberrations in treated *Drosophila* sperm [2,3,8]. F_1 mutant flies produced from x-irradiated sperm are usually whole body mutants in which all cells carry the mutation or structural rearrangement. In contrast, F_1 mutant flies produced from sperm exposed to S-dependent chemical mutagens frequently are mosaic mutants in which only a fraction of the cells carry the mutation. These findings reflect the fact that x-irradiation of the unreplicated sperm nucleus will produce damage leading to chromosome-type aberrations, so that any viable structural change would be present in all descendant cells from the fertilized egg. Exposure to chemical mutagens results in chromatid-type changes that develop when the sperm DNA replicates in the egg and, with the exception of isochromatid aberrations, segregate to only one of the two first cleavage cells.

Another important difference between the effects of x-rays and chemical agents on *Drosophila* sperm is that with chemical agents the frequency of chromosome structural changes increases as the posttreatment storage time (days or weeks) prior to fertilization increases. Similar results have been reported for mutations in fungal spores [4] and chromosome damage in plant cells [24]. Does such a storage effect exist in human cells?

To answer this question, we have taken advantage of the fact that it is possible to store human peripheral blood lymphocytes in culture for a week or so before adding PHA (phytohemagglutinin) to stimulate the cells into mitosis. In our experiments [19], we exposed lymphocytes for 1 h to a 6×10^{-7} M solution of the alkylating agent mitomycin C (MMC) at zero time, transferred the cells into culture medium containing bromodeoxyuridine (BUdR), and then added PHA to the cultures 1, 5, or 8 d later. The cells were fixed 72 h after the addition of PHA, and those in the first mitosis in culture (cells with nonharlequin-stained chromosomes) were scored for aberrations. The results, which are summarized in Figure 2, illustrate two important conclusions. First, the storage of human lymphocytes for a week or so after exposure to an alkylating agent

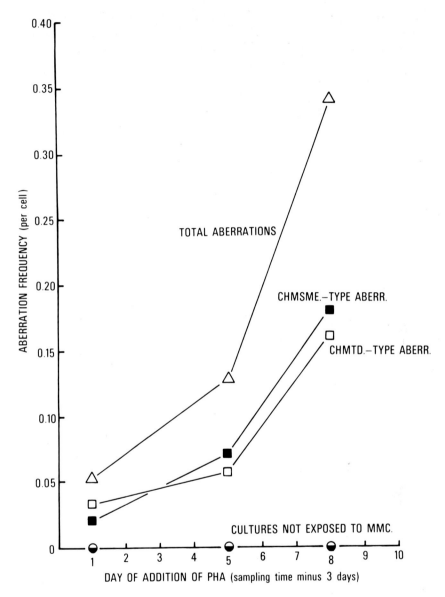

Fig. 2. Frequency of chromosomal aberrations in human lymphocytes exposed for 1 h to mitomycin C (6×10^{-7} M) immediately prior to transfer to culture medium and stored in the medium for various time periods before the addition to the medium of the mitotic stimulant, PHA. All cells were cultured for a further 3 days following the addition of PHA. Note the large increase in aberration frequency subsequent to storage for 8 days and the fact that a significant part of this increase is due to the presence of chromosome-type aberrations (an aberration type that is virtually absent in cells exposed to PHA immediately after MMC exposure and analyzed at their first mitosis in culture).

results in an almost tenfold increase in the frequency of induced chromosome damage. Second, although the aberrations seen in unstored MMC-treated G_0 cells are almost certainly all chromatid-type changes, after storage the cells also yield chromosome-type changes; that finding has been reported for plant cells also [25]. These results indicate a time-dependent alteration in the induced DNA lesions after storage, so that one must be cautious in extrapolating from the results obtained in rapidly proliferating cell systems to those exposed in tissues with slow cell-turnover rates.

In addition to, and often as a consequence of, structural rearrangements, a second type of chromosome change that follows exposure to carcinogens involves the addition or loss of one or more chromosomes in exposed cells. Of particular interest in the context of carcinogenesis are the aneuploidies that are due to chromosome losses, for these may uncover normally unexpressed or recessive loci, which are then free to be activated or expressed in the hemizygous state. The possible importance of induced hemizygosity as one of several steps resulting in transformation to a neoplastic state has been widely discussed (e.g., Ohno [54]).

THE SPONTANEOUS FREQUENCIES OF CHROMOSOME CHANGES IN HUMAN SOMATIC CELLS

How frequent are these chromosome changes? Chromosome structural changes, losses, and gains occur spontaneously. The structural changes are largely consequences of misreplication; and the aneuploidies may derive from the structural changes or from errors in the processes of chromatid separation and segregation at mitosis. The incidence of these events in humans can be estimated from analyses of chromosomes in human peripheral blood lymphocytes cultured *in vitro* for 48 h. In cultures from individuals who have not knowingly been exposed to ionizing radiations or to chemical mutagens we find that exchange aberrations are present at a frequency of about one per 1,000 cells, and deletions at about 1 per few hundred cells. The frequencies of these aberrations show small but consistent increases with age and marked increases if individuals have been exposed to mutagens.

Chromosome losses or gains are much more frequent events than are structural changes. X chromosomes in both males and females, and particularly the inactive X in females, are prone to nondisjunction; when additional copies are present in a cell, they often show premature centromere separation and have the appearance of "large acentric fragments" [23]. Neither the loss of one copy of chromosome 21 nor the loss of the Y chromosome in male cells is uncommon. The incidence of all these aneuploidies is heavily age-dependent, and in some aging females, up to 10% of the cells may lack an X chromosome [21,23]. Because the additional or missing chromosome in lymphocytes often is relatively inactive genetically, this kind of variation may be tolerated well in these cells. It should be emphasized that these aneuploidy variations are present in normal individuals who have not been exposed to known mutagens.

THE INDUCTION OF ABERRATIONS IN HUMAN CHROMOSOMES FOLLOWING IN VITRO AND IN VIVO EXPOSURE TO IONIZING RADIATIONS

A wealth of information exists on the response of human chromosomes to ionizing radiations, both in vitro and in vivo. In vitro studies on human blood lymphocytes exposed to x- or γ-rays have shown that the frequency of induced aberrations is a function of dose and, moreover, that low doses — of the order of 10-20 rads — produce increases in frequency that are easily demonstrable if sufficient cells are analyzed (Fig. 3). Our earlier studies on chromosome damage in lymphocytes of cancer patients who were exposed to low levels of whole body x-irradiation show that the response of cells irradiated in vitro was the same as if they were exposed in vivo (Table I). These findings lead us to expect that exposure of an individual to an acute dose of about 5 rads of x- or γ-rays will produce a three-to-fourfold increase in aberration frequency above background, which should be detectable if sufficient cells are counted. This dose level is equal to the internationally agreed maximum that is permitted for occupational exposure of radiation workers. In a further series of studies, we have examined chromosome aberration incidence in blood lymphocytes of patients with ankylosing spondylitis who had received therapeutic partial-body exposure to x-rays. We have sampled blood cells hours, weeks, months, and years after exposure; the results (Fig. 4) show that the aberration frequency declines with time, but that significantly elevated frequencies are apparent even many years after exposure.

In the light of the findings from these three kinds of experiment, we set up a long-term study on the incidence of aberrations in a population of nuclear dockyard workers (submarine refitters) who had been exposed to γ-radiation; their exposure was well within the occupational limits of 5 rads per yr over a period of up to 10 yr [18]. Our results are summarized in Table II and Figure 5. They show that (1) an increasing aberration frequency with increasing exposure can be detected — at the highest dose levels we see a fourfold increase in aberration frequency; (2) the incidence of aberrations increases linearly with increasing dose and there is no evidence of threshold; (3) the rate of increase is a function not only of dose

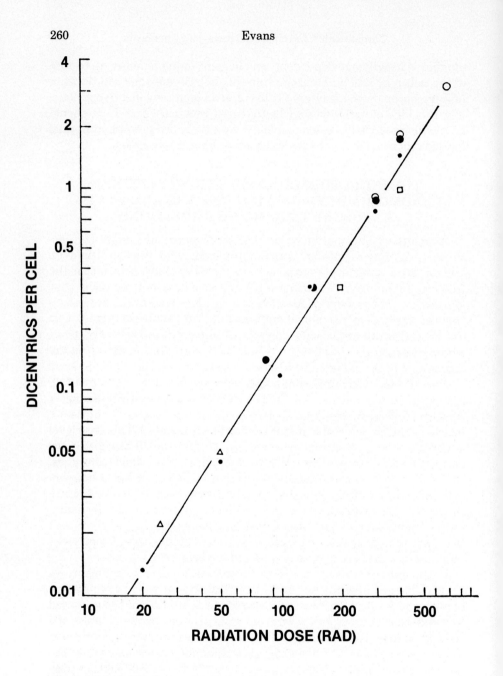

Fig. 3. Relationship between radiation dose and incidence of dicentric aberrations in human blood lymphocytes exposed to x-rays *in vitro* [59]. Note that the incidence of dicentrics in unirradiated samples is around 0.001 per cell.

TABLE I. Comparison of the Dicentric and Ring Aberrations (D+R) Frequencies in Blood Lymphocytes of Cancer Patients Prior to and Following Exposure to X-radiation In Vivo (Whole Body) and In Vitro

Patient no. & age (yr)	Control sample		Dose (rads)	In vitro sample		In vivo sample	
	Cells scored	D+R /cell		Cells scored	D+R /cell	Cells scored	D+R /cell
I, 63	200	0	30	200	0.025	200	0.03
II, 69	100	0	36	200	0.03	200	0.02
III, 56	100	0.02	40	100	0.03	100	0.04
IV, 64	100	0.03	44	200	0.03	200	0.03
V, 67	–	–	45	200	0.02	300	0.05
VI, 75	100	0	50	100	0.08	100	0.01

but also of age and time of exposure. If we consider the data over the entire 10-yr period, the total aberration frequency increases at a rate of 2.4 ± 0.8 aberrations/cell/rem $\times 10^{-4}$. Inasmuch as the aberration frequency declines with time after exposure, it may be more instructive to look at the response in relation to the dose received in the 12 months prior to blood sampling. In this case, the rate of increase in total aberrations is 4.4 ± 1.4/cell/rem $\times 10^{-4}$.

Our results have recently been confirmed by studies of atomic energy radiation workers exposed to chronic low level γ-radiation [42]. It is clear that we can detect radiation-induced genetic damage in the somatic cells in people who have been exposed to levels of ionizing radiation that are within the currently accepted levels. Our results say nothing about the possible consequences of such chromosome damage, although there currently is much debate over possible, but disputed, increased cancer rates in nuclear dockyard [50] and atomic energy workers [43] in the U.S.A.

THE INDUCTION OF ABERRATIONS IN HUMAN CHROMOSOMES AFTER EXPOSURE IN VITRO AND IN VIVO TO CHEMICAL MUTAGENS

Studies on the mutagenicity in bacteria and eukaryotic cells of chemical agents that are carcinogenic in mammals have shown that there is a strong correlation between the propensity for a substance to produce DNA damage and chromosome aberrations, and its potential as a carcinogen [30,34,46]. Therefore, interest in the actions of chemical mutagens in inducing chromosome damage stems not only from the possibility that the presence of a chemical mutagen in the environment could result in an increased incidence of cancer but also from the fact that exposure to these agents may result in an increased incidence of transmitted genetic disease.

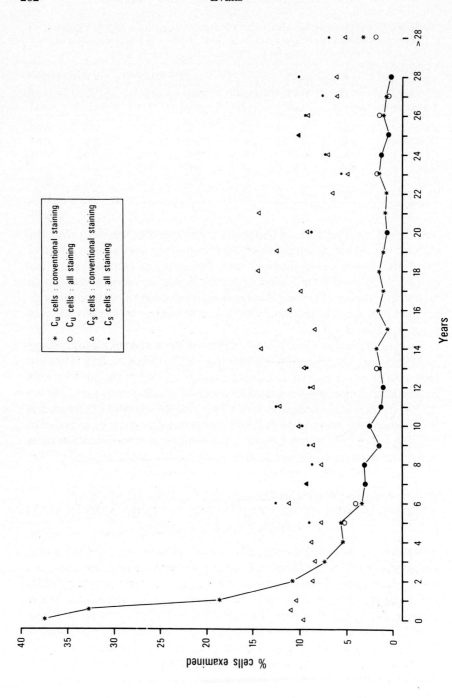

It has been estimated that about 70,000 man-made chemicals have been introduced into our environment and that 1,000 new compounds are added annually. Chemical agents are present in the household and working environment; they are added to food, and used as hair dyes, solvents, adhesives, herbicides, pesticides, and medicines. When added to human lymphocytes in cell culture, many of these chemicals are potent inducers of chromosome aberrations. The list of chromosome damaging substances is extensive, and scores of substances have been tested for their effects in inducing chromosome damage in human peripheral blood cells cultured *in vitro* [9]. I should like to refer to two examples.

A few years ago, it was shown that the occurrence of angiosarcomas in certain workers in the plastics industry, particularly the autoclave workers, was associated with exposure to vinyl chloride monomer (VCM). This led to the demonstration [12,22,58] that the incidence of chromosome aberrations in peripheral blood cells of workers exposed to VCM was considerably higher than it was in workers who were not exposed (Table III); this finding has been confirmed repeatedly [38]. In follow-up studies [28,60], it became evident that the aberration frequencies in blood cells of exposed workers have declined with time, and that the measures taken to limit VCM levels in the workplace have been effective. In collaboration with colleagues at Imperial Chemical Industries Ltd., we have also studied the incidence of sister-chromatid exchanges (SCEs) in cells of exposed workers and examined the response of human lymphocyte chromosomes to SCE induction by VCM *in vitro* [1]. As the properties of SCEs are discussed elsewhere in this volume [40], here it is sufficient to note that these exchange events are readily visualized with special staining techniques, but that they are not aberrations in the conventional sense for they do not alter chromosome morphology. Nevertheless, SCEs may reflect induced damage in DNA [15], and they are readily induced by exposure of cells to low doses of many different chemical mutagens or carcinogens [56,57]. When we expose human lymphocytes to VCM gas *in vitro*, the gas does not induce SCEs or chromosome aberrations. However, if we expose the cells in the presence of mixed-function oxidases from rat liver (S9-mix), then VCM (or rather a metabolite of it) is a potent inducer of DNA damage and chromosome aberrations (Fig. 6).

My second example concerns one of the most widespread carcinogens to which humans are exposed at high concentrations, namely, cigarette

---

Fig. 4. Frequencies of cells with unstable (C_u) and stable (C_s) aberrations, at various time intervals following irradiation, in the peripheral bloods of patients treated with x-ray therapy for ankylosing spondylitis.

TABLE II. Frequencies of Aberrations in Peripheral Blood Lymphocytes of Nuclear Dockyard Workers Exposed for Periods up to 10y to γ-Ray Doses Within Permissible Occupational Limits

Dose range (rem)	Mean dose	Mean age	Number of People sampled	Number of Cells scored	Number of Dicentrics	Number of Rings	Number of Acentrics	Percentage of cells with Unstable aberrations	Percentage of cells with Abnormal monocentrics
0– 0.9	0.2	32.6	80	8,700	14	2	14	0.30	0.11
1– 4.9	2.5	31.7	80	8,500	23	6	17	0.47	0.15
5– 9.9	7.7	37.4	71	9,200	21	5	30	0.54	0.23
10–14.9	12.2	39.8	68	9,075	32	4	37	0.68	0.21
15–19.9	17.4	39.4	29	3,700	21	0	18	0.81	0.16
20–24.9	21.8	41.3	22	2,900	15	3	7	0.76	0.24
25–29.9	26.7	42.3	9	1,040	6	1	7	1.05	0.38
>30	32.9	41.3	6	600	2	0	6	1.33	0.33
Total	8.65	36.4	197	43,715	134	21	136	0.57	0.19

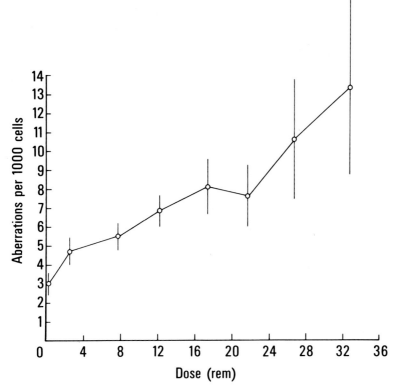

Fig. 5. Total unstable aberrations per 1,000 cells (means with standard errors) in blood lymphocytes of nuclear dockyard workers exposed to accumulated doses of γ-rays over a 10-yr period.

smoke. Obe and Herha [52] in Berlin and my colleagues and I in Edinburgh have compared the incidence of aberrations in the peripheral blood lymphocytes of cigarette smokers who inhale with that of nonsmokers. The Berlin smokers were very heavy smokers, 40–60 cigarettes per day, whereas the majority of the Edinburgh smokers used fewer than 20 cigarettes per day. In both groups, the incidence of exchange aberrations is significantly higher in smokers (Table IV). Studies of the incidence of sister-chromatid exchanges (SCEs) in the blood cells of smokers [39] indicate that in general, but depending on the numbers of cigarettes smoked, cigarette smokers show small but consistently increased frequencies of SCEs relative to nonsmokers. Inhaling cigarette smoke clearly leads to genetic damage in the somatic cells of the smoker.

The increased incidence of DNA damage and chromosome aberrations in blood cells of cigarette smokers, and the relative ease with which we

TABLE III. Chromosomal Aberrations in Vinyl-Chloride-Exposed Workers and Controls Not Exposed to Vinyl Chloride [58]

Exposed category	No.	% C_u cells	% C_s cells
Exposed	56	1.45	0.38
Non-exposed	24	0.46	0.09

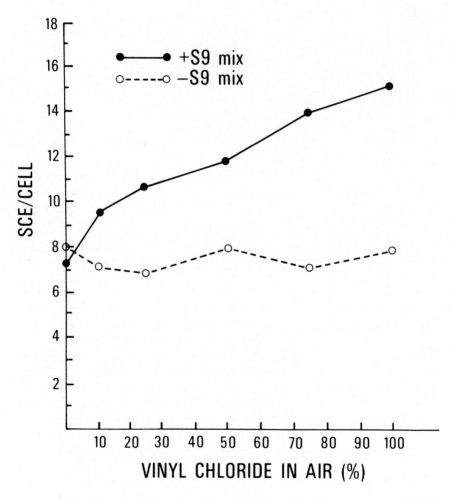

Fig. 6. The induction of SCEs in human lymphocytes exposed to vinyl chloride monomer gas (in air) in the presence or absence of rat liver mixed function oxidases (S9 mix).

TABLE IV. Incidence of Exchange Aberrations in Lymphocyte Chromosomes in Two Populations (Berlin and Edinburgh) of Inhaling Cigarette Smokers and Matched Nonsmokers

	Berlin	Edinburgh
Control	11.3×10^{-4}	32.5×10^{-4}
	(16 in 14,164)	(14 in 4,300)
Smokers	68.0×10^{-4}	63.6×10^{-4}
	(26 in 3,823)	(35 in 5,500)

can measure SCE induction in human lymphocyte chromosomes *in vitro* stimulated us to examine the response of cultured human cells to cigarette tar condensate *in vitro*. We exposed cultures of human lymphocytes (10^6 lymphocytes in 10 ml of culture medium) to small quantities of cigarette tobacco tar collected in a smoking machine and diluted with dimethyl sulfoxide (DMSO). Condensates from low-, mid-, and high-tar cigarettes were found to be potent inducers of SCEs, and although the different types of cigarettes yielded different amounts of condensate, equivalent amounts from the different types of cigarettes gave the same SCE yield [31].

This kind of experiment has been repeated using a number of cigarette tars and on blood cells from a range of different individuals. In all cases the cigarette tars are potent inducers of SCEs; in general, we find that the SCE frequency is more than doubled by the addition of 0.5 mg of condensate to a culture (Fig. 7), i.e., a dose representing 0.0125 of a high-tar or 0.05 of a low-tar cigarette — in other words, in the smoke condensing from less than one puff of a cigarette!

There are about 1,200 different compounds in cigarette smoke [20], including various precarcinogens, procarcinogens, carcinogens, and promotors. At least until recently, the polycyclic aromatic hydrocarbons — in particular the eventual metabolites of benzo(a)pyrene (BP) — were regarded as the most potent carcinogens. We examined the potency of BP in inducing SCEs in our human-lymphocyte system (Fig. 8) and found that about 2.5 μg of BP (equal to the amount in the smoke of 50–250 cigarettes) is required to produce a frequency of SCEs equivalent to that produced by 0.5 mg of smoke condensate (that is 0.0125 of a high-tar cigarette.) To put it another way: Weight for weight, cigarette-smoke condensate is between 4,000 and 20,000 times more potent than the BP it contains in inducing SCEs in human lymphocytes *in vitro*. Our conclusion that BP was not responsible for the increased SCEs in human lymphocytes exposed to cigarette smoke condensate was further reinforced by studies with Chinese hamster cells [33].

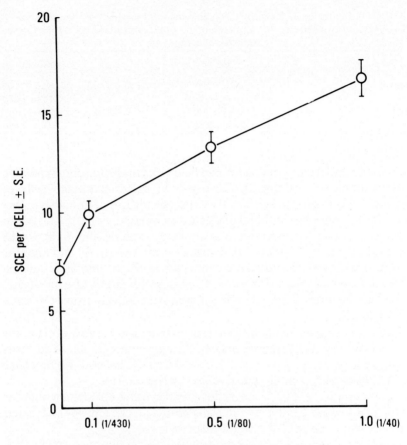

Fig. 7. Relationship between cigarette-smoke condensate (in milligrams of condensate per culture) and SCE induction in human lymphocytes exposed to condensate *in vitro*. The figures in parenthesis refer to the fraction of high-tar cigarette that would have to be smoked to give the stated amounts of condensate.

Chinese hamster cells are relatively insensitive to induction of SCEs by compounds such as BP that require microsomal activation. However, if we add S9, BP becomes a potent SCE inducer, but its activation can be blocked if α-naphtho-flavone (ANF) is present in the culture medium. By contrast, cigarette-smoke condensate is potent in inducing SCEs in the presence or absence of S9, and the addition of ANF to cigarette smoke condensate has no effect whatever on its potency (Fig. 9). These findings, (1) that smoke condensate is active in the absence of S9-mix, and (2) that

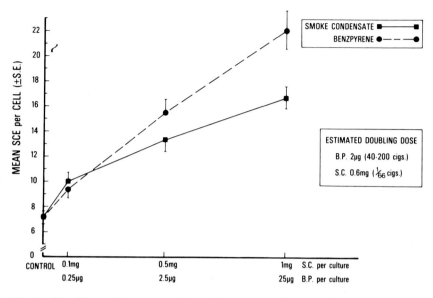

Fig. 8. The effects of cigarette-smoke condensate (mg/culture) and benzo(a)pyrene (BP) (µg/culture) in inducing SCE in cultured human lymphocytes.

its action is not diminished in the presence of ANF, indicate that the induction of SCEs by cigarette smoke condensate is not due to its BP content, or to any other polycyclic hydrocarbon that requires activation by the microsome system.

These findings parallel earlier work that showed that the polycyclic aromatic hydrocarbons present in very small quantities in cigarette-smoke condensate could not, by themselves, account for the carcinogenic effect of the condensate. Recent studies in Japan by Matsumoto et al. [44,45] and by Nagao et al. [48,49] have shown that an important group of mutagens in tobacco tars and in burned organic materials consists of the aromatic amines that are pyrolysis products of amino acids and proteins. In particular, this work has demonstrated that the major, active, substances that are mutagenic in *Salmonella* are two pyrolysates from tryptophan labeled TRP-P-1 and TRP-P-2, and one from glutamic acid, GLU-P-1. Sugimura and his group [63] have shown that Syrian hamster fibroblasts *in vitro* are transformed after treatment with these compounds, and that tumors are produced when the transformed cells are transferred to hamster cheek pouches. Moreover, the injection of these compounds into Syrian hamsters results in malignant fibrosarcomas at the sites of injection, an indication that these substances from the pyrolysates of amino acids and proteins are both carcinogens and mutagens. Tohda et al. [63] examined the effects of these pyrolysates in

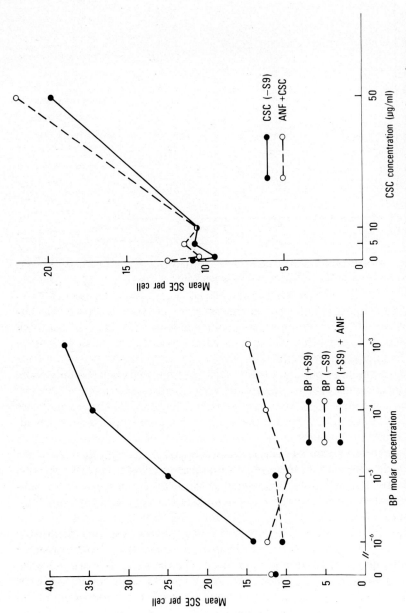

Fig. 9. SCE induction in Chinese hamster cells exposed to benzo(a)pyrene (BP) and cigarette-smoke condensate (CSC) in the presence or absence of S9. Note the requirement for S9 to activate BP and its inactivation in the presence of α-naphtho-flavone (ANF), whereas the action of CSC is independent of both S9 and ANF.

inducing SCEs in a line of human lymphoblastoid cells and found that they were active inducers. Comparative studies on their potency relative to other known SCE inducers show that, weight for weight, TRP-P-2 is as potent as aflatoxin B_1 and three times as effective as TRP-P-1, which in turn, is at least as effective as BP. However, in contrast to cigarette smoke condensate, these pyrolysates are active only in the presence of S9 mix and must be metabolized to an active form.

The interaction of the various components in cigarette smoke condensate to give DNA damage is complex and will not be pursued here. However, the suggestion that BP and related molecules may not be the important tar constituents with regard to chromosome damage inspired us to undertake some further work. Kellerman et al. [35,36] reported that different individuals had different degrees of inducibility for the enzyme complex aryl hydrocarbon hydroxylase (AHH). This AHH complex is involved in the metabolism and activation of precarcinogenic polycyclic hydrocarbons such as BP; and the data of Kellerman et al. [36] indicated that populations of individuals with lung cancer appeared to contain a much higher incidence of high-level inducers than did controls without the disease. There is some evidence for a genetic factor in lung cancer [64], and the inducibility of AHH in the mouse is controlled by a single genetic locus [51]. In humans, however, AHH inducibility probably is controlled by more than one locus [10] and, indeed, the original association between AHH inducibility and cigarette-smoke-induced lung cancer has not been convincingly confirmed [41,55]. Our results on SCE induction by cigarette smoke condensate suggested that differences between the responses of cells from different individuals to the activation of polycyclic aromatic hydrocarbons might be less relevant than differences in the responses to SCE induction *in vitro* by tobacco-tar condensates. We [32], therefore, set up a study to look at the responses of lymphocyte chromosomes from four different groups of people:

Group A: ten healthy cigarette smokers, none of whom (to our knowledge) had been exposed unduly to mutagens other than cigarette smoke.

Group B: ten healthy nonsmokers who were examined in pairs with the people in group (A) and were matched with them on the basis of age and sex.

Group C: 12 patients with histologically proven, but untreated, nondisseminated anaplastic or squamous cell bronchial carcinomas, in whom diagnostic biopsies had been obtained by fiber-optic bronchoscopy. Ten of these 12 patients were continuing to smoke cigarettes at the time of testing, and were labeled C1; two had stopped smoking (2 and 5 yr previously), and were labeled C2.

Group D: 10 patients who were selected as controls for group C. None of the group D controls had any form of cancer, on average they were 10 yr older and had smoked a similar number of cigarettes per day, but for 10 yr longer than our lung-cancer patients. All the people in groups D and C were smokers who inhaled and smoked their cigarettes to comparable butt lengths.

Blood lymphocytes from each of these individuals were cultured both in the absence and in the presence of various concentrations of cigarette-tobacco tar condensate, and dose-response curves for SCE induction were obtained. The results from the healthy smokers (group A) and healthy nonsmokers (group B) showed typical dose-response curves for SCEs against cigarette-smoke condensate in milligrams per culture (Fig. 10). Each point in Figure 10 represents a determination from a single individual; although there is considerable scatter, it is evident that the SCE levels in smokers are higher than in nonsmokers. This difference is true for the basal levels, i.e., the levels observed in blood cells not exposed to smoke condensate in culture, as well as for the levels observed after exposure to tar condensate *in vitro*. The results from the lung-cancer patients and their matched controls (groups C and D) again show a reasonable scatter (Fig. 11), but the cancer subjects show a higher incidence, at least in their response to smoke condensate in culture, than do the smoking matched controls.

The pooled data, summarized in Figure 12, show that the basal SCE frequencies of the three groups of smokers, B, C, and D, do not differ but are higher than the basal SCE frequencies in healthy nonsmokers. The almost parallel dose-response curves following exposure to condensate *in vitro* might mean that the increased SCE rates after *in vitro* induction simply emphasize *in vivo* differences, with similar increments being added to all four groups for each *in vitro* unit dose. This interpretation cannot be excluded, but at lower doses of condensate, the response is certainly not uniform among groups. What is clear is that the *in vitro* induction highlights the differences among the four groups: Highest responses are found for smokers with lung cancer (C1). Among smokers, the lowest response was found in the older, heavy-smoking, noncancer patients group (D); this is in keeping with our selection of these people as possibly representing a relatively low-risk group. The unselected group of healthy and younger smokers (A), which might be expected to contain individuals of varying risk for lung cancer, showed a mean response intermediate between those of the lung cancer subjects (C1) and their controls (D). This demonstrates an association of risk with the extent of DNA damage and is consistent with the possibility that somatic mutation might be important in initiating malignant transformation.

Our results also indicate that the number and type of cigarettes smoked affect SCE rates significantly. For example, the heaviest

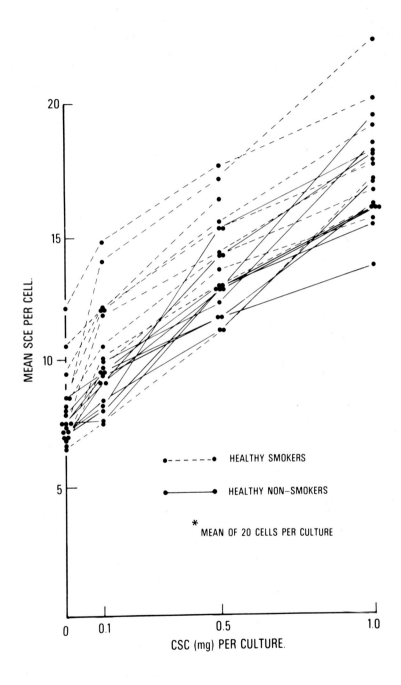

Fig. 10. Mean SCE frequencies (20 cells per point) in lymphocytes of healthy non-smokers and healthy smokers exposed to cigarette smoke condensate *in vitro*.

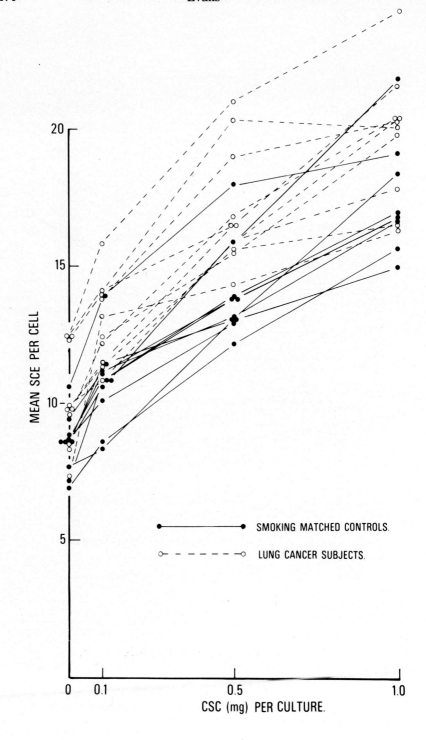

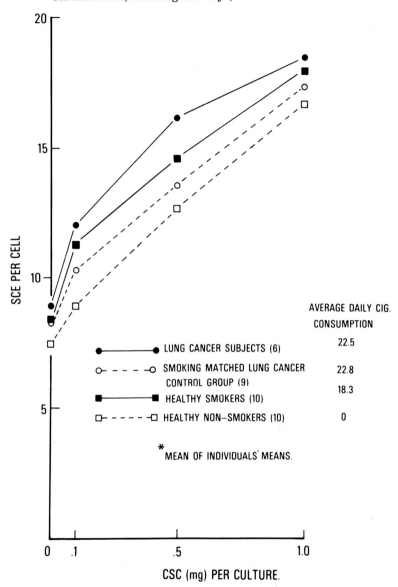

Fig. 12. Pooled "within group" data on SCE frequencies in lymphocytes of nonsmokers, smokers, and lung-cancer patients exposed to cigarette-smoke condensate *in vitro*.

Fig. 11. Mean SCE frequencies (20 cells per point) in lymphocytes of lung cancer subjects and older smoking matched controls exposed to cigarette-smoke condensate *in vitro*.

smokers show the highest frequencies. In addition, that smokers who apparently are exposed to the same amounts of cigarette smoke *in vivo* can have quite different SCE rates is evident from the results for groups C1 and D. These differences may stem from discrepancies in the actual amounts of cigarettes smoked — smoking histories, after all, describe only an alleged amount. Or, they may stem from varying patterns of inhalation, or of air flow into the bronchial tree, both of which would influence the dose of carcinogen delivered to the bronchial mucosa and to the peripheral lymphocytes. Alternatively, if real smoke dosage *in vivo* was equivalent in both groups (and there is no way of determining this precisely), then the differences may reflect the influence of other factors on the extent of measured DNA damage. Tokuhata and Lilienfeld [64] have shown that there is a significant familial aggregation of lung cancer which does not extend to spouses; thus genetic factors appear to be involved.

CONCLUDING COMMENTS

Our discussion of the effects on chromosomes of carcinogenic rays and chemicals has focused on the readily visible, gross changes in chromosome structure and number that are produced by exposure to mutagens or carcinogens, or both, and has not addressed the possible relevance of these changes in the initiation, maintenance, and development of the neoplastic state. These latter aspects will be dealt with elsewhere in this volume, but one point should be emphasized here. There is ample evidence for the association of gross structural changes, and especially translocations, with various malignancies — for example, chronic myeloid leukemia, renal cell carcinoma, and certain lymphomas [17]. It is equally clear that in many cases, no obvious chromosome aberration is present in the early stages of malignancy development. If the loss or suppression of activity of one or a few genetic loci, is an important event in the initiation or early development of a malignancy, then we must expect that such a loss will often not be detectable by standard cytogenetic methods, unless it is associated with a gross chromosomal rearrangement. However, the increased resolution afforded by new cytogenetic techniques is expected to uncover more subtle chromosome alterations and these, together with the application of molecular methods to define chromosome regions in terms of their DNA content rather than their gross morphology, will surely reveal further correlations between chromosome changes and cancer.

LITERATURE CITED

1. ANDERSON, D., RICHARDSON, C.R., PURCHASE, I.F.H., EVANS, H.J., O'RIORDAN, M.L.: Chromosomal analysis in vinyl chloride exposed workers: Comparison of the standard technique with the sister-chromatid exchange technique. Mutat. Res. 83:137-144, 1981.
2. AUERBACH, C.: Problems in chemical mutagenesis. Cold Spring Harbor Symp. Quant. Biol. 16:199-213, 1951.
3. AUERBACH, C.: Mutation Research: Problems Results and Perspectives. London: Chapman and Hall, 1976.
4. AUERBACH, C., RAMSAY, D.: Analysis of the storage effect of diepoxibutate (DEB) in Neurospora crassa. Mutat. Res. 18:129-141, 1973.
5. AUERBACH, C., ROBSON, J.M.: Mutation chemically induced. Product of mutations by allyl isothio-cyanite. Nature 154:81, 1944.
6. BOVERI, T.: Über mehrpolige Mitosen als Mittel zur Analyse des Zellkerns. Verh. Phys.-Med. Ges. (Würzburg) 35:67-90, 1902.
7. BOVERI, T.: Zur Frage der Entstehung maligner Tumoren. Jena; G. Fischer, 1914.
8. BROWNING, L.S., ALTENBURG, E.: Gonadal mosaicism as a factor in determining the ratio of visible to lethal mutations in Drosophila. Genetics 46:1317-1321, 1961.
9. BUCKTON, K.E., EVANS, H.J.: Human peripheral blood lymphocyte cultures: An in vitro assay for the cytogenetic effects of environmental mutagens. In Cytogenetic Assays of Environmental Mutagens (Hsu, T.C., ed.), Totowa: Allanheld, Osmun, 183-202, 1982.
10. COOMES, M.L., MASON, W.A., MUIJSSON, I.E., CANTRELL, E.T., ANDERSON, D.E., BUSBEE, D.L.: Aryl hydrocarbon hydroxylase and 16α-hydroxylase in cultured human lymphocytes. Biochem. Genet. 14:671-685, 1976.
11. DARLINGTON, C.D., KOLLER, P.C.: The chemical breakage of chromosomes. Heredity 1:187-222, 1947.
12. DUCATMAN, A., HIRSCHHORN, K., SELIKOFF, I.J.: Vinyl chloride exposure and human chromosome aberrations. Mutat. Res. 31:163-168, 1975.
13. EVANS, H.J.: Chromosome aberrations induced by ionizing radiations. Int. Rev. Cytol. 13:221-321, 1962.
14. EVANS, H.J.: Effects of ionizing radiations on mammalian chromosomes. In Chromosomes and Cancer (German, J., ed.), New York: John Wiley & Sons, 191-237, 1974.
15. EVANS, H.J.: What are sister chromatid exchanges? In Chromosomes Today (de la Chapelle, A., Sorsa, M., eds.), vol. 6, Amsterdam: Elsevier/North-Holland Biomedical Press, 315-326, 1977.
16. EVANS, H.J.: Molecular mechanisms in the induction of chromosome aberrations. In Progress in Genetic Toxicology (Scott, D., Bridges, B.A., Sobels, F.H., eds.), Amsterdam: Elsevier/North-Holland Biomedical Press, 57-74, 1977.
17. EVANS, H.J.: Genes, chromosomes and neoplasia: An overview. In Proceeding of the 33rd Annual Symposium on Fundamental Cancer Research. Genes, Chromosomes, and Neoplasia (Arrighi, F.E., Rao, P.N., Stubblefield, E., eds.), New York: Raven, 511-527, 1981.
18. EVANS, H.J., BUCKTON, K.E., HAMILTON, G.E., CAROTHERS, A.: Radiation-induced chromosome aberrations in nuclear-dockyard workers. Nature 277:531-534, 1979.
19. EVAN, H.J., VIJAYALAXMI: Storage enhances chromosome damage after exposure of human leukocytes to mitomycin C. Nature 284:370-372, 1980.
20. FALK, N.L.: Chemical agents in cigarette smoke. In Handbook of Physiology: Reaction to Environmental Agents (Lee, D.H.K., ed.), Baltimore: Waverly Press, 199-211, 1977.

21. FITZGERALD, P.H.: A mechanism of X chromosome aneuploidy in ageing women. *Humangenetik* 28:153-158, 1975.
22. FUNES-CRAVIOTO, F., LAMBERT, B., LINDSTEN, J., EHRENBERG, L., NATARAJAN, A.T., OSTERMAN-GOLKAR, S.: Chromosome aberrations in workers exposed to vinyl chloride. *Lancet* 1:459, 1975.
23. GALLOWAY, S.M., BUCKTON, K.E.: Aneuploidy and ageing: Chromosome studies on a random sample of the population using G-banding. *Cytogenet. Cell Genet.* 20:78-95, 1978.
24. GICHNER, T., VELEMINSKY, J.: Differential response of mutagenic effects to storage of barley seeds treated with propyl methanesulphonate and isopropyl methanesulphonate. *Mutat. Res.* 16:35-40, 1972.
25. GICHNER, T., VELEMINSKY, J., ONDREN, M.: Liquid-holding mediated enhancement of the frequency of chromosomal aberrations induced by ethyleneimine in barley embryos cultivated in vitro. *Mutat. Res.* 71:101-107, 1980.
26. GRIGGS, H.G., BENDER, M.A.: Photoreactivation of ultraviolet-induced chromosomal aberrations. *Science* 179:86-88, 1973.
27. HANSEMANN, D.: Über asymmetrische Zellteilung in Epithelkrebsen und deren biologische Bedeutung. *Virchow's Arch. Pathol. Anat.* 119:299-326, 1890.
28. HANSTEEN, I.-L., HILLESTAD, L., THIIS-EVENSEN, E., HELDAAS, S.S.: Effects of vinyl chloride in man. A cytogenetic follow-up study. *Mutat. Res.* 51:271-278, 1978.
29. HERTWIG, O., HERTWIG, R.: Über den Befruchtungs – und Teilungsvorgang des tierischen Eies unter dem Einfluss äusserer Agentien. Jena: G. Fischer, 1887.
30. HOLLSTEIN, M.J., MCCANN, J., ANGELOSANTO, F.A., NICHOLS, W.W.: Short term tests for carcinogens and mutagens. *Mutat. Res.* 65:133-226, 1979.
31. HOPKIN, J.M., EVANS, H.J.: Cigarette smoke condensates damage DNA in human lymphocytes. *Nature* 279:241-242, 1979.
32. HOPKIN, J.M., EVANS, H.J.: Cigarette smoke-induced DNA damage and lung cancer risks. *Nature* 283:388-390, 1980.
33. HOPKIN, J.M., PERRY, P.E.: Benzo(a)pyrene does not contribute to the SCEs induced by cigarette smoke condensate. *Mutat. Res.* 77:377-381, 1980.
34. ISHIDATE, M., ODASHIMA, S.: Chromosome tests with 134 compounds on Chinese hamster cells *in vitro* – a screening for chemical carcinogens. *Mutat. Res.* 48:337-354, 1977.
35. KELLERMAN, G., LUYTEN-KELLERMAN, M., SHAW, C.R.: Genetic variation of aryl hydrocarbon hydroxylase in human lymphocytes. *Am. J. Hum. Genet.* 25:327-331, 1973.
36. KELLERMAN, G., SHAW, C.R., LUYTEN-KELLERMAN, M.: Aryl hydrocarbon hydroxylase inducibility and bronchogenic carcinoma. *N. Engl. J. Med.* 289:934-937, 1973.
37. KOERNICKE, M.: Über die Wirkung von Röentgen – und Radiumstrahlen auf pflanzliche Gewebe und Zellen. *Ber. Dtsch. Bot. Ges.* 23:404-415, 1905.
38. KUCEROVA, M., POLIKOVA, Z., BATORA, J.: Comparative evaluation of the frequency of chromosomal aberrations and the sister chromatid exchange numbers in peripheral lymphocytes of workers occupationally exposed to vinyl chloride monomer. *Mutat. Res.* 67:97-100, 1979.
39. LAMBERT, B., LINDBLAD, A., NORDENSKJOLD, M., WERELIUS, B.: Increased frequency of sister chromatid exchanges in cigarette smokers. *Hereditas* 88:147-149, 1978.
40. LATT, S.: This volume.
41. LIEBERMAN, J: Aryl hydrocarbon hydroxylase in bronchogenic carcinoma. *N. Engl. J. Med.* 298:686-687, 1978.
42. LLOYD, D.C., PURROTT, R.J., REEDER, E.J.: The incidence of unstable chromosome aberrations in peripheral blood lymphocytes from unirradiated and occupationally exposed people. *Mutat. Res.* 72:523-532, 1980.

43. MANCUSO, T.F., STEWART, A., KNEALE, G.: Radiation exposures of Hanford workers dying from cancer and other causes. *Health Phys.* 33:369-384, 1977.
44. MATSUMOTO, T., YOSHIDA, D., MIZUSAKI, S., OKAMOTO, H.: Mutagenic activity of amino acid pyrolyzates in *Salmonella typhimurium* TA98. *Mutat. Res.* 48:279-286, 1977.
45. MATSUMOTO, T., YOSHIDA, D., MIZUSAKI, S., OKAMOTO, H.: Mutagenicities of the pyrolyzates of peptides and proteins. *Mutat. Res.* 56:281-288, 1978.
46. MILLER, E.C.: Some current perspectives on chemical carcinogenesis in humans and experimental animals. *Cancer Res.* 38:1479-1496, 1978.
47. MULLER, H.J.: Artificial transmutation of the gene. *Science* 66:84-87, 1927.
48. NAGAO, M., HONDA, M., SEINO, Y., YAHAGI, T., KAWACHI, T., SUGIMURA, T.: Mutagenicities of protein pyrolysates. *Cancer Lett.* 2:335-340, 1977.
49. NAGAO, M., HONDA, M., SEINO, Y., YAHAGI, T., SUGIMURA, T.: Mutagenicities of smoke condensates and the charred surface of fish and meat. *Cancer Lett.* 2:221-226, 1977.
50. NAJARIAN, T., COLTON, T.: Mortality from leukaemia and cancer in shipyard nuclear workers. *Lancet* 1:1018-1020, 1978.
51. NEBERT, D.W., GOUJON, F.M., GIELEN, J.E.: Aryl hydrocarbon hydroxylase induction by polycyclic hydrocarbons: Simple autosomal dominant trait in the mouse. *Nature* 236:107-110, 1972.
52. OBE, G., HERHA, J.: Chromosomal aberrations in heavy smokers. *Hum. Genet.* 41:259-263, 1978.
53. OEHLKERS, F.: Die Auslösung von Chromosomen-mutationen on der Meiosis durch Einwirkung von Chemikalien. *Z. Indukt. Abstamm. Vererbungsl.* 81:313-341, 1943.
54. OHNO, S.: Aneuploidy as a possible means employed by malignant cells to express recessive phenotypes. *In* Chromosomes and Cancer (German, J., ed.), New York: John Wiley & Sons, 77-94, 1974.
55. PAIGEN, B., GURTOO, H.L., MINOWADA, J., HOUTEN, L., VINCENT, R., PAIGEN, K., PARKER, N.B., WARD, E., HAYNES, N.T.: Questionable relation of aryl hydrocarbon hydroxylase to lung-cancer risk. *N. Engl. J. Med.* 297:346-350, 1977.
56. PERRY, P.E.: Chemical mutagens and sister-chromatid exchange. *In* Chemical Mutagens (de Serres, F.J., Hollaender, A., eds.), vol. 6, New York: Plenum 1-39, 1980.
57. PERRY, P., EVANS, H.J.: Cytological detection of mutagen-carcinogen exposure by sister chromatid exchange. *Nature* 258:121-125, 1975.
58. PURCHASE, I.F.H., RICHARDSON, C.R., ANDERSON, D.: Chromosomal and dominant lethal effects of vinyl chloride. *Lancet* 2:410-411, 1975.
59. REPORT OF THE UNITED NATIONS SCIENTIFIC COMMITTEE ON THE EFFECTS OF ATOMIC RADIATION. 24th Session, Supplement No. 13 (A/7613), New York; United Nations, 1969.
60. ROSSNER, P., SRAM, R.J., NOVAKOVA, J., LAMBL, V.: Cytogenetic analysis in workers occupationally exposed to vinyl chloride. *Mutat. Res.* 73:425-427, 1980.
61. STADLER, L.J.: Genetic effects of X-rays in maize. *Proc. Natl. Acad. Sci. USA* 14:69-75, 1928.
62. STADLER, L.J., SPRAGUE, G.F.: Genetic effects of unfiltered ultraviolet light on maize. *Proc. Natl. Acad. Sci. USA* 22:572-578, 1936.
63. TOHDA, H., OIKAWA, A., KAWACHI, T., SUGIMURA, T.: Induction of sister-chromatid exchanges by mutagens from amino acid and protein pyrolysates. *Mutat. Res.* 77:65-69, 1980.
64. TOKUHATA, G.K., LILIENFELD, A.M.: Familial aggregation of lung cancer in humans. *J. Natl. Cancer Inst.* 30:289-312, 1963.
65. VIG, B.K., LEWIS, R.: Genetic toxicology of bleomycin. *Mutat. Res.* 55:121-145, 1978.

MOLECULAR ALTERATIONS IN DNA ASSOCIATED WITH MUTATION AND CHROMOSOME REARRANGEMENTS

BERNARD STRAUSS, KATHLEEN AYRES, KALLOL BOSE, PETER MOORE, SAMUEL RABKIN, ROBERT SKLAR, AND VALERIE LINDGREN

INTRODUCTION

Discussions of the molecular mechanisms of mutagenesis often focus on the structure of nucleotides altered by reaction with different mutagens and on the mispairing errors that can occur when such altered nucleotides are replicated [20,21]. This approach may be adequate for studies of mutation in some bacterial viruses, but many investigations with bacteria, notably with *Escherichia coli*, demonstrate the role of metabolic systems, most of which are inducible, in the mutagenesis process [96]. In the bacteria, the signal for induction of the mutagenic process—the "SOS repair" system—seems to be inhibition of DNA synthesis, inasmuch as treatments from thymine starvation to UV-irradiation are inducers and their one common denominator is an effect on DNA synthesis [70]. The inducible process may involve formation of wrong base pairs, but the replicative system that makes these mistakes is apparently altered from the normal. One can therefore ask whether mammalian cells respond to mutagenic agents as a result of "simple" mispairing during replication or whether some additional process, set in motion by the inhibition of DNA synthesis, is required. In this paper we wish to argue (1) that treatments that are carcinogenic are ones that damage DNA and inhibit DNA synthesis (as suggested by Painter [61,62]) and (2) that the phenomena of mutagenesis and carcinogenesis result from the cellular response required to bypass the DNA damage. Chromosome aberrations result in large part from similar processes in which DNA synthesis and its inhibition play a key role.

REACTION OF DNA WITH MUTAGENS

Carcinogen treatment of cells results in the alteration of DNA mainly, but not invariably, by the addition of particular groups (adducts) to the bases. Depending on the treatment, almost any reactive site in DNA can be altered. Our studies generally have used two mutagenic agents: N-methyl-N'-nitro-N-nitrosoguanidine (MNNG) and N-acetoxy-N-2-acetylaminofluorene (AAAF). MNNG is a highly mutagenic methylating agent, and its reaction with cells can result in the addition of methyl groups to 16 different sites distributed among the four nucleotides (Fig. 1) [83]. AAAF, by contrast, is much more restricted and reacts primarily at the C-8 of guanine with only a minor reaction product at the N-2 position of guanine when either native DNA or cells are treated [42,95]. Treatment of single-stranded ΦX174 with AAAF results largely in the C-8 guanine adduct with only about 4% of reaction at other sites (Fig. 2). We use the acetylaminofluorene-DNA (AAF-DNA) reaction product as a paradigm of DNA reacted with a bulky hydrocarbon or other bulky group, and use the MNNG as an example of naturally occurring groups (e.g., CH_3) that are mutagenic when transferred to unusual positions in the cell.

A chemical group can have very different structural effects which are dependent on its position. Methyl groups at the N-7 position of guanine are usually ignored by cells [46,67] or are removed slowly [43,54,84]. However, reaction at the O^6 position of guanine has immediate and serious biological consequences. The presence of AAF adducts in the double-stranded DNA structure may result in the rotation of the guanine from the *anti* to *cis* configuration and the insertion or intercalation of AAF rings into the DNA structure [28]. In cells, AAF may be deacylated, and the dimensions of this compound then permit it to sit within the standard DNA structure without deformation [22a,42,73a,95]. The exact configuration of a reacted DNA may be particularly important in determining what happens when a replicating polymerase approaches the altered nucleotide (see below), but we have very little knowledge of the structural changes resulting from most combinations.

The loss of a purine or pyrimidine base from DNA without breakage of the phosphodiester backbone produces an apurinic or apyrimidinic site (AP site) depending on the base removed. AP sites arise continually in DNA, ordinarily by spontaneous depurination [50] but also as a re-

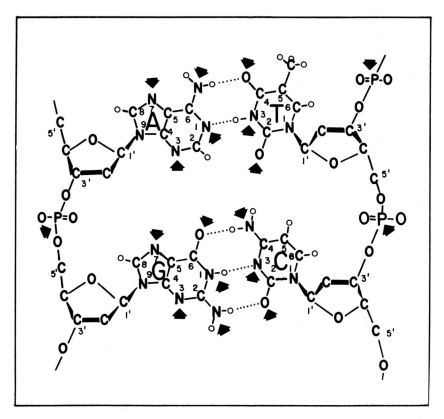

Fig. 1. Sites for methylation in DNA. The arrows indicate those positions in DNA at which alkylation adducts have been reported [83].

sult of enhanced degradation after treatment of DNA or cells with alkylating agents [44]. Alkylation of any of the ring nitrogens, for example by the production of 7-methylguanine or 3-methyladenine, creates a quaternary nitrogen that unstabilizes the glycosidic bond holding the base and deoxyribose moieties together. The rate of spontaneous depurination depends on the site of alkylation and on the nature of the attached group [44]. Large groups may depurinate relatively rapidly and, for example, one of the activated aflatoxin products with DNA at the N-7 of guanine is very unstable [10,55]. Some alterations in DNA as a result of treatment of cells with mutagenic agents have little *initial* biological effect (e.g., the formation of 7-methylguanine) because addition products at this position do not interfere with hydrogen bonding and

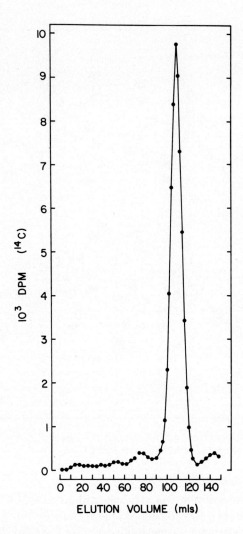

Fig. 2. Reaction products of single-stranded ΦX174 DNA with ^{14}C AAAF. ΦX174 DNA (35.8 μg) containing 1449 pmol ^{14}C AAAF(158,000 dpm) in 100 μl 10 mM Tris HCl (pH 8.0), 100 mM NaCl, and 1 mM $MgCl_2$ was sequentially digested at 37°C with *N. crassa* endonuclease (25 units, 16 h), DNAse I (200μg, 3 h), alkaline phosphatase (10 units, 3h), spleen phosphodiesterase (6 units, 1 h), and venom phosphodiesterase (500 μg, 16 h). The digest was loaded onto an LH20 column (0.8 × 20 cm) and eluted with 20 ml of 10 mM $(NH_4)_2CO_3$ followed by 35% ethanol containing 10 mM $(NH_4)_2CO_3$ and 2 ml fractions were collected. One milliliter of each fraction was counted for radioactivity in Aquasol scintillant. Recovery of radioactivity from the column was 90%, and the major peak (at 110 ml) was identified as *N*-acetyl-N-(guan-8-yl)-2-aminofluorene by thin-layer chromatography.

have little effect on the structure. However, spontaneous or enzymatic removal of the alkylated base to give an AP site may result in a profound effect. Production of AP sites may destroy the biological activity of DNA [46] because bacteriophage that contain such sites are no longer infective. (It is not clear whether this occurs because the AP site *per se* is a block to replication or because there is endonucleolytic cleavage of the DNA as it enters the host cell.) AP sites, if replicated, would be expected to be mutagenic because the directing influence of a complementary base is missing [82]. We have now shown using our *in vitro* system (see below) that AP sites are blocks to DNA synthesis (unpublished data) and Schaaper and Loeb [*Proc.Natl.Acad.Sci.USA* 78:1773–1777, 1981] observe that AP DNA yields mutations only when present in SOS-induced *Escherichia coli* as might be expected for a treatment which blocks replication.

Cells have developed mechanisms to recognize and remove unusual bases from the DNA, and particular mechanisms remove unusually methylated bases. We do not know why the methylated bases are recognized so specifically. Among the DNA N-glycosylases that specifically recognize altered bases is one that recognizes and removes 3-methyladenine from DNA [49]. These enzymes form AP sites which are subject to rapid degradation by endonucleolytic enzymes to give broken DNA chains. The enzymology of the breakage reaction is complex because there are different ways of attacking and splitting a chain with an AP site [15]. Only one way produces an end group with an intact nucleotide containing a 3'OH group (Fig. 3). As only such nucleotides (with free 3'OH groups) can serve as substrates for the DNA polymerase that are required to repair the damage, some "cleaning up" of the broken AP chain often is required, for example [see 91]; as a result, the detailed enzymology of the repair reactions at such sites is complex.

The AP endonucleases can be used as probes for the presence of AP sites produced as intermediates in the degradation of mutagen-treated DNA, and much of the recent activity in study of excision repair mechanisms has centered around the possibility that AP sites may be necessary intermediates [15,30,69,79]. One of the major classes of antitumor antibiotics, neocarzinostatin (NCS), which has the interesting property of releasing primarily Ts but also As from opposite strands [12,31], acts with AP sites as intermediates [6] (Fig. 4). Although the compound itself will produce breaks in the DNA chain by a two-stage reaction [12], the intermediate existence of AP sites permits endonucleases to attack the structure and produce substrates which may be more amenable to cellular repair reactions. The compound bleomycin acts in a similar manner [12,71] and is reported to be an efficient chromosome-breaking

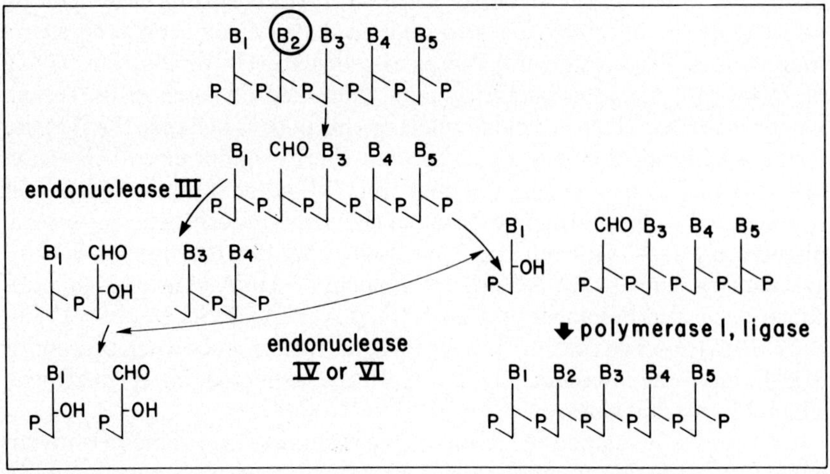

Fig. 3. Reaction products resulting from hydrolysis of apurinic/apyrimidinic (AP) sites [15].

agent [34]. NCS itself induces numerous DNA double-strand breaks *in vivo* [2,92].

CELLULAR REPAIR MECHANISMS

Cells remove a variety of lesions induced in their DNA by mutagens and carcinogens; when the DNA is restored to its original state, we speak of "excision repair." Although extensive excision repair can occur in a cell, mutation or death can still result if some lesions are left unexcised. Cells also bypass lesions by the use of different mechanisms. Such bypass can result in survival; thus the mechanisms are also — but I believe incorrectly — referred to as repair even though the lesions remain in the DNA. Several very distinct mechanisms can account for the removal of different lesions [29]. Perhaps the most straightforward (although one of the most recently recognized) is the system that adds a purine into an AP site [17,18,51]. A second mechanism involves the sequential action of a DNA N-glycosylase, a polymerase and exonuclease, and finally a ligase to cut out the damaged base and resynthesize a patch of new DNA by use of the undamaged strand as a template (Fig. 5). This series of reactions is used for the repair of UV-induced pyrimidine dimers in T4 bacteriophage and in *Micrococcus luteus*. As recently shown, the first step in this chain of reactions is a splitting of the pyrimidine dimer from the DNA by a glycosylase action [15,30,69,79]. It is not

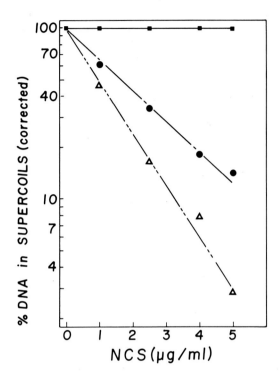

Fig. 4. Effect of apurinic endonuclease on the degradation of DNA treated with neocarzinostatin (NCS). ColEl DNA was treated with NCS alone (■), in the presence of mercaptoethanol (●), or with mercaptoethanol plus AP endonuclease (△) [5]. The radioactive DNA substrate was analyzed by gel electrophoresis [6].

known whether the same sequence of reactions is involved in the repair of UV damage in E. coli, for at least three gene products and ATP are required for the incision step in this organism [80,81]. The mechanism of the excision repair sequence in human cells is unknown. It has been widely reported that xeroderma pigmentosum (XP) cells are unable to remove not only pyrimidine dimers but also a wide range of polyaromatic hydrocarbon adducts [8,53,98]. We do not understand the structural feature that is common to these different lesion-ions and permits the same system to recognize that they possess a common component. Our inability to specify the mechanism(s) of excision repair in human cells means that we cannot identify the exact molecular defect in XP cells. The large number of complementation groups in this disease [40] may be viewed either as a mystery that will be understood when we know

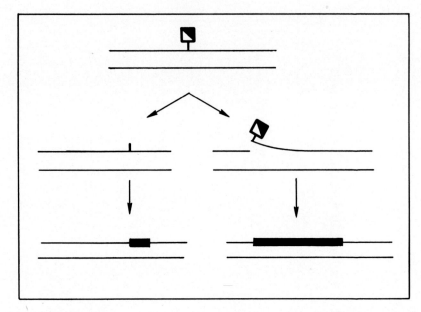

Fig. 5. Schemes of excision repair. The drawing illustrates two mechanisms: (1) Right side. Incision at the site of a lesion (□) followed by excision, insertion of a repair patch, and ligase action. (2) Left half. Removal of the lesion by a DNA-N-glycosylase producing an AP site followed by action of an AP endonuclease, insertion of a repair patch, and ligase action.

the mechanism or as tools that will permit us to unravel the mechanism by analysis of the differences between these groups.

Damage can be removed without the removal of bases or the need for repair synthesis (i.e., insertion of a "patch"). For example, in the process of photoreactivation, UV-induced pyrimidine dimers are split by an enzyme that utilizes light energy for the reaction [89,90]. Although not yet absolutely clear, it seems that the system that removes the O^6-methylguanine (O^6MeG) adduct from DNA does so by a direct transfer of the methylated base from alkylated DNA to a cysteine residue in an acceptor protein, possibly without removal of the base itself [39,59a]. O^6-methylguanine (and the other O^6 alkyl derivatives) is important because of the evidence that this adduct is a premutagenic [78,85] and precarcinogenic lesion [26]. The finding that cells have developed a rather special mechanism for removing the O^6-methyl lesion also seems to testify to its special importance.

TABLE I. Effect of Chloramphenicol on Mutation and the Removal of O⁶ Methylguanine From the DNA of *E. coli* (ABAA)*

MNNG µg/ml	Chlorampenicol	7MeG dpm/mg DNA $\times 10^{-3}$	O⁶MeG dpm/mg DNA $\times 10^{-7}$	Mutation frequency $\times 10^{-2}$
1.25	0	4.7	4.3	0.08
1.25	+	6.6	70.5	1.5
2.5	0	15.4	13.3	1.0
2.5	+	20.6	255.0	12.2

*Data from Sklar and Strauss [85]. Mutation frequency to *ara*, *aceBA*, *arg*, and *met* determined by replica plating.

TABLE II. MNNG-Induced Mutation Does Not Require *recA* or *lexA* Gene Products*

	Freq. rpoB $\times 10^6$		
Strain of *E. coli*	Control	MNNG	MNNG plus chloramphenicol
ABAA	0.06	4.9	130
recA12		5.4	270
recA13		2.1	27
uvrE	14	17.0	78
lexA		5.8	130

*Cells were treated with 2.5 µg MNNG/for 40 min in the presence or absence of chloramphenicol. They were resuspended in fresh medium, grown to stationary phase and plated on rifampin-containing medium [85]. The number of resistant colonies was scored to give the frequency of rifampin resistant (rpoB) bacteria.

The system for removal of O⁶MeG in bacteria has been extensively investigated, particularly by the group including Cairns, Schendel, Jego, Samson, and others, and most of our ideas as to how this adduct is handled in bacteria derive from their studies [7,14,78]. In *E. coli*, O⁶MeG is removed by an inducible system of limited capacity. The protein involved, which is only now being purified by Lindahl and his collaborators [39], has the peculiar property of being "used up" in the process, probably because of the transfer of the methyl of O⁶MeG to an amino acid on the acceptor protein. The key role of O⁶MeG in bacterial mutation can be demonstrated by preventing induction of the removal system, for example by addition of chloramphenicol at the time of MNNG treatment (Table I), which increases mutation frequency and prevents O⁶MeG removal [78,85]. As neither the *recA* nor *lexA* gene products are involved in the mutagenesis (Table II), it appears that

O^6MeG-induced mutations in *E. coli* result directly from mispairing during replication, which implies in turn that O^6MeG at lower concentrations is *not* a block to replication in this organism. Contrast these results with those obtained with neocarzinostatin in which *recA* [93] and *umuC* genes are required, an indication that "the NCS lesions in DNA are not mutagenic per se but rather are substrates for processes that generate mutational events at the sites of lesions" [22].

The susceptibility of different mammalian tissues to agents that induce O^6MeG is likely to vary because of the different capacities of various cell types to remove the alkyl groups [26]. Even different cell types within an organ may have different capacities; for example, nonparenchymal cells of the liver are deficient in their ability to remove O^6MeG as compared to hepatocytes [48]. This difference in capability between different cells or organs of the same animal implies that a developmental mechanism is involved. We have observed that human lymphoblastoid lines derived from different individuals differ in their ability to remove O^6MeG. Some lines are deficient in their removal capability, and we have designated such strains as mex⁻ in contrast to the removal proficient mex⁺ strains (Fig. 6). There appears to be no correlation between the mex⁻ characteristic and two of the repair-deficiency diseases, XP and ataxia-telangiectasia (AT): Lymphoblastoid lines derived from XP patients of several complementation groups may be either mex⁺ or mex⁻ (Table III), although all retain their excision-repair deficiency for AAAF-induced damage; and Shiloh and Becker [personal communication] report that lines derived from AT patients also may be either mex⁺ or mex⁻. The capacity for removal of O^6MeG is limited and readily saturated, particularly when compared with the ability to remove 3-methyladenine adducts, an addition product removed by a specific glycosylase (Fig. 7). Treatment with approximately 0.2 μg/ml of MNNG saturates the O^6MeG removal system of cells of the human lymphoma line Raji, provided that cell density is less than 4×10^7 cells/ml. Relatively long periods are required for the recovery of the ability to remove O^6MeG in mex⁺ Raji cells, so that a pretreatment with nonlabeled MNNG results in the loss of ability to remove O^6MeG produced by a second dose of (labeled) MNNG (Fig. 8). Pretreatment with *N*-ethyl-*N'*-nitro-*N*-nitrosoguanidine inhibits removal of a challenge dose of MNNG, an indication that methyl- and ethyl-derivatives can be removed by the same system. It should be pointed out that the capacity for O^6MeG removal in lymphoblastoid cells is very limited as compared with an induced *E. coli*. We calculate that Raji cells can remove about 1.6×10^{12} methyl groups per mg DNA as compared to Cairn's calculation of about 4.5×10^{14} methyl

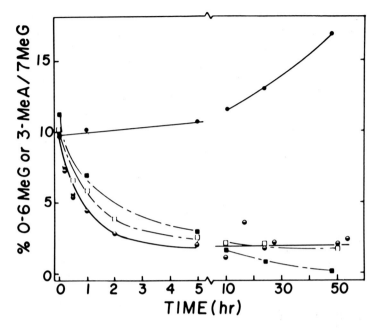

Fig. 6. Removal of O^6methylguanine by the mex⁺ strain, Raji, or the mex⁻ H2BT [56]. Cells were treated with 0.5 µg/ml ³H MNNG for 2 min in saline-citrate containing N-acetyl-L-cysteine. The cells were then suspended in medium and incubated. DNA was isolated and the purines analyzed by thin-layer chromatography [86]. Squares represent the ratio of 3-methyladenine to 7-methylguanine × 100; circles, ratio of O^6methylguanine to 7-methylguanine × 100; completely filled symbols, spherocytosis derived lymphoblastoid line H2BT; open squares and half-filled circles, lymphoma line Raji.

groups/mg DNA for induced bacteria. We have no data to indicate whether either mex⁺ or mex⁻ cells will respond to lower doses of MNNG by the induction of a removal system [73].

The suggestion that developmental mechanisms may affect the ability of particular cells to remove O^6MeG does not preclude the possibility that genetic mechanisms may be superimposed. We are interested in understanding the mechanism(s) that determines the mex⁺ characteristic of our cells, particularly in light of the suggestion that the mechanism may be related to cellular transformation [13]. In order to study the question we have hybridized the mex⁺ lymphoma line Raji with the mex⁻ L33-6-1, first by isolating a thioguanine-resistant, ouabain-resistant Raji line and then fusing with polyethylene glycol and selecting an HAT medium containing ouabain. Since thioguanine resistance (THG^R) is recessive and ouabain resistance (OUA^R) is dominant, only

TABLE III. Removal of O^6 Methylguanine by Lymphoblastoid Lines Derived From Xeroderma Pigmentosum (XP) Patients*

Strain of cells	Derivation	Time incubated (h)	7-MeG dpm/µg DNA	$\dfrac{O^6MeG}{7\text{-MeG}} \times 100$
GM2500	XP group A	0	11.5	10.0
		24	11.6	2.5
GM2246	XP group C	0	12.7	9.4
		24	7.6	1.6
GM2498	XP group C	0	8.1	10.4
		24	5.1	14.0
GM2253	XP group D	0.5	–	3.8
		3	–	2.0
GM2473	XP group D heterozygote†	0.5	–	9.8
		3	–	9.9

*Cells were treated as described in the legend to Figure 6 (data from Sklar and Strauss [86]). Cultures from the Human Genetic Mutant Cell Repository.
†Mother of donor of GM2253.

hybrids of a $THG^R\ OUA^R \times THG^S\ OUA^S$ cross will grow when plated in medium with hypoxanthine, aminopterin, thymidine, and ouabain. The two strains selected had modal chromosome numbers of 88 and 92 and karyotype analysis indicated a Y chromosome coming from Raji and some chromosomes characteristic of the L33 parental cells. Although L33-6-1 is one of the most deficient strains in terms of capacity to remove O^6MeG, one of the hybrids, hybrid A, does have some ability to remove the adduct (Table IV). We are also testing the ability of mex⁻ survivors of MNNG treatment to remove O^6MeG. Although such experiments alone will not decide between genetic or developmental mechanisms, they are compatible with some sort of chromosomal control.

ADDUCTS WHICH ARE NOT REMOVED

DNA replication may begin before all adducts are removed from the DNA if initiation itself has not been inhibited. Damage may remain either because the excision-repair systems are unable to deal with the number of adducts present or because the cells are genetically incompetent for excision, as is the case with XP cells and pyrimidine dimers [66]. Such replication may have at least three outcomes: (1) It can proceed without damage to the cell; (2) it can result in mutation; or (3) it can result in cell death. It is likely that any of these outcomes can occur as a result of the same lesion(s) in the DNA. Evidence comes from a series of

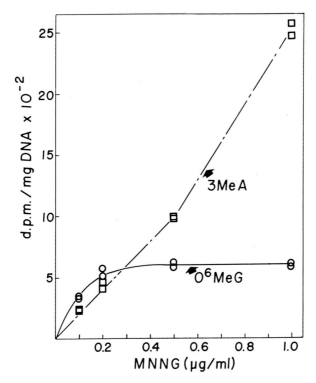

Fig. 7. Capacity for O^6methylguanine and 3-methyladenine removal. Raji cells (2 × 10^7/ml) were treated with varying concentrations of ^{3}H MNNG (112 mCi/mM) as described in the legend to Figure 6. The cells were transferred to medium and incubated for 2 h. The total content of O^6methylguanine and 3-methyladenine in the DNA was determined and the amount removed was calculated based on an estimate of an initial amount of methylation of 10% of the 7-methylguanine radioactivity for both 3-methyladenine and O^6methylguanine.

"holding experiments" in which treated fibroblasts are held for various times in a state of confluence, at which DNA synthesis does not occur, before DNA synthesis is released by trypsinization and (replating) of the cells. The frequency of mutation declines and survival increases the longer the confluent cultures are incubated, and this increase in survival and decrease in mutation frequency is accompanied by removal of the induced lesion [53,98]. Furthermore, the effect for UV- and bulky chemical lesions is not seen in XP cells. Maher and her co-workers [53,98] conclude that the same lesion is both mutagenic and inactivating. However, given that an inactivated (dead) cell cannot yield a mu-

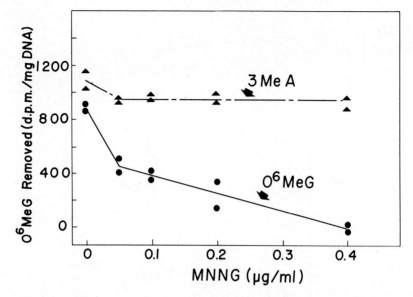

Fig. 8. Effect of pretreatment with MNNG on the ability of cells of the Raji lymphoma line to remove O^6methylguanine or 3-methyladenine. Cells were treated with unlabeled MNNG as indicated and then, after 2 h, they were given a second treatment with ^{3}H MNNG (1.0 μg/ml; 112 mCi/mM). The cells were incubated an additional 2 h, then harvested, and the amount of radioactive O^6methylguanine or 3-methyladenine was determined as described in the legend to Figure 7.

TABLE IV. Removal of O^6 Methylguanine From Two Hybrid Clones Derived From Fusion of the Mex+, Clone Rot 5 of Raji With the Mex Clone L6 of Strain L33-6-1*

Strain of cells	7-MeG μg DNA	O^6MeG 7-MeG	O^6MeG removed μg DNA
rot 5	10.24	0.028	0.794
rot 5	10.43	0.028	0.809
L6	10.7	0.1199	−0.160
L6	9.49	0.011	−0.148
Hybrid A	8.13	0.074	0.248
Hybrid A	7.64	0.080	0.188
Hybrid B	8.21	0.111	−0.050
Hybrid B	9.39	0.102	0.024

*Cultures were treated with 0.2 μg/ml of MNNG for 10 min with no N-acetyl cysteine added and then incubated 2 h. Hybrid B was in culture for several months; hybrid A was from a frozen culture.

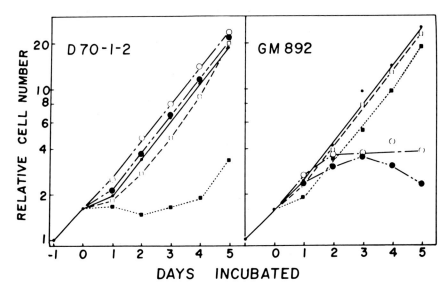

Fig. 9. Effect of MNNG and AAAF on the growth of the mex⁻ line GM892 and the mex⁺ (XP heterozygote) line D70-1-2 [86]. Rapidly growing cells were treated for 1 h with either MNNG or AAAF and then placed in fresh medium and incubated at 37°C. The cultures were counted and diluted to 4×10^5 cells/ml at daily intervals. Small filled circles, represent control; large open circles, MNNG (0.1 μg/ml); large closed circles, MNNG (0.2 μg/ml); open squares, AAAF (4 μg/ml), closed squares, AAAF (8 μg/ml).

tant clone, cells must make some decision as to which lesions are to be inactivating and which mutagenic. The major question is, "How is this decision made?" We translate this as "What is the mechanism of DNA replication on a template containing lesions?" This is directly related to the question of how chromosome breaks are induced because of the observations, which date back at least to the studies of Evans and Scott [23], that DNA replication is required for the conversion of chemical damage into chromatid breaks. Furthermore, although the causes of cell death after treatment with alkylating agents is not certain, one guess is that improper chromosome distribution resulting from breakage plays a large role [68]. This hypothesis fits our observation that after treatment with either methyl methanesulfonate (MMS) or MNNG (but not with AAAF, the cell growth continues at an undiminished rate for about one division before the inhibitory effect of the treatment becomes evident (Fig. 9).

A major problem in determining the molecular changes responsible for chromosomal events and mutations is to identify the particular le-

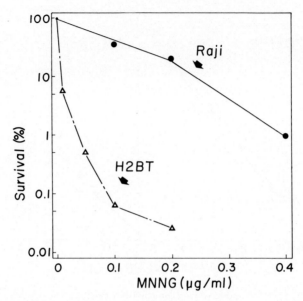

Fig. 10. Inactivation by MNNG of the mex$^+$ line Raji and the mex$^-$ line H2Bt. Cell survival was determined by the microwell-plating method [41].

sion responsible for any particular effect. With AAAF for example, one cannot automatically be sure that the biological effects are due to formation of the C-8 acetylated adduct rather than to the deacylated product or to the N-2 addition product. It is known that these are excised at different rates [97] but we know nothing about their relative *in vivo* effects on mutation and DNA synthesis. However, it can be argued that the O^{6}MeG adduct *per se* is lethal for mammalian cells. Mex$^-$ cells, which are deficient in their ability to remove O^{6}MeG, are highly sensitive to cell killing by MNNG (Fig. 10). Although it could be maintained that mex$^-$ cells are also deficient in the ability to remove some other lesion, experiments by Bodell et al. [4] indicate that there is real specificity in the O^{6}alkylG removal deficiency. Mex$^-$ lines are also more sensitive to MMS than are mex$^+$ [33], but we believe this differential sensitivity (which is less than to MNNG) is due to the small but sufficient quantities of O^{6}MeG made by MMS treatment. We are not sure whether AP sites, which are produced in quantity by MNNG and MMS, are also blocks to replication; however, the differential sensitivity of the mex$^+$ and mex$^-$ lines to alkylating agents, despite their ability to respond to MMS or MNNG treatment with increased repair synthesis, suggests that O^6MeG is lethal for lymphoid cells, perhaps because it blocks DNA replication. As indicated above, O^6MeG does not block replication in bacterial cells (or does not require the SOS system at low levels of reaction), and so there may be a difference in the replication systems of bacterial and animal cells with respect to response to this adduct.[1]

INHIBITION OF DNA SYNTHESIS

Insofar as this analysis is correct, knowledge of how chromosome breaks occur and how mutations are produced requires an understanding of the behavior of the DNA synthetic machinery when a block to DNA synthesis is reached. We have therefore begun an examination of this process that uses a simple *in vitro* model (Fig. 11) based on a system developed for the study of *in vitro* mutagenesis [25,32]. The single-stranded DNA bacteriophage ΦX174, primed with a single restriction fragment prepared from double-stranded ΦX174 Replicative Form (RF), forms a substrate for DNA synthesis at which initiation must occur at the 3′OH terminus of the primer. The site of termination of DNA synthesis can be located by determining the size of the newly synthesized fragment after treatment with the original restriction enzyme. The template can be treated with a variety of mutagens, different polymerases can be tested, and the ionic environment can be manipulated. The effect of protein cofactors can be studied. Although the system is deficient in some respects — for example, agents affecting the initiation of DNA synthesis cannot be studied — it does permit isolation of some of the factors involved in the termination of DNA synthesis.

DNA synthesis is blocked by the introduction of either pyrimidine dimers, or AAF or benzpyrene adducts, and the blocks occur, when DNA polymerase I is used, at the expected positions, so that differences in the specificity of reaction can be seen [59] (Fig. 12). The data for AAF-reacted DNA (the only case so analyzed) support the hypothesis that each lesion is an absolute block to DNA synthesis catalyzed by *E. coli* polymerase I (Fig. 13) [58]. Analysis of the results for synthesis catalyzed by pol I, pol III (holoenzyme), or T4 DNA polymerase leads us to suppose that termination occurs one base prior to the reacted nucleotide on the template (Fig. 14). The enzyme AMV reverse transcriptase, acting on AAF-reacted but not UV-treated DNA, inserts a nucleo-

---

[1]We have recently obtained lymphoblastoid lines which can remove some O^6MeG but which are still very sensitive to MNNG-induced killing. O^6MeG removal capacity may therefore be necessary but not sufficient to protect against MNNG-induced cytotoxicity.

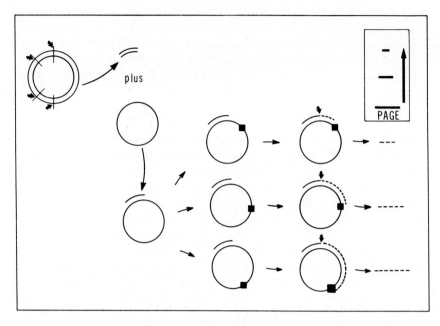

Fig. 11. Use of ΦX174 DNA to locate the site of a block to DNA synthesis. Single circles represent ΦX174; double circles, replicative form; solid squares, sites of lesions; dashed line, newly synthesized DNA; PAGE, polyacrylamide gel electrophoresis. Upper left: preparation of restriction fragments from ΦX174 replicative form. Center left: annealing of an isolated restriction fragment to single-stranded ΦX174 DNA circles and reaction with mutagen/carcinogen. Center right: *in vitro* DNA synthesis with labeled deoxynucleoside triphosphates using the prepared ΦX174 template + primer followed by restriction enzyme digestion of the fragments. Top right: polyacrylamide gel electrophoresis of the denatured fragments followed by autoradiography. The restriction fragments are separated by size with differences in length of a single nucleotide being apparent.

tide opposite the reacted nucleotide. DNA polymerase-α, isolated either from human lymphoma cells or from calf thymus, appears to act in an "in between" manner, inserting bases at some positions as does AMV reverse transcriptase and at others as do the bacterial polymerases [57]. Because of the reports that substitution of Mn^{2+} for Mg^{2+} in reaction mixtures results in an accumulation of errors in *in vitro* systems (for example see [32]), we tested the effect of reaction mixtures containing Mn^{2+} on the termination site of pol I reaction mixtures. We found (Fig. 15) that such substitution results in a change in termination site so that a base appears to be inserted at a nucleotide level opposite a lesion. This insertion occurs for both templates containing UV and AAF lesions.

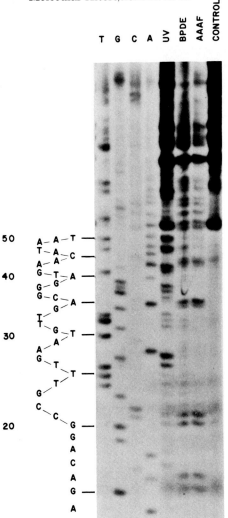

Fig. 12. Polyacrylamide gel analysis of the products synthesized by pol I on DNA templates reacted with AAF on *anti*benzpyrene diolepoxide (BPDE) or irradiated with UV. About 265 AAF adducts or 140 *anti*-BPDE adducts were present per DNA molecule. DNA was irradiated with 100 J/m^2 of UV. The primer was HaeII fragment 5. Lanes A, C, G, and T are sequence standards, and numbering is from the center of the HaeII recognition site (from [59]).

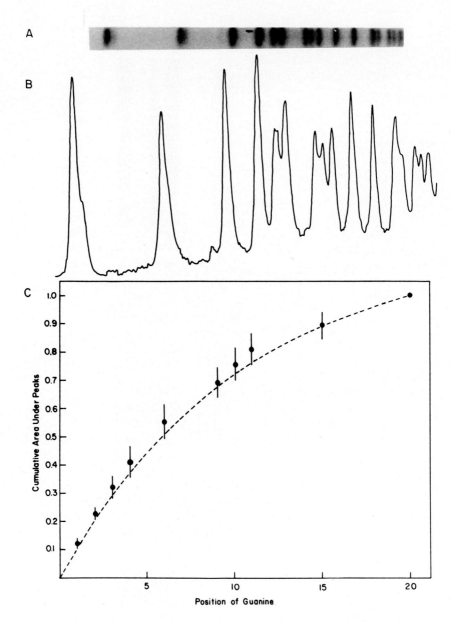

Fig. 13. Quantitation of termination by pol I using sequence gels. A. Channel from 20% polyacrylamide sequence gel on which have been run the products of synthesis by pol I on a template containing 113 AAF adducts. Each fragment terminates one base before the position of a guanine in the template. The smallest fragments are on the left, so that synthesis is proceeding from left to right, and the smallest and largest termination frag-

One major question relates to the nature of the base that is inserted opposite the lesion (Fig. 16). To answer it, one of us (P.M.) reacted an AAF-containing template with nucleotides by using T4 DNA polymerase and, after isolating the DNA product, he used this as a template for an AMV reverse transcriptase reaction and added one nucleoside triphosphate at a time. It appears that only deoxycytidine triphosphate (C) will convert the T4 polymerase pattern to the AMV reverse transcriptase pattern, an indication that only C is incorporated; that result is difficult to understand, given the lack of coding specificity expected for an AAF-G (N-acetyl-N-(guan-8-yl)-2-aminoflurorene) in a double-stranded structure [27,28].

This work, even at its present preliminary stage, permits a number of conclusions. Inhibition at the nucleotide level of DNA synthesis is variable and determined by a number of factors. Certainly the nature of the lesion is important, but also important are the nature of the enzyme or enzyme system, the ionic environment, and, possibly (as judged by the results with polymerase α) the place in the sequence at which the lesion occurs.

THE BYPASS OF LESIONS

The *in vitro* data show that a variety of mutagenic and clastogenic adducts are absolute blocks to DNA synthesis. The *in vivo* evidence indicates that at least some of these adducts can be replicated past, or "bypassed." Mammalian cells will survive and replicate with a limited number of adducts still present in their DNA. Such replication has been demonstrated for DNA that carries pyrimidine dimers [8], polyaromatic hydrocarbons [19], and O^6MeG [1], and this list is not exhaustive. It is not known how or why certain adducts escape the excision processes. Certainly excision is slower when alterations occur in those portions of the DNA covered by the nucleosomes [36,87]. Because the adducts that do remain should still be capable of blocking DNA synthesis, either they are altered and cannot do so or, most likely, the *in vitro* situation does not represent fairly the possible *in vivo* responses.

---

ments in the section of the gel shown are 33 and 118 nucleotides long, respectively. B. Densitometric scan of gel shown in A. C. Cumulative summation of peak sizes from scan through 20 successive guanines (total area under peaks set as 1.0) Data show mean and standard error of five separate experiments. The dashed line is calculated on the assumption that each AAF adduct is an absolute block to elongation by pol I using the relationship $T = 1 - p^n/1 - p^N$, where T is the proportion of chains terminating, N is the total number of guanines in the sequence, p is the probability of continuing past any guanine calculated from the average number of adducts, and n is the number of guanines up to a particular position in the sequence [58].

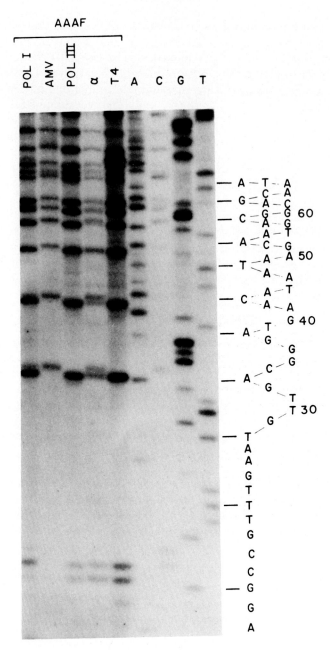

Fig. 14. Polyacrylamide gel analysis of the products synthesized on an AAF-reacted template by pol I; AMV reverse transcriptase, pol III; DNA polymerase α, and T4 DNA polymerase. The DNA template was primed with HaeII restriction fragment 5 and contained 113 AAF residues per ΦX174 molecule [57].

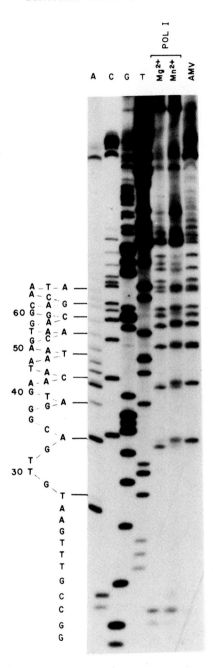

Fig. 15. The effect of divalent cation on the sites of termination by pol I on AAF-treated DNA. Where indicated $MnCl_2$ (0.5 mM) was substituted for $MgCl_2$ (8 mM) [57].

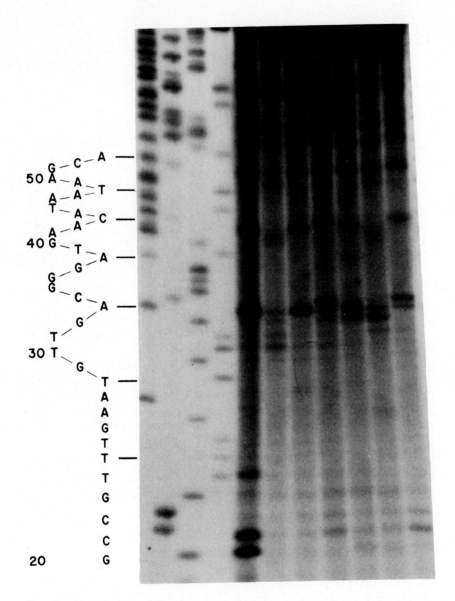

Fig. 16. Identification of the nucleotide inserted by AMV reverse transcriptase opposite AAF adducts in the DNA. One microgram of an AAF-reacted template, primed with HaeII fragment 5, was incubated for 40 min at 37 °C with 12.5 units T4 DNA polymerase in the presence of [^{32}P]-dATP, dCTP, dGTP, and TTP (each 2.5 μM, 410 Ci/mmol). The la-

Bacteria solve the problem of lesions in their DNA by at least two mechanisms. The first involves a recombinational insertion of a segment of parental DNA into the gap produced as a result of the block in DNA synthesis at the site of a lesion and its reinitiation distal to the lesion. The parental DNA is then replaced by a repair-type synthesis [24,72] (Fig. 17). Bacteria also have developed an inducible mechanism, SOS repair, that permits replication of DNA that contains pyrimidine dimers and other lesions. Because mutations that block the induction of this system make cells sensitive to radiation and simultaneously prevent the induction of mutations (such strains are called "mutagen stable"), it is supposed that the induced mode of DNA synthesis in SOS repair is in some way "error prone" [70,96].

It is not yet known how bypass occurs in mammalian cells. The search for recombinational repair mechanisms has, on the whole, been unsuccessful [47], although some limited recombination has been suggested [56]. The presence of an inducible mechanism has not been demonstrated in an unequivocal fashion. There may be technical flaws in the observation [11] that pretreatment promotes the ability of cells to synthesize DNA past a pyrimidine dimer [63]; however the observation that pretreatment of cells with chemical mutagens promotes the mutability of SV40 virus that has been treated with mutagen and used to infect these cells seems to indicate the operation of an inducible process, at least for SV40 replicating in monkey kidney cells [74-76]. Certainly, the existence of a process that is induced by cell damage and permits replication past blocks in DNA has not been excluded.

To understand both the ways in which mammalian cells bypass lesions and the quantitative limits to bypass reactions probably requires a knowledge of the special features of DNA replication in eukaryotic cells. DNA replication occurs in many independent units, or replicons, and in any cell in the S period of the cell-division cycle numerous replicons may be synthesizing DNA simultaneously [16]. Synthesis is bidirectional, starting from an initiation point and proceeding outward [35]. The biochemistry of the termination step, when two replicons join and the daughter DNA molecules separate, is for the most part unknown

---

beled DNA was separated from unincorporated nucleotides by alcohol precipitation followed by gel filtration through a 1.5-ml Sephadex G50 column. Aliquots of the DNA were then incubated for 15 min at 37°C in 50 mM Tris HCL (pH 8.0), 8 mM $MgCl_2$, and 5 mM DTT either with no enzyme (lane 1) or with 19 units of AMV reverse transcriptase (lanes 2-7). Nucleotides were added as follows: no addition (lane 2), 50 μM dATP (lane 3), 50 μM dCTP (lane 4), 50 μM dGTP (lane 5), 50 μM TTP (lane 6), 50 μM each dATP, dCTP, dGTP, and TTP (lane 7). The samples then were digested with HaeII and prepared for loading as usual. Lanes A, C, G, and T are the sequences standards.

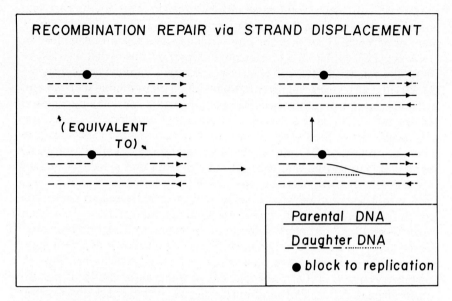

Fig. 17. Model of recombination repair *via* strand displacement. Solid line represents parental DNA; dashed and dotted lines, daughter DNA; circle, block to replication.

[16]. It is not even certain whether there are fixed termination points at which a completed replicon awaits the arrival of a second, or whether replicons keep extending until they meet an approaching replicon in a head-on collision. Evidence indicates clusters of replicating units that synthesize simultaneously and whose initiation is coordinately controlled [37,38,53a,64,64a].

The nondeterminate nature of termination of replicons in undamaged DNA [16] fits the suggestion by Park and Cleaver [65] that pyrimidine dimers might act as termination sites for synthesis in UV-irradiated DNA. They then proposed that the probability of synthesizing past the terminated site would increase with time, but did not suggest a particular mechanism to account for such a change. Understanding the mechanism of termination may permit us to suggest a mechanism. If DNA synthesis proceeds in two directions, then a gap of at least one nucleotide may always exist opposite the carcinogen-reacted base, either because synthesis does not ordinarily proceed beyond one base prior to the reacted nucleotide or because synthesis of an Okazaki fragment will not start at this position. However, if conditions in the cell change so that a nucleotide is inserted opposite the reacted base, as in the action of AMV reverse transcriptase on AAF-reacted templates, then, given the apparent stochastic nature of the initiation of Okazaki fragments, one would expect that, regardless of the polarities, a continuous sequence of nucleotides eventually would be provided and then ligate into a complete chain. This is only a speculation, for the hypothesis requires insertion of bases opposite both nucleotides in a pyrimidine dimer and in our experiments (pol I with Mn^{2-} substituted for Mg^{2+}) we have seen only insertion opposite the first pyrimidine of a dimer[2].

The scheme proposed by Park and Cleaver [65] also provides some limit on the number of lesions that might tolerated (called the "magic number" by B. Weinstein in conversation). With one or fewer lesions per average replicon, we would expect DNA synthesis to proceed without the presence of any gaps because synthesis from two directions would terminate at the lesion. However (Fig. 18), if there were more than one lesion per replicon, a gap of nonsynthesized DNA would occur between two stalled DNA growing points. Presumably then, organisms with smaller replicons could tolerate a higher concentration of adducts (a suggestion made by H.J. Evans in the discussion after presentation of this paper). In addition, a Poisson analysis of the amount of DNA synthesis as a function of the amount of reaction would make it possible to decide whether the hypothesis has any quantitative validity; but, such data are not available.

MUTATIONS AND CHROMOSOME ABBERATIONS

The speculation above makes it evident how both mutation and chromosome aberrations could occur as a result of DNA synthesis. Although we have observed (see above) that the "correct" base is inserted by AMV reverse transcriptase opposite an AAF-lesion in DNA, the data (Fig. 16) do not preclude 5% or even 10% of misincorporation. Induction of a system that inserted a base at the site of the reacted nucleotide might, at the same time, induce a system that produced substitution mutations. Any "slippage" of synthesis at the site of a block would produce frame-shift mutation. Although AMV reverse transcriptase inserts the correct base opposite the AAF-G lesion, we find that the specificity of insertion by pol I with Mn^{2+} is more complex. For example, deoxyadenosine is most commonly inserted by this enzyme opposite G-AAF. A complete discussion of these phenomena is now being prepared.

---

[2]In recent experiments we have observed insertion of nucleotides opposite the second base in a pyrimidine dimer with Mn^{2+} as the metal ion at high (500 μM) concentrations of single deoxynucleoside triphosphates.

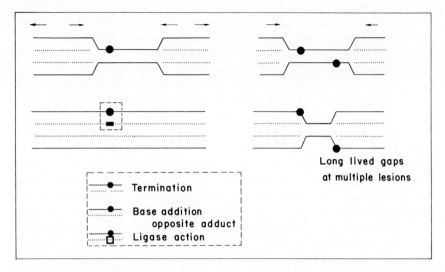

Fig. 18. Model of "bypass" as a result of termination of replicon progression at the site of lesions. Modified from Park and Cleaver [65]. Left: one lesion per replicon. Right: two lesions per replicon showing "frozen" growing points.

Chromosome damage occurs after treatment with agents that produce breaks in the DNA strands as a result of chemical or enzymatic action as well as with agents such as UV or AAF that (as far as we know) do not produce unstable lesions in the DNA. Chromatid aberrations produced by such agents as MNNG or methyl methanesulfonate (MMS) are readily understood as resulting from DNA synthesis proceeding on a DNA template containing a single-strand break. One expects that such single-strand breaks would be converted to double-strand or chromatid breaks as a direct result of DNA synthesis [3]. It is more difficult to understand how agents that are blocks to DNA synthesis but do not cause single-strand breaks may act to produce chromosome aberrations. We suppose that if there is only one lesion per replicon and if the cell is able to insert a nucleotide opposite the lesion, then there will be no significant gaps in the newly synthesized DNA. If, however, there is more than one lesion per replicon, then there will be separated growing points which remain "frozen" for a period of time and are likely to be targets for endonuclease action, for example, the endonucleases involved in unwinding cellular DNA [9]. Such endonuclease action could result in chromatid breaks.

Although this speculation clearly is incomplete, it can account for the correlation of the inhibition of DNA synthesis with mutation and clastogenic action. We suppose that the ability of cells with lesions in their DNA to replicate means that the enzyme systems present *in vivo*, either normally or as a result of induction and in contrast to the *in vitro* behavior of some polymerases, can add nucleotides opposite the point of a DNA lesion. Our data suggest that it is the nature of the polymerase rather than the proofreading exonuclease that determines the site at which nucleotides are inserted. Therefore, the production of chromosome breaks and of mutations is determined by the characteristics of the polymerases and nucleases that are present after a halt in DNA synthesis at the point of a lesion. The hypothesis implies that we should look at the characteristics of the replication system found in carcinogen-treated organisms, a suggestion already in the literature in an attempt to account for the increased variability of precarcinogenic cells [52].

SUMMARY

Cellular repair mechanisms recognize specific chemical adducts in DNA. One of the more specific repair mechanisms is that which removes O^6methylguanine adducts produced by treating cells with N-methyl-N'-nitro-N-nitrosoguanidine. In bacteria, O^6methylguanine is removed by an inducible system. In mammalian cells the system concerned with removal of this adduct is limited in amount and is "used up," in contrast to most enzyme reactions. Cells may be either proficient or deficient in their ability to remove O^6methylguanine, and hybrids between deficient and proficient strains seem to have an intermediate activity.

Mutagenic and clastogenic agents almost invariably are inhibitors of DNA synthesis. An *in vitro* system for the study of inhibition based on ΦX174 DNA is described. By use of this system, it has been shown that DNA synthesis usually stops one nucleotide before the affected template nucleotide in carcinogen-treated DNA. Under some circumstances, however, a nucleotide is inserted opposite the carcinogen-reacted template base. A hypothesis that explains how cells can replicate DNA that contains lesions is developed, and is based on the induction of systems that can replicate damaged DNA up to the point of the damaged nucleotide.

Acknowledgments. Original work from this laboratory was supported by grants from the National Institutes of Health (GM 07816, CA 19265) and contracts NO1 CP-85669 from the National Cancer Institute and DE-AC02-76EV02040 from the Department of

Energy. R.S. is a trainee of a program on Enviromental Biology (Mutagenesis and Carcinogenesis) supported by the National Cancer Institute (5T32CA09273). We thank the Cancer Research Program, National Cancer Institute, Division of Cancer Cause and Prevention, Bethesda, Maryland, for providing ^{14}C AAAF.

LITERATURE CITED

1. ABANOBI, S., COLUMBANO, A., MULIVOR, R., RAJALAKSHMI, SARMA, D.: In vivo replication of hepatic deoxyribonucleic acid of rats treated with dimethylnitrosamine: Presence of dimethylnitrosamine-induced O^6-methylguanine, N^7-methylguanine and N^3-methyladenine in the replicated hybrid deoxyribonucleic acid. *Biochemistry* 19:1382-1387, 1980.
2. BEERMAN, T., GOLDBERG, I.: DNA strand scission by the antitumor protein, neocarzinostatin. *Biochem. Biophys. Res. Commun.* 59:1254-1260, 1974.
3. BENDER, M.: Relationship of DNA lesions and their repair to chromosomal aberration production. *Basic Life Sci.* 15:245-265, 1980.
4. BODELL, W., SINGER, B., THOMAS, G., CLEAVER, J.: Evidence for removal at different rates of O-ethyl pyrimidines and ethylphosphotriesters in two human fibroblast lines. *Nucleic Acids Res.* 6:2819-2829, 1979.
5. BOSE, K., KARRAN, P., STRAUSS, B.: Repair of depurinated DNA in vitro by enzymes purified from human lymphoblasts. *Proc. Natl. Acad. Sci. USA* 75:794-798, 1978.
6. BOSE, K., TATSUMI, K., STRAUSS, B.: Apurinic/apyrimidinic endonuclease sensitive sites as intermediates in the in vitro degradation of deoxyribonucleic acid by Neocarzinostatin. *Biochemistry* 19:4761-4766, 1980.
7. CAIRNS, J.: Efficiency of the adaptive response of *Escherichia coli* to alkylating agents. *Nature* 286:176-178, 1980.
8. CLEAVER, J., BOOTSMA, D.: Xeroderma pigmentosum: Biochemical and genetic characteristics. *Annu. Rev. Genet.* 9:19-38, 1975.
9. COZZARELLI, N.: DNA gyrase and the supercoiling of DNA. *Science* 207:953-960, 1980.
10. CROY, R., WOGAN, G.: Temporal patterns of covalent DNA adducts in rat liver after single and multiple doses of Aflotoxin B_1. *Cancer Res.* 41:197-203, 1981.
11. D'AMBROSIO, S., AEBERSOLD, P., SETLOW, R.: Enhancement of post-replication repair in ultraviolet light irradiated Chinese hamster cells by irradiation in G_2 or S phase. *Biophys. J.* 23:71-78, 1978.
12. D'ANDREA, A., HASELTINE, W.: Sequence specific cleavage of DNA by the antitumor antibiotics neocarizinostatin and bleomycin. *Proc. Natl. Acad. Sci. USA* 75: 3608-3612, 1978.
13. DAY, R., ZIOLKOWSKI, C., SCUDIERO, D., MEYER, S., LUBINIECKI, A., GIRARDI, A., GALLOWAY, S., BYNUM, G.: Defective repair of alkylated DNA by human tumour and SV40-transformed human cell strains. *Nature* 288:724-727, 1980.
14. DEFAIS, M., JEGGO, P., SAMSON, L., SCHENDEL, P.: Effect of the adaptive response on the induction of the SOS pathway in *E. coli* K-12. *Mol. Gen. Genet.* 177:653-660, 1980.
15. DEMPLE, B., LINN, S.: DNA N-glycosylases and UV repair. *Nature* 287:203-307, 1980.

16. DePamphilis, M., Wasserman, P.: Replication of eukaryotic chromosomes: A close up of the replication fork. *Annu. Rev. Biochem.* 49:627–666, 1980.
17. Deutsch, W., Linn, S.: DNA binding protein from cultured human fibroblasts that is specific for partially depurinated DNA and that inserts purines into apurinic sites. *Proc. Natl. Acad. Sci. USA* 76:141–144, 1979.
18. Deutsch, W., Linn, S.: Further characterization of a depurinated DNA-purine base insertion activity from cultured human fibroblasts. *J. Biol. Chem.* 254:12099–12103, 1979.
19. Dipple, A., Roberts, J.: Excision of 7-bromomethylbenz(a)anthracene DNA adducts in replicating mammalian cells. *Biochemistry* 16:1499–1503, 1977.
20. Drake, J.W.: The Molecular Basis of Mutation. San Francisco: Holden-Day, 1970.
21. Drake, J., Baltz, R.: The biochemistry of mutagenesis. *Annu. Rev. Biochem.* 45:11–37, 1976.
22. Eisenstadt, E., Wolf, M., Goldberg, I.: Mutagenesis by Neocarzinostatin in *Escherichia coli* and *Salmonella typhimurium*: Requirement for $umuC^+$ or plasmid pKM 101. *J. Bacteriol.* 144:656–660, 1980.
22a. Evans, F., Miller, D., Beland, F.: Sensitivity of the conformation of deoxyguanosine to binding at the C-8 position by N-acetylated and unacetylated 2-aminofluorene. *Carcinogenesis* 1:955–959, 1980.
23. Evans, H., Scott, D.: The induction of chromosome aberrations by nitrogen mustard and its dependence on DNA synthesis. *Proc. R. Soc. Lond.* [Biol.] 173:491–512, 1969.
24. Ganesan, A.: Persistence of pyrimidine dimers during post replication repair in ultraviolet light irradiated *Escherichia coli* K12. *J. Mol. Biol.* 87:103–119, 1974.
25. Gopinathan, K., Weymouth, L., Kunkel, T., Loeb, L.: Mutagenesis *in vitro* by DNA polymerase from an RNA tumor virus. *Nature* 278:857–859, 1979.
26. Goth, R., Rajewsky, M.: Persistence of O^6-ethylguanine in rat brain DNA: Correlation with nervous system-specific carcinogenesis by ethylnitrosourea. *Proc. Natl. Acad. Sci. USA* 71:639–643, 1974.
27. Grunberger, D., Weinstein, I.: The base displacement model: An explanation for the conformational and functional changes in nucleic acids modified by chemical carcinogens. *In* Biology of Radiation Carcinogenesis (Yuhas, J., Tennant, R., Regan, J., eds.), New York: Raven, 175–187, 1976.
28. Grunberger, D., Weinstein, I.B.: Biochemical effects of the modification of nucleic acids by certain polycyclic aromatic carcinogens. *Prog. Nucleic Acid Res. Mol. Biol.* 23:105–149, 1979.
29. Hanawalt, P., Cooper, P., Ganesan, A., Smith, C.: DNA repair in bacteria and mammalian cells. *Annu. Rev. Biochem.* 48:783–836, 1979.
30. Haseltine, W., Gordon, L., Lindan, C., Grafstrom, R., Shaper, N., Grossman, L.: Cleavage of pyrimidine dimers in specific DNA sequences by a pyrimidine dimer DNA-glycosylase of *M. luteus*. *Nature* 285:634–641, 1980.
31. Hatayama, T., Goldberg, I., Takeshita, M., Grollman, A.: Nucleotide specificity in DNA scission by neocarzinostatin. *Proc. Natl. Acad. Sci. USA* 75:3603–3607, 1978.
32. Hibner, U., Alberts, B.: Fidelity of DNA replication catalyzed *in vitro* on a natural DNA template by the T4 bacteriophage multi-enzyme complex. *Nature* 285:300–305, 1980.
33. Higgins, N.P., Strauss, B.: Differences in the ability of human lymphoblastoid cells to exclude bromodeoxyuridine and in their sensitivity to methyl methanesulfonate and to incorporated ^{3}H thymidine. *Cancer Res.* 39:312–320, 1979.

34. HITTELMAN, W., RAO, P.: Bleomycin induced damage in prematurely condensed chromosomes and its relationship to cell cycle progression in CHO cells. *Cancer Res.* 34:3433-3439, 1974.
35. HUBERMAN, J., RIGGS, A.: On the mechanism of DNA replication in mammalian chromosomes. *J. Mol. Biol.* 32:327-341, 1968.
36. KANEKO, M., CERRUTI, P.: Excision of N-acetoxy-2-acetylaminofluorene induced DNA adducts from chromatin fractions of human fibroblasts. *Cancer Res.* 40: 4313-4319, 1980.
37. KAPP, L., PAINTER, R.: Replicon sizes in mammalian cells as estimated by an X-ray plus bromodeoxyuridine photolysis method. *Biophys. J.* 24:739-748, 1978.
38. KAPP, L., PAINTER, R.: DNA fork displacement rates in synchronous aneuploid and diploid mammalian cells. *Biochim. Biophys. Acta* 562:222-230, 1979.
39. KARRAN, P., LINDAHL, T., GRIFFIN, B.: Adaptive response to alkylating agents involves alteration *in situ* of O^6-methylguanine residues in DNA. *Nature* 280:76-77, 1979.
40. KEYZER, W., JASPERS, N., ABRAHAM, P., TAYLOR, A., ARLETT, C., ZELLE, B., TAKEBE, H., KINMONT, P., BOOTSMA, D.: A seventh complementation group in excision-deficient xeroderma pigmentosum. *Mutat. Res.* 62:183-190, 1979.
41. KRAEMER, K., WATERS, H., BUCHANAN, J.: Survival of human lymphoblastoid cell after DNA damage measured by growth in microtiter wells. *Mutat. Res.* 72: 285-294, 1980.
42. KRIEK, E.: Persistent binding of a new reaction product of the carcinogen N-hydroxy-N-2-acetylaminofluorene with guanine in rat liver DNA *in vivo*. *Cancer Res.* 32:2042-2048, 1972.
43. LAVAL, J., PIERRE, J., LAVAL, P.: Release of 7-methylguanine residues from alkylated DNA by extracts of Micrococcus luteus and Escherichia coli. *Proc. Natl. Acad. Sci. USA* 78:852-855.
44. LAWLEY, P., BROOKES, P.: Further studies on the alkylation of nucleic acids and their constituent nucleotides. *Biochem. J.* 89:127-138, 1963.
45. LAWLEY, P., MARTIN, C.: Molecular mechanisms in alkylation mutagenesis. Induced reversion of bacteriophage T4rII AP72 by ethyl methanesulphonate in relation to extent and mode of ethylation of purines in bacteriophage DNA. *Biochem. J.* 145:85-91, 1975.
46. LAWLEY, P., ORR, D.: Specific excision of methylation products from DNA of *Escherichia coli* treated with N-methyl-N'-nitro-N-nitrosoguanidine. *Chem. Biol. Interactions* 2:154-157, 1970.
47. LEHMANN, A., KIRK-BELL, S.: Pyrimidine dimer sites associated with the daughter DNA strands in UV-irradiated human fibroblasts. *Photochem. Photobiol.* 27: 297-308, 1978.
48. LEWIS, J., SWENBERG, J.: Differential repair of O^6-methylguanine in DNA of rat hepatocytes and nonparenchymal cells. *Nature* 288:185-187, 1980.
49. LINDAHL, T.: DNA-glycosylases, endonucleases for apurinic-apyrimidinic sites, and base excision repair. *Prog. Nucleic Acid Res. Mol. Biol.* 22:135-192, 1979.
50. LINDAHL, T., NYBERG, B.: Rate of depurination of native DNA. *Biochemistry* 11: 610-618, 1972.
51. LIVNEH, Z., ELAD, D., SPERLING, J.: Enzymatic insertion of purine bases into depurinated DNA *in vitro*. *Proc. Natl. Acad. Sci. USA* 76:1089-1093, 1979.
52. LOEB, L., SPRINGGATE, C., BATTULA, N.: Errors in DNA replication as a basis of malignant change. *Cancer Res.* 34:2311-2321, 1974.
53. MAHER, V., DORNEY, D., MENDRALA, A., KONZE-THOMAS, B., MCCORMICK, J.: DNA

excision repair processes in human cells eliminate the cytotoxic and mutagenic consequences of ultraviolet irradiation. *Mutat. Res.* 62:311–323, 1979.
53a. MAKINO, F., OKADA, S.: Effects of ionizing radiations on DNA replication in cultured mammalian cells. *Radiat. Res.* 83:668–676, 1975.
54. MARGISON, G., PEGG, A.: Enzymatic release of 7-methylguanine from methylated DNA by rodent liver extracts. *Proc. Natl. Acad. Sci. USA* 78:861–865, 1981.
55. MARTIN, C., GARNER, R.: Aflotoxin B-oxide generated by chemical or enzymic oxidation of aflotoxin B_1 causes guanine substitution in nucleic acids. *Nature* 267:863–865, 1977.
56. MENIGHINI, R., HANAWALT, P.: T4-endonuclease V-sensitive sites in DNA from ultraviolet-irradiated human cells. *Biochim. Biophys. Acta* 425:428–437, 1976.
57. MOORE, P., BOSE, K., RABKIN, S., STRAUSS, B.: Sites of termination of *in vitro* DNA synthesis on UV- and AAF-treated ΦX174 templates with prokaryotic and eukaryotic DNA polymerases. *Proc. Natl. Acad. Sci. USA* 78:110–114, 1981.
58. MOORE, P., RABKIN, S., STRAUSS, B.: Termination of in vitro DNA synthesis at AAF adducts AAF in the DNA. *Nucleic Acids Res.* 8:4473–4484, 1980.
59. MOORE, P., STRAUSS, B.: Sites of inhibition of *in vitro* DNA synthesis in carcinogen- and UV-treated ΦX174 DNA. *Nature* 278:664–666, 1979.
59a. OLSSON, M., LINDAHL, T.: Repair of alkylated DNA in *Escherichia coli.* Methyl group transfer from O^6-methylguanine to a protein cysteine residue. *J. Biol. Chem.* 255:10569–10571, 1980.
60. PAIKA, K., KRISHAN, A.: Bleomycin-induced chromosomal aberrations in cultured mammalian cells. *Cancer Res.* 33:961–965, 1973.
61. PAINTER, R.: Rapid test to detect agents that damage human DNA. *Nature* 265:650–651, 1977.
62. PAINTER, R.: DNA synthesis inhibition in HeLa cells as a simple test for agents that damage DNA. *J. Environ. Pathol. Toxicol.* 2:65–78, 1978.
63. PAINTER, R.: Does ultraviolet light enhance post replication repair in mammalian cells? *Nature* 275:243–245, 1978.
64. PAINTER, R.B.: Role of DNA damage and repair in cell-killing induced by ionizing radiation. *In* Radiation Biology in Cancer Research (Meyn, R., Withers, H., eds.), New York: Raven, 59–68, 1980.
64a. PAINTER, R., YOUNG, B.: Formation of nascent DNA molecules during inhibition of replication initiation in mammalian cells. *Biochim. Biophys. Acta* 418:146–153, 1976.
65. PARK, S., Cleaver, J.: Postreplication repair: Questions of its definition and possible alteration in xeroderma pigmentosum cell strains. *Proc. Natl. Acad. Sci. USA* 76:3927–3931, 1979.
66. PARK, S., CLEAVER, J.: Recovery of DNA synthesis after ultraviolet irradiation of xeroderma pigmentosum cells depends on excision repair and is blocked by caffeine. *Nucleic Acids Res.* 6:1151–1160, 1979.
67. PRAKASH, L., STRAUSS, B.: Repair of alkylation damage: Stability of methyl groups in *Bacillus subtilis* treated with methyl methanesulfonate. *J. Bacteriol.* 102:760–766, 1970.
68. PUCK, T.: The Mammalian cell as a Microorganism. San Francisco: Holden-Day, 1972.
69. RADANY, E., FRIEDBERG, E.: A pyrimidine dimer-DNA glycosylase activity associated with the v gene product of bacteriophage T4. *Nature* 286:182–185, 1980.
70. RADMAN, M., VILLANI, G., BOITEAUX, S., DEFAIS, M., CAILLET-FAUQUET, P., SPADARI, S.: On the mechanism and genetic control of mutagenesis induced by car-

cinogenic mutagens. *In* Origins of Human Cancer (Hiatt, H., Watson, J., Winsten, J., eds.), Book B, Cold Spring Harbor: Cold Spring Harbor Laboratory, 903–922, 1977.
71. Ross, S., Moses, S.: Two actions of bleomycin on superhelical DNA. *Biochemistry* 17:581–586, 1978.
72. Rupp, W., Wilde, C., Reno, D., Howard-Flanders, P.: Exchanges between DNA strands in ultraviolet-irradiated *Escherichia coli*. *J. Mol. Biol.* 61:25–44, 1971.
73. Samson, L., Schwartz, J.: Evidence for an adaptive DNA repair pathway in CHO and human skin fibroblast cell lines. *Nature* 287:861–863, 1980.
73a. Santella, R., Kriek, E., Grunberger, D.: Circular dichroism and proton magnetic resonance studies of dApdG modified with 2-aminofluorene and 2-acetylaminofluorene. *Carcinogenesis* 1:897–902, 1980.
74. Sarasin, A., Benoit, A.: Induction of an error-prone mode of DNA repair in UV-irradiated monkey kidney cells. *Mutat. Res.* 70:71–81, 1980.
75. Sarasin, A., Hanawalt, P.: Carcinogens enhance survival of UV-irradiated simian virus 40 in treated monkey kidney cells: Induction of a recovery pathway? *Proc. Natl. Acad. Sci. USA* 75:346–350, 1978.
76. Sarasin, A., Hanawalt, P.: Replication of ultraviolet-irradiated simian virus 40 in monkey kidney cells. *J. Mol. Biol.* 138:299–320, 1980.
77. Schendel, P., Defais, M., Jeggo, P., Samson, L., Cairns, J.: Pathways of mutagenesis and repair in *Escherichia coli* exposed to low levels of simple alkylating agents. *J. Bacteriol.* 135:466–475, 1978.
78. Schendel, P., Robins, P.: Repair of O^6-methylguanine in adapted *Escherichia coli*. *Proc. Natl. Acad. Sci. USA* 75:6017–6020, 1978.
79. Seawell, P., Smith, C., Ganesan, A.: *denV* gene of bacteriophage T4 determines a DNA glycosylase specific for pyrimidine dimers in DNA. *J. Virol.* 35:790–797, 1980.
80. Seeberg, E.: Reconstitution of an *Escherichia coli* repair endonuclease activity from the separated $uvrA^+$ and $uvrB^+/uvrC^+$ gene products. *Proc. Natl. Acad. Sci. USA* 75:2569–2573, 1978.
81. Seeberg, E., Rupp, W.D., Strike, P.: Impaired incision of ultraviolet-irradiated deoxyribonucleic acid in *uvrC* mutants of *Escherichia coli*. *J. Bacteriol.* 144:97–104, 1980.
82. Shearman, C., Loeb, L.: Effects of depurination on the fidelity of DNA synthesis. *J. Mol. Biol.* 128:197–218, 1979.
83. Singer,B.: *N*-Nitroso alkylating agents: Formation and persistence of alky derivatives in mammalian nucleic acids as contributing factors in carcinogenesis. *J. Natl. Cancer Inst.* 62:1329–1339, 1979.
84. Singer, B., Brent, T.: Human lymphoblasts contain DNA glycosylase activity excising N-3 and N-7 methyl and ethyl purines but not O^6-alkylguanine or 1-alkyladenine. *Proc. Natl. Acad. Sci. USA* 78:856–860, 1981.
85. Sklar, R., Strauss, B.: The role of the *uvrE* gene product and of inducible O^6methylguanine removal in the induction of mutations by *N*-methyl-*N'*-nitro-*N*-nitrosoguanidine in *Escherichia coli*. *J. Mol. Biol.* 143:343–362, 1980.
86. Sklar, R., Strauss, B.: Removal of O^6-methylguanine from DNA of normal and xeroderma pigmentosum-derived lymphoblastoid cell lines. *Nature* 78:856–860, 1981.
87. Smerdon, M., Lieberman, M.: Distribution within chromatin of deoxyribonucleic acid repair synthesis occurring at different times after ultraviolet radiation. *Biochemistry* 19:2992–3000, 1980.

88. STRAUSS, B., BOSE, K., ALTAMIRANO, M., SKLAR, R., TATSUMI, K.: Response of mammalian cells to chemical damage. *ICN-UCLA Symp. Mol. Cell. Biol.* 9: 621-624, 1978.
89. SUTHERLAND, B., CHAMBERLIN, M., SUTHERLAND, J.: Deoxyribonucleic acid photoreactivating enzyme from *Escherichia coli. J. Biol. Chem.* 248:4200-4205, 1974.
90. SUTHERLAND, B., HARBER, L., KOCHEVAR, I.: Pyrimidine dimer formation and repair in human skin. *Cancer Res.* 40:3181-3185, 1980.
91. TATSUMI, K., BOSE, K., AYRES, K., STRAUSS, B.: Repair of Neocarzinostatin-induced deoxyribonucleic acid in human lymphoblastoid cells: Possible involvement of apurinic/apyrimidinic sites as intermediates. *Biochemistry* 19:4767-4772, 1980.
92. TATSUMI, K., NAKAMURA, T., WAKISAKA, G.: Damage of mammalian cell DNA *in vivo* and *in vitro* induced by neocarzinostatin. *Gann* 65:459-461, 1974.
93. TATSUMI, K., NISHIOKA, H.: Effect of DNA repair systems on antibacterial and mutagenic activity of an antitumor protein, neocarzinostatin. *Mutat. Res.* 40: 195-204, 1977.
94. Warner, H., Demple, B., DEUTSCH, W., KANE, C., LINN, S.: Apurinic-apyrimidinic endonucleases in the repair of pyrimidine dimers and other lesions in DNA. *Proc. Natl. Acad. Sci. USA* 77:4602-4606, 1980.
95. WESTRA, J., KRIEK, E., HITTENHAUSEN, H.: Identification of the persistently bound form of the carcinogen *N*-acetyl-2-aminofluorene to rat liver DNA in vitro. *Chem. Biol. Inter.* 15:149-164, 1976.
96. WITKIN, E.: Ultraviolet mutagenesis and inducible DNA repair in *Escherichia coli. Bacteriol. Rev.* 40:869-907, 1976.
97. YAMASAKI, H., PULKRABEK, P., GRUNBERGER, D., WEINSTEIN, I.B.: Differential excision from DNA of the -8 and N^2 guanosine adducts of *N*-acetyl-2-aminofluorene by single strand-specific endonucleases. *Cancer Res.* 37:3756-3760, 1977.
98. YANG, L., Maher, V., McCORMICK, J.: Error-free excision of the cytotoxic mutagenic N^2-deoxyguanosine DNA adduct formed in human fibroblasts by ($\pm$)-7β, 8α-dihydroxy-9α, 10α-epoxy-7,8,9,10-tetrahydrobenzo(a)pyrene. *Proc. Natl. Acad. Sci. USA* 77:5933-5937, 1980.

VIRAL INTERACTIONS WITH THE MAMMALIAN GENOME RELEVANT TO NEOPLASIA

WARREN W. NICHOLS

Interactions of viruses with host cell genomes were first reported in the late 1940s and early 1950s in bacterial systems with phage. The first observations of these interactions in mammalian cells were made in the 1960s. In mammalian systems, the primary interest in viral effects on the host genome was in relation to cancer, both from the standpoint of the role and mechanism of virus induction of cancer and as a general model for all types of carcinogenesis. The first reported viral effects on the mammalian cell genome were those of induced chromosomal change, and these reports were followed by observations on viral insertion, induced gene mutations, and gene transfer. Many viruses now have been shown to have cytogenetic effects. Not all of these will be enumerated here as several extensive reviews are available [20,37,46,47]. It should be pointed out, however, that cytogenetic effects are produced by both RNA and DNA viruses, by both transforming and tumorigenic viruses, and by both nontransforming and nontumorigenic viruses.

TYPES OF VIRUS-INDUCED GENETIC CHANGE

Among the chromosome abnormalities produced are open chromosome breaks (Fig. 1), which are discontinuities in chromosomes and chromatids that are most likely to result when DNA and, or, protein synthesis are inadequate for repair processes. Also produced are stable and unstable chromosomal rearrangements, including chromosome and chromatid intrachanges and interchanges (Fig. 2). Premature chromosome condensation (PCC), sometimes referred to as chromosome pulverization, is another effect (Fig. 3). This usually is considered to be an indirect effect of the virus, the result of cell fusion between a mitotic cell and an

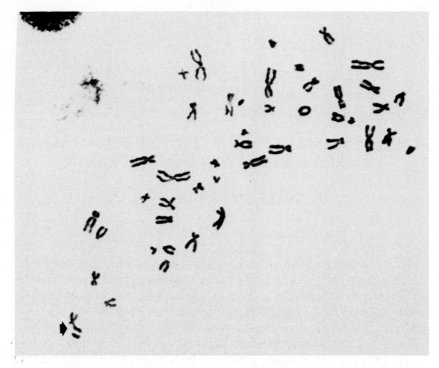

Fig. 1. Metaphase chromosomes from human leukocyte treated in vitro with Schmidt Ruppin Rous sarcoma virus showing open chromatid break at arrow (from *Symp. Int. Soc. Cell Biol.* 3:255–271, 1964, Academic Press).

interphase cell. In this case, the mitotic cell produces a protein substance that induces chromatin condensation in the interphase cell, and the characteristic appearance depends on which part of interphase the cell is in [40]. The apparent fragmentation seen in Figure 3 occurs when the interphase nucleus is in the S phase of the cell cycle. When the interphase nucleus is in G1, the PCC has the appearance of a long, attenuated single-stranded unit, whereas in the G2, the PCC has the appearance of an elongated double-stranded unit. PCC is also seen, however, in diploid cells, and there the mechanism is not as clearly understood. There are also alterations in chromosome number: most frequently, increases in the number of polyploid cells, but occasionally, changes around the diploid number, including hypodiploidy. An increased frequency of sister-chromatid exchange (SCE) events has been described in association with SV40 transformation [47,67,68] and with transformation with Rauscher leukemia virus [10] (Fig. 4).

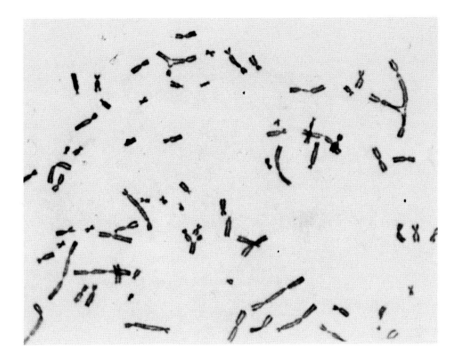

Fig. 2. Human diploid fibroblasts treated with SV40 virus demonstrating a variety of chromosome and chromatid interchanges and an increase in chromosome number.

With respect to chromosomal changes in lytic and transforming virus infections, in the former, the effects are usually seen very rapidly, whereas in the latter, there is often a long latent period between the addition of virus and the first appearance of chromosome defects. In the case of SV40, this period can be several weeks; and, in Epstein-Barr virus (EBV) transformation of human lymphocytes, after infection it can be as long as a year until chromosomal abnormalities are seen.

In some infections, the viral DNA or a DNA copy of the viral RNA are inserted covalently into the host cell DNA. This was observed in SV40 by Sambrook in 1968 [54], and in RNA viruses, both by Temin and Mizutani [57] and by Baltimore [4] in 1970. While many of these inserted viruses alter cell growth by means of a viral "onc" gene contained in their genome, recent work exemplified by Hayward et al. [21] demonstrated that altered growth patterns can be produced by viruses that alter cellular gene regulation without a viral "onc" gene. In this "viral promoter mechanism" of cell transformation there are long terminal repeats (LTRs)

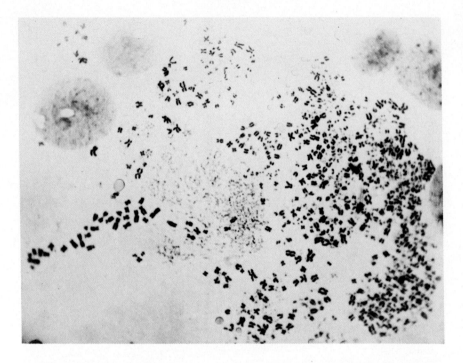

Fig. 3. Syncytium produced by cell fusion after addition of measles virus in vitro. Premature chromosome condensation (also referred to as chromosome pulverization) is seen in center and to a lesser degree in other nuclear groups (from *Hereditas* 5:380–382, 1964).

at the ends of the virus that initiate transcription of the viral genome, and these rarely integrate next to and cause transcription of a cellular "onc" gene.

In recent years, evidence also has been growing that viruses can induce gene mutations. These data have been observed by using somatic cell genetic techniques. Taylor [55] supplied precedent for this effect in 1963 by demonstrating that Mu-1 phage produced mutations in bacteria; he also demonstrated that this phage was inserted randomly throughout the bacterial genome rather than at a single, specific insertion site, as is true with most other bacteriophages. With SV40 virus in a variety of cell systems, gene mutations have been found by Marshak and his group in the USSR [38], by T.L. Theile and his colleagues in Germany [58–60], and in our laboratory.

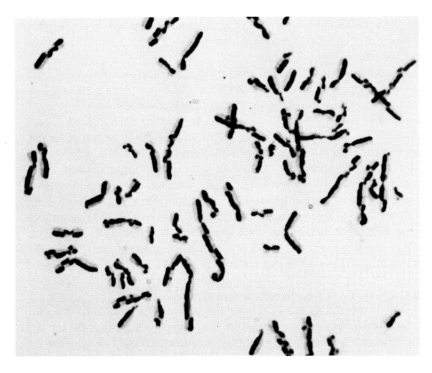

Fig. 4. Increased frequency of sister-chromatid exchanges (SCEs) induced by SV40 virus transformation in human diploid fibroblasts (from *Cancer Res.* 38:960-964, 1978).

The final type of interaction to mention is gene transfer. This is exemplified by the integration of the herpes simplex virus-thymidine kinase gene into thymidine kinase deficient (TK−) cells which are then shown to be able to grow in HAT medium; this was demonstrated by Bacchetti and Graham [2,3], as well as by others.

OBSERVATIONS ON VIRUS-INDUCED CELL TRANSFORMATION

Evidence for the role of somatic mutation in cancer has been growing steadily; it includes the demonstration that most, if not all, chemical carcinogens are mutagens [1,41,43]; epidemiologic observations [29,31]; observations of the increased risk of cancer in chromosome-fragility syndromes and abnormalities of DNA repair [7,11,16,19,52]; specific chromosomal changes within tumor cells of clinical and experimental malignancies [45,51]; and specific prezygotic chromosomal defects that predispose to specific cancers [12,30,50,66].

The epidemiologic work of Knudson [29,31] indicates the possibility of a two- or multiple-hit mechanism for cancer. On the basis of several observations, it seems probable that viral transformation also is a multistep process. In human diploid cells, SV40 transformation is seen as a morphologic and kinetic alteration that is followed by a period of "crisis" in which most cells die. This crisis is followed by a period in which one or a few clones repopulate the culture to give rise to a permanent transformed line. Girardi [17] found that crisis occurred at a regular interval after phase 3 (or senescent phase-out) was observed in uninfected diploid fibroblasts; the interval was approximately 9 wk. These observations suggest that the first stage of transformation, which precedes crisis, only delays the regularly observed senescence of uninfected diploid cells. This possibility is supported by the observations of Gotoh [18]; he carried 91 clones of transformed WI-38 cells, all of which died. This also indicates the rather low rate of survival of crisis.

Other observations that indicate the probability of a multistep process in transformation are found in the work with SV40 deletion mutants. In a series of deletion mutants developed by Paul Berg and his colleagues, mutants of early viral genes were used to infect several animal cell lines, and these failed to transform when an assay system based on growth in soft agar was used. The same virus-cell combinations, however, did transform when a monolayer assay system was used [9,42]. The work of Martin and others [39] indicates that this difference may be based on a requirement for cell division for stable transformation and that the virus deletion mutants lack the ability to stimulate mitosis but can produce transformation with the addition of suitable growth factors.

There is no doubt about the importance of the inserted viral genome in transformation. In most cases it can be shown that the inserted viral genetic material is functional, and that turn-off of the viral genome via temperature-sensitive viral mutants leads to a reversion of many of the acquired phenotype characteristics of transformation [56]. It is not clear, however, whether the induction of transformation or malignancy is produced by the insertion and function of the viral material alone or whether cellular participation or alterations also are necessary. A steadily growing number of investigators believe that cellular alterations are necessary. Dulbecco [15], who has commented that the expression of the viral genome is controlled by cellular genes, believes that the viral genome can be transcribed only if it is included in a transcribed segment of cellular

DNA. Thus, a viral genome that is inserted in cellular heterochromatin could not be responsible for transformation because it would not be transcribed and, therefore, would not be expressed. Dulbecco also suggests that the reason some cells infected by transforming viruses are not successfully transformed may be that transformation can be accomplished only when the cellular gene balance has been altered in specific ways by rare, virus-independent, genetic events.

Croce and his colleagues [35] recently have reported very interesting work on murine teratocarcinoma cells. Teratocarcinoma stem cells were shown not to support the growth of SV40 or polyoma virus; however, when the stem cells were permitted to differentiate, they became susceptible to the viruses. The resistance of the stem cells was not due to blockage at the level of virus adsorption, or to penetration, uncoating, or transport to the nucleus. Croce and co-workers used a recombinant plasmid that was linked both to the thymidine kinase gene of herpes simplex virus and an SV40 genome, and inserted these into TK-stem cells. They found that the TK gene is expressed in stem cells whereas the SV 40 is not. If differentiation of the teratocarcinoma cells is permitted, then both the TK and SV40 genome are expressed, indicating not only that cellular regulation occurs but also that it differs with the state of differentiation.

Klein [27,28] has compared the EBV transformation of human lymphocytes with Burkitt's lymphoma; he believes there is an initial step in the EBV transformation that immortalizes the cells, but that cytogenetic change is required for malignant growth in vivo. Thus, lymphocytes that are infected with EB virus and grow continually will not form colonies in agarose or produce tumors in nude mice until cytogenetic changes have taken place, usually months or a year after infection. In Burkitt's lymphoma, which also is associated with EBV, there is a characteristic translocation between chromosomes 8 and 14.

The concept of a cytogenetic requirement after initiation also is supported by the observation that T-cell lymphomas in mice are found to be trisomic for chromosome 15, whether they were initiated by chemical, x-ray, or viral means. Thus, no matter what was the initiating agent, the same chromosomal alteration accompanied the development of the tumor. Similar observations have now been extended to B-cell tumors. In addition, Sager [53] has reported high frequencies of rearrangements in mouse cell DNA adjacent to the integrated sites of inserted SV40 genome, and hypothesizes that these may influence the expression of transformation and tumorigenic potential. Further, Whitehouse [65] has proposed that proviral genes in a cell may have selective advantages for both the cell and the virus, and that the proviral genes are kept in check by cellular regulatory genes that are capable of initiating host-DNA synthesis. Any alteration to these suppressor genes, either directly or through mutation, would alter the cell-provirus balance toward transformation.

From studies on various chromosomal abnormalities, it is clear that the cellular genome is important in the susceptibility of cells to virus transformation. However, there is uncertainty about the ways in which chromosomal abnormalities influence virus transformation. For instance, in Fanconi's anemia [33,61], in which increased chromosome breakage occurs, and also in chromosomal trisomies [33,49,62], there is an increased susceptibility to SV40 transformation, as shown by both the numbers of transformed foci produced at specific multiplicities of infection and by the percentage of cells that express virus-induced T antigen 72 h after infection. Ataxia-telangiectasia however, also a condition with an increased spontaneous rate of chromosome abnormalities, has been shown to be more susceptible to SV40-induced chromosome breakage [64], but the transformation rate is the same as that found in normal cells [33,64]. Bloom's syndrome also has an increased level of chromosome breakage; in one study, it has been reported to have a decreased transformation rate when infected with SV40 virus, but in another study to have a transformation rate similar to that of normal cells [34,63]. Cells from patients with xeroderma pigmentosum show no increase of transformation frequency either with SV40 virus alone or with SV40 virus combined with ultraviolet [23,25,48].

VIRUS-INDUCED GENE MUTATIONS

I will now return to virus-induced chromosome and gene mutations in order to compare the time sequence of their appearance and some characteristics of the gene-mutational events. When cells are examined sequentially after infection, chromosomal mutations associated with viral transformation are observed at the time the cells become T-antigen positive. As shown in Figure 5, the frequencies of both the SCEs per cell and the anaphase abnormalities parallel the increase in T-antigen-positive cells. The time at which cells become T-antigen-positive varies with the multiplicity of infection used.

SV40 also has been associated with induced gene mutations. These seem to be produced earlier than the chromosomal mutations, and in Marshak's [38] experiments, mutations were observed 24 to 96 h after infection. Induced gene mutations seem to be related to the viral genome: Theile could produce the mutations by using extracted SV40 DNA, but found that noninfected DNA did not induce an increase in mutation frequency [59]. The mutations produced also seem, at least in part, to be point mutations rather than small deletions. In most of the systems examined, the studies utilized mutations to thioguanine or 8-azaguanine re-

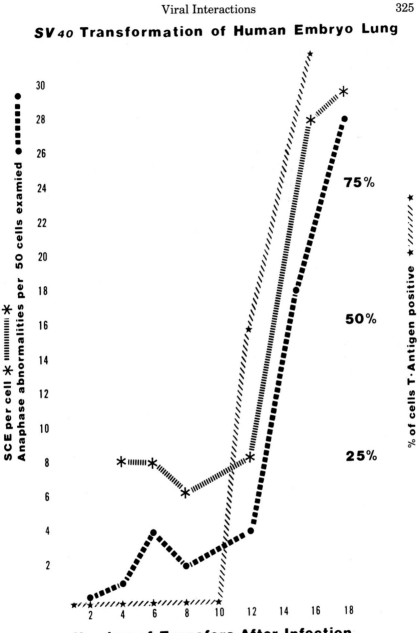

Fig. 5. Chart showing time relationship of increasing frequencies of sister-chromatid exchange (SCE), anaphase chromosome abnormalities, and T antigen-positive cells after addition of SV40 virus (from *Cancer Res.* 38:960–964, 1978).

sistance, and these could be either true gene mutations or small deletions; but Luebbe and colleagues [36] have reported that SV40 can induce revertants of Lesch-Nyhan cells which could not be explained on the basis of deletion events. The viral-induced mutations are increased in a synergistic manner when chemical mutagen treatment is added to the viral treatment [60]. It is noteworthy that some chemical mutagens and tumor promoters also increase the rate of SV40 transformation [22,44,69]. Most studies of viral-induced mutations have been done in animal cell systems or in heteroploid human cells. In our own laboratory, when studying mutation frequency in diploid cells, we saw an increased mutation frequency at the hpt locus in three out of five cultures examined (Table I). In one culture of human embryo lung, no increase was noted on six successive transformations. The reasons for these differences are not yet apparent, but several possibilities will be investigated, such as the timing of the mutational analysis after infection.

TABLE I. Frequencies of Gene Mutations in Control and Virus-Transformed Culture*

Cell line	Freq. of 8-azaguanine resistance	No. of cells examined
HEL 67 (normal lung)	2.75×10^{-6}	2.8×10^{7}
HEL 67 VA	1.25×10^{-5}	3.1×10^{7}
HEL 76 (normal lung)	2.0×10^{-6}	3.31×10^{7}
VAI	1.5×10^{-6}	5.8×10^{6}
VAII	2.6×10^{-6}	1.44×10^{7}
VAIII	6.5×10^{-7}	1.07×10^{7}
VAIV	5.4×10^{-7}	2.04×10^{7}
VAV	1.3×10^{-6}	2.69×10^{7}
VAVI	$< 1.9 \times 10^{-7}$	4.9×10^{6}
GM 37 (normal skin)	4.14×10^{-6}	2.4×10^{7}
GM 37 VA	1.81×10^{-6}	5.0×10^{7}
GM 54 (galactosemia)	$< 1.4 \times 10^{-7}$	7.2×10^{6}
GM 54 VA	3.6×10^{-4}	1.27×10^{7}
GM 1131 (retinoblastoma)	$< 8.3 \times 10^{-7}$	1.2×10^{6}
GM 1131 VA	4.9×10^{-5}	1.7×10^{6}

*Frequencies of mutations at hpt locus in control and SV40 transformed (VA) pre-crisis human diploid fibroblasts in five different cultures. HEL 76 did not exhibit a detectable increase in mutation frequency in six separate transformations.

THE TWO-HIT HYPOTHESIS OF VIRAL TRANSFORMATION

At this point, it can be said that there is a possibility that viral transformation is a multistep event; the inserted viral genome into the cell genome is an important event in transformation, and it is possible that the cell genome may be involved in the transformation process, and that transforming viruses have the ability to produce chromosomal and gene mutations. In the context of these observations, Knudson's two-hit mutational model of carcinogenesis has interesting implications. His studies indicate that for many or all tumor types, two subgroups exist: those that are sporadic and those that have a hereditary predisposition. In the development of a sporadic tumor, two specific mutations must occur, which is rare. In the hereditary subgroup, one mutation already is present in a genome through inheritance from the germline and only one somatic mutation is necessary for tumor production [29]. The requirement for only a single somatic mutation in the hereditary group makes this a much more common event and produces tumors at a younger age that are multifocal and bilateral when applicable.

Cell transformation of normal human diploid cells rarely has been achieved with chemical carcinogens or mutagens. If two specific mutations are necessary and mutation rates are 10^{-5} or 10^{-6}, transformation would be expected only once in 10^{10} or 10^{12} cells per generation. This event would be so rare that it would be difficult to detect it in a cell culture system, and this may explain why transformation of diploid cells with chemicals is accomplished so rarely. I would propose that viral transformation seen so regularly with human diploid cells is comparable to Knudson's hereditary subgroup of cancers. In the case of viral transformation, the effect of two mutational events is produced by the insertion and function of the viral genome as one event, and then by the induction of chromosomal or gene mutation as the second event. According to this hypothesis, the virus would insert randomly in the genome of infected cells, and it would both produce viral messenger RNA and induce chromosome and gene mutations. The functioning viral genome would be responsible for the altered cell phenotype, including the increased growth rate. At the time of crisis, most of these cells would degenerate or die, and only those in which an appropriate second mutation is produced would survive and propagate with the full malignant phenotype (Fig. 6).

It is also conceivable that, in rare cases, the complete transformation could occur without viral genome function by induction of the two necessary cellular mutations, as in the case of sporadic tumors. Bishop [6] has reported one mouse line, which is transformed by Rous sarcoma virus, that does not have detectable levels of virus-specific RNA and thus could

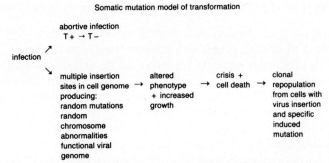

Fig. 6. Diagram of two-hit mutational model of virus in transformation of human cells.

be comparable with this second type of process. Other observations that are compatible with this hypothesis are the differences in the number of viral insertion sites observed by Croce and his colleagues [13,14,26] and by Kucherlapati et al [32], who found by chromosome mapping techniques that SV40 virus is limited to one chromosome in the transformed cell: chromosome 7, 17, or 8. When restriction endonuclease techniques and Southern blots were used to look at insertion sites, multiple sites of SV40 insertion were found [5,8,24]. It is likely that these differences are due to the utilizaztion of postcrisis cells when only one site was found, and of precrisis cells when multiple sites were found. This would be due to the loss by death of most cells with random mutations and the selection and clonal proliferation of the rare cell with the specific second mutation. In our own laboratory, Dr. Arnold Dion's preliminary work with restriction endonuclease and Southern blots on pre- and postcrisis cells indicates a large reduction in the number of insertion sites after crisis. Whether or not this is reduced to one site is not yet clear because of uncertainties introduced by free virus in the preparations.

It is likely that either a chromosome mutation or gene mutation could effect the final stage to full transformation or malignancy, much as is seen in hereditary retinoblastoma [30]. Some of these retinoblastoma cases appear to have a gene mutation that predisposes to tumor formation, whereas others have a specific deletion in chromosome 13 that leads to tumor formation. In our own laboratory we are attempting to elucidate this hypothesis by examining induced chromosomal and gene mutations in relation to transformation by wild-type virus, and by mutant deleted and temperature-sensitive viruses.

Acknowledgments. Dr. Nichols is the S. Emlen Stokes Professor of Genetics at the Institute for Medical Research. This work was supported by grant No. 2P01-Ag-00378-09 AGE from the National Institute of Aging, National Institutes of Health, Bethesda, Maryland.

LITERATURE CITED

1. AMES B.N., SIMS P., GROVER P.L.: Epoxides of carcinogenic polycyclic hydrocarbons are frameshift mutagens. *Science* 176:47-49, 1972.
2. BACCHETTI S., GRAHAM F.L.: Transfer of the gene for thymidine kinase to thymidine kinase-deficient human cells by purified herpes simplex viral DNA. *Proc. Natl. Acad. Sci. USA* 74:1590-1594, 1977.
3. BACCHETTI S., GRAHAM F.L.: DNA mediated transfer of herpes simplex virus TK gene to human TK- cells: Properties of the transformed lines. *IARC Sci. Publ.* 24(pt. 1):495-499, 1978.
4. BALTIMORE D.: RNA dependent DNA polymerase in virions of RNA tumor viruses. *Nature* 226:1209, 1970.
5. BATTULA M., TEMIN H.: Infectious DNA of spleen necrosis virus is integrated at a single site in the DNA of chronically infected chicken fibroblasts. *Proc. Natl. Acad. Sci. USA* 74:281-285, 1977.
6. BISHOP J.M., JACKSON H., QUINTRELL N., VARMUS H.E.: Transcription of RNA tumor virus genes in normal and infected cells. *In* Silvestri L.G. (ed.): "Possible Episomes in Eukaryotes 6." Amsterdam and London: North Holland, pp. 61-72, 1973.
7. BLOOM A.D.: Induced chromosomal aberrations in man. *In* Harris H., Hirschhorn K. (eds): "Advances in Human Genetics Vol. 3." New York: Plenum Press, pp. 99-172, 1972.
8. BOTCHAN H., TOPP W., SAMBROOK J.: The arrangement of simian virus 40 sequences in the DNA of transformed cells. *Cell* 9:269-287, 1976.
9. BOUCK N., BEALES N., SHENK T., BERG P., DI MAYORCA G.: New region of the simian virus 40 genome required for efficient viral transformation. *Proc. Natl. Acad. Sci. USA* 75:2473-2477, 1978.
10. BROWN R.L., CROSSEN P.E.: Increased incidence of sister chromatid exchanges in Rauscher leukaemia virus infected mouse embryo fibroblasts. *Exp. Cell Res.* 103:418-420, 1976.
11. CLEAVER J.E.: Defective repair replication of DNA in Xeroderma pigmentosum. *Nature* 218:652-656, 1968.
12. COHEN A.J., LI F.P., BERG S., MARCHETTO D.J., TSAI S., JACOBS S.C., BROWN R.S.: Hereditary renal-cell carcinoma associated with a chromosomal translocation. *N. Engl. J. Med.* 301:592-595, 1979.
13. CROCE C.M.: Assignment of the integration site for simian virus 40 to chromosome 17 in GM54VA, a human cell line transformed by simian virus 40. *Proc. Natl. Acad. Sci. USA* 74:315-318, 1977.
14. CROCE C.M., GIRARDI A.J., KOPROWSKI H.: Assignment of the T-antigen gene of simian virus 40 to human chromosome C-7. *Proc. Natl. Acad. Sci. USA* 70:3617-3620, 1973.
15. DULBECCO R.: Cell transformation by viruses and the role of viruses in cancer. *J. Gen. Microbiol.* 79:7-17, 1973.
16. FRAUMENI J.F., MILLER R.W.: Epidemiology of human leukemias. Recent observations. *J. Natl. Cancer Inst.* 38:593-605, 1967.
17. GIRARDI A.J., JENSEN F.C., KOPROWSKI H.: SV40 induced transformation of human diploid cells: Crisis and recovery. *J. Cell. Comp. Physiol.* 65:69-84, 1965.
18. GOTOH S., GELB L., SCHLESSINGER D.: SV40-transformed human diploid cells that remain transformed throughout their limited lifespan. *J. Gen. Virol.* 42:409-414, 1979.

19. HAMERTON J.L.: Chromosomes and neoplastic disease. *In* Hamerton J.L. (ed.): "Human Cytogenetics, Clinical Cytogenetics Vol.11." New York: Academic Press, pp. 407-441, 1971.
20. HARNDEN D.G.: Viruses, chromosomes and tumors. The interaction between viruses and chromosomes. *In* German J. (ed.): "Chromosomes and Cancer." New York: John Wiley and Sons, pp. 151-190, 1974.
21. HAYWARD W.S., NEEL B.G., ASTRIN S.M.: Activation of a cellular *onc* gene by promoter insertion in ALV-induced lymphoid leukosis. *Nature* 290:475-480, 1981.
22. HENDERSON E.E., RIBECKY R.: Transformation of human leukocytes with Epstein-Barr virus after cellular exposure to chemical or physical mutagens. *J. Natl. Cancer Inst.* 1:33-39, 1980.
23. KERSEY J.H., GATTI R.A., GOOD R.A., ARRONSON S.A., TODARO G.J.: Susceptibility of cells from patients with primary immunodeficiency diseases to transformation by simian virus 40. *Proc. Natl. Acad. Sci. USA* 69:980-982, 1972.
24. KETNER G., KELLY T.J.: Integrated simian virus 40 sequences in transformed cell DNA: Analysis using restriction endonucleases. *Proc. Natl. Acad. Sci. USA* 73:1102-1106, 1976.
25. KEY D.J., TODARO G.J.: Xeroderma pigmentosum cell susceptibility to SV40 virus transformation: Lack of effect of low dosage ultraviolet radiation in enhancing viral-induced transformation. *J. Invest. Dermatol.* 62:7-10, 1974.
26. KHOURY G., CROCE C.M.: Quantitation of the viral DNA present in somatic cell hybrids between mouse and SV40 transformed human cells. *Cell* 6:535-542, 1975.
27. KLEIN G.: Lymphoma development in mice and humans: Diversity of initiation is followed by convergent cytogenetic evolution. *Proc. Natl. Acad. Sci. USA* 76:2442-2446, 1979.
28. KLEIN G.: The role of viral transformation and cytogenetic changes in viral oncogenesis. *Ciba Found. Symp.* 66:335-358, 1979.
29. KNUDSON A.G.: Mutation and cancer: Statistical study of retinoblastoma. *Proc. Natl. Acad. Sci. USA* 68:820-823, 1971.
30. KNUDSON A.G., MEADOWS A.T., NICHOLS W.W., HILL R.: Chromosomal deletion and retinoblastoma. *N. Engl. J. Med.* 295:1120-1123, 1976.
31. KNUDSON A.G., STRONG L.C.: Mutation and cancer: Neuroblastoma and phenochromocytoma. *Am. J. Hum. Genet.* 24:513-532, 1972.
32. KUCHERLAPATI R., HWANG S.P., SHIMIZU N., MCDOUGALL J.L., BOTCHAN M.R.: Another chromosomal assignment for a simian virus 40 integration site in human cells. *Proc. Natl. Acad. Sci. USA* 75:4460-4464, 1978.
33. LEVIN S., HAHN T.: Transformation of cells in culture from children with chromosomal and immune deficiency disorders. *Isr. J. Med. Sci.* 8:133-136, 1972.
34. LIN M.S., ALFI O.S.: Chromosome fragility and susceptibility of Bloom's syndrome fibroblasts to SV40 transformation. *Experientia* 36:296-297, 1980.
35. LINNENBACH A., HUEBNER K., CROCE C.M.: DNA-transformed murine teratocarcinoma cells: Regulation of expression of simian virus 40 tumor antigen in stem versus differentiated cells. *Proc. Natl. Acad. Sci. USA* 77:4875-4879, 1980.
36. LUEBBE L., STRAUSS M., GEISSLER E.: Reversion of HGPRT deficiency of skin fibroblasts from Lesch-Nyhan patients. *FEBS Proc.* 56:67, 1979.
37. MAKINO S.: "Human Chromosomes." Amsterdam and New York: Elsevier/North-Holland Biomedical Press, 1975.
38. MARSHAK M.A., VARSHANER N.B., SHAPIRO N.I.: Induction of gene mutations and chromosomal aberrations by simian virus 40 in cultured mammalian cells. *Mutat. Res.* 30:383-396, 1975.

39. MARTIN R.G.: The transformation of cell growth and transmogrification of DNA synthesis by simian virus 40. *In* Klein G., Weinhouse S. (eds.): "Advances in Cancer Research 34." New York: Academic Press, pp. 1-68, 1981.
40. MATSUI W., WINFELD H., SANDBERG A.A.: Dependence of chromosome pulverization in virus-fused cells on events in the G2 period. *J. Natl. Cancer Inst.* 47:401-411, 1971.
41. MCCANN J., CHOI E., HAMASAKI E., AMES B.N.: Detection of carcinogens as mutagens in the Salmonella microsome test: Assay of 300 chemicals. *Proc. Natl. Acad. Sci. USA* 72:5135-5139, 1975.
42. MERTZ J.E., CARBON J., HERZBERG M., DAVIS R.W., BERG P.: Isolation and characterization of individual clones of simian virus 40 mutants containing deletions, duplication and insertions in their DNA. *Cold Spring Harbor Symp. Quant. Biol.* 39(pt. 1):69-84, 1975.
43. MILLER J.A.: Carcinogenesis by chemicals: An overview. *Cancer Res.* 30:559-576, 1970.
44. MILO G.E., BLAKESLESS J.R., HART R., YOHN D.S.: Chemical carcinogen alteration of SV40 virus induced transformation of normal human cell populations in vitro. *Chem. Biol. Interact.* 22:185-197, 1978.
45. MITELMAN F., MARK J., LEVAN G., LEVAN A.: Tumor etiology and chromosome pattern. *Science* 176:1340-1341, 1972.
46. NICHOLS W.W.: Viruses and chromosomes. *In* Busch H. (ed.): "The Cell Nucleus, Vol. 2." New York: Academic Press, pp. 437-458, 1974.
47. NICHOLS W.W., BRADT C.I., TOJI L.H., GODLEY M., SEGAWA M.: Induction of sister chromatid exchanges by transformation with simian virus 40. *Cancer Res.* 38:960-964, 1978.
48. PARRINGTON J.M., DELHANTY J.D.A., BADEN H.P.: Unscheduled DNA synthesis, i.v.-induced chromosome aberrations and SV40 transformation in cultured cells from xeroderma pigmentosum. *Ann. Hum. Genet.* 35:149-160, 1971.
49. PAYNE F.E., SCHMICKEL R.D.: Susceptibility of trisomic and of triploid human fibroblasts to simian virus 40 (SV40). *Nature* 230:190, 1971.
50. RICCARDI V.M., SUJANSKY E., SMITH A.C., FRANCKE U.: Chromosomal imbalance in the aniridia-Wilm's tumor association. I/P interstitial deletion. *Pediatrics* 61:604-610, 1978.
51. ROWLEY J.D.: Chromosome abnormalities in cancer. *Cancer Genet. Cytol.* 2:175-198, 1980.
52. RYSER H.J.P.: Chemical carcinogenesis. *N. Engl. J. Med.* 285:721-734, 1971.
53. SAGER R., ANISOWICZ A., HOWELL N.: Genomic rearrangements and tumor-forming potential in an SV40 transformed mouse cell line and its hybrid and cybrid progeny. *Cold Spring Harbor Symp. Quant. Biol.* 45:747-754, 1980.
54. SAMBROOK J., WESTPHAL H., SRINIVASSAN T.R., DULBECCO R.: The integrated state of viral DNA in SV40 transformed cells. *Proc. Natl. Acad. Sci. USA* 60:1288-1295, 1968.
55. TAYLOR A.L.: Bacteriophage-induced mutation in Escherichia coli. *Proc. Natl. Acad. Sci. USA* 50:1043-1051, 1963.
56. TEGTMEYER P.: Function of simian virus 40 geneA in transforming infection. *J. Virol.* 15:613-618, 1975.
57. TEMIN H.M., MIZUTANI S.: RNA-dependent DNA polymerase in virions of Rous sarcoma virus. *Nature* 226:1211, 1970.
58. THEILE M., KRAUSE H.: The combined mutagenic action of simian virus 40 and other carcinogens in Chinese hamster cells. *Stud. Biophys.* 76:45-46, 1979.

59. THEILE M., SCHERNECK S., GEISSLER E.: DNA of simian virus 40 mutates Chinese hamster cells. *Arch. Virol.* 65(¾):293-310, 1980.
60. THEILE M., STRAUSS M., LUEBBE L., SCHERNECK S., KRAUSE H., GEISSLER E.: SV40 induced somatic mutations: Possible relevance to viral transformation. *Cold Spring Harbor Symp. Quant. Biol.* 44:377-382, 1980.
61. TODARO G.J., GREEN H., SWIFT M.R.: Susceptibility of human diploid fibroblast strains to transformation by SV40 virus. *Science* 153:1252-1254, 1966.
62. TODARO G.J., MARTIN G.M.: Increased susceptibility of Down's syndrome fibroblasts to transformation by SV40. *Proc. Soc. Exp. Biol. Med.* 124:1232-1236, 1967.
63. WEBB T., HARDING M.: Chromosome complement and SV40 transformation of cells from patients susceptible to malignant disease. *Br. J. Cancer* 36:583-591, 1977.
64. WEBB T., HARNDEN D.G., HARDING M.: The chromosome analysis and susceptibility to transformation by simian virus 40 of fibroblasts from ataxia-telangiectasia. *Cancer Res.* 37:997-1002, 1977.
65. WHITEHOUSE H.L.K.: Chromosome integration of viral DNA: The open-replicon hypothesis of carcinogenesis. *In* German J. (ed.): "Chromosomes and Cancer." New York: John Wiley and Sons, pp. 42-76, 1974.
66. WILSON M.G., TOWNER J.W., FUJIMOTO A.: Retinoblastoma and D-chromosome deletions. *Am. J. Hum. Genet.* 25:57-61, 1973.
67. WOLFF S., BODYCOTE J., THOMAS G.H., CLEAVER J.E.: Sister chromatid exchange in Xeroderma pigmentosum cells that are defective in DNA excision repair of postreplication repair. *Genetics* 81:349-355, 1975.
68. WOLFF S., RODIN B., CLEAVER J.E.: Sister chromatid exchanges induced by mutagenic carcinogens in normal and xeroderma pigmentosum cells. *Nature* 265:347-349, 1977.
69. YAMAMOTO N., ZUR HAUSEN H.: Tumour promoter TPA enhances transformation of human leukocytes by Epstein-Barr virus. *Nature* 280:244-245, 1979.

GENOMIC REARRANGEMENTS AND THE ORIGIN OF CANCER

RUTH SAGER

REDISCOVERING BOVERI: THE PROBLEM OF CAUSALITY

Boveri's contribution to clear thinking about cancer [5] ranks nearly with Mendel's contribution to clear thinking about genes. Written close to 70 years ago, Boveri's view of the origin of cancer anticipates much of what we know today. Some of the principal points are summarized in Table I.

It has been known ever since Boveri's time that chromosomes in tumor cells are often grossly rearranged in addition to being aneuploid [8]. In the days before chromosome banding (i.e., before ca. 1970), only a small fraction of the rearrangements that were present could be identified, and many of the complex changes seen could not be analyzed. Within the past 10 years, as banding techniques have improved [14,56], a new body of knowledge has been developing, documenting the occurrence of chromosome changes that include very small deletions as well as more obvious rearrangements and fragmentation seen in cells of tumor origin [39,48,57]. Most significantly, particular translocations, often with identical breakpoints in cells from numerous patients, have been found in specific kinds of cancer [22,33,42,48]. Thus, new support for Boveri's hypothesis is coming, remarkably enough in this molecular age, from improved techniques of light microscopy.

Some apparently diploid tumor-derived cells have also been reported [48], and their existence, as well as the complex and diverse chromosome changes seen in advanced tumors, together have limited the enthusiasm of some investigators for Boveri's hypothesis. For example, it has been widely accepted that no chromosome changes are found in approximately 50% of acute leukemias. Currently, however, with the application of

TABLE I. Boveri's Hypothesis (1914)*

1. Malignant cells arise from normal cells
2. Tumors arise from single malignant cells; i.e., cancer is of clonal origin
3. Cells of malignant tumors have an abnormal chromatin content
4. The effect of the abnormality is to bring about the capacity for unlimited proliferation characteristic of tumor cells
5. Any process which leads to this abnormal chromatin constitution may result in the origin of a malignant tumor. The probability is increased by:
   Heavy cell proliferation
   Aging
   Exposure to x-rays or certain chemicals
   Genetic predisposition

*Summarized from Theodor Boveri: "Zur Frage der Entstehung maligner Tumoren" [5].

banding techniques to prophase chromosomes about ten times as many deletions and other small rearrangements are being detected in patients with acute nonlymphocytic leukemia (ANLL), as were seen by conventional banding methods [57]. Solid tumors, previously refractory to growth in culture, are being grown with increasing success, providing new material for karyotypic analysis. Biopsies provide samples from early stages of tumor growth, in which regularities may be discerned in chromosome rearrangement which are later observed by further genomic reshuffling. We may anticipate, therefore, that the detailed association of particular chromosome changes with early tumors will be further specified in the near future, as available techniques are more widely applied.

Pursuing Boveri's hypothesis, we come then to the question of causality. Do genomic rearrangements precede the acquisition of tumor-forming ability? How do rearrangements arise? Are the rearrangements causally related to the disease? That is, do specific rearrangements give rise to specific gene products that are required for tumor formation? If so, why are different rearrangements found in different kinds of cancer. If genomic rearrangements are necessary for the development of tumor-forming ability, then what determines the conversion of a normal cell with its stable genome into a cell which is genomically unstable and undergoing rearrangements?

In the rediscovery of Mendel's laws in 1900, the physical identity of the unit factor was totally unknown. But causality was established—between the hypothetical factor and the altered phenotype. In current reexamination of Boveri's hypothesis that chromosome breakage and aneuploidy underlie the origin of cancer, causality remains unproven, because the mechanisms of genomic rearrangement are not understood and the specific genomic changes that convert normal cells to tumor cells have not yet been identified. Nonetheless, new molecular methods are now becoming available to examine both rearrangement mechanisms and the resulting changes in DNA sequence and gene expression.

The modern era in the analysis of genomic rearrangement was ushered in by McClintock's epochal discovery of *transposable elements* in the corn plant, *Zea mays* [23–27]. Subsequently, analogous genetic elements capable of transposition were found in bacteria [50], in *Drosophila* [12], and in yeast [9]. Suggestive evidence for the existence of similar elements is also being found in mammalian cells [16].

On the basis of these studies, a new picture has been emerging of what McClintock has called "the plasticity of the genome" [27]. Transposable elements have been shown to act as built-in mutators and genomic reorganizers. In addition, programmed rearrangements have been discovered, as in immunoglobulin synthesis [52] and in yeast mating type control [13]. The rearrangement potentials undoubtedly extend beyond what has been discovered so far, and it is entirely reasonable to anticipate that abnormal or scrambled rearrangements of developmental programs built into genomic structure will play a central role in carcinogenesis. The potential role of controlling elements in carcinogenesis was explored in depth at an International Cancer Research Workshop [46] and subsequently discussed by Cairns [7].

Support for this concept has also come from recent molecular studies of tumor viruses. In the retroviruses (RNA tumor viruses) interest has been focused on the products of single viral genes, the *src* and *onc* genes, which have been shown to act like growth factors, playing some role as yet unknown in turning on and maintaining DNA synthesis and cell division and inducing tumorigenicity in otherwise quiescent cells. In addition to the gene products that retroviruses introduce into infected cells, the viruses also induce chromosome breaks [32]. In recent molecular studies, terminal, inverted, and repeated sequences have been identified in retroviral RNA, and the genomic copies of these sequences may act as agents of transposition, as they do in all other transposable elements that have been studies. In cells from retrovirus-induced tumors, presumed promotor sequences of viral origin have been found proximal to host genes whose transcription rates have been greatly elevated as a consequence [30,34]. Although the functions of these genes are not known, the consistency with which the same or similar sequences have been found in tumors of independent origin strongly suggests that the genes play some role in tumor formation [3]. Thus, the current retrovirus studies support the hypothesis that specific host genes, transcribed at abnormally high rates as a result of viral-induced genomic rearrangement, can contribute to tumor development. DNA tumor viruses also induce chromosome breakage [29], but they appear not to contain terminal repeats, and their modes of integration and excision are unknown [53].

With the exception of host sequences related to or identified by viral sequences, no specific genes involved in any essential steps in tumorigenesis have been identified. Attempts to identify such genes by methods of DNA transfection, microinjection, and molecular cloning are in progress in numerous laboratories [e.g., 10,49]. Their chances for success will depend, *inter alia*, upon how complex a series of genetic changes are involved in the transformation of a normal cell into a cancer cell.

In this chapter, some of the current evidence linking chromosome changes with tumorigenicity will be reviewed briefly, to indicate both the plausibility of the rearrangement hypothesis and the availability of material for further analysis. Mechanisms of genomic rearrangement known from studies with other organisms will be considered in relation to the origin of cancers in mammalian cells.

GENOMIC REARRANGEMENTS AS CAUSAL AGENTS IN THE ORIGIN OF CANCER

Evidence supporting a causal role for genomic rearrangements in the origin and progression of malignancy has come from diverse lines of investigation. Probably the most compelling evidence comes from the cancer-prone conditions of patients with hereditary diseases that are responsible for damage to DNA, diseases discussed extensively in this volume. In none of these genetic diseases is the enzymatic basis well understood, but the essential feature that they have in common, the fact that cancers arise following damage to DNA, is well documented.

The correlation of tumor induction with the exposure to ionizing radiation has been studied since the times of the Curies; here, too, there is extensive evidence of the role of radiation in chromosome damage [54]. In addition, chemicals such as alkylating agents have been shown to induce single- and double-strand chromosome breaks. Although the data are better developed for some agents than others, the overall correlation is strong between exposure to particular chemicals, chromosome changes, and tumorigenicity [35].

Karyotypic studies using advanced banding techniques of cells recovered from clinical cancers are now providing more focused material for causality studies than anything previously available. The most extensive studies are of leukemias and lymphomas, in which cells at early stages of the disease are much easier to obtain than with most solid tumors. Beginning with the discovery that the Philadelphia chromosome (a translocation between human chromosomes 9 and 22) is found in most cases of chronic granulocytic leukemia, numerous correlations have been established between particular forms of cancer and specific nonrandom chromosome rearrangements or breaks [22,33,40,42,48]. The important points to stress are (1) that these rearrangements are postzygotic, occurring during the life of the individual; and, (2) that the breakpoints are often identical (at the level of the light microscopy) in specimens from different patients. Thus, the breaks and rearrangements found in an increasing number of cancers of diverse origin clearly have a nonrandom component.

Nonrandom chromosomal rearrangements have also been associated with certain hereditary diseases that predispose to cancer, of which retinoblastoma is one that has been most extensively investigated from the genetic point of view [21]. The presence of an interstitial deletion in chromosome 13 associated with some cases of hereditary retinoblastoma is a recent finding [55], as is the presence of a similar deletion in chromosome 11 in some patients with Wilms tumor [38].

It is apparent that particular genomic rearrangements favor the expression of tumor-forming ability in a histotype-specific way, in that different rearrangements are found in cancers of different histotype origin. Why histotype-specific? We do not know. We can speculate that the rearrangements are necessary to disrupt specific differentiation programs built into genomic organization, and to bring histotype-specific genes under a different regulatory control. The frequent breakage of chromosome 14 in certain lymphomas and multiple myeloma, for example, may relate to the location of gene programs for lymphatic differentiation on chromosome 14 [20]. On the other hand, the frequent presence of chromosome 8 in translocations in several leukemias and lymphomas suggests a role for this chromosome as well.

A different approach to the problem of identifying specific chromosome changes associated with tumorigenicity was taken in my laboratory. We studied two closely related cell lines that differ greatly in chromosome stability and in ease of tumor induction, but nonetheless show a striking regularity, namely, changes in chromosome 3 in most of the cell lines we established from tumors that arose in these experiments. Cell lines CHEF/18 and CHEF/16 both arose from a single Chinese hamster embryo [44] and both have remained stably diploid in cell culture [17,45]. However, CHEF/18 cells are anchorage dependent and nontumorigenic in the nude mouse, whereas CHEF/16 cells are anchorage independent and highly tumorigenic [44]. Virtually every CHEF/

and they can transport genes. When a transposon or an IS element happens to integrate in a position that interferes with normal transcription, the phenotypic consequence is like a mutation. Thus, the term "mutation," although operationally correct, is not an adequate description of the phenomenon. However, when further transposition occurs, it may leave the vacated gene back in its wild-type state, if excision is precise. If excision is imprecise, loss or partial alteration of gene function may result, depending on the specific sequence that was affected.

Thus, bacterial mobile elements have two principal effects: (1) upon the expression of any genes into which they happen to integrate; and (2) upon chromosome organization per se. They are both mutators and rearrangers.

Although the molecular analysis of the maize material is just getting under way, McClintock's classical cytogenetic analysis has provided a wealth of detailed information about the behavior of the transposable elements and of the genes they affect. Some molecular interpretations of her data have already been proposed, first by Fincham and Sastry in a comprehensive review [11], and subsequently by Nevers and Saedler [31] in a more specific comparison of controlling elements of maize with IS elements of bacteria. In both accounts the authors strongly support the view that transposable elements alter gene expression by integration at or near the affected gene, and that the somatic reversions to wild-type result from excision. The first experimental support in maize for this interpretation has been published recently [6].

To facilitate discussion, a brief résumé of salient features of the controlling elements of maize will be presented.

(1) Several nonoverlapping classes of controlling elements have been identified (e.g., *Ac, Spm, Dt*), each of which exhibits specificity with respect to the alleles it can affect. This specificity appears unrelated to gene function inasmuch as different alleles at the same locus may be affected by different classes of elements.

(2) Each class of controlling elements includes both *cis-* and *trans-*acting elements, both of which are capable of transposition.

(3) The consequences of transposition include a wide range of alterations at the DNA level, including chromosome breaks, deletions, inversions, duplications, and, at the phenotypic level, changes in gene expression.

(4) These elements were discovered in cells undergoing chromosome breakage and rearrangement induced by x-ray treatment, but they also have been detected in field corn in which their aberrant behavior is presumably of spontaneous origin.

Each class of controlling elements is probably analogous to a transposon with a specific pair of IS elements, and each class is probably present in multiple copies per genome. In most of McClintock's studies, the behavior of two elements of the same class was followed simultaneously, one located *cis* to the "mutable" gene under observation and the second located elsewhere in the genome. The *cis* element was often defective, requiring the presence of the *trans* element for transpositional activity. The molecular basis for the *trans* recognition and activation is not yet known, but it seems likely that the *cis*- and *trans*-acting elements are related structurally.

Maize elements share with bacterial elements the ability both to induce mutationlike effects on gene expression and to mediate extensive genomic rearrangements.

While the bacterial systems are unparalleled for rapid and detailed molecular analysis, the studies with maize have contributed added parameters to our knowledge by virtue both of the sophistication of maize genetics and of the complexities of higher eukaryotic nuclear organization and differentiation. Certain findings with particular relevance to the origin of cancer will be noted here.

Chromosome breakage and gross rearrangement. At the time of excision, complicated recombinational events may be triggered that can lead to a variety of rearrangements. In McClintock's studies, a particular element called *Ds* (dissociator) was identified that left in its wake double-strand breaks, deletions, duplications, and inversions as well as transpositions. Other elements exhibited more restricted behavior, some of them excising cleanly, with or without subsequent transposition, others causing small changes that led to altered gene expression associated with their imprecise excision.

Genetic control of time and frequency of transposition. Some of the controlling elements of maize are strongly controlled in the frequency at which transpositions occur, as well as in the time during development when they occur. The precision of this control is similar to that of normal development, suggesting that some controlling elements may come under the regulation of normal developmental programs [26], perhaps as a result of a particular genomic location. This kind of control has not been observed in microbial systems and provides one example of the maize system as a more instructive model than bacteria or yeast for anticipating the effects of transposable elements in mammalian systems. In studies of rearrangements induced by SV40 viral integration, we observed different rates of rearrangement from one clone to another, reminiscent of the genetic controls found in maize [47].

Reversible suppression of gene expression. McClintock described a set of controlling elements, called the *Spm* system (suppressor-

16 cell is capable of giving rise to a tumor in the nude mouse when tested by coinjection with 10^{-7} x-irradiated normal CHEF/18 cells [44].

Though diploid in culture, CHEF/16 cells undergo extensive rearrangements during tumor growth [18]. In tumors which were excised at early times in their development, stem lines could be readily inferred from the chromosome-rearrangement patterns, and clonality of tumor origin was evident. Suggestive evidence that chromosome 3 may be involved in tumorigenicity comes from the fact that some change in chromosome 3 was noted in all but two of the 14 stem lines studied, but the changes were not uniform. Nine stem lines had an increase in chromosome 3 material, usually an extra 3q, whereas two other stem lines lost a copy of either a long or short arm; and, one carried a 1,3 translocation.

In contrast to the tumorigenicity of CHEF/16 cells, CHEF/18 cells are stably nontumorigenic. Even after treatment with EMS, MNNG, or 4-NQO at effectively mutagenic dosages, very few cells became tumorigenic [52]. In a comprehensive study [52], 21 anchorage-independent and 13 low-serum mutants from CHEF/18 cells, either spontaneous or mutagen-induced, were recovered and characterized. Only four of the anchorage mutants and one of the LS mutants proved to be tumorigenic in nude mice, although after a second round of mutagenesis several of the lines did give rise to tumors. Chromosome studies of all the mutants and their tumor-derived products [19] revealed far fewer rearrangements than had been found in tumor-derived CHEF/16 cells. Nonetheless, alterations affecting chromosome 3 were present in most, but not all, of the tumor-derived material, supporting further the view that this chromosome may carry genes involved in tumor formation.

Bloch-Shachter and Sachs [4] reported, in studies of SV40- and chemically transformed Chinese hamster cells, that extra chromosome 3 material was present in malignant cells and subsequently lost in nonmalignant revertants. However, their cells were near-triploids, and changes in the dosages of genes on the other chromosomes could not be ruled out as contributing factors in the phenotypic changes that were observed. Nonetheless, the possibility that particular regions on chromosome 3 may be involved in tumorigenicity in the Chinese hamster merits further study.

TRANSPOSITION AS A MECHANISM FOR GENOMIC REARRANGEMENT

The nonrandom arrangements present prezygotically in hereditary diseases that predispose to cancer, such as retinoblastoma, and postzygotically in various leukemias, lymphomas, and other cancers suggest that some kind of inborn genetic engineering is at work. What mechanisms are known that could perform these feats? On the basis of present-day knowledge, the most likely mechanism involves transposable genetic elements [28], which were discovered by Barbara McClintock in her work with the corn plant, Zea mays [23-27], and first comprehensively described by her in the early 1950s. Their potential for a key role in carcinogenesis has been pointed out by McClintock [27] and will be discussed below.

A word first about transposable elements in other organisms. In the mid-1960s, the first examples in bacteria of transposable elements were described; they were called insertion sequences or IS elements [51]. Current research with these elements and the larger transposons of which they are a component is revealing their importance in bacterial variation and evolution [15]. Analogous elements now have been found in yeast and in *Drosophila* [28]. All of these elements bear an astonishing functional similarity to the controlling elements in maize. Detailed molecular information is available for bacterial IS elements and transposons, and so their properties of particular relevance to our theme will be reviewed first. This summary is based on recent reviews [8,28].

All transposable elements described to date carry inverted repeated sequences at their termini. When the elements are examined *in vitro*, they can be shown by strand separation and reannealing to form stem-and-loop structures, the stems being the complementary base-paired repeats. In addition, some transposons have been shown to contain sequences that code for an enzyme, called a transposase, which is responsible for their excision. In the excision process, the cleavage occurs in the host DNA at the junction and is asymmetrical, leaving a short, complementary, single strand free at each end. When integration occurs, these ends are filled in, resulting in the origin of new, short (usually less than 12 base pairs), inverted repeats at the junctions. In contrast to the long, inverted repeats which are constant components of each transposon, the specific sequences in the short ends differ from one integrated site to another, because the integration process is nonhomologous. As yet, there is no evidence of site specificity of integration: any sequence, or one of many, will do.

Transposon integration, therefore, is mediated by enzymes different from those that mediate the homologous recombination associated with phage integration and from those associated with meiotic recombination in eukaryotes. Any DNA sequence can be inserted between a pair of IS elements. Thus, these elements have aptly been called "portable regions of homology"; they can join together unrelated sequences, mutator), which showed important differences from the *Ac-Ds* system described above. The *Spm* elements regulate the expression of a gene to be reversibly turned on and off in a cyclical manner. This mode of action resembles that described for mating-type control in yeast, in which expression of alternative mating types is regulated by a precise and specific transpositional event. The possibility that analogous switches are present in mammalian cells deserves consideration, especially in premalignant or tumor cells that are quiescent for some time and suddenly switch on new rounds of cell division.

In summary, transposable elements of maize, yeast, and bacteria provide examples of the kind of activity that could play a decisive role in the reprogramming of cellular functions that occurs in cancer cells. Not only can these elements mediate chromosome rearrangements of many kinds as well as mutations, but—of unique importance—they also can respond to normal genetic and developmental programs.

ORIGIN OF TUMORIGENICITY: A MULTISTAGE GENETIC PROCESS

In this final section, I shall sketch the hypothesis that underlies my thinking about the origin of malignancy.

The initial event is damage to DNA. This damage may be induced by radiation, by chemicals, by viral integration, by defective DNA repair, or possibly by as yet unidentified processes. The ability of external agents to damage DNA is heavily documented. Relative efficiencies and differences in the mode of attack, and in the repair enzymes induced, are topics of current research beyond the scope of this discussion. One point of special interest, however, is the possibility that double-strand breaks are particularly effective in tumorigenesis [1].

One class of hereditary diseases, those that lead to defects in DNA synthesis, predispose to cancer with high frequency. Studies of these diseases and their molecular consequences have provided powerful evidence for the role of DNA damage in the induction of cancer.

A feature of radiation-induced damage, and to some extent of chemical-induced damage, is that the lesions are distributed randomly, not randomly at the level of the immediate nucleotide sequence, but at the level of the functional genetic unit that is disrupted. Clearly, then, different lesions will have different consequences. The consequences of importance for tumorigenicity are those that lead to further chromosomal changes, as discussed below.

Tumor viruses break chromosomes, as documented in the early studies of viral infection and reviewed in this volume [32]. In recent years, tumor virologists have focused on the growth-stimulating functions of virally coded proteins, but the fact that viruses are capable of integrating into chromosomal DNA suggests that they carry some nuclease activity, although no integration-excision functions have been demonstrated yet for the DNA tumor viruses [53]. The recent discovery of specific repeat sequences at the ends of RNA tumor-virus transcripts in DNA suggests that they integrate by mechanisms similar to those described for bacterial transposons, and if that is correct, they may also be able to transpose. However, no retrovirus-induced transpositions have been described yet, although rearrangements have been inferred [3,10].

Chromosome breakage and rearrangement. Controlling elements may play their role in normal developmental processes without transposition, in that *trans*-acting elements may send signals to *cis*-acting elements without actually moving. What is so striking about tumor cells and about the aberrant control systems described by McClintock is the high frequency of dramatic chromosome rearrangements, readily seen with the light microscope, as well as submicroscopic changes inferred from genetic data and now being substantiated by molecular analysis.

The inherited rearrangements that predispose to cancer, and are thus present prezygotically, lead one to suggest that analogous rearrangements are engineered in the origin of postzygotic cancer. Thus, the DNA damage, of whatever origin, contributes to the generation of cancer-prone cells if it leads to further rearrangements. One mechanism known to achieve this effect is initiating aberrant transposition as in maize. Therefore, I suggest that the second stage in the origin of tumor-forming cells is characterized by the cascading of chromosome breaks and rearrangements, initiated by DNA damage and perpetuated by transpositional events. These events generate a heterogeneous mixture of phenotypes, as is seen in tumor growth, and provide the genetic diversity for operation of the process of selection. Tumor promotors may act by inducing continuing rounds of cell division, thereby permitting the cascading of rearrangements. [2].

Selection of successfully growing mutant cells. The process of natural selection, operating during tumor growth, is probably the most essential visualized feature of this hypothesis. Everyone has seen instances of selection occurring in mixed cell populations in the laboratory, and there is no reason to doubt that similar exercises in differential growth rate and cell-cell interaction occur in tumor progression. Only the earliest-arising tumor cells, in which relatively little chromosome rearrangement has occurred, are sufficiently homogeneous to provide clear evidence of clonality. The most malignant cells are also sometimes rather homogeneous, as they approach an endpoint of selection. In between these extremes, heterogeneity is the rule, providing at the same time the basis for rapid evolution of anticancer-drug resistance.

In summary, this hypothesis has three stages: first, the initial DNA damage, which may be of diverse origins; second, the generation of genomic and phenotypic diversity engineered by transpositional elements that were activated by the damage to the DNA; and, third, the selection from among this diversity of cellular phenotypes, those capable of further growth, tumor formation, and progression.

This hypothesis has two novel features that distinguish it from previous proposals. One novel feature is the concept that DNA damage is carcinogenic not primarily by inducing simple gene mutations but rather by initiating a process of genomic instability; the second is that transposable elements provide the instability that is required for the transformation of normal cells to cancer cells. Both of these features are being evaluated experimentally.

LITERATURE CITED

1. Auerbach, C.: Mutation Research. New York: John Wiley & Sons, 1976.
2. Berenblum, I.: Sequential aspects of chemical carcinogenesis: Skin. *In* Cancer (Becker, F.F., ed.), New York: Plenum Publishing Corp., Vol. 1, 323-344, 1975.
3. Blair, D.G., Oskarsson, M., Wood, T.G., McClements, W.L., Fischinger, P.J., Van de Woude, G.G.: Activation of the transforming potential of a normal cell sequence: A molecular model for oncogenesis. *Science* 212:941-942, 1981.
4. Bloch-Shachter, N., Sachs, I.: Identification of a chromosome that controls malignancy in Chinese hamster cells. *J. Cell. Physiol.* 93:205-212, 1977.
5. Boveri, T.: The Origin of Malignant Tumors. Baltimore: The Williams & Wilkins Co., 1929 (Translated from the German volume of 1914).
6. Burr, B., Burr, F.A.: Detection of changes in maize DNA at the shrunken locus due to the intervention of Ds elements. *Cold Spring Harbor Symp. Quant. Biol.* 14:463-465, 1981.
7. Cairns, J.: The origin of human cancers. *Nature* 289:353-357, 1981.
8. Calos, M.P., Miller, J.H.: Transposable elements. *Cell* 20:579-595, 1980.
9. Cameron, J.R., Loh, E.Y., Davis, P.W.: Evidence for transposition of dispersed repetitive DNA families in yeast. *Cell* 16:739-751, 1979.
10. Cooper, G.M., Okenquist, S., Silverman, L.: Transforming activity of DNA of chemically transformed and normal cells. *Nature* 284:418-421, 1980.
11. Fincham, J.R.S., Sastry, G.R.K.: Controlling elements in maize. *Annu. Rev. Genet.* 8:15-50, 1974.
12. Green, M.M.: Transposable elements in Drosophila and other Diptera. *Annu. Rev. Genet.* 14:109-120, 1980.

13. HERSKOWITZ, I., OSHIMA, Y.: Control of cell type in *Saccharomyces cerevisiae:* Mating type and mating type interconversion. *In* The Yeast *Saccharomyces cerevisiae* (Strathern, J.N., Jones, E.W., Broach, J., eds.), Cold Spring Harbor, New York: Cold Spring Harbor Laboratory, in press, 1981.
14. HSU, T.C.: Longitudinal differentiation of chromosomes. *Annu. Rev. Genet.* 7:153-176, 1973.
15. IIDA, S., MEYER, J., ARBER, W.: Genesis and natural history of IS-mediated transposons. *Cold Spring Harbor Symp. Quant. Biol.* 45:27-37, 1981.
16. JAGADEESWARAN, P., FORGET, B.G., WEISSMAN, S.M.: Short interspersed repetitive DNA elements in eukaryotes. *Cell* (in press) 1981.
17. KITCHIN, R., SAGER, R.: Genetic analysis of tumorigenesis. V. Chromosomal analysis of tumorigenic and non-tumorigenic diploid Chinese hamster cell lines. *Somat. Cell Genet.* 6:75-87, 1980.
18. KITCHIN, R., SAGER, R.: Genetic analysis of tumorigenesis. VI. Chromosome rearrangements in tumors derived from diploid premalignant Chinese hamster cells in nude mice. *Somat. Cell Genet.* 6:615-629, 1980.
19. KITCHIN, R., GADI, I.K., SMITH, B.I., SAGER, R.: Genetic analysis of tumorigenesis: XII. *Somat. Cell Genet.*, in press.
20. KLEIN, G.: Lymphoma development in mice and humans: Diversity of initiation is followed by convergent evolution. *Proc. Natl. Acad. Sci. USA* 76:2442-2446, 1979.
21. KNUDSON, A.G., HETHCOTE, H.W., BROWN, B.W.: Mutation and childhood cancer: A probabilistic model for the incidence of retinoblastoma. *Proc. Natl. Acad. Sci. USA* 72:5116-5120, 1975.
22. MARK, J.: Chromosomal abnormalities and their specificity in human neoplasms. *Adv. Cancer Res.* 24:165-222, 1977.
23. McCLINTOCK, B.: Chromosome organization and gene expression. *Cold Spring Harbor Symp. Quant. Biol.* 16:13-47, 1951.
24. McCLINTOCK, B.: Controlling elements and the gene. *Cold Spring Harbor Symp. Quant. Biol.* 21:197-216, 1956.
25. McCLINTOCK, B.: The control of gene action in maize. *Brookhaven Symp. Biol.* 18: 162-184, 1965.
26. McCLINTOCK, B.: Genetic systems regulating gene expression during development. *Dev. Biol.* 1:84-112, 1967.
27. McCLINTOCK, B.: Mechanism that rapidly reorganizes the genome. *Stadler Genet. Symp.* 10:25-48, 1978.
28. Movable Genetic Elements, New York: Cold Spring Harbor Laboratory, 1981, Vol. 45.
29. MOORHEAD, P.S., SAKSELA, E.: Sequence of chromosome alterations during SV40 transformation of a human diploid cell strain. *Hereditas* 52:271-284, 1965.
30. NEEL, B.G., HAYWARD, W.S., ROBINSON, H.L., FANG, J., ASTRIN, S.M.: Avian leukosis virus-induced tumors have common proviral integration sites and synthesize discrete new RNAs: Oncogenesis by promotor insertion. *Cell* 23:323-334, 1981.
31. NEVERS, P., SAEDLER, H.: Transposable genetic elements as agents of gene instability and chromosomal rearrangements. *Nature* 268:109-115, 1977.
32. NICHOLS, W.: This volume.
33. NOWELL, P.C., HUNGERFORD, D.A.: A minute chromosome in human chronic granulocytic leukemia. *Science* 132:1497, 1960.
34. PAYNE, G.S., COURNEIDGE, S.A., CRITTENDEN, L.B., FADLY, A.M., BISHOP, J.M., VARMUS, H.E.: Analysis of avian leukosis virus DNA and RNA in bursal tumors: Viral gene expression is not required for maintenance of the tumor state. *Cell* 23: 311-322, 1981.

35. PULLMAN, B., Ts'O, P., GELBOIN, H., (eds.): Carcinogenesis: Fundamental Mechanisms and Environmental Effects. Hingham, Mass.: D. Reidel Publishing Co., 1981.
36. RAY, J.H., GERMAN, J.: The chromosome changes in Bloom's syndrome, ataxia-telangiectasia, and Fanconi's anemia. In Genes, Chromosomes, and Neoplasia (Arrighi, F.E., Rao, P.N., Stubblefield, E. eds.), New York: Raven Press, 351–378, 1981.
37. RHOADES, M.M.: The genetic control of mutability in maize. Cold Spring Harbor Symp. Quant. Biol. 9:138–144, 1941.
38. RICCARDI, V.M., SUJANSKY, E., SMITH, A.C., FRANKE, U.: Pediatrics 61:604–610, 1978.
39. ROWLEY, J.D.: Chromosome abnormalities in cancer. Cancer Genet. Cytogenet. 2:175–198, 1980.
40. ROWLEY, J.D.: Chromosomes in human leukemia. Annu. Rev. Genet. 14:17–39, 1980.
41. ROWLEY, J.D.: Do all leukemic cells have an abnormal karyotype? N. Engl. J. Med. 305:164–166, 1981.
42. ROWLEY, J.D.: A possible role for nonrandom chromosomal changes in human hematologic malignancies. In Chromosomes Today (De la Chapelle, A., Sorsa, M. eds.), Amsterdam: Elsevier/North-Holland, 1977.
43. SAGER, R.: On the mutability of the waxy locus in maize. Genetics 36:510–540, 1951.
44. SAGER, R., KOVAC, R.: Genetic analysis of tumorigenesis: I. Properties of a pair of tumorigenic and non-tumorigenic hamster cell lines. Somat. Cell Genet. 4:375–392, 1978.
45. SAGER, R., KOVAC, P.: Genetic analysis of tumorigenesis: IV. Chromosome reduction and marker segregation in progeny clones from Chinese hamster cell hybrids. Somat. Cell Genet. 5:491–502, 1979.
46. SAGER, R.: Report on workshop: Chromosomal transpositions as causal agents in cancer. Nature 282:447–448, 1979.
47. SAGER, R., ANISOWICZ, A., HOWELL, N.: Genomic rearrangements and tumor-forming potential in an SV40-transformed mouse cell line and its hybrid and cybrid progeny. Cold Spring Harbor Symp. Quant. Biol. 45:747–754, 1981.
48. SANDBERG, A.A.: Chromosomes in Human Cancer and Leukemia. New York: Elsevier North-Holland, 1979.
49. SHIH, C., SHILO, B-Z., GOLDFARB, M.P., DANNENBERG, A., WEINBERG, R.A.: Passage of phenotypes of chemically transformed cells via transfection of DNA and chromatin. Proc Natl. Acad. Sci. USA 76:5714–5718, 1979.
50. SMITH, B.L., SAGER, R.: The multistep origin of tumor-forming ability in CHEF cells. Cancer Res. 42:389–396, 1982.
51. STARLINGER, P., SAEDLER, H.: IS-elements in microorganisms. Curr. Top. Microbiol. Immuno. 75:111–152, 1976.
52. TONEGAWA, S., HOZUMI, N., MATTHYSSENS, G., SCHULLER, R.: Somatic changes in the content and context of immunoglobulin genes. Cold Spring Harbor Symp. Quant. Biol. 41:877–889, 1977.
53. TOOZE, J. (ed.): DNA Tumor Viruses. New York: Cold Spring Harbor Laboratory, 1980.
54. YUHAS, J.M., TENNANT, R.W., REGAN, J.D. (eds.): Biology of Radiation Carcinogenesis. New York: Raven Press, 1976.
55. YUNIS, J.J., RAMSAY, N.: Retinoblastoma and sub-band deletion of chromosome 13. Am. J. Dis. Child. 132:161–163, 1978.
56. YUNIS, J.J., SAWYER, J.R., BALL, D.W. The characterization of high-resolution G-banded chromosomes of man. Chromosoma 67:293–307, 1978.
57. YUNIS, J.J. BLOOMFIELD, C.D., ENSRUD, K.: All patients with acute nonlymphocytic leukemia may have a chromosomal defect. N. Engl. J. Med. 305:135–139, 1981.

BLOOM'S SYNDROME. X. THE CANCER PRONENESS POINTS TO CHROMOSOME MUTATION AS A CRUCIAL EVENT IN HUMAN NEOPLASIA

JAMES GERMAN

Clinical Bloom's syndrome [6,18] consists of growth deficiency, a sun-sensitive skin lesion, and a poor immune response that predisposes to severe infections. The biochemical defect is unidentified [9], but evidence exists for disturbed DNA synthesis, a manifestation of which is a striking degree of chromosome instability of a specific type (Fig. 1) [19,20]. The several interesting aspects of this rare, recessively transmitted disorder have been under investigation in my laboratory since 1960.

NEOPLASIA IN BLOOM'S SYNDROME

One of the most significant features of Bloom's syndrome is a predisposition to neoplasia. Tabulated elsewhere in this volume [8] are the 29 neoplasms (26 malignant) that have been documented in the approximately 100 persons known to have had this rare condition; the data are depicted in an abbreviated way in Figure 2. Here, the possible significance of the enormously increased cancer incidence in Bloom's syndrome is discussed.

The Bloom's syndrome population is a young one, no person past 50 yet having been recognized. It is for this reason that the occurrence of 29 neoplasms in 100 people is remarkable. Figure 2 shows that, at all ages, neoplasia is diagnosed clinically much more often than in the general population. Leukemias and lymphoid neoplasms are relatively more

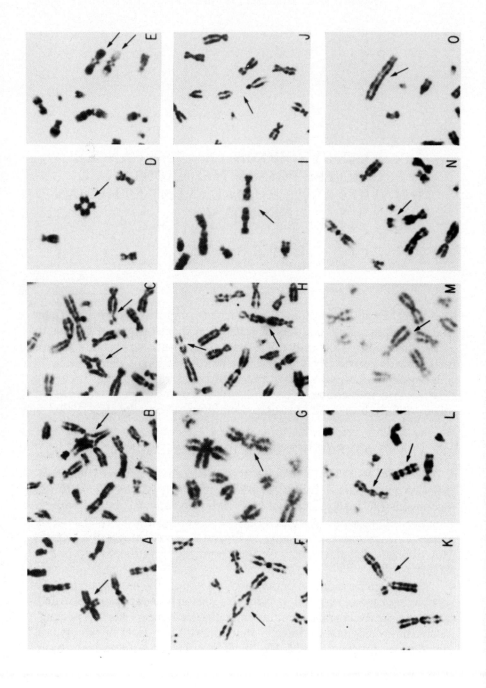

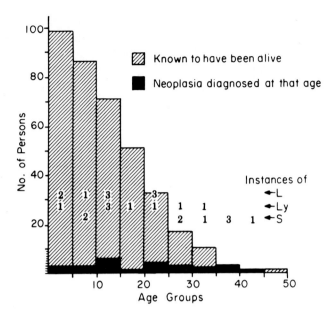

Fig. 2. Diagnosis of neoplasia in Bloom's syndrome at different ages. L, leukemia; Ly, lymphoma, lymphosarcoma, Hodgkin's disease; S (solid), carcinoma, Wilms' tumor, meningioma. Not included are colon polyps, a benign skin tumor, and a histologically unconfirmed metastatic carcinoma.

Fig. 1. Chromatid abberations of various types from cultured Bloom's syndrome cells. Although each of the aberrations displayed will be found occasionally in cells from normal people, their frequency in Bloom's syndrome cells is increased greatly. Lesions of the type seen in J–O, C (right arrow), and H (upper arrow) – gaps, breaks, interchange figures affecting nonhomologous chromosomes, dicentrics, and acentrics – are increased in frequency also in Fanconi's anemia and ataxia-telangiectasia, rare genetic disorders that, like Bloom's syndrome, predispose to neoplasia. However, the aberrations in A–I are highly characteristic of just Bloom's syndrome cells; they probably are on the same basis as the constitutionally increased sister-chromatid exchange frequency that is found only in Bloom's syndrome (not pictured here). Each quadriradial configuration (Qr) in A–E and G indicates that during the preceding interphase an interchange had occurred between two of the four chromatids of two chromosomes. When staining techniques are used that permit chromosome identification, the Qr characteristic of Bloom's syndrome, the Qr-I [20], is shown to affect homologous chromosomes (e.g., the two Nos. 6 in A, the two Nos. 1 in B). Qr-Is indicate that homologous chromosomes had exchanged segments at what appear to be the same sites of two nonsister chromatids. Close association of the tips of homologous chromosome arms also is abnormally increased in frequency in Bloom's syndrome (H, I, and probably F), and this probably has the same significance as the Qr-I, i.e., homologous chromatid interchange, but of short terminal segments (telomeres) of the two chromosomes affected (cf. A, B, and D in which the points of exchange are at or near centromeres).

common below the age of 25, epithelial neoplasms thereafter. The sites at which the carcinomata have occurred are not unusual — base of the tongue, epiglottis, esophagus, gastroesophageal junction, transverse colon, sigmoid colon, and rectosigmoid colon. What is unusual is the young ages at which they were diagnosed — several decades earlier than the mean ages of occurrence of such cancers in the general population. The only brain tumor, a meningioma, was diagnosed at the age of 9, exceptionally early for that benign neoplasm. The lymphoid tumors have been of various types (lymphomata of several types, lymphosarcomata in several areas of the body, and Hodgkin's disease), and the acute leukemias have been classed nonlymphocytic, myelomonocytic, and lymphocytic. In three persons, neoplasms have originated at multiple sites: laryngeal carcinoma plus lymphoma; esophageal carcinoma plus colon carcinoma; and, gastroesophageal carcinoma plus colon carcinoma plus multiple benign tumors.

An important question, not yet answerable because the data are so few, is whether the pattern of neoplasia produced in Bloom's syndrome, i.e., the distribution of types of neoplasms, is similar to or different from that of the general population. The three most common carcinomata in the general population are those affecting the lung, the colon, and the breast. Neither lung nor breast as yet has been recorded as the primary site of neoplasia in Bloom's syndrome, but three examples of colon cancer, just mentioned, have occurred. Even when adequate data will have been accrued — necessarily a slow process — it can be anticipated that the early ages at which carcinomata are diagnosed in Bloom's syndrome will create serious problems in making the comparison that will answer the question.

POSSIBLE BASES FOR THE PRONENESS TO NEOPLASIA IN BLOOM'S SYNDROME

The variety of types of neoplasms that occurs in Bloom's syndrome is noteworthy. It differs in this respect from certain other recessive genetic disorders that feature chromosome instability and cancer proneness [8]; e.g., xeroderma pigmentosum predisposes predominantly to cancer of sun-exposed tissues, Werner's syndrome particularly to sarcoma, and Fanconi's anemia predominantly to leukemia and hepatic neoplasia. The striking increase in Bloom's syndrome of cancers of multiple types, affecting multiple tissue types, suggests that *an event of paramount importance in the etiology of neoplastic transformation occurs much more frequently than in most other people.* The postulated event, which occurs commonly in Bloom's syndrome but only rarely in other persons, probably results mainly from exposure to environmental carcinogens in the general population.

It is important to identify the postulated crucial event in the etiology of neoplastic transformation. Results from the cytological investigation of Bloom's syndrome advance somatic recombination and mutation (1 and 2 below) as leading candidates, but immunodeficiency (3 below) must be considered also. Possibly, more than one of these three contribute to the remarkable neoplasia predisposition of persons homozygous for the Bloom's syndrome gene.

(1) Somatic Recombination

A cell that contains a mutation in just one chromosome can, through somatic recombination, become homozygous for the mutation. Cytological evidence for increased somatic recombination does exist in Bloom's syndrome cells both in vitro and in vivo [2,4,23] (Fig. 1, A-I), as does genetic evidence for increased recombination in viruses infecting such cells in vitro [28]. The frequency of sister-chromatid exchange is increased greatly also [2]. Somatic recombination, therefore, is a candidate for being the crucial event postulated to occur in neoplastic transformation [7].

(2) Mutation

The basis for tumor initiation generally is believed to be mutation. Increased numbers of cytologically detectable types of chromosome mutations (Fig. 1, J-O) long have been known to occur spontaneously in the various types of Bloom's syndrome cells cultured in vitro [19,20]. Recently, an increased rate of spontaneous mutation at specific loci has been reported for cultured Bloom's syndrome fibroblasts [10,26,27], and evidence for such in vivo is now available also [24]. The event postulated above, therefore, may be some type of mutation.

One type of abnormal chromosome behavior pertains to both of the mechanisms mentioned in the two preceding paragraphs. *Unequal* chromatid exchange, either between two chromatids of homologous chromosomes or between sister chromatids of a single chromosome, can result in duplication and deletion of short chromosome segments [7]. Furthermore, such segmental realignments could have as their consequence so-called position effect. (At least some of the mutations resulting from unequal inter- and intrachromosomal exchange would be detectable by high-resolution chromosome banding.)

(3) Defective Immunity

Immune defenses are impaired severely in most persons with Bloom's syndrome. In fact, this syndrome may be grouped properly with the better known genetically determined immunodeficiencies, almost all of which predispose to cancer [22]. Although the role in clinical cancer prevention of immune surveillance for neoplastic clones, if any, remains undefined, the defective immune function in Bloom's syndrome must be considered as a possible explanation for, or contributory factor to, the elevated cancer incidence. (This also is the case for ataxia-telangiectasia, the only other "chromosome-breakage syndrome" in which immunodeficiency is a major clinical feature.)

THE CHROMOSOME-MUTATION THEORY OF CARCINOGENESIS

Which of these, if any, is decisive in neoplastic transformation? Because of two pieces of circumstantial evidence that I cited and emphasized a decade ago [5], a mutational event of the type that usually is associated with microscopically detectable chromosome rearrangement — a chromosome mutation — remains in my opinion the most probable. The evidence consists of the following: First, many human cancers are clones of cells with mutant genomes, the mutations being demonstrable cytologically [21,29]; second, either excessive exposure to one of the environmental causes of chromosome mutation or having one of the several human genetic disorders that feature chromosome instability predisposes a person to cancer. Cairns [1], using this and additional evidence, expressed a similar view, "that the steps leading to human cancer are more likely to be transpositions than localized changes in sequence."

Somatic crossing-over, if it were exact (as it is in the germ line during meiosis), would not produce the microscopically detectable changes in the genome demanded by the actual observations in human cancers [21], so it seems not to qualify as the crucial event. Furthermore, the several other genetically determined cancer-prone disorders in which chromosome instability has been demonstrated feature neither increased sister-chromatid exchange (SCE) nor increased homologous chromatid interchange. That chromatid gaps, breaks, and unbalanced rearrangements do occur with increased frequency in Bloom's syndrome cells (Fig. 1, J–O) suggests that the chromosome mechanism responsible for the dramatically increased chromatid exchange rate is error prone. Thus, Bloom's syndrome would be comparable to the environmental causes of cancer and the other chromosome-breakage syndromes in providing excessive instability of the genome, from which a microscopically detectable rearrangement that transforms the normal cell into a neoplastic one has an increased chance of emerging.

A visible chromosome mutation, whether it is an apparently balanced or a clearly unbalanced translocation or an inversion, deletion, or duplication of a chromosome segment, does not itself specify the accompanying significant functional change at the molecular level that may be brought about when the rearrangement takes place. Among the conceivable functional consequences of chromosome mutation are the following: inactivation of a locus controlling cellular proliferation, either through total deletion of the DNA or by interruption of its nucleotide sequence; duplication of such a locus; alteration of the activity of such a locus as result of a new and aberrant positioning in relation to heterochromatin (i.e., position effect) or to some genetic determinant that itself can control the locus; and, activation of a so-called movable genetic element that then can either induce additional mutations or alter the activity of the chromatin it adjoins [14].

If the hypothesis is made that chromosome mutation is the crucial event occurring in Bloom's syndrome cells before they become neoplastic — and therefore as argued above, the crucial event occurring in cells of persons from the general population as well — a schema for the emergence of at least a proportion of clinical cancers can be devised. This is presented in Figure 3.

By this concept, the chromosome constitution of neoplasms that the cytogeneticist observes can be expected to vary, depending on when the cells are sampled. Ordinarily, leukemias are sampled relatively earlier in their progression than are epithelial tumors, which may explain why the changes in their chromosome complements usually are less drastic than those in "solid" tumors. Thus, some of the changes detectable in leukemic clones — and frequently just a single, simple chromosome rearrangement will be present — may well be crucial ones that can divert a differentiated cell and its progeny into a course of unscheduled proliferation. As cytogenetic techniques have improved, a degree of specificity of the chromosome mutation sites — "breakpoints" — for different neoplasms has become evident [13], the first examples of specificity being recognized in certain human leukemias and Burkitt's lymphoma. How does this observation of specificity fit into the model proposed here, especially in view of the fact that in Bloom's syndrome, sister-chromatid exchanges, homologous chromatid exchanges, and gaps, breaks, and rearrangements occur at essentially all regions of the genome? Specificity with respect to the sites of the postulated chromosome mutation would

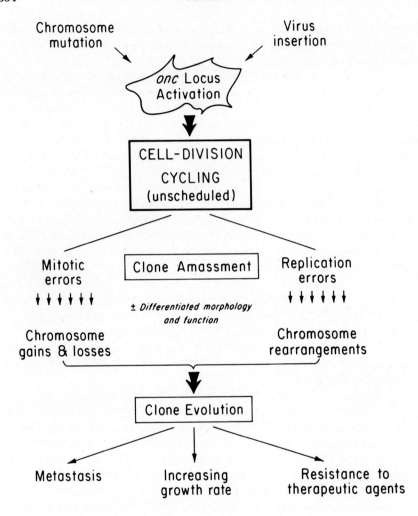

Fig. 3. Proposed schema for the origin and progression of malignant neoplasia.

A nonterminally differentiated cell that is to become the progenitor of a clinical cancer is represented near the top of the figure. An "oncogene" (onc) [3,12,25] in that cell is activated as result of one of the two events depicted, either a chromosome mutation or, as suggested by Hayward et al. [11], the integration near the locus of a viral genome. The two events are not mutually exclusive in that the proposed viral integration might occur at the time of a chromosome rearrangement. A chromosome mutation, if it is to accomplish this alone, would have to be one that affects the activity of chromatin near it, including the onc locus, by one of the mechanisms mentioned in the text; or, the newly integrated viral genome would activate the onc gene it has come to adjoin. (Another be expected if relatively few loci in any particular differentiated cell type can serve the role of an oncogene, i.e., trigger autonomous cell cycling; of the many breaks and rearrangements produced, the few that activate the oncogene would be selected for. The chromosome rearrangements that would be capable of converting a normal into a neoplastic cell could be those, as hypothesized by Hayward et al. [11], "that join c-*onc* (cellular oncogene) coding sequences to transcriptionally active elements. The different types of cancer may reflect the particular c-*onc* gene activated and the type of cell in which activation occurs." Further breakpoint specificity could be determined by which structural loci are active in the particular type of cell affected, and that actively transcribed regions of chromatin are hypersusceptible to chromosome rearrangement because of the chemical and physical modulation they undergo in order to be transcribed. In this second case, the specificity would relate directly to the differentiated state of the cell that was the progenitor of the neoplastic clone, the cell in which the mutation had occurred.

---

possibility for carcinogenesis, not represented in the figure, is the infection of the cell by a retrovirus that bears its own *onc* gene.) The essence of the matter is that a cell and its progeny no longer are able to retire either intermittently or permanently from cell-division cycling, as they had been programmed to do (uppermost box in the figure).

Consequences of the continual cellular proliferation would ensue. These as yet are poorly defined and doubtless vary, depending on the locus activated and the cell type affected: (1) The number of cells in the neoplastic clone would increase (center box in the figure), the rate of the increase depending mainly on the duration of the cell-division cycle. (2) The differentiated morphology and function that the now-neoplastic clone might display would depend on what had been the prior commitment of the progenitor of the clone, i.e., on the course of differentiation that had already been established in that one cell before its neoplastic conversion. Such specialized functions the clone might thus be obliged to perform would be expected to be less manifest than normal because of its new, and abnormal, preoccupation with passing repeatedly through cell-divison cycles.

(3) Genetic instability in the clone of cells derived from this cell is another consequence [15,16], manifested cytologically as an unusually large number of cells with extra or missing chromosomes, the appearance from time to time of new chromosome rearrangements, or both. Abnormalities of mitosis are well known as part of the cancer-cell phenotype [17] and account easily for the numerical chromosome abnormalities that abound in far-advanced epithelial cancers. (The basis remains unknown for both the abnormal mitoses and the appearance from time to time of new structural chromosome rearrangements that characterize malignant neoplastic clones.)

Cells with new genetic constitutions would be generated in abundance as result of the genomic instability in the neoplastic clone, and some of these subclones would be expected to have proliferative advantages over the parent clone [15]. (4) When, and if, selection occurs for some of these new genomes (lowest box in the figure), not only would the clinical character of the neoplasm change (e.g., its growth rate, its ability to metastasize, the emergence of drug tolerance) but a new karyotype also would be recognized by the cytogeneticist as the predominating one, perhaps with a new modal number and, or, additional marker chromosomes.

CONCLUSION

In the genetic disturbance of growth known as Bloom's syndrome the incidence of several common types of malignant neoplasms is greatly increased. The rationale behind a 22-year study by this laboratory of affected persons, their families, and their cells is the following. The significant event(s) responsible for neoplastic transformation of normal cells, leading to what is an important and very common disease in the general population (i.e., clinical cancer), should occur with an increased frequency in persons with this very rare disorder; consequently, the event(s) should be more readily identifiable and analyzable in Bloom's syndrome. Now, the knowledge gained during this study, when considered along with information from other fields, particularly tumor virology [12], permits formulation of a unified concept of the origin and progression of neoplasia (Fig. 3).

The initial event postulated to occur if a normal cell is to become neoplastic is one that alters the genome, and the alteration is heritable. This mutation as a rule results in a microscopically visible change in the genome. Such a chromosome mutation might come about because of constitutional (i.e., genetically determined) instability of the genome, as in Bloom's syndrome, or it might result from some environmental insult; possibly it can occur spontaneously. A series of cellular changes then follow, the essential one being the diversion of the cell and its progeny into continued cellular proliferation, an abnormal course for the affected cell because of its earlier commitment to some normal differentiated function.

LITERATURE CITED

1. CAIRNS, J.: The origin of human cancers. *Nature* 289:353–357, 1981.
2. CHAGANTI, R.S.K., SCHONBERG, S., GERMAN, J.: A manyfold increase in sister chromatid exchanges in Bloom's syndrome lymphocytes. *Proc. Natl. Acad. Sci. USA* 71:4508–4512, 1974.
3. DALLA FAVERA, R., GELMANN, E.P., GALLO, R.C., WONG-STAAL, F.: A human *onc* gene homologous to the transforming gene *(v-sis)* of simian sarcoma virus. *Nature* 292:31–35, 1981.
4. GERMAN, J.: Cytological evidence for crossing-over *in vitro* in human lymphoid cells. *Science* 144:298–301, 1964.
5. GERMAN, J.: Genes which increase chromosomal instability in somatic cells and predispose to cancer. *In* Progress in Medical Genetics. Vol. VIII (Steinberg, A.G., Bearn, A.G., eds.) New York: Grune and Stratton, pp. 61–101, 1972.
6. GERMAN, J.: Bloom's syndrome. II. The prototype of human genetic disorders predisposing to chromosome instability and cancer. *In* Chromosomes and Cancer. (German, J., ed.), New York: John Wiley and Sons, pp. 601–617, 1974.
7. GERMAN, J.: A biological role for chromatid exchange in mammalian somatic cells? *In:* Gene Amplification. (Schimke, R., ed.), Cold Spring Harbor, New York: Cold Spring Harbor laboratory, pp. 307–312, 1982.

8. GERMAN, J.: Patterns of neoplasia associated with the chromosome-breakage syndromes. (This volume)
9. GERMAN, J., SCHONBERG, S.: Bloom's syndrome. IX. Review of cytological and biochemical aspects. *In:* Genetic and Environmental Factors in Experimental and Human Cancer. (Gelboin, H.V., et al., eds.), Tokyo: Japan Scientific Societies Press, pp. 181-186, 1980.
10. GUPTA, R.S., GOLDSTEIN, S.: Diphtheria toxin resistance in human fibroblast cell strains from normal and cancer-prone individuals. *Mutat. Res.* 73:331-338, 1980.
11. HAYWARD, W.S., NEEL, B.G., ASTRIN, S.M.: Activation of a cellular *onc* gene by promoter insertion in ALV-induced lymphoid leukosis. *Nature* 290:475-480, 1981.
12. HAYWARD, W.S., NEEL, B.G., ASTRIN, S.M.: Avian leukosis viruses: Activation of cellular "oncogenes." *In* Advances in Viral Oncology. (Klein, G., ed.) New York: Raven Press, pp. 207-233, 1982.
13. Human Gene Mapping 6. *Cytogenet. Cell Genet.* Volume 32, 1982.
14. Movable Genetic Elements. *Cold Spring Harbor Symp. Quant. Biol. Vol. 45*, Cold Spring Harbor, New York: Cold Spring Harbor Laboratory, 1,025 pages, 1981.
15. NOWELL, P.C.: Chromosome changes and the clonal evolution of cancer. *In* Chromosomes and Cancer. (German, J., ed.), New York: John Wiley and Sons, pp. 267-285, 1974.
16. NOWELL, P.C.: Tumor progression and clonal evolution: The role of genetic instability. (This volume.)
17. OKSALA, T., THERMAN, E.: Mitotic abnormalities and cancer. *In* Chromosomes and Cancer. (German, J., ed.), New York: John Wiley and Sons, pp. 239-263, 1974.
18. PASSARGE, E.: Bloom's syndrome. (This volume.)
19. RAY, J.H., GERMAN, J.: The chromosome changes in Bloom's syndrome, ataxia-telangiectasia, and Fanconi's anemia. *In* Genes, Chromosomes, and Neoplasia. (Arrighi, F.E., Rao, P.N., Stubblefield, E., eds.), New York: Raven Press, pp. 351-378, 1981.
20. RAY, J.H., GERMAN, J.: The cytogenetics of the chromosome-breakage syndromes. (This volume.)
21. ROWLEY, J.D.: Do all leukemia cells have an abnormal karyotype? *N. Engl. J. Med.* 305:164-166, 1981.
22. SPECTOR, B.D.: Immunodeficiency-Cancer Registry: 1975 update. *In* Genetics of Human Cancer. (Mulvihill, J.J., Miller, R.W., Fraumeni, J.F., Jr., eds.), New York: Raven Press, pp 339-342, 1977.
23. THERMAN, E., OTTO, P.G., SHAHIDI, N.T.: Mitotic recombination and segregation of satellites in Bloom's syndrome. *Chromosoma* 82:627-636, 1981.
24. VIJAYALAXMI, EVANS, H.J., RAY, J.H., GERMAN, J.: Bloom's syndrome. XI. Elevated incidence of 6-thioguanine resistant lymphocytes. (In preparation)
25. WHITEHOUSE, H.L.K.: Chromosome integration of viral DNA: The open-replicon hypothesis of carcinogenesis. *In* Chromosomes and Cancer. (German, J., ed.), New York: John Wiley and Sons, pp. 41-76, 1974.
26. WARREN, S.T., SCHULTZ, R.A., CHANG, C.C., TROSKO, J.E.: Elevated spontaneous mutation rate in Bloom syndrome fibroblasts.*Am. J. Hum. Genet.* 32:361A, 1980.
27. WARREN, S.T., SCHULTZ, R.A., CHANG, C.C., WADE, M.H., TROSKO, J.E.: Elevated spontaneous mutation rate in Bloom syndrome fibroblasts. *Proc. Natl. Acad. Sci. USA* 78:3133-3137, 1981.
28. YOUNG, C.S.H., FISHER, P.B.: Adenovirus recombination in normal and repair-deficient human fibroblasts. *Virology* 100:179-184, 1980.
29. YUNIS, J.J., BLOOMFIELD, C.D., ENSRUD, K.: All patients with acute nonlymphocytic leukemia may have a chromosomal defect. *N. Engl. J. Med.* 305:135-139, 1981.

THE SIGNIFICANCE OF CHROMOSOME CHANGE TO NEOPLASTIC DEVELOPMENT

R.S.K. CHAGANTI

Hypotheses non fingo
I. Newton: Philosophiae Naturalis Principia Mathematica, 2nd Edition, London, 1713.

INTRODUCTION

Abnormal mitoses often seen in cancer cells have puzzled biologists ever since they were described by 19th century pathologists. A number of important ideas emerged during the late 19th and early 20th centuries about the significance of these abnormalities for the origin of cancer cells. In 1890 von Hansemann [65] suggested that mitotic irregularities were responsible for disordered growth. Boveri's hypothesis [28] that mitotic errors are directly responsible for the origin of neoplasia drew attention to the importance of somatic cell genotypes in the origin of tumor cells. The stem-line concept of tumor evolution proposed by Winge in 1930 [190] and elaborated by the cytological studies of transplanted rodent tumors by Levan, Hauschka, and others [70,106,107] established the importance of selection in the evolution of cell types in a population of proliferating tumor cells which are genotypically heterogeneous. In 1960 Nowell and Hungerford [137] discovered the first specific chromosome abnormality to be recognized in a human tumor, namely, the Philadelphia (Ph¹) chromosome in chronic myelogenous leukemia (CML) cells; this discovery laid the foundation for the view that tumors are clonal in

---

I wish to dedicate this article to my mentor and friend Charles E. Ford, D.Sc., F.R.S., in admiration of his many pioneering contributions to human and mammalian cytogenetics and to transplantation biology.

nature. Cytogenetic analysis of Chinese hamster sarcoma and human leukemia enabled C.E. Ford and associates [52] to define the now universally recognized properties of neoplastic cells in the *primary host:* (1) a population—though related, cell types may arise within it; and (3) one of such variant cell types may replace all others, becoming the dominant component of the types may replace all others, becoming the dominant component of the population. Soon after, Fialkow et al. [49] demonstrated that, in female patients with Ph¹ chromosome-positive CML who are also heterozygous for the enzyme marker glucose-6-phosphate dehydrogenase (G6PD), the leukemic granulocytes exhibited only one of the two isozyme types, thereby completing the evidence for the clonal nature of this disorder. Subsequently, a number of other neoplasms have been shown to be clonal in the G6PD system [48]. Thus, clonal origin and genotypic evolution of tumors, anticipated in the hypotheses of von Hansemann, Boveri, and Winge, were shown to be essential components of neoplastic growth; indeed, by the end of the 1960s they had come to be regarded as the central dogma of tumor cytogenetics.

The next series of major advances in cancer cytogenetics came about in the early 1970s and coincided with the introduction of new methods of study such as banding analysis of chromosomes and somatic cell hybridization. The study of inherited human disorders with increased predisposition to neoplastic disease and spontaneous chromosome instability progressed rapidly during the past decade, generating new ideas about the significance of chromosome changes to neoplastic development and new approaches to study it. The role of chromosome change in the origin and evolution of neoplastic cells will be evaluated in this chapter based on a review of three sets of pertinent data: (1) chromosome changes in preneoplastic and neoplastic cells; (2) chromosome control of neoplastic proliferation as revealed by somatic cell hybridization experiments; and (3) the natural history of neoplastic proliferation in the chromosome-breakage syndromes.

CHROMOSOME CHANGES IN NEOPLASTIC CELLS

According to currently held views, carcinogen-induced tumorigenesis consists of two major steps. The first, *initiation*, appears to comprise a specific local change at the DNA level (i.e., mutation), whereas the second, *promotion*, is complex and probably comprises many changes, some of which also may be genetic in nature [14,15,182,183]. Two recognizable endpoints in the history of the initiated cell are the *transformed* and the *tumorigenic* states. The transformed state, inducible in cultured normal cells treated with carcinogens, characterizes a cell that has escaped constitutive restriction on life-span and acquired the ability to proliferate indefinitely, given appropriate circumstances [93]. The tumorigenic state characterizes a transformed cell that has gained the properties of selective proliferation, and colonization and expansion in distant sites in the body of the host. The problem before the cytogeneticist is to place the chromosome change seen in tumor cells in the context of the model of tumorigenesis outlined above. It can be best approached by asking two questions: (1) How do these chromosome changes arise? and (2) What role do they play in the emergence of neoplastic cells? The relevant data and their bearing on the above questions will be discussed below.

Specificity of Chromosome Changes in Human Neoplasia

Recent years have witnessed an explosion in the literature pertaining to banding analysis of tumors, especially of the human lymphohematopoietic system. These studies, which have recently been reviewed extensively [126,128, 159,160,162,164], have established that the chromosome changes seen in neoplastic cells are not only nonrandom but are sometimes quite specific. Mitelman's recent summary of chromosome changes in 1,871 human tumors belonging to 15 histologic types, and including myeloid, lymphoid, epithelial, mesenchymal, and neurologic tumors, revealed that only chromosomes Nos. 1, 3, 5, 6, 7, 8, 9, 11, 12, 13, 14, 17, 20, 21, and 22 become preferentially involved in their chromosome variation; of these, chromosomes Nos. 1, 8, and 14 predominate [108, 129, 130]. The variation itself is expressed as either translocations and other rearrangements or gains and losses of chromosomes or chromosome regions, or both.

Although nonrandom gain and loss of specific chromosomes and chromosome regions has been well documented in leukemias, lymphomas, and other tumors [159,162], recent observations of a rare group of acute lymphoblastic leukemia patients provide a dramatic example of the association between chromosome loss and neoplastic proliferation. These leukemias, of which less than a dozen cases have been described, are characterized by extreme reduction in chromosome number [30,34]. Unaccompanied by rearrangement, they can reach haploidy of all autosomes (Fig. 1). The abnormal proliferation of these near-haploid cells may be viewed as resulting from the loss of an entire set of genes which normally are required to be present in two doses in order to modulate functions related to normal growth and differentiation of target cells[1]; elsewhere, I have called the situation leukemogenesis by default [34].

---

[1]Target cells are defined here as cells in given tissues or organs that will undergo neoplastic transformation under given circumstances.

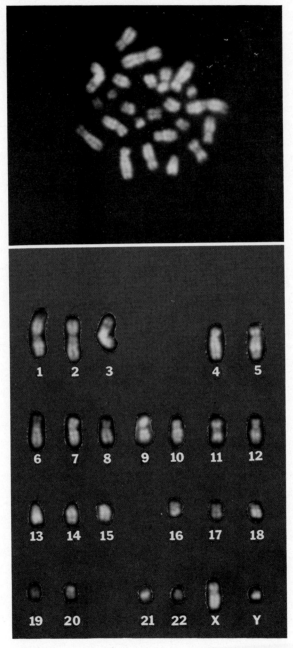

Fig. 1. Q-banded karyotype of a near-haploid leukemia cell. The acute lymphoblastic leukemia in a 7-year-old boy was characterized by monosomy for all chromosomes.

Specific translocations characterize a number of neoplastic cell types. Some of the well-known examples are the following: t(9;22)(q34;q11) in CML; t(8;22)(q25;q22) in the M2 subset of acute myelogenous leukemia (French-American-British classification [12]); t(15;17)(q25;q22) in acute promyelocytic leukemia; t(8;14)(q23;q32) in Burkitt's lymphoma (BL); and t(6;14)(q21;q24) in ovarian cystadenocarcinoma [159,164]. Those seen in CML and non-Hodgkin's lymphoma (NHL) can serve as models in considering the significance of rearrangement to emergence of tumor cells. The Ph[1] chromosome in CML and the 14q+ marker chromosome in endemic BL and other forms of NHL are derived in the majority of cases from the two translocations, t(9;22)(q34;q11) and t(8;14)(q23;q32), respectively [115,116,157]. In about 10% of Ph[1]-positive CML cases the translocation is between 22q and a chromosome other than the No. 9 [158]. Similarly, in a proportion of BL cases, the translocation is between 8q on the one hand and 2q or 22q on the other [20,95,127,131]. These observations suggest that perturbations of the distal regions of chromosomes Nos. 22 and 8 are of special and specific significance to the development of CML and NHL, respectively. Similar deduction(s) can be made about other chromosome regions found nonrandomly involved in rearrangement in tumors.

The significance of nonrandomness of chromosome changes seen in tumor cells to their emergence and evolution has been addressed by several investigators [128,156]. One hypothesis is that certain chromosomes of the human genome carry genetic information of importance to cellular proliferation. If products of such genes control functions related to cell proliferation, then their perturbation—due to position effect (as a result of rearrangement) or to alterations in gene dosage (as a result of loss or gain of chromosome material)—might lead to disruption of normal genetic regulation and to acquisition of a new proliferative advantage by the cell. The idea of position effect finds some support in the observation that the staining property of the segment of 8q (23→terminus) may be altered when it is translocated to 14q, indicating possible changes in chromatin function [76,116]. An examination of the human gene map suggested to Rowley [156] that most of the chromosomes that are nonrandomly involved in change in tumor cells also carry genes for nucleic acid biosynthesis; from this she postulated that the genes in question might be those that control DNA metabolism.

The recent work of Hayward and his associates [71,133], which elucidated the mechanism of activation of the cellular *onc* gene responsible for avian lymphoid leukosis, provides a different and intriguing model in considering the role of chromosome rearrangement in the origin of neoplastic proliferation. Avian leukosis virus (ALV) induces neoplastic transformation by activating the cellular *onc* gene, *c-myc*, itself a counterpart of the transforming gene of MC29 virus. The AVL provirus integrates randomly in the cellular genome. Its occasional integration adjacent to the c-myc gene causes initiation of transcription by the viral promotor and enhanced expression of c-myc, leading to neoplastic transformation. By analogy, some of the specific chromosome rearrangements seen in human tumors may be viewed as recombining cellular *onc* genes with endogenous or viral promotors, thereby leading to neoplastic transformation.

The above hypotheses address the consequences of chromosome change for the establishment and maintenance of neoplastic proliferation but not the question of the origin of the abnormalities themselves. Two main hypotheses have been proposed to explain the possible mechanisms by which chromosome changes arise in the first place in cells destined to become neoplastic. One suggests that carcinogens act directly on specific chromosomal sites leading to abnormality which is seen to be specific to the inducing agents [128]. The other suggests that genetic instability of initiated preneoplastic cells promotes fixation of specific chromosome abnormalities in specific target cells, due, most likely, to the fact that these chromosomes carry genes that control the function or proliferation, or both, of the target cells [93].

Inducing Agent-Specific Chromosome Changes

During the early and middle 1970s Mitelman, Levan, and their associates performed extensive cytogenetic analyses of tumors in the Chinese hamster and the rat that were induced by a variety of carcinogens. These studies demonstrated that chromosome changes seen in such tumors not only were nonrandom but also exhibited a relationship to the carcinogen. The following summary of these data is based on Mitelman's recent reviews [126,128]. Primary sarcomas of Chinese hamster induced by Rous sarcoma virus (RSV) and 7,12-dimethyl-benz(α)anthracene (DMBA) were characterized by nonrandom involvement of chromosomes Nos. 5, 6, and 10 in the former and chromosome No. 11 in the latter. RSV-induced sarcomas in the rat had normal diploid complements in the primary tumors and exhibited a stepwise evolution leading to nonrandom aneuploidy of chromosome Nos. 7, 12, and 13 during serial transplantation. By contrast, in the rat, primary leukemia and sarcoma induced by a variety of chemical carcinogens (6,8,12- and 7,8,12-trimethyl-benz (α) anthracene (TMBA), 20-methylcholanthrene, 3,4-benzpyrene, and N-nitroso-N-butylurea) were predominantly trisomic for the No. 2 chromosome. The No. 2 chromosome of the rat was further observed to exhibit increased breakage and sister-chromatid exchange (SCE) when treated with DMBA and TMBA *in vitro* as well as *in vivo*, which suggests that the carcinogens act directly on specific chromosomes or chromosomal regions. Recently, geographic differences have been observed in the clustering of nonrandom chromosomal abnormalities in human tumors. Mitelman interpreted these differences as suggesting relationships to inducing agents [29,126,129].

From these observations Mitelman [126] has postulated two levels of chromosome change associated with tumor-cell origin. The first (primary) results from direct interaction between the carcinogen and the specific gene or chromosome locus which controls normal cellular proliferation. This interaction may in turn result in chromosome change, which, depending upon the nature of rearrangement and size of chromosome segment involved, may or may not be recognized through the light microscope. Nevertheless, this primary change usually leads to the establishment of a preneoplastic cell whose proliferation is no longer under its own control; however, on rare occasions it can lead to the generation of fully neoplastic cells. The second (secondary) level of change is random and comprises the raw material for the karyotypic evolution which is associated with the further evolution of the neoplastic cell. This evolution may be governed by such phenomena as gene amplification and elimination.

Target-Cell-Specific Chromosome Changes

The best examples of target-cell specificity of chromosome changes are seen in the neoplastic proliferations of the murine lymphoid system and of human lymphoid and myeloid systems. In the case of the mouse, thymomas that arise spontaneously in the AKR strain [46], as well as the T-cell leukemias that arise in non-AKR mice following treatment with diverse carcinogens (x-rays, viruses, chemicals) have been shown to be trisomic for the No. 15 chromosome [36,188]. Further cytogenetic studies of carcinogen-induced T- as well as B-cell leukemias in mice showed that murine leukemogenesis, irrespective of the inducing agent, is characterized by trisomy of the No. 15 chromosome (Fig. 2) [189]. Furthermore, by using chromosomally marked mice in leukemia induction experiments, Wiener et al. [186] were able to demonstrate that the duplication of the distal region of the No. 15 chromosome is the sole cytogenetic abnormality that is critical to leukemia development. The significance of this chromosome region to the emergence of murine lymphoid malignancies has been further underscored in studies of primary plasmacytomas induced in BALBc mice by Pristane oil treatment [138]. A translocation which involves either chromosome Nos. 6 and 15 or 12 and 15 was seen in each of seven tumors studied; in both translocations, the breakpoint in the No. 15 chromosome was the same and was situated in the D3 band. Because the determinants for the immunoglobulin light and heavy chains in the mouse are known to be localized on chromosome Nos. 6 and 12, respectively [78], it was suggested that breaks in these chromosomes are possibly related to either gene rearrangements or immunoglobulin heavy-chain switching that occur during differentiation of immunoglobulin

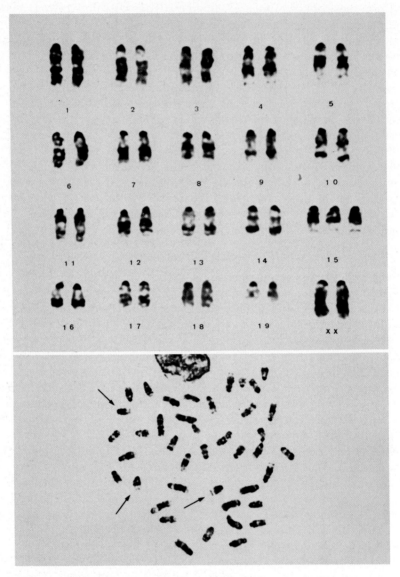

Fig. 2. G-banded karyotype of a mouse T-cell leukemia cell showing trisomy of the No. 15 chromosome (karyotype courtesy of Dr. Francis Wiener, Stockholm).

structural genes [*138*]. Further studies of plasmacytomas by Wiener et al. [*187*] showed that the translocation t(12;15)(F2;D3) invariably was present in all cells of early transfer generations of λ- as well as ϰ-chain-producing tumors. Analysis of breakpoints indicated that the translocated end of the No. 15 chromosome was directly joined to band F1 of the No. 12 chromosome, with the F2 region possibly being deleted; the locus for heavy chain has been recently localized proximal to band 12F2 [*121*]. Arguing that the distal segment of the No. 15 chromosome contains a "supergene" region that controls normal differentiation and responsiveness of the lymphohematopoietic system to growth control, these authors postulated that the joining together of the site that codes for the heavy chain with the site that controls proliferation might play a key role in the origin of these tumors [*187*].

Parallel to the situation in the mouse, the human immunoglobulin heavy-chain determinants have been assigned to chromosome No. 14 [*42,169*]. As already pointed out, band q32 of this chromosome frequently engages in translocation in BL, NHL, and plasma cell dyscrasias [*127*]. The human light-chain determinants have recently been assigned to the short arm of the No. 2 chromosome and the long arm of the No. 22 chromosome [*47,114,117*], both of which engage in translocation with the No. 14 chromosome at band q32 [*18,127,131*]. Therefore, as discussed previously, it is perturbation of chromosome No. 8 that seems to be important in the development of B-cell neoplasia. Indeed, evidence of its nonrandom involvement in abnormality, either as trisomy or as translocation in diverse neoplasms, points to the presence on this chromosome of genetic information of importance to cellular proliferation, especially of the lymphohematopoietic system [*73,155*]. In this respect, the distal region of human chromosome No. 8 appears to be analogous to the distal region of murine chromosome No. 15; perhaps they contain one or more homologous *onc* genes that control cellular proliferation.

Recent studies of Ph^1 chromosome-positive leukemia further illustrate the nature of target-cell-related chromosome change. Until a few years ago the Ph^1 chromosome served as the paradigm for a specific chromosome change associated with neoplastic proliferation of a specific cell type, namely, a granulocyte. The discovery that the leukemic blasts of occasional CML patients, who were presenting in the blastic phase of their disease, can express in their cytoplasm terminal deoxynucleotidyl transferase (TdT), an enzyme marker of developing thymic lymphocytes [*27*], cast doubt on the strictness of the specificity [*165*]. Subsequently, it has been shown that 25–30% of adult acute lymphoblastic leukemia patients with non-T, non-B leukemic cells also are Ph^1-chromosome-positive [*23,33,163*]. The translocation between chromosome Nos. 8 and 22 found in some patients with BL, discussed earlier, [18,127,131] also involves breakage and rearrangement at 22q11, leading to the generation of the Ph[1] chromosome. We have observed a Ph[1] chromosome derived from a translocation involving chromosomes Nos. 8, 14, and 22 in the leukemic phase of a BL patient; the leukemic cells also exhibited two 14q+ chromosomes [Chaganti et al., unpublished] (Fig. 3). It is of additional interest to note here that in a very few cases of CML, the Ph[1] chromosome was derived from a translocation between chromosomes Nos. 14 and 22 with breakpoints in bands q32 and q22, respectively [83,144; Chaganti et al., unpublished] (Fig. 4). Finally, the Ph[1] chromosome has been observed in rare cases of acute myeloblastic leukemia [23]. This ubiquity of the Ph[1] chromosome in neoplasia of the lymphohematopoietic system, as well as the engagement of 22q in translocation with both 8q and 14q with breaks in bands classically seen perturbed in lymphoid malignancies, clearly indicates that the target cell for the Ph[1]-positive leukemia is a pluripotential hematopoietic stem cell that is capable of undergoing lymphoid as well as myeloid differentiation in the neoplastic phase, although it predominantly follows the myeloid path.

In proposing a model for lymphoma development in man and mouse based on observations of target-cell-related chromosome change, Klein [93] had argued that carcinogen-induced initiation brings about, by local changes in DNA, the establishment of "long-lived preneoplastic cells" which, frozen in their state of differentiation, are capable of continued division in response to appropriate stimulation, akin to *in vitro*-transformed cells. The human B lymphocyte immortalized *in vitro* by the Epstein-Barr virus infection is suggested as an example of such a preneoplastic cell. Conversion of preneoplastic cells to the fully neoplastic state is said to be contingent upon the occurrence of specific genetic changes which, in some systems, can consist of detectable chromosome abnormalities. Such abnormalities, therefore, can be expected to be present in the majority of tumors that originate from the same cell type. The abnormalities themselves are considered to arise as part of random genetic changes but become selectively fixed in the emergent neoplastic cells due to the proliferative advantage of the cells that carry them; thus, the specific chromosome change in these systems is due to convergent evolution [93]. Such a scheme would require the genetic system of the preneoplas-

---

Fig. 3. Q-banded karyotype of a leukemic marrow cell from a patient with L3 (Burkitt-type) leukemia. A complex three-way translocation involving chromosomes Nos. 8, 14, and 22 most probably was responsible for the origin of the abnormal chromosomes 8q, 14q+, and 22q− (Ph[1]) (arrows) seen in this clone. Also note a second 14q+ chromosome and several other abnormal chromosomes (e.g., 1q+, 3q−).

Chromosome Change to Neoplastic Development 369

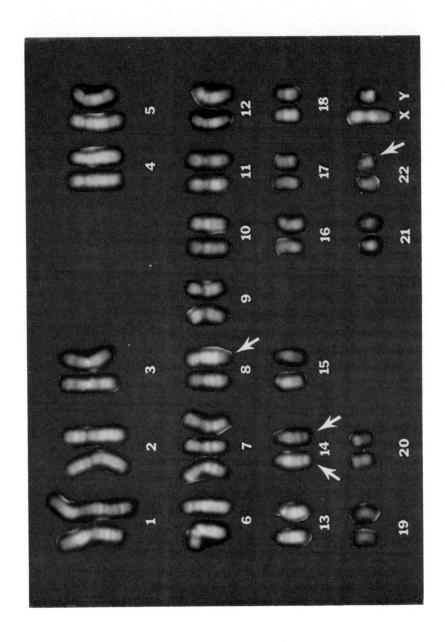

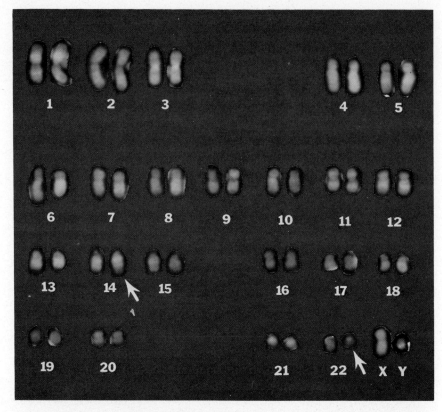

Fig. 4. Q-banded karyotype of a CML patient showing an unusual derivation of the Ph[1] chromosome, from a translocation between 14q and 22q, with breakpoints in bands q32 and q11, respectively.

tic cells to be unstable; indeed, Nowell [136] previously had advanced the view that the genomes of early neoplastic (preneoplastic) cells are unstable and hence promote evolution. The concept of convergent evolution is in good agreement with the evolutionary models of neoplastic origin and progression and most likely explains much of the nonrandomness of chromosome changes seen in tumor cells.

The Question of Normal Karyotypes in Tumor Cells

A discussion of the significance of chromosome changes in tumor cells will be incomplete without considering those tumors in which cytogenetic analysis reveals only normal chromosome complements. This has generally been reported to be the case in as many as 50% of acute leukemias when studied with convential banding methods. Several reasons can be advanced for the observation of normal diploid karyotypes in cytogenetic preparations of leukemic tissue. The first is that the resolution achieved by currently used methods of metaphase chromosome preparation and banding is limited and does not permit recognition of small structural changes. Indeed, Yunis et al. [191] very recently observed that when studied with high-resolution banding methods the leukemic cells of all 24 ANLL patients studied had abnormal chromosome complements. The second has to do with the proportions and rates of proliferation of normal and leukemic cells in the marrow of patients. Usually only about a milliliter or less of aspirated marrow is available for cytogenetic studies, and a chromosomally abnormal clone is easily missed if its proportion in the marrow cells was small, or if its proliferation was slower than that of the normal cells. A method such as fluorescence-activated cell sorting based on DNA content of nondividing cells sometimes reveals the presence of aneuploid leukemic cells in cases where cytogenetic methods failed to do so. The third possibility is reactive proliferation of host cells, which can be extensive enough to mask the presence of tumor cells with chromosome abnormalities [188].

Rare instances of chromosomally normal cells that have been found by cytochemical or immunological methods to be leukemic have been of great interest because they seem to point to involvement of special etiologic agents such as viruses that cause direct transformation. One of the best examples of such a tumor is mouse leukemia caused by the Abelson murine leukemia virus, a leukemia with an extremely rapid onset. The leukemic cells in this case remain diploid with normal chromosomes. The virus presumably integrates directly within the host DNA and causes leukemia onset without a latency period, thus leaving no time for karyotypic evolution [96]. A possible counterpart to this in humans is the rare instance of leukemic relapse in cells of donor origin in patients who have received allogeneic marrow transplantation for the treatment of their original leukemia. Three out of five such leukemias reported in the literature were diploid [50,135,180]. In the case most recently described, the leukemic nature of the chromosomally normal diploid cells was established by cytochemical and immunological analyses [135].

In summary, studies of chromosome changes in human, murine, and rodent tumors performed during the past decade applying the banding methods yielded a wealth of information which enabled new speculation concerning the role of chromosome change in the emergence of neoplastic cells. The major merit of these hypotheses has been their attempts to integrate cytogenetic data with currently held molecular, genetic, developmental, and evolutionary views of neoplastic origin and proliferation.

The testing of these hypotheses necessarily has to be in experimental systems. The recent dramatic successes in the analysis of vertebrate DNA sequences associated with transformation [40,71,104] suggest that before long these hypotheses will be subjected to verification by the powerful methods of molecular genetics. As far as cancer cytogenetics is concerned, the next decade promises to be as stimulating and rewarding as the past two have been, if not more so!

GENETIC ANALYSIS OF NEOPLASTIC PROLIFERATION BY SOMATIC CELL HYBRIDIZATION

During the past decade several groups have used somatic cell hybridization to address the question of chromosome control of malignant proliferation. The experimental strategy of these studies has consisted of fusing neoplastic cells with other neoplastic, or with nonneoplastic (normal) cells, and evaluating the tumorigenic potential of ensuing hybrids. Although the initial aim of these studies has been to determine whether the neoplastic trait behaves as dominant or recessive, the view that developed from the analyses of the chromosome constitutions and the tumorigenic phenotypes of the hybrids was that the classical dominance-recessivity complementation relationship does not apply to this property when analyzed in this fashion [97]. The work of Harris, Klein, and their collaborators [69, 185] clearly indicated that the expression or nonexpression by hybrids of the neoplastic phenotype was related to the number of chromosomes from the nonneoplastic parent retained in them; those that retained most or all tended to be nontumorigenic, while those that retained few tended to be tumorigenic. The conclusion from these results was that genetic information present on certain chromosomes of the normal parental cells suppressed the neoplastic trait, acting in a dominant fashion, the implication being that the initial (genetic) change that took place in the neoplastic cell must have been recessive in nature [69,185]. Further evidence for this view derives from the studies of Harris, which showed that when a range of neoplastic cells were fused with each other, the resulting hybrids usually were neoplastic as well, indicating absence of complementing alleles [69].

Chromosome suppression of tumorigenicity has been studied in considerable detail in intraspecific as well as interspecific cell hybrids (diploid human × tumorigenic Chinese hamster; diploid human × tumorigenic mouse; diploid human × tumorigenic human). In a recent series of experiments Klinger has shown that human chromosome Nos. 9 and 11 suppress tumorigenicity of human × hamster hybrids and that human chromosome Nos. 13 and 17, either in combination with each other or with chromosome Nos. 9 and 11, can act as alternate suppressors in these hybrids [97–99]. In the case of human × mouse hybrids, human chromosome No. 8 in combination with either chromosome No. 9 or No. 11, or with both, acts as the suppressor. Doubtless, further analysis will bring to light specific chromosome regions that control the tumorigenic ability of hybrid clones derived from the fusion of normal cells with tumor cells of diverse kinds. It has been suggested that the suppressive effect of normal chromosomes most probably is based on the restoration of genetic information lost or rendered nonfunctional in the tumorigenic cell rather than, as mentioned earlier, complementation of mutant genes [98]. According to this view, transformation to tumorigenic state is the end result of one or more mutational events which led to the loss of regulation of proliferation, a view not unlike the one arrived at from the study of chromosome changes in tumor cells, discussed earlier. In this connection, it is interesting to note that some of the human chromosomes found involved nonrandomly in change in tumor cells (e.g., Nos. 8, 9, 11, 13, 17) also exhibit suppressive properties in normal × tumorigenic cell hybrids. Thus, the conclusion concerning chromosome control of neoplastic proliferation inferred from studies of chromosome changes in tumor cells finds support in the results of genetic analysis by somatic cell hybridization.

THE CHROMOSOME-BREAKAGE SYNDROMES

The chromosome-breakage syndromes, the main topic of this symposium, are discussed extensively elsewhere in this volume [3, 55, 61, 140]. Therefore, I shall restrict myself to a consideration of the natural history of neoplasia in the three classical disorders, ataxia-telangiectasia (AT), Bloom's syndrome (BS), and Fanconi's anemia (FA). Several investigators have addressed this question previously and the main set of hypotheses, put forward by German [57–60] and others [67, 168], can be summarized in the following way. Most human neoplasms consist of clonal proliferations of cells with chromosome abnormalities. Environmental agents (e.g. x-ray) that cause cancer also cause chromosome rearrangement and the proliferation of clones of cells with chromosome abnormalities. In the chromosome-breakage syndromes, an extraordinarily high rate of *in vivo* transformation takes place against a background of spontaneous chromosome instability brought about by germ-line mutations. Therefore, these mutations can be grouped along with environmental carcinogens when considering human carcinogenesis. Pathways to neoplastic transformation in these syndromes may be related to either the presence of DNA-repair defects (which themselves may be etiologically related to chromosome breakage) or the generation of clones of cells with chromosome abnormalities which have gained a proliferative advantage, or both. Origin of skin cancer in xeroderma pigmentosum (XP), an autosomal recessive disorder characterized by inability of cells to remove damage to DNA induced by ultraviolet (UV) radiation, may be considered as the paradigm for the first pathway; the origin of lymphocytic leukemia in a proliferating T-cell clone of AT, which also carries a specific translocation between the two No. 14 chromosomes, may be considered as the paradigm for the second pathway, above. A third pathway sometimes suggested recognizes that gene mutations and visible chromosome mutations represent two ends of a common spectrum of genetic instability in somatic cells induced by carcinogens and, therefore, also can represent the same spectrum produced by germ-line mutations. Thus, these pathways perceive a central role for chromosome breakage in the tumorigenesis that takes place in these disorders, and by extension, in tumorigenesis per se. By this view, the origin of neoplasia in the chromosome-breakage syndromes, especially that in BS, can serve as a model for the origin of neoplasia in general [58].

Recently, Cairns [32] has challenged the conventional view of neoplastic initiation by gene mutation and proposed instead that transposition of major chromosome segments may be the mechanism by which common cancers are initiated in humans. He also chose BS with its 100-fold increase in the incidence of common cancers as an example for this model of cancer cell origin. Attractive in its simplicity and directness, the view of initiation through chromosome rearrangement presents several difficulties in connection with its acceptance as a general model for carcinogenesis. First, being based on a series of correlations, it does not lend itself to experimental verification. Second, in disregarding the need for the occurrence of a local change or changes prior to the appearance of recognizable chromosome rearrangements, it advances a cataclysmic view of carcinogenesis; the bulk of currently available experimental evidence, however, points to an evolutionary nature of the process [53, 93, 183]. Third, as discussed earlier, some malignant tumors do retain apparently normal karyotypes. Fourth, the chromosome-breakage syndromes AT and FA are disorders of multiple developmental defects, and, as will be shown below, carcinogeneis in them takes place predominantly in organs that also exhibit profound developmental defects, a fact not easily explained by the chromosome-breakage hypothesis.

I shall take the position that the pathways to neoplasia in these syndromes are multiple and complex rather than unitary and simple, and that they relate to developmental defects in target organs and specific, rather than common, cellular defects. I shall propose that cells of these syndromes, unlike those of the dominantly inherited cancer syndromes such as polyposis coli and Gardner's syndrome [100,101], are not in a state of constitutive initiation brought about by the germ-line mutation. Finally, I shall attempt to define the pathways of malignancy in the chromosome-breakage syndromes, as far as current evidence permits, in terms of their neoplasms and neoplasia-associated traits.

The Neoplasms

The frequencies with which the various neoplasia have been reported to occur in the three syndromes, classified as to the tissue type from which they arise, are summarized in Table I. The predominance of lymphoid neoplasms to the complete exclusion of myeloid leukemia in AT, the predominance of myeloid leukemia and hepatic tumors to the complete exclusion of lymphoid neoplasms in FA, and the presence of myeloid, lymphoid and other tumors in BS become immediately apparent from this table. The main conclusion from these data is that the bulk of carcinogenesis in AT and FA is restricted to certain target cells while no such target-cell restriction is apparent in BS. A primary requisite of any hypothesis that attempts to explain carcinogenesis in these syndromes, therefore, is to explain their unique patterns of neoplasia.

Neoplasia-Associated Traits

Individuals affected with the three syndromes manifest one or more of six features classically associated with neoplastic proliferation: namely, immunodeficiency, clonal proliferation, developmental defects, carcinogen hypersensitivity, increased mutation rate, and increased chromosome instability [5,57,59]. An analysis of the relationship between the above traits exhibited by these disorders and the neoplasms seen in af-

TABLE I. Neoplasia in Chromosome-Breakage Syndromes*

Syndrome	Tumor types and their frequencies†				
	Myeloid leukemias	Non-Hodgkin's lymphoma and lymphocytic leukemia	Hodgkin's disease	Nervous system tumors	Epithelial and mesenchymal tumors
AT	0	68	10	3	19
FA	50	0	0	0	50‡
BS	19	37	4	4	37

*Source of data: [4, 58, 60].
†Expressed as percentage of total neoplasms reported in each syndrome.
‡Predominantly in the liver.

fected individuals can be expected to shed light on the natural history of neoplasms themselves.

Immunodeficiency. AT patients often suffer from profound deficiencies in both T- and B-cell functions [21,24,119,143]. Because of defective T-cell function they are often unable to mount successful cell-mediated immunity. The defective B-cell function manifests itself as absence or low production of IgA and other immunoglobulins. Affected persons suffer from frequent infections, especially of the respiratory system. Their rudimentary or embryonic thymus, noted by Boder and Sedgwick [25] and others, must play a role in the defective T-cell development and function, which in turn presumably leads to a failure of the homeostasis needed for proper B-cell function. BS patients also suffer from frequent childhood infections, although their immunodeficiency does not seem to be as severe as that in AT. In BS, the nature of the immune defect, including the status of the thymus, is not well understood. Clinical information as well as available laboratory data suggest defective T-cell function and T-B-cell interaction [62]. So far, immunodeficiency has not been noted in FA patients.

Immunodeficiency of primary (genetic) as well as secondary (acquired through immunosuppressive treatment) origin and autoimmune states such as systemic lupus erythematosis are associated with an increased incidence of lymphoid malignancies [1,54,56,64,89,142,171]. A deficient or defective immune system traditionally has been credited with setting the stage for the origin and expansion of neoplastic lymphoid cells by two mechanisms. The first relates to origin of neoplastic cells through breakdown of normal homeostasis in cell interactions leading to chronic, unregulated antigenic stimulation, which in turn leads to chronic proliferation. In support of this view are data from a recent review of 35 lymphoreticular lesions that developed in patients with primary immunodeficiency disorders [54]. Of 21 lesions among those classified as NHL, 50% consisted of abnormal proliferative processes (three which could not be characterized as either benign or malignant and eight which were classified as immunoblastic sarcomas, commonly regarded as polyclonal proliferations). The second mechanism for cancer development in immunodeficient individuals relates to establishment of a permissive state for neoplastic development through an inability of the host to mount proper surveillance against tumor cells [92,109].

The classical immune surveillance concept [31,179], which required a cytotoxic T-cell capable of recognizing tumor cells by their antigenic properties, has been criticized by a number of investigators [134,145,146,174,175], although other investigators maintained the validity of the concept, especially in the context of virally derived tumors

[92,109]. Recently, a family of naturally occurring lymphoid cells has been discovered that can lyse virally infected cells as well as tumor cells without reference to their antigenic properties and which, in addition, can induce interferon production. These cells, designated as a group as natural killer (NK) cells, have been suggested to be the effectors of natural immunity against tumor cells and virally infected cells [79–81,90]. Evidence is rapidly accumulating which demonstrates that defective NK-cell function characterizes primary immunodeficiency disorders such as X-linked lymphoproliferative syndrome (XLP) [177], the Wiskott-Aldrich syndrome [111,112], the Chediak-Highashi syndrome [102], and severe combined immunodeficiency disease [102,111]. Lipinski et al. [111] showed that two AT patients had decreasd NK-cell function. The status of NK-cell function in BS at present is unknown. Clearly, the status of "natural" cell-mediated immune competence of AT and BS needs to be studied in detail.

An important pathway in lymphoid malignancy in these disorders seems to be related to potentially oncogenic virus infections. Purtilo and his associates showed that in the case of XLP syndrome, hyposensitivity to Epstein-Barr virus (EBV), presumably based on immunodeficiency, often leads to polyclonal neoplastic proliferation in the lymphoid system [148]. Recent studies have indicated that lymphoid neoplasms that arise in XLP patients, and also in renal transplant recipients, contain many copies of the EBV genome, implicating the virus in the origin of these tumors [66,149]. AT patients exhibit evidence of a chronic EBV infection [19,85], and one case of fatal disseminated herpes simplex virus (HSV) infection has also been reported, suggesting a general susceptibility of these patients to infection by herpes viruses [13]. Recently, Saemundsen et al. [161] have shown that in one lymphoma from an AT patient, 50–70 EBV-genome equivalents per cell were present. The Wiskott-Aldrich syndrome constitutes another example of hyposensitivity to infections by HSV. Thus, susceptibility to viruses with oncogenic potential increasingly is recognized as an important factor in lymphoma induction in immunodeficient individuals [94].

In summary, malignancies that arise in target lymphoid cells of immunodeficient patients appear to be rooted in the consequences of abnormal regulation of the immune system stemming from either defective development (primary) or therapeutic intervention (secondary). Lymphoid neoplasms of AT may, therefore, be viewed as arising through the same mechanisms by which they arise in other immunodeficient states.

Hodgkin's disease (HD), especially of the lymphocyte depletion type, is a frequent component of the spectrum of neoplasia associated with some primary immunodeficiency disorders; 10% of neoplasms reported in AT and 4% of those reported in BS are HD [54,60,171]. Although the identity and nature of the proliferating cell in HD remains a mystery, recent immunologic studies of HD patients demonstrate that their cell-mediated immunity is severely impaired [45,87,88]. In particular, the response of T-lymphocytes to stimulation by phytohemagglutinin (PHA) and other lectins has been reported to be depressed in patients, including asymptomatic ones with localized disease [88]. These observations, when considered in conjunction with the high incidence of AT and BS, both immunodeficiencies with defects in T-cell development and function, raise the possibility of an etiologic relationship between abnormal cell-mediated immune response and the development of HD.

Clonal proliferation. AT and FA patients exhibit spontaneous clonal proliferation *in vivo* of cells carrying chromosome mutations in the lymphoid and myeloid systems, respectively; BS patients do not exhibit a comparable proliferation [152,153]. Nearly a decade ago, Hecht and his associates [74] reported the appearance and progressive and advantageous expansion in the peripheral circulation of AT patients of PHA-responsive lymphocyte clones carrying a specific abnormality, namely, a 14q+ chromosome derived from the translocation t(14;14)(q12;q32). The No. 14 chromosome has subsequently been shown to engage frequently in translocation not only with its homolog but also with the No. 7 chromosome in these clones, breakpoints being situated in bands 7p13, 7q15, 14q12, and 14q32. Chromosomes other than No. 7 (e.g., 6,X) have also been seen to participate in translocation with No. 14, although less frequently than the No. 7 [118,153]. Cells of the clone with the 14-14 translocation, on occasion, (but, so far, not of those with 7-14 or other translocation) have undergone change to a neoplastic state, manifesting as lymphocytic leukemias [110,118,167,170]. The progression of a cell with the 14–14 translocation to a fully neoplastic state has often been cited as evidence for a chromosome abnormality predisposing to neoplastic transformation, the implication being that the constitutional chromosome instability had lead to the establishment of the clone in the first place [72,118]. The assumption that these cells, which may be considered as preneoplastic, are on the path to becoming fully neoplastic appears valid; at issue is their origin. The clonal cells in the peripheral circulation do not divide spontaneously. They respond to mitogenic stimulation by PHA and, at least in two cases, have been shown to carry receptors for the Fc segment of IgG and IgM; they therefore are differentiated T-lymphocytes [167; Chaganti, Auerbach and Gupta, unpublished].

The origin and nature of the mutant T-lymphocyte clones in AT can be evaluated best by considering the origin and normal sequence of migration and differentiation of thymic lymphocytes. The classical experiments of

Ford and his associates [51,124], which used the T6 marker chromosome to study the migration of injected bone-marrow stem cells in irradiated syngeneic mice, established almost 20 years ago that the traffic of hematopoietic cells in the lymphoid system is unidirectional; stem cells migrate from bone marrow to the peripheral lymphoid organs (lymph node, spleen) *invariably journeying* through the thymus. Based on the results of his mouse studies using the chromosome marker technique, Stutman [172,173,176] has proposed a model of lymphocyte maturation according to which prethymic precursors (immigrant stem cells from bone marrow) are allowed to emigrate into the periphery, only after acquisition of an appropriate immunologic repertoire during intrathymic "maturation." The emigrant cells, called postthymic precursor cells, home to the peripheral lymphoid organs where they continue their differentiation into the various subclasses of immunocompetent T-lymphocytes, again under thymic influence, a process described as postthymic maturation. The acquisition of the repertoire and clonal expansion of diverse lymphocyte genotypes during thymic maturation is itself suggested to be based on somatic mutation, selection, and clonal proliferation [84]. Supporting such a view is the observation of an extraordinarily high rate of cellular proliferation and death in the murine thymus [84,122,123]. Thus, intrathymic maturation is contingent upon a viable and accessible thymic stroma, while extrathymic maturation is contingent upon the appropriate thymic influence [for extensive review of T-cell differentiation, see references 154,176,193].

Relevant to the above view is the observation that a low level of spontaneous chromosome mutation appears to be associated with human T-lymphocyte differentiation in general. A common experience in clinical cytogenetics laboratories is the detection of sporadic cells with chromosome rearrangements in PHA-stimulated blood lymphocytes of immunologically normal individuals under study for various reasons. Among such mutations, chromosomes Nos. 7 and 14 have been reported to engage in nonrandom rearrangement at the rate of approximately 2.5×10^{-3} to 1×10^{-3} cells, with breakpoints situated in bands 7p13, 7q32-5, 14q12, and 14q23, respectively [9,11,75,86,184,192]. Among 22,191 PHA-stimulated blood lymphocytes from 843 individuals examined during a 3½-year period, we have encountered 69 cells with chromosome rearrangements, their frequency being 3×10^{-3} (Fig. 5) [Chaganti, Alonso, and Hew, in preparation]. The rearrangements in 25 of the aberrant cells involved chromosomes Nos. 7 and (or) 14, the breakpoints again being situated in 7q13, 7q22, 7q32-4, 14q12, and 14q32 (Table II). Pertinent to this observation is the report by Hustinx et al. [82] of a patient with IgA deficiency, mental deficiency, and developmental defects, but without

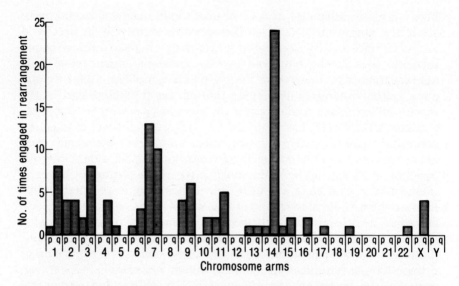

Fig. 5. Frequency of involvement of chromosome arms in rearrangement seen in 69 aberrant cells detected among 22,191 PHA-stimulated blood lymphocytes from 843 individuals studied with G-banding during the period January 1, 1978 to June 30, 1981 in the Cytogenetics Laboratory of The New York Hospital. Cell culture and harvest, chromosome preparation, banding, and analysis of all specimens were performed by the same technician using the same methods throughout (see also Table 1) [Chaganti, Alonso, and Hew, in preparation].

TABLE II. Distribution of Rearrangement Types That Involved Chromosomes No. 7 and 14*

Rearrangement	No. of times encountered
Translocation	
t(1;14)(q12;p11)	1
t(4;14)(q26;q24)	1
t(6;14)(q25;q12)	1
t(7;14)(p13;q12)	12
t(7;14)(q22;q32)	1
t(7;14)(q32-4;q12)	7
t(14;22)(q32;q12)	1
Inversion	
inv(7)(p12;q34)	1

*These rearrangements were encountered in 22,191 PHA-stimulated blood lymphocytes from immunologically normal individuals studied at The New York Hospital as described in the text.

either ataxia or telangiectasia, in whom 23 out of 96 PHA-stimulated lymphocytes exhibited 7-7 or 7-14 translocations with breakpoints situated in bands 7p13, 7q32, and 14q11. Several of the translocation types seen were present in more than one cell, indicating clonal origin. Thus, the breakpoints at which No. 7 and other chromosomes exchange segments with chromosome No. 14 in proliferating clones of AT and other immunodeficient patients, as well as in sporadic lymphocytes of normal persons, are the same, which, in turn, suggests that these identical nonrandom rearrangements in T-lymphocytes from individuals with diverse genetic backgrounds arise through related mechanisms. The occurrence of these cells in blood of normal persons is high enough (2.5×10^{-3} to 1×10^{-3}) to suggest that most normal individuals carry minor clones of such cells. These cells may therefore be viewed as arising in AT as well as in normal persons as a result of the mutation associated with thymic differentiation discussed above. It can therefore be suggested that while clonal expansion of these cells is restricted severely in the competent immune system of normal individuals, it is permitted in the improperly developed and incompetent thymus-dependent immune system of AT and other immunodeficient patients. Cells with the 14-14 translocation have not been observed so far among lymphocytes from normal persons. In our study, band 14q32 was found to be engaged in translocation far less frequently than band 14q12, suggesting that this band is less labile than the other three in thymic lymphocytes. However, rearrangement involving this band, as discussed earlier, is associated with neoplastic transformation of lymphoid cells irrespective of whether the host genotype is AT or normal. Further evidence that this fragility of chromosome Nos. 7 and 14 is restricted to thymocytes derives from the following observations. In three AT patients studied, the bone marrow cells were found to be chromosomally normal at a time when all T lymphocytes in the peripheral circulation comprised clones with the 14-14 translocation [2,167; Chaganti, unpublished]. Likewise, in the patient reported by Hustinx et al. [82], discussed above, no evidence of rearrangements involving chromosomes No. 7 and 14 was seen in 50 metaphases from bone marrow cells while blood lymphocytes, as discussed earlier, exhibited 7-7 and 7-14 translocations. Whether the postulated random and nonrandom breakage of the thymocyte chromosomes occurs spontaneously or takes place in response to the action of an intrathymic mutagen currently is unknown; at least one intracellular molecule unique to the thymocyte, namely TdT, has been suggested to be a mutagen [10].

Recent *in vitro* studies of bone marrow strongly suggest that the hematopoietic system of FA patients is characterized by an intrinsic stem-cell defect that manifests itself as a marked decrease in colony growth in the preanemic, anemic, and leukemic phases of the disorder [7,37,44,113,147,166]. Proliferation of clones with chromosome abnormalities in the bone marrow of FA patients has also been reported by a number of investigators in the anemic and leukemic phases of the disease; at the same time, there is little evidence to indicate that T-lymphocyte clones with chromosome abnormalities proliferate in this disorder [7]. Unfortunately, banding methods were not applied in the analysis of most of the marrow clones described; therefore, the nature of chromosome change that takes place in them can not be properly evaluated. In one well-studied case, the chromosomally abnormal clone detected during the anemic phase that preceded the onset of clinical leukemia was shown to undergo clonal evolution concomitant with the appearance of the leukemia, the chromosomes involved in aberration being Nos. 1, 3, 7, and 12 [17]. In a second carefully documented case, Auerbach et al. [7] observed loss of chromosome Nos. 7 and 8 in the leukemic cells, indicating that the chromosome change seen in these leukemias is probably similar to that seen generally in preleukemic and leukemic states of myeloid origin. Thus, leukemogenesis in these patients probably follows a path similar to that encountered in non-FA patients with preleukemic cytopenias, and some forms of idiopathic aplastic anemia, which are characterized by stem-cell defects and which predispose to myeloid leukemia [6, also see reference No. 4 for a review of leukemia in bone marrow-failure syndromes].

Developmental defects. That organs or tissues whose development is defective act as targets for neoplastic development is well recognized, although the mechanisms of neoplastic transformation in developmentally defective organs are unknown [26,125,132]. The relationship in AT and FA between abnormal development of lymphoid and myeloid systems on the one hand and the origin of the predominant neoplasia on the other has already been discussed. In AT, the development of the central nervous and the female reproductive system are abnormal, and in both systems neoplasms develop with increased frequency [67,171]. Other target-organ-related neoplasms can be expected to be discovered in these two disorders as more information becomes available.

Carcinogen sensitivity. Cellular as well as clinical hypersensitivity to selected carcinogens is exhibited by all three syndromes [77]; Arlett and Lehmann [5] and Harnden [68] have summarized these data. AT cells are hypersensitive to γ-rays [141,178] and the radiomimetic drug bleomycin [39,105], whereas AT patients are hypersensitive to x-ray treatment [43]. FA cells are hypersensitive to DNA crosslinking and alkylating agents [8,16], and clinically FA patients are hypersensitive to drugs, such as cyclophosphamide, which are alkylating agents [7,63]. BS cells are hypersensitive to ethyl methanesulfonate [*103*]. Of all the hypersensitivities exhibited by these disorders, only the γ-ray sensitivity of AT is of potential significance to carcinogenesis; it is difficult to envisage consistent exposure of FA and BS patients to DNA crosslinking and alkylating agents even at low doses during their normal course of life. In the case of AT, natural radiation may be considered a predisposing factor similar to the situation in XP, wherein exposure to solar UV radiation precipitates carcinogenesis [*38*]. The relationship between x-ray exposure and human carcinogenesis has been studied extensively in diverse kinds of exposure. The spectrum of neoplasia seen in children and adults exposed to x-rays in the general population has been well documented [*22,41*]. The main neoplasms that occur are leukemia (especially of myeloid origin) and cancer of the thyroid, breast, lung, and bone [*68,151*]. The spectrum of neoplasia seen in AT, as discussed earlier, is quite dissimilar to this; in AT, myeloid leukemia and thyroid cancer are conspicuous by their absence. Therefore, natural γ-ray radiation can be ruled out as a special carcinogen for AT patients. Thus, it is difficult to recognize a direct relationship between hypersensitivity and neoplastic transformation for any of the three syndromes. Indeed, hypersensitivity of the type expressed by these syndromes leads to cytotoxicity, cell death, and increased chromosome breakage which itself is usually lethal to the cell, but not, as far as is known, to transformation. FA patients with pancytopenia are invariably treated with androgenic steroids. The possible role of these drugs in the induction of liver cancer has been raised by several investigators with reference to FA [*4,60*], as well as to non-FA patients [*120*].

Somatic mutation. If the chromosome breakage seen in these syndromes represents one end of a spectrum of mutation that includes both gene and chromosome mutation, then spontaneous as well as induced gene-mutation rates can be expected to be elevated in cultured cells of patients with all three chromosome-breakage syndromes; such a hypothesis would imply also that the basic cellular defect(s) of the three syndromes must be related in some way, perhaps as DNA-repair defects. Available data from measurement of mutation rates in cultured fibroblast cells indicate that AT cells are either immutable or hypomutable, and FA cells are hypomutable [reviewed in *5*]; therefore, the above correlation seems invalid. Recently, Warren et al. [*181*] have shown that BS cells are hypermutable. Although all of the data reported on mutation rates of chromosome-breakage syndrome cells pertain to loci unrelated to transformation, they may be considered to reflect the general mutability status of cells from these syndromes. Thus, enhanced initiation by somatic mutation driven by a germ-line mutation is unlikely in FA and AT, although such a mechanism of initiation may well take place in BS. The latter syndrome is not characterized by pronounced developmental defects in any target organs; growth retardation and disturbance in immune function being the developmental defects noted in these patients. Yet, the spectrum of neoplastic disease seen in this disorder is similar to that found in the general population. In view of these data, a model of enhanced initiation by mutation in a hypermutable system becomes a reasonable pathway to malignancy in the case of BS, just as it is in the case of XP.

Chromosome breakage. The spectrum of chromosome breakage seen in these disorders has been reviewed previously [152,153]. Although many hypotheses have been proposed, the molecular basis of chromosome breakage of AT and FA cells remains to be elucidated; evidence for the postulated DNA-repair defects in these two syndromes has been equivocal as well [59,139].

In the case of BS, many careful studies have ruled out defects in known pathways of DNA repair [comprehensive review in 62]. A consistent disturbance in DNA metabolism has, however, been described in BS cells, which consists of a slower-than-normal rate of progression of DNA-replication fork when studied by DNA-fiber autoradiography and by pulse-chase experiments with tritiated thymidine. This cellular defect is most probably the underlying basis for the dramatic increase in SCE and quadriradial (Qr) and other exchanges seen in BS cells. Symmetrical Qrs that involve homologous chromosomes represent exchanges of nonsister chromatids, i.e., somatic recombination [35]. It is noteworthy in this connection that Radman and Kinsella [91,150] have suggested that one mechanism of promotion of carcinogenesis may consist of attainment of homozygosity of mutations responsible for initiation through somatic recombination. Therefore, if recombinational promotion exists, then BS would have the distinction of being self-sufficient for highly efficient intrinsic mechanisms for initiation as well as promotion. The spectrum of neoplasia and their incidence in this disorder may very well be pointing to such a situation.

Thus, analysis of the circumstances leading to neoplastic development in the three disorders suggests strongly that the germ-line mutation in none of them acts as the first change in the path to malignancy; the pathways, as proposed here, are complex. Those of AT and FA, rooted in defective development, comprise an immense challenge, whereas those in BS are rooted in a specific cellular defect and hence should be amenable to verification by appropriate experimentation *in vitro*.

SUMMARY

In this review, three sets of data pertaining to the question of the significance of chromosome changes to neoplastic development have been reviewed and evaluated; namely, (1) the nature of chromosome change in preneoplastic and neoplastic cells, (2) chromosome control of neoplastic proliferation as revealed by somatic cell hybridization experiments, and (3) the natural history of neoplastic disease in the chromosome-breakage syndromes.

(1) Chromosome changes in neoplastic cells have been studied for over 50 years. The experimental studies of the early 1950s emphasized the evolution of karyotypes in cultured tumor cells and led to the establishment of the stem-line concept. Subsequently, the emphasis shifted to the study of murine, rodent, and human primary neoplasms. The main outcome of these investigations was the establishment, by the end of the 1960s, of what may be regarded as the central dogma of tumor cytogenetics which can be stated as follows: neoplastic cells often exhibit chromosome changes; they are clonal in nature, and their evolution to increasing levels of malignancy is associated with further evolution in the chromosome complement. The application of the banding techniques to the analysis of tumor chromosomes in the 1970s yielded extensive new data concerning the types of chromosome changes that take place in spontaneous as well as carcinogen-induced neoplasms in a variety of mammalian species. Chromosome changes were found to characterize most of these tumors, and the chromosomes that participated in these changes exhibited either specificity or a high degree of nonrandomness. A complex set of relationships emerged from these data. In some tumors, chromosome changes appeared to be specified by the tumor-inducing agents, although in others they appeared to be specified by target cells of tumorigenesis, and still other tumors exhibited no chromosome changes. Although evidence for neoplastic initiation resulting from major chromosome changes is lacking, available data suggest strongly that chromosome change plays an important, even critical, role in the progression of initiated (preneoplastic) cells to fully neoplastic states. The completion of the elucidation of this role appears to lie within the domain of the rapidly expanding new field of the molecular genetics of neoplastic transformation.

(2) Genetic analysis of malignancy by somatic cell hybridization established the fact that neoplastic proliferation is under chromosome control; the chromosomes that effect control again tend to be nonrandomly distributed in the relevant genomes, a conclusion in agreement with the one derived from direct observation of chromosome changes in tumor cells.

(3) The so-called classical chromosome-breakage syndromes (AT, FA, and BS), as a group, are characterized by autosomal recessive modes of inheritance, spontaneous chromosome instability, and increased predisposition to neoplastic disease. Two of these, AT and FA, also exhibit pro-

found defects in the development of certain organs, and neoplasias in these cases arise mainly in the developmentally defective organs. Analysis of the circumstances leading to neoplastic development in the three disorders suggests strongly that the germ-line mutation is in no case the first change in the path to malignancy. The pathways to malignancy seen here are complex; those in AT and FA appear to be rooted in defective development, whereas those in BS are more likely rooted in a specific cellular defect. Therefore, these disorders provide major challenges to understanding the relationship between defective development and neoplastic tranformation on the one hand and between the consequence of abnormal DNA metabolism and transformation on the other.

Acknowledgments. I acknowledge with pleasure a debt of gratitude to James German, friend and colleague, who 10 years ago introduced me to the study of human and cancer genetics. Over the years, he has sustained and excited my interest in these areas through countless discussions, and given generously of his support and encouragement. I would like to express my appreciation to my associates Arleen Auerbach and Suresh Jhanwar, my past and present graduate students Mira Wiegensberg, Nancy Schneider, Jacqueline Burns, and Edward Chaum, and my former associate Steven Schonberg for the many stimulating discussions that I had with them about the role of genes and chromosomes in the origin and evolution of human cancer. I am also grateful to Peter Benn for critically reading the manuscript. The views expressed in this article, however, are my own responsibility.

This review was initiated at the Department of Tumor Biology, Karolinska Institutet, Stockholm, Sweden. I am grateful to the Cancer Research Institute, Inc., of New York for the award of a fellowship which made my visit to Stockholm possible. My sincere thanks are due to George Klein and Francis Wiener for their kind hospitality and for the many stimulating discussions during my stay in their laboratory. Research work of my laboratory is supported by the NCI contact CP-85665, NCI grants CA-23766 and CA-20194, and the Hecksecher Foundation.

LITERATURE CITED

1. AGUDELO, C.A., SCHUMACHER, H.R., GLICK, J.H., MOLINA, J.: Non-Hodgkin's lymphoma in systematic lupus erythematosus. *J. Rheumatol.* 8:69-78, 1981.
2. AL SAADI, A., PALUTKE, M., KRISHNA KUMAR, G.: Evolution of chromosomal abnormalities in sequential cytogenetic studies of ataxia telangiectasia. *Hum. Genet.* 55:23-29, 1980.
3. ALTER, B.P., POTTER, N.U.: Long-term outcome in Fanconi's anemia: description of 26 cases and reviews of the literature. This volume, pp. 43-62.
4. ALTER, B.P., RAPPEPORT, J.M., PARKMAN, R.: The bone marrow failure syndromes. *In* Hematology of Infancy and Childhood (Nathan, D.G., Oski, F.A., eds.), 2nd ed., Philadelphia: W.B. Saunders, 168-249, 1981.
5. ARLETT, C.F., LEHMAN, A.R.: Human disorders showing increased sensitivity to the induction of genetic damage. *Annu. Rev. Genet.* 12:95-115, 1978.
6. AUERBACH, A.D., ADLER, B., CHAGANTI, R.S.K.: Effect of a carcinogen and a protease inhibitor on chromosome breakage in Fanconi anemia. *J. Supramol. Biol. Suppl.* 5:202, 1981.

7. AUERBACH, A.D., WEINER, M.A., WARBURTON, D., YEBOA, K., LU, L., BROXMEYER, H.: Acute myeloid leukemia as the first hematologic manifestation of Fanconi anemia. *Am. J. Hematol.* 12:289-300, 1982.
8. AUERBACH, A.D., WOLMAN, S.R.: Carcinogen-induced chromosome breakage in chromosome instability syndromes. *Cancer Genet. Cytogenet.* 1:21-28, 1979.
9. AYME, S., MATTEI, J.F., MATTEI, M.G., AURRAN, Y., GIRAUD, F.: Nonrandom distribution of chromosome breaks in cultured lymphocytes of normal subjects. *Hum. Genet.* 31:161-175, 1976.
10. BALTIMORE, D.: Is terminal deoxynucleotidyl transferase a somatic mutagen in lymphocytes? *Nature* 248:409-411, 1974.
11. BEATTY-DE SANA, J.W., HOGGARD, M.J., COOLEDGE, J.W.: Non-random occurrence of 7-14 translocations in human lymphocyte cultures. *Nature* 255:242-243, 1975.
12. BENNETT, J.M., CATOVSKY, D., DANIEL, M-T., FLANDRIN, G., GALTON, D.A.G., GARLNICK, R., SULTAN, C.: Proposals for the classification of the acute leukemias. *Br. J. Haematol.* 33:451-458, 1976.
13. BEN-ZVI, A., SOFFER, D., YATZIV, S.: Disseminated herpes simplex virus infection in ataxia-telangiectasia. *Acta Paediatr. Scand.* 67:667-670, 1978.
14. BERENBLUM, I: Carcinogenesis and tumor pathogenesis. *Adv. Cancer Res.* 2:129-175, 1954.
15. BERENBLUM, I: Sequential aspects of chemical carcinogenesis. In Cancer: A Comprehensive Treatise (Becker F.F., ed.), New York: Plenum Press, 323-344, 1975.
16. BERGER, R., BERNHEIM, A., GLUCKMAN, E., GISSELBRECHT, C.: In vitro effect of cyclophosphamide metabolites on chromosomes of Fanconi anemia patients. *Br. J. Haematol.* 45:565-568, 1980.
17. BERGER, R., BERNHEIM, A., LE CONIAT, M., VECCHIONE, D., SCHAISON, G.: Chromosomal studies of leukemic and preleukemic Fanconi's anemia patients. *Hum. Genet.* 56:59-62, 1980.
18. BERGER, R., BERNHEIM, A., WEH, H.J., FLANDRIN, G., DANIEL, M.T., BROUET, J.C., COLBERT, N.: A new translocation in Burkitt's tumor cells. *Hum. Genet.* 53:111-112,1979.
19. BERKEL, A.I., HENLE, W., HENLE, G., KLEIN, G., ERSOY, F., SANAL, O.: Epstein-Barr virus-related antibody patterns in ataxia-telangiectasia. *Clin. Exp. Immunol.* 35:196-201, 1979.
20. BERNHEIM, A., BERGER, R., LENOIR, G.: Cytogenetic studies on African Burkitt's lymphoma cell lines: t(8;14), t(2;8) and t(8;22) translocations. *Cancer Genet. Cytogenet.* 3:307-315, 1981.
21. BIGGER, D.W., GOOD, R.A.: Immunodeficiency in ataxia-telangiectasia. In Immunodeficiency in Man and Animals. Birth Defects: Original Article Series, Vol. 11, No. 1 (Bergsma, D., ed.), New York: National Foundation-March of Dimes, 271-276, 1975.
22. BIZZOZERO, O.J., JOHNSON, K.G., CIOCCO, A., HOSHINO, T., ITOGA, T., TOYODA, S., KAWASAKI, S.: Radiation-related leukemia in Hiroshima and Nagasaki, 1946-1964. *N. Engl. J. Med.* 274:1095-1101, 1976.
23. BLOOMFIELD, C.D., LINDQUIST, L.L., BRUNNING, R.D., YUNIS, J.J. COCCIA, P.F.: The Philadelphia chromosome in acute leukemia. *Virchows Arch. B. Cell. Pathol.* 29:81- 91, 1978.
24. BODER E.: Ataxia-telangiectasia: Some historic, clinical and pathologic observations. In Proceedings of the 2nd International Workshop on Immunodeficiency Diseases in Man (Bergsma, D., Good, R.A., eds.), Sunderland, Mass.: Sinauer Associates, 225-270, 1975.

25. BODER, E., SEDGWICK, R.P.: Ataxia-telangiectasia. A familial syndrome of progressive cerebellar ataxia, oculocutaneous telangiectasia and frequent pulmonary infection. *Pediatrics* 21:526-554, 1958.
26. BOLANDE, R.P.: Childhood tumors and their relationship to birth defects. *In* Genetics of Human Cancer (Mulvihill, J.J., Miller, R.W., Fraumeni, J.F. Jr., eds.), New York: Raven Press, 43-75, 1977.
27. BOLLUM, F.J.: Terminal deoxynucleotidyl transferase as a hematopoietic cell marker. *Blood* 54:1203-1215, 1979.
28. BOVERI, T.: Zur Frage der Entstechung maligner Tumoren. Jena: Fischer, 1914.
29. BRANDT, L., NILSSON, P.G., MITELMAN, F.: Trends in incidence of acute leukemia. *Lancet* ii:1069. 1979.
30. BRODEUR, M.B., WILLIAMS, D.L., LOOK, A.T., BOWMAN, W.P. KALWINSKY, D.K.: Near-haploid acute lymphoblastic leukemia: A unique subgroup with a poor prognosis? *Blood* 58:14-19, 1981.
31. BURNET, M.: Immunological Surveillance. New York: Pergamon Press, 1970.
32. CAIRNS, J.: The origin of human cancers. *Nature* 289:353-357, 1981.
33. CATOVSKY, D.: Ph^1-positive acute leukemia and chronic granulocytic leukemia: One or two diseases. *Br. J. Haematol.* 42:493-498, 1979.
34. CHAGANTI, R.S.K.: Near-haploid chromosome numbers in leukemia: Significance to leukemogenesis. *Am. J. Hum. Genet.* 33:61A,1981.
35. CHAGANTI, R.S.K., SCHONBERG, S., GERMAN, J.: A manifold increase in sister chromatid exchanges in Bloom's syndrome lymphocytes. *Proc. Natl. Acad. Sci. USA* 71:4508-4512, 1974.
36. CHANG, T.D., BIEDLER, J.L., STOCKERT, E., OLD, L.J.: Trisomy of chromosome 15 in X-ray-induced mouse leukemia. *Cancer Res.* 18:225, 1977.
37. CHU, J-Y.: Granulopoiesis in Fanconi's aplastic anemia. *Proc. Soc. Exp. Biol. Med.* 161:609-612, 1979.
38. CLEAVER, J.E.: Human diseases with in vitro manifestations of altered repair and replication of DNA. *In* Genetics of Human Cancer (Mulvihill, J.J., Miller, R.W., Fraumini, J.F. Jr., eds.), New York: Raven Press, 355-363, 1977.
39. COHEN, M.M., SIMPSON, S.J., PAZOS, L.: Specificity of bleomycin-induced cytotoxic effects on ataxia telangiectasia lymphoid cell lines. *Cancer Res.* 41:1817-1823, 1981.
40. COOPER, G.M., OKENQUIST, S., SILVERMAN, L.: Transforming activity of DNA of chemically transformed and normal cells. *Nature* 284:418-421, 1980.
41. COURT BROWN, W.M., DOLL, R.: Mortality from cancer and other causes after radiotherapy for ankylosing spondylitis. *Br. Med. J.* 2:1327-1332, 1965.
42. CROCE, C.M., SHANDER, M., MARTINIS, J., CICUREL, L., D'ANCONA G., DOLBY, T.W., KOPROWSKI, H.: Chromosomal location of genes for human immunoglobulin heavy chains. *Proc. Natl. Acad. Sci. USA* 76:3416-3419, 1979.
43. CUNLIFFE, P.M., MANN, J.R., CAMERON, A.H., ROBERTS, K.D., WARD, H.W.C.: Radiosensitivity in ataxia telangiectasia. *Br. J. Radiol.* 48:374-376, 1975.
44. DANESHBOD-SKIBA, G., SHAHIDI, M.J.: Myeloid and erythroid colony growth in nonanemic patients with Fanconi's anaemia. *Br. J. Haematol.* 44:33-38, 1980.
45. DESFORGES, J.F., RUTHERFORD, C.J., PIRO, A.: HODGKIN'S DISEASE. *N. Engl. J. Med.* 301:1212-1222, 1979.
46. DOFUKU, R., BIEDLER, J.L., SPENGLER, B.A., OLD, L.J.: Trisomy of chromosome 15 in spontaneous leukemia of AKR mice. *Proc. Natl. Acad. Sci. USA* 72:1515-1517, 1975.

47. ERIKSON, J., MARTINIS, J., CROCE, C.M.: Assignment of the genes for human λ immunoglobulin chains to chromosome 22. *Nature* 294:173-175, 1981.
48. FIALKOW, P.J.: Clonal origin and stem cell evolution of human tumors. *In* Genetics of Human Cancer (Mulvihill, J.J., Miller, R.W., Fraumeni, J.F., Jr., eds.), New York:Raven Press, 439-453, 1977.
49. FIALKOW, P.J., GARTLER, S.M., YOSHIDA, A.: Clonal origin of chronic myelocytic leukemia in man. *Proc. Natl. Acad. Sci. USA* 58:1468-1471, 1967.
50. FIALKOW, P.J., THOMAS, E.D., BRYANT, J.I., NEIMAN, P.E.: Leukaemic transformation of engrafted human marrow cells in vivo. *Lancet* i:251-255, 1971.
51. FORD, C.E.: Traffic of lymphoid cells in the body. *In* The Thymus: Experimental and Clinical Studies (Wolstenholme, G.E.W., Porter, R., eds.), Boston: Little, Brown & Co, 131-152, 1966.
52. FORD, C.E., CLARKE, C.M.: Cytogenetic evidence of clonal proliferation in primary reticular neoplasms. *Can. Cancer Conf.* 5:129-146, 1963.
53. FOULDS, L.: The natural history of cancer. *J. Chronic Dis.* 8:2-37, 1958.
54. FRIZZERA, G., ROSAI, J., DEHNER, L.P., SPECTOR, B.D., KERSEY, J.H.: Lymphoreticular disorders in primary immunodeficiencies: New findings based on an up-to-date histologic classification of 35 cases. *Cancer* 46:692-699, 1980.
55. GATTI, R.A., HALL, K.: Ataxia-telangiectasia: search for a central hypothesis. This volume, pp. 23-41.
56. GATTI, R.A., GOOD, R.A.: Occurrence of malignancy in immunodeficiency diseases. *Cancer* 28:89-98, 1971.
57. GERMAN, J.: Genes which increase chromosomal instability in somatic cells and predispose to cancer. *Prog. Med. Genet.* 8:61-101, 1972.
58. GERMAN, J.: Bloom's syndrome. II. The prototype of human genetic disorders predisposing to chromosome instability and cancer. *In* Chromosomes and Cancer (German, J., ed.), New York: John Wiley and Sons, 601-617, 1974.
59. GERMAN, J.: The association of chromosome instability, defective DNA repair, and cancer in some rare human genetic diseases. *In* Human Genetics (Armandares, S., Lisker, R., Ebling, F.J.G., Henderson, I.W., eds.), Amsterdam: Excerpta Medica, 64-68, 1978.
60. GERMAN, J.: The cancers in chromosome-breakage syndromes. *In* Radiation Research. Proc. Sixth Intl. Congr. Radiat. Res. (Okada, S., Imamura, M., Terashima, T., Yamaguchi, H., eds.), Tokyo: Jpn. Assoc. Radiat. Res., Univ. of Tokyo, 496-505, 1979.
61. GERMAN, J.: Patterns of neoplasia associated with the chromosome-breakage syndromes. This volume.
62. GERMAN, J., SCHONBERG, S.: Bloom's syndrome. IX: Review of cytological and biochemical aspects. *In* Genetic and Environmental Factors in Experimental and Human Cancer (Gelboin, H.V., Matsushima, T., Sugimura, T., Sugimura, T., Takayama, S., Takebe, H., eds.), Tokyo: Jpn. Sci. Soc. Press, 175-186, 1980.
63. GLUCKMAN, E., DEVERGIE, A., SCHAISON, G., BUSSEL, A., BERGER, R., SOHIER, J., BERNARD, J.: Bone marrow transplantation in Fanconi anaemia. *Br. J. Haematol.* 45:557-564, 1980.
64. GOOD, R.A.: Relations between immunity and malignancy. *Proc. Natl. Acad. Sci. USA* 69:1026-1032, 1972.
65. HANSEMANN, D. VON: Über asymmetrische Zellteilung in Epithelkrebsen und der biologische Bedeutung. *Virchow's Arch. Pathol. Anat. Physiol.* 119:299-326, 1980.
66. HANTO, D.W., FRIZZERA, G., PURTILO, D.T., SAKAMOTO, K., SULLIVAN, J., SAEMUND-

SEN, A.K., KLEIN, G., SIMMONS, R.L., NAJARIAN, J.J.: Clinical spectrum of lymphoproliferative disorders in renal transplant recipients and evidence for the role of the Epstein-Barr virus. *Cancer Res.* 41:4253-4261, 1981.
67. HARNDEN, D.G.: Cytogenetics of high risk groups. *Excerpta Med. Int. Congr. Ser.* 484:33-41, 1980.
68. HARNDEN, D.G.: Mechanisms of genetic susceptibility to cancer. *In* Carcinogenesis: Fundamental Mechanisms and Environmental Effects (Pullman, B., Ts'o, P.O.P., Gelboin, H., eds.), Boston: D. Reidel Publishing Co. 235-244, 1980.
69. HARRIS, H.: Some thoughts about genetics, differentiation, and malignancy. *Somatic Cell Genet.* 5:923-930, 1979.
70. HAUSCHKA, T.S.: The chromosomes in ontogeny and oncogeny. *Cancer Res.* 21: 957-974, 1961.
71. HAYWARD, W.S., NEEL, B.G., ASTRIN, S.M.: Activation of a cellular *onc* gene by promotor insertion in ALV-induced lymphoid leukosis. *Nature* 290:475-480, 1981..
72. HECHT, F., McCAW, B.K.: Chromosome instability syndromes. *In* Genetics of Human Cancer (Mulvihill, J.J., Miller, R.W., Fraumeni, J.F. Jr., eds.), New York: Raven press, 105-113, 1977.
73. HECHT, F., KAISER-McCAW, B., SANDBERG, A.: Chromosome translocations in cancer. *N. Engl. J. Med.* 304:1493, 1981.
74. HECHT, F., McCAW, B.K., KOLER, R.D.: Ataxia-telangiectasia clonal growth of translocation lymphocytes. *N. Engl. J. Med.* 289:286-291, 1973.
75. HECHT, F., KAISER-McCAW, B., PEAKMAN, D., ROBINSON, A.: Non-random occurrence of 7-14 translocations in human lymphocyte cultures. *Nature* 255:243-244, 1975.
76. HECTH, F., KAISER-McCAW, B., PATIL, S., WYANDT, H.: Are balanced translocations really balanced? Preliminary evidence for position effect in man. *Birth Defects: Original Article Series* Vol. XIV 6C:281-286, 1978.
77. HEDDLE, J.A., KREPINSKY, A.B., MARSHALL, R.R.: Cellular sensitivity to mutagens and carcinogens in the chromosome-breakage and other cancer-prone syndromes. This volume, pp. 203-234.
78. HENGARTNER, H., MEO, T., MULLER, E.: Assignment of genes for immunoglobulin *k* and heavy chains to chromosomes 6 and 12 in mouse. *Proc. Natl. Acad. Sci. USA* 75:4494-4498, 1978.
79. HERBERMAN, R.B.: Natural Cell Mediated Immunity Against Tumors. New York: Academic Press, 1980.
80. HERBERMAN, R.B., ORTALDO, J.R.: Natural killer cells: Their role in defenses against disease. *Science* 214:24-30, 1981.
81. HERBERMAN, R.B., NUNN, M.E., LAVRIN, D.H.: Natural cytotoxic reactivity of mouse lymphoid cells against syngeneic and allogeneic tumors. I. Distribution of reactivity and specificity. *Int. J. Cancer* 16:216-229, 1975.
82. HUSTINX, T.W.J., SCHERES, J.M.J.C., WEEMAES, C.M.R., TER HAAR B.G.A., JANSSEN, A.H.: Karyotype instability with multiple 7/14 and 7/7 rearrangements. *Hum. Genet.* 49:199-208, 1979.
83. ISHIHARA, T., KOHNO, S.I., MINAMIHISAMATSU, M. KUMATORI, T.: Banding analysis of Ph¹ chromosome translocation: A hypothetic Ph¹ region relating to the development of CML. *Natl. Inst. Radiol. Sci. Annu. Rep.* P. 50, NIRS-15, 1975.
84. JERNE, N.K.: The somatic generation of immune recognition. *Eur. J. Immunol.* 1:1-9, 1971.
85. JONCAS, J., LAPOINTE, N., GERVAIS, F., LEYRITZ, M., WILLS, A.: Unusual prevalence

of antibodies to Epstein-Barr virus early antigen in ataxia-telangiectasia. *Lancet* i:1160, 1977.
86. KAISER-MCCAW, B., PEAKMAN, D., HECHT, F., ROBINSON, A.: Recurrent somatic 7;14 translocation in lymphocytes. *Am. J. Hum. Genet.* 31:99A, 1979.
87. KAPLAN, H.S.: Hodgkin's Disease. Cambridge, Mass.: Harvard University Press, 1980.
88. KAPLAN, H.: Hodgkin's disease. Biology, treatment, prognosis. *Blood* 57:813–822, 1981.
89. KERSEY, J.H., SPECTOR, B.D.: Immunodeficiency diseases. *In* Persons at High Risk of Cancer (Fraumeni, J.F., ed.), New York: Academic Press, 55–67, 1975.
90. KIESSLING, R.E., KLEIN, E., WIGZELL, H.: "Natural" killer cells in the mouse. I. Cytotoxic cells with specificity for mouse moloney leukemia cells. Specificity and distribution acording to genotype. *Eur. J. Immunol.* 5:112–117, 1975.
91. KINSELLA, A.R., RADMAN, M.: Tumor promoter induces sister chromatid exchanges: Relevance to mechanisms of carcinogenesis. *Proc. Natl. Acad. Sci. USA* 75:6149–6153, 1978.
92. KLEIN, G.: Immunological surveillance against neoplasia. *Harvey Lect.* 69:71–102, 1975.
93. KLEIN, G.: Lymphoma development in mice and humans: Diversity of initiation is followed by convergent cytogenetic evolution. *Proc. Natl. Acad. Sci. USA* 76:2442–2446, 1979.
94. KLEIN, G.: Viruses and cancer. *In* Cancer Achievements, Challenges and Prospects for the 1980s. New York: Grune & Stratton, 81–100, 1981.
95. KLEIN, G.: The role of gene dosage and genetic transpositions in carcinogenesis. *Nature* 294:313–318, 1981.
96. KLEIN, G., OHNO, S., ROSENBERG, N., WIENER, F., SPIRA, J., BALTIMORE, D.: Cytogenetic studies on Abelson-virus-induced mouse leukemias. *Int. J. Cancer* 25:805–811, 1980.
97. KLINGER, H.P.: Genetic analysis of cell malignancy: Evidence from somatic cell genetics. *Bull. Schweiz. Akad. Med. Wiss.* 34:377–388, 1978.
98. KLINGER, H.P.: Suppression of tumorigenicity in somatic cell hybrids. I. Suppression and reexpression of tumorigenicity in diploid human X D98 AH2 hybrids and independent segregation of tumorigenicity from other cell phenotypes. *Cytogenet. Cell. Genet.* 27:254–266, 1980.
99. KLINGER, H.P.: The role of chromosomes in tumorigenicity. *In* Chromosomes Today. Volume 7. London: George Allen & Unwin, 220–224, 1981.
100. KNUDSEN, A.G. JR.: Genetics and etiology of human cancer. *Adv. Hum. Genet.* 8:1–51, 1977.
101. KOPELOVICH, L.: Hereditary adenomatosis of the colon and rectum: Recent studies on the nature of cancer promotion and cancer progression *in vitro*. *In* Colorectal Cancer: Prevention, Epidemiology, and Screening (Winawer, S., Schottenfeld, D., Sherlock, P., eds.), New York: Raven Press, 97–108, 1980.
102. KOREN, H.S., AMOS, D.B., BUCKLEY, R.H.: Natural killing in immunodeficient patients. *J. Immunol.* 120:796–799, 1978.
103. KREPINSKY, A.B., HEDDLE, J.A., GERMAN, J.: Sensitivity of Bloom's syndrome lymphocytes to ethyl methanesulfonate. *Hum. Genet.* 50:151–156, 1979.
104. KRONTIRIS, T.G., COOPER, G.M.: Transforming activity of human tumor DNAs. *Proc. Natl. Acad. Sci. USA* 78:1181–1184, 1981.
105. LEHMANN, A.R., STEVENS, S.: The response of ataxia telangiectasia cells to bleo-

mycin. *Neucleic Acids Res.* 6:1953-1960, 1979.
106. LEVAN, A.: Relation of chromosome status to the origin and progression of tumors: The evidence of chromosome numbers. *In* Genetics and Cancer. Austin: University of Texas Press, 151-182, 1959.
107. LEVAN, A., BIESELE, J.J.: Role of chromosomes in carcinogenesis, as studied in serial tissue cultures of mammalian cells. *Ann. N.Y. Acad. Sci.* 71:1022-1053, 1958.
108. LEVAN, G., MITELMAN, F.: What is the significance of chromosome aberrations in malignant cells? *In* International Cell Biology 1980-1981 (Schweiger, H.G., ed.), Berlin: Springer-Verlag, 467-476, 1981.
109. LEVANTHAL, B.G., KAIZER, H.: Etiologic factors in non-Hodgkin's lymphoma. *In* Non-Hodgkin's Lymphoma in Children (Graham-Polo, J., ed.), New York: Masson Publishing, 1-11, 1980.
110. LEVITT, R., PIERRE, R.V., WHITE, W.L., SIEKERT, R.G.: Atypical lymphoid leukemia in ataxia-telangiectasia. *Blood* 52:1003-1011, 1978.
111. LIPINSKI, M., VIREHZER, J.L., TURSZ, T., GRISCELLI, C.: Natural killer and killer cell activities in patients with primary immunodeficiencies or defects in immune interferon production. *Eur. J. Immunol.* 10:246-249, 1980.
112. LOPEZ, C.: Resistance to herpes simplex virus-type 1 (HSV-1). *Curr. Top. Microbiol. Immunol.* 92:15-24, 1981.
113. LUI, V.K., RAGAB, A.H., FINDLEY, H.S., FRAUEN, B.T.: Bone marrow cultures in children with Fanconi anemia and the TAR syndrome. *J. Pediatr.* 91:952-954, 1977.
114. MALCOM, S., BARTON, P., BENTLY, D.L., FERGUSON-SMITH, M.A., MURPHY, C.S., RABBITTS, T.H.: Assignment of a V_k locus for immunoglobulin light chains to the short arm of chromosome 2 (2cen→p13) by *in situ* hybridization using a cRNA probe of $H_k101\lambda Ch4A$. *Human Gene Mapping 5 (1981): Fifth International Workshop on Human Gene Mapping. Cytogenet. Cell. Genet.* (in press).
115. MANOLOV, G., MANOLOVA, Y.: Marker band in one chromosome 14 from Burkitt lymphomas. *Nature* 237:33-34, 1972.
116. MANOLOVA, Y., MANOLOV, G., KIELER, J., LEVAN, A., KLEIN, G.: Genesis of the 14q+ marker in Burkitt's lymphoma. *Hereditas* 90:5-10, 1979.
117. McBRIDE, O.W., SWAN, D., LEDER, P., HIETER, P., HOLLIS, G.: Chromosomal location of human immunoglobulin light chain constant region genes. *Human Gene Mapping 5 (1981): Fifth International Workshop on Human Gene Mapping. Cytogenet. Cell Genet.* (in press).
118. McCAW, B.K., HECHT, F., HARNDEN, D.G., TEPLITZ, R.L.: Somatic rearrangement of chromosome 14 in human lymphocytes. *Proc. Natl. Acad. Sci. USA* 72:2071-2075, 1975.
119. McFARLIN, D.E., STROBER, W., WALDMAN, T.A.: Ataxia telangiectasia. *Medicine (Baltimore)* 51:281-314, 1972.
120. MEADOWS, A.T., NAIMAN, J.L., VALDESDAPENA, M.: Hepatoma associated with androgen therapy for aplastic anemia. *J. Pediatr.* 84:109-110, 1974.
121. MEO, T., JOHNSON, J.J., BEECHEY, C.V., ANDREWS, S.J., PETERS, J., SEARLE, A.G.: Linkage analysis of murine immunoglobulin heavy chain and serum prealbumin genes establish their location on chromosome 12 proximal to the t(5;12) 31H breakpoint in band 12F1. *Proc. Natl. Acad. Sci. USA* 77:550-553, 1980.
122. METCALF, D.: The nature and regulation of lymphopoiesis in the normal and neoplastic thymus. *In* The Thymus: Experimental and Clinical Studies (Wolstenholme, G.E.W., Porter, R., eds.), Boston: Little Brown & Company, 242-263, 1966.
123. METCALF, D.: Relation of the thymus to the formation of immunologically reactive

cells. *Cold Spring Harbor Symp. Quant. Biol.* 32:583-590, 1967.
124. MICKLEM, H.S., FORD, C.E., EVANS, E.P., GRAY, J.: Interrelationships of myeloid and lymphoid cells: Studies with chromosome-marked cells transfused into lethally irradiated mice. *Proc. R. Soc. Lond. Biol.* 165:78-102, 1966.
125. MILLER, R.W.: Cancer and congenital malformations: Another view. *In* Genetics of Human Cancer (Mulvihill, J.J., Miller, R.W., Fraumeni, J.F. Jr., eds.), New York: Raven Press, 77-81, 1977.
126. MITELMAN, F.: Cytogenetics of experimental neoplasms and non-random chromosome correlations in man. *Clin. Hematol.* 9:195-219, 1980.
127. MITELMAN, F.: Marker chromosome 14q+ in human cancer and leukemia. *Adv. Cancer Res.* 34:141-170, 1981.
128. MITELMAN, F.: Tumor etiology and chromosome pattern: Evidence from human and experimental neoplasms. *In* Genes, Chromosomes, and Neoplasia (Arrighi, F.E., Rao, P.N., Stubblefield, E., eds.), New York: Raven Press, 335-350, 1981.
129. MITELMAN, F., LEVAN, G.: Clustering of aberrations to specific chromosomes in human neoplasia. III. Incidence and geographic distribution of chromosome aberrations in 856 cases. *Hereditas* 89:207-232, 1978.
130. MITELMAN, F., LEVAN, G.: Clustering of aberrations to specific chromosomes in human neoplasms IV. A survey of 1,871 cases. *Hereditas* 95:79-139, 1981.
131. MIYOSHI, I., HAMASAKI, K., MIYAMOTO, K., NAGUSE, K., NARAHARA K., KOICHI, K., KIMURA, I., SATO, J.: Chromosome translocations in Burkitt's lymphoma. *N. Engl. J. Med.* 304:734, 1981.
132. MULVIHILL, J.: Congenital and Genetic Disease. *In* Persons at High Risk of Cancer (Fraumeni, J.F., ed.), New York: Academic Press, 3-37, 1975.
133. NEEL, B.G., HAYWARD, W.S., ROBINSON, H.L., FANG, J.M., ASTRIN, S.M.: Avian leukosis virus-induced tumors have common proviral integration sites and synthesize discrete new RNAs: Oncogenesis by promoter insertion. *Cell* 23:323-334, 1981.
134. NEHLSEN, S.L.: Prolonged administration of antithymocyte serum in mice. 1. Observations on cellular and humoral immunity. *Clin. Exp. Immunol.* 9:63-77, 1971.
135. NEWBURGER, P.E., LATT, S.A., PESANDO, J.M., GUSTASHAW, K., POWERS, M., CHAGANTI, R.S.K., O'REILLY, R.J.: Leukemia relapse in donor cells after allogeneic bone-marrow transplantation. *N. Engl. J. Med.* 304:712-714, 1981.
136. NOWELL, P.C.: The clonal evolution of tumor cell populations. *Science* 194:23-28, 1976.
137. NOWELL, P.C., HUNGERFORD, D.A.: A minute chromosome in human chronic granulocytic leukemia. *Science* 132:1497, 1960.
138. OHNO, S., BABONTIS, M., WIENER, F., SPIRA, J., KLEIN, G., POTTER, M.: Nonrandom chromosome changes involving the distal end of chromosome 15 and the Ig-gene carrying chromosomes (Nos. 12 and 6) in pristane-induced mouse plasmacytomas. *Cell* 18:1001-1107, 1979.
139. PAINTER, R.B., YOUNG, B.R.: Radiosensitivity in ataxia-telangiectasia: A new explanation. *Proc. Natl. Acad. Sci. USA* 77:7315-7317, 1980.
140. PASSARGE, E.: Bloom syndrome. This volume.
141. PATERSON, M.C., SMITH, P.J.: Ataxia telangiectasia: An inherited human disorder involving hypersensitivity to ionizing radiation and related DNA-damaging chemicals. *Annu. Rev. Genet.* 13:291-318, 1979.
142. PENN, I.: Tumors arising in organ transplant recipients. *Adv. Cancer Res.* 28:31-61, 1978.

143. PETERSON, R.D.A., KELLY, W.D., GOOD, R.A.: Ataxia telangiectasia: Its association with a defective thymus, immunological deficiency disease, and malignancy. *Lancet* i:1189–1193, 1964.
144. POTTER, A.M., SHARP, J.C., BROWN, M.J., SOKOL, R.J.: Structural rearrangements associated with the Ph^1 chromosome in chronic granulocytic leukemia. *Humangenetik* 29:223–228, 1975.
145. PREHN, R.T.: Immunological surveillance: Pro and con. *Clin. Immunobiol.* 2:191–203, 1974.
146. PREHN, R.T.: Tumor progression and homeostasis. *Adv. Cancer Res.* 23:203–236, 1975.
147. PRINDULL, G., JENTSCH, E., HANSMANN, I.: Fanconi's anaemia developing erythroleukemia. *Scand. J. Haematol.* 23:59–63, 1979.
148. PURTILO, D.T.: Immune deficiency predisposing to Epstein-Barr virus-induced lymphoproliferative disease: The X-linked lymphoproliferative syndrome as a model. *Adv. Cancer Res.* 34:279–312, 1981.
149. PURTILO, D.T., SAKAMOTO, K., SAEMUNDSEN, A.K., SULLIVAN, J.L., SYNNERHOLM, A-C., ANVERT, M., PRITCHARD, J., SLOPER, C., SIEFT, C., PINCOTT, J., PACHMAN, L., RICH, K., CRUZI, F., CORNET, J.A., COLLINS, R., BARNES, N., KNIGHT, J., SANDSTED, B., KLEIN, G.: Documentation of Epstein-Barr virus (EBV) infection in immunodeficient patients with life-threatening lymphoproliferative disease by clinical, virological, and immunopathological studies. *Cancer Res.* 41:4226–4236.
150. RADMAN, M., KINSELLA, A.R.: Chromosomal events in carcinogenic initiation and promotion: Implications for carcinogenicity testing and cancer prevention strategies. *In* Molecular and Cellular Aspects of Carcinogen Screening Tests (Montesano, R., Beutsch, H., Tomatis, L., eds.), Lyons: IARC Scientific Publications No. 27, 75–90, 1980.
151. RATANEN, J.: Radiation carcinogenesis. *J. Toxicol. Environ. Health.* 6:971–976, 1980.
152. RAY, J.H., GERMAN H.: The cytogenetics of the "chromosome-breakage syndromes." This volume, pp. 135–167.
153. RAY, J.H., GERMAN, J.: The chromosome changes in Bloom's syndrome, ataxiatelangiectasia, and Fanconi's anemia. *In* Genes, Chromosomes, and Neoplasia (Arrighi, F.E., Rao, P.N., Stubblefield, E., eds.), New York: Raven Press, 351–378, 1981.
154. REINHERZ, E.L., SCHLOSSMAN, S.F.: Regulation of immune response: Inducer and suppressor T-lymphocyte subsets in human beings. *N. Engl. J. Med.* 303:370–373, 1980.
155. RICCARDI, V., FORGASON, J.: Chromosome 8 abnormalities as components of neoplastic and hematologic disorders. *Clin. Genet.* 15:317–326, 1979.
156. ROWLEY, J.D.: Mapping human chromosomal regions related to neoplasia: Evidence from chromosomes 1 and 17. *Proc. Natl. Acad. Sci. USA* 74:5729–5733, 1977.
157. ROWLEY, J.D.: A new and consistent chromosomal abnormality in chronic myelogenous leukemia identified by quinacrine fluorescent and Giemsa staining. *Nature* 243:290–293, 1973.
158. ROWLEY, J.D.: Ph^1 positive leukemia, including chronic myelogenous leukemia. *Clin. Hematol.* 9:55–86, 1980.
159. ROWLEY, J.D.: Chromosome abnormalities in cancer. *Cancer Genet. Cytogenet.* 2:175–198, 1980.
160. ROWLEY, J.D.: Nonrandom chromosome changes in leukemia. *In* Genes, Chromosomes, and Neoplasia (Arrighi, F.E., Rao, P.N., Stubblefield, E., eds.), New

York: Raven Press, 273-296, 1981.
161. SAEMUNDSEN, A.K., BERKEL, A.I., HENLE, W., HENLE, G., ANVERET, M., SANAL, O., ERSOY, F., CAGLAR, M., KLEIN, G.: Epstein-Barr-virus carrying lymphoma in a patient with ataxia-telangiectasia. Br. Med. J. 282:425-427, 1981.
162. SANDBERG, A.A.: The Chromosomes in Human Cancer and Leukemia. New York: Elsevier, 1980.
163. SANDBERG, A.A., KOHNO, S., WAKE, N., MINOWANDA, J.: Chromosomes and causation of human cancer and leukemia. XLII. Ph¹-positive ALL: An entity within myeloproliferative disorders? Cancer Genet. Cytogenet. 2:145-174, 1980.
164. SANDBERG, A.A., WAKE, N.: Chromosomal changes in primary and metastatic tumors and in lymphoma: Their nonrandomness and significance. In Genes, Chromosomes, and Neoplasia (Arrighi, F.E., Rao, P.N., Stubblefield, E., eds.), New York: Raven Press, 297-333, 1981.
165. SARIN, P.S., ANDERSON, P.N., GALLO, R.D.: Terminal deoxynucleotidyl transferase activities in human blood leukocytes and lymphoblast cell lines: High levels of some patients with chronic myelogenous leukemia in acute phase. Blood 47:11-20, 1976.
166. SAUNDERS, E.F., FREEDMAN, M.H.: Constitutional aplastic anaemia: Defective haematopoietic stem cell growth in vitro. Br. J. Haematol. 40:277-287, 1978.
167. SAXON, A.R., STEVENS, R.H., GOLDE, D.W.: Helper and suppressor T-lymphocyte leukemia in ataxia telangiectasia. N. Engl. J. Med. 300:700-704, 1979.
168. SETLOW, R.B.: Repair deficient human disorders and cancer. Nature 271:713-717, 1978.
169. SHANDER, M., CROCE, C.M.: Genetics of human immunoglobulins: Assignment of the genes for μ, α, and γ immunoglobulin chains to human chromosome 14. Transplant. Proc. 12:417-420, 1981.
170. SPARKES, R.S., COMO, R., GOLDE, D.W.: Cytogenetic abnormalities in ataxia telangiectasia with T-cell chronic lymphocytic leukemia. Cancer Genet. Cytogenet. 1:329-336, 1980.
171. SPECTOR, B.D., PERRY, G.S. III, KERSEY, J.H.: Genetically determined immunodeficiency diseases (GDID) and malignancy: Report from the immunodeficiency-cancer registry. Clin. Immunol. Immunopathol. 11:12-29, 1978.
172. STUTMAN, O.: Hematopoietic origin of cells responding to phytohemagglutinin in mouse lymph nodes. In Fifth Leukocyte Culture Conference (Harris, J., ed.), New York: Academic Press, 671-681, 1970.
173. STUTMAN, O.: Traffic of cells and development of immunity. In Membranes and Viruses in Immunopathology (Day, S.B., Good, R.A., eds.), New York: Academic Press, 437-450, 1972.
174. STUTMAN, O.: Tumor development after 3-methylcholanthrene in immunologically deficient athymic nude mice. Science 183:534-536, 1974.
175. STUTMAN, O.: Immunodepression and malignancy. Adv. Cancer Res. 22:261-422, 1975.
176. STUTMAN, O.: Intrathymic and extrathymic T cell maturation. Immunol. Rev. 42:138-184, 1978.
177. SULLIVAN, J.L., BYRON, K.S., BREWSTER, F.E., PURTILO, D.T.: Deficient natural cell activity in X-linked lymphoproliferative syndrome. Science 210:543-545, 1980.
178. TAYLOR, A.M.R., HARNDEN, D.G., ARLETT, C.F., HARCOURT, S.A., LEHMANN, A.R., STEVENS, S., BRIDGES, B.A.: Ataxia telangiectasia: A human mutant with abnormal radiation sensitivity. Nature 258:427-429, 1975.
179. THOMAS, L.: Discussion. In Cellular and Humoral Aspects the Hypersensitive

States (Lawrence, H.S., ed.), New York: Hoeber-Harper, 529-532, 1959.
180. THOMAS, E.D., BRYANT, J.I., BUCKNER, C.D.: Leukemic transformation of engrafted human marrow cells in vivo. *Lancet* i:1310-1313, 1972.
181. WARREN, S.T., SCHULTZ, R.A., CHANG, C.C., WADE, M.H., TROSKO, J.E.: Elevated spontaneous mutation rate in Bloom syndrome fibroblasts. *Proc. Natl. Acad. Sci. USA* 78:3133-3137, 1981.
182. WEINSTEIN, I.B.: Cell regulation and cancer. *Differentiation* 13:65-66, 1979.
183. WEINSTEIN, I.B., YAMASAKI, H., WIGLER, M., LEE, L-S., FISHER, P., JEFFREY, A., GRUNBERGER, D.: Molecular and cellular events associated with the action of initiating carcinogens and tumor promoters. *In* Carcinogens: Identification and Mechanisms of Action (Griffin, A.C., Shaw, C.R., eds.), New York: Raven Press, 399-418, 1979.
184 WELCH, J.P., LEE, C.L.Y.: Non-random occurrence of 7-14 translocations in human lymphocyte cultures. *Nature* 255:241-242, 1975.
185. WIENER, F.: Studies on the Chromosomal Control of Malignant Behavior by Cell Hybridization. Stockholm: Karolinska Institutet, 1974.
186. WIENER, F., OHNO, S., SPIRA, J.: Cytogenetic mapping of the trisomic segment of chromosome 15 in murine T-cell leukemia. *Nature* 275:658-660, 1978.
187. WIENER, F., BABONTIS, M., SPIRA, J., KLEIN, G., POTTER, M.: Cytogenetic studies of IgA/lambda-producing murine plasmacytomas: Regular occurrence of a t(12;15) translocation. *Somatic Cell. Genet.* 6:731-738, 1980.
188. WIENER, F., OHNO, S., SPIRA, J., HARAN-GHERA, N., KLEIN, G.: Chromosomal changes (trisomy 15 and 17) associated with tumor progression in leukemias induced by radiation leukemia virus (RadLV). *J. Natl. Cancer Inst.* 60:227-237, 1978.
189. WIENER, F., BABONTIS, M., SPIRA, J., BREGULA, U., KLEIN, G., MERWINS, R.M., ASOFSKY, R., LYNES, M, HAUGHTON, G.: Chromosome 15 trisomy in spontaneous and carcinogen-induced murine lymphomas of B-cell origin. *Int. J. Cancer.* 27:51-58, 1981.
190. WINGE, Ö.: Zytologische Untersuchungen über die Natur maligner Tumoren. II. Teerkarzinome bei Mauzen. *Z. Zellforsch.* 10:683-735, 1930.
191. YUNIS, J.J., BLOOMFIELD, C.D., ENSRUD, K.: All patients with acute nonlymphocytic leukemia may have a chromosomal defect. *N. Engl. J. Med.* 305:135-139, 1981.
192. ZECH, L., HAGLUND, U.: A recurrent structural aberration, t(7;14), in phytohemagglutinin-stimulated lymphocytes. *Hereditas* 89:69-73, 1978.
193. ZINKERNAGEL, R.M.: Thymus and lymphohemopoietic cells: Their role in T cell maturation in selection of T cells' H-2-restriction-specificity and in H-2 linked Ir gene control. *Immunol. Rev.* 42:224-270, 1978.

ONCOGENES, CHROMOSOME MUTATION, AND THE DEVELOPMENT OF NEOPLASIA

R.S.K. CHAGANTI AND SURESH C. JHANWAR

INTRODUCTION

The possible significance of acquired and inherited chromosome change in neoplastic development has been discussed at length in the preceding chapter in this volume [6]. A number of highly significant data pertaining to transformation-inducing retroviral and cellular genes — the so-called oncogenes — have been published recently. Localization of the positions of some of these genes on human chromosomes provided support for some of the previously proposed hypotheses regarding the role of chromosome change in the etiology of cancer [6,29]. We shall review these developments briefly in this chapter.

ONCOGENES

Knowledge about the nature of oncogenes has accelerated in a dramatic fashion during the past few years as a result of two types of investigations: (1) the analysis of transformation caused by RNA tumor viruses (retroviruses) and (2) the analysis of transformation induced in the mouse NIH3T3 fibroblast cells by transfection with DNA sequences derived from normal as well as tumor cells.

Viral Oncogenes

Retroviruses that cause rapid transformation in vertebrate cells typically contain within their genomes two types of function-related genes — those that determine the oncogenic property and those that regulate virus replication. The viral genome is considered to be a recombination product because it is made up partly of a large segment that is replication competent and viral in origin and a small segment that is homologous to a gene of the host species. The cell-related gene in the virus is responsible for its transforming property and is termed the v-*onc* gene to distinguish it from its cellular homolog; the latter in turn is termed the c-*onc* gene [2, 3,9,56,58]. A number of v-*onc* genes now have been identified from rapidly transforming viruses [25] (Table I). The c-*onc* genes as a class are DNA sequences which have been highly conserved throughout vertebrate evolution. Their function is unknown, although they have been suggested to be part of a cell's essential genetic machinery responsible for the control of proliferation, development, and differentiation [9,58]; they have also been called proto-oncogenes [3]. Various lines of evidence suggest that proto-oncogenes are capable of inducing transformation under appropriate conditions. For instance, clones of the normal cellular *mos* and *ras* genes, when activated by ligation to transcriptional promoter sequences of viral origin, induce neoplastic transformation efficiently in NIH3T3 cells [4,17,40]. One hypothesis of the origin of retroviruses states that they arose from eukaryotic cells as movable genetic elements [55].

TABLE I. Certain Characteristics of Known Oncogenes of Rapidly Transforming Retroviruses [3,9]

v-*onc* Gene	Virus	Host species	Tumor type in host species	c-*onc* Gene detected in vertebrate genomes
abl	Abelson leukemia virus	Mouse	Leukemia	Yes
bas	BALB-murine sarcoma virus	Mouse	Sarcoma	Yes
erb	Avian erythroblastosis virus	Chicken	Leukemia, sarcoma	Yes
fes	Snyder-Theilin feline sarcoma virus	Cat	Sarcoma	Yes
fms	McDonough feline sarcoma virus	Cat	Sarcoma	Yes
fps	Fujinami sarcoma virus	Chicken	Sarcoma	Yes
mos	Moloney sarcoma virus	Mouse	Sarcoma	Yes
myb	Avian myeloblastosis virus	Chicken	Leukemia	Yes
myc	Myelocytomatosis virus, strain MC29	Chicken	Carcinoma, sarcoma, leukemia	Yes
ras	Harvey sarcoma virus	Rat, mouse	Sarcoma, leukemia	Yes
rel	Reticuloendotheliosis virus, strain T	Turkey	Lymphoma	Yes
ros	UR2 sarcoma virus	Chicken	Sarcoma	Yes
sis	Simian sarcoma virus	Woolly monkey	Sarcoma	Yes
src	Rous sarcoma virus	Chicken	Sarcoma	Yes
yes	Y73 sarcoma virus	Chicken	Sarcoma	Yes

The best-studied viral oncogene, src, the transforming gene of Rous sarcoma virus (RSV), serves as the model for v-onc genes of rapidly transforming retroviruses. v-src, which has a homologous c-src gene in the host species (chicken), encodes a 60K phosphoprotein (pp 60^{src}), a tyrosine-specific protein kinase [21]. In chicken cells transformed by RSV, viral pp 60^{src} levels are found to be about 100-fold higher than that of cellular pp 60^{src} [9]. Thus, the basis for retroviral transformation is most probably the overproduction of v-onc gene product. Because the v-onc gene is homologous to the c-onc gene, transformation can be said to result from abnormal expression of a normal cellular gene, the so-called dosage effect.

That activation of a cellular oncogene also leads to transformation has been shown in B-cell lymphomas in the chicken induced by the slowly transforming avian leukosis virus (ALV). In this case the cellular oncogene myc, the counterpart of the transforming gene of the rapidly transforming retrovirus MC29 (v-myc), is activated when the ALV long terminal repeat (LTR) sequences containing the viral promoter integrate upstream from it. Under the influence of the inserted promoter, a 30–100-fold increase in c-myc activity has been detected in transformed cells [26]. ALV, unlike the rapidly transforming viruses, lacks the oncogene, and neoplastic development, being dependent upon the random integration of the promoter at the appropriate site, requires a latency period [26]. Although a number of transforming genes derived from retroviruses are recognized now, the mechanisms of activation and the products of all these genes have not been studied in detail; the activation mechanisms of src and myc genes described above serve as good models.

Oncogenes From Transformed Cells

Transfection of DNA isolated from normal as well as neoplastic human and other mammalian cells into NIH3T3 cells has yielded an array of results which are of significance to the understanding of cellular oncogenes. High molecular weight DNA (>30 kilobases) from normal cells derived from human and other vertebrate species fails to induce neoplastic transformation of such cells [12]. However, the same DNA, when fragmented into smaller pieces (0.5–5 kilobases), induces transformation, albeit at low frequencies [12]. The DNA isolated from these transformed cells, on the other hand, causes transformation with high efficiency when used in secondary transfection experiments [12]. The interpretation of these results was that transformed cells contain activated oncogenes which could be transmitted by transfection, the activation itself being brought about by intragenomic rearrangements involving the gene [9, 12]. These results, then, would predict that spontaneous and induced tumors contain activated oncogenes which can be transmitted in transfection assays in a dominant fashion. Indeed, such has been found to be the case with DNAs derived from a number of tumor cell types that originated in diverse species, including the human [57,58].

Analysis of the sensitivity of transforming activity of tumor-derived oncogenes to digestion with restriction endonucleases showed that (1) different oncogenes are activated in different tumors and (2) independent tumors belonging to the same cell type exhibit the same activated oncogene irrespective of the mode of induction of the tumor [9]. This apparent specificity of relationship between oncogene and tumor cell type strongly suggests that the oncogenes identified with transformation indeed may be the very genes involved normally in the control of proliferation and differentiation of the particular target-cell types.

The question as to whether any of these transforming genes derived from tumor cells are related to the v-*onc* or c-*onc* genes discussed earlier has been investigated by a number of laboratories. Molecular hybridization of clones of a number of v-*onc* genes to restriction endonuclease-cleaved DNAs extracted from NIH3T3 cells (which have previously been transformed by human tumor DNA) showed that the c-*onc* genes from bladder and lung carcinomas were homologous to the *ras* genes, respectively, of Harvey (ras^H) and Kirsten (ras^K) sarcoma virus [8,20,42,51]. Initial molecular analysis of these activated *ras* genes isolated from human tumors indicated that they were structurally similar to their homologs from normal cells, suggesting that transformation in these cases involved activation of normal c-*onc* genes [20,23,46]. Very recently, however, the transforming *ras* gene isolated from two bladder carcinoma cell lines was shown to differ from its normal counterpart by a point mutation which comprised replacement of a guanosine by thymidine [41,47].

Oncogene Mapping

The final approach to the study of oncogenes to be discussed here relates to mapping their positions on the chromosomes of both normal and neoplastic cells of human and other species. The relevance of this to chromosome changes encountered in tumor cells is obvious [6,29]. During the past year several investigators have pursued these studies employing two methods: (1) somatic cell hybridization and (2) in situ molecular hybridization. In the first method, human cells are fused with rodent cells carrying appropriate genetic markers that permit selection of hybrid cells; in ensuing clones of such hybrid cells correlation of retention of specific chromosomes with the presence of c-*onc* gene sequences detected by molecular hybridization constitutes evidence for synteny between the two. In the second method, the oncogene DNA is amplified in an appropriate vector, radiolabeled with ^3H or ^{125}I, and hybridized in situ to chro-

TABLE II. ASSIGNMENT OF ONCOGENES TO HUMAN AND MURINE CHROMOSOMES

Oncogene	Assignment to chromosome*			
	Human		Murine	
	S	L	S	L
c-abl	9 [18,27]			
c-fes	15 [11,27]			
c-mos	8 [45]	8q22 [37]	4 [54]	
c-myb	6 [16]			
c-myc	8 [15]	8q24 [37,54]		15D3 [35,54]
c-sis	22 [14,53]			

*S, synteny by somatic cell hybridization; L, localization by in situ hybridization or other molecular method; 8q24 refers to long arm, region 2, band 4 of human chromosome No. 8 [43]; 15D3 refers to region D, band 3 of murine chromosome No. 15 [38].

mosome preparation on a slide. The position of the gene is ascertained from analysis of autoradiographs of hybridized preparations. To date, synteny with specific human chromosomes of somatic cells has been established for six c-onc genes—abl, fes, mos, myb, myc, sis [14–16,18,27,45, 52,54]—while localization to a band of a somatic chromosome as well as to a chromomere of a germ-line (pachytene) chromosome has been established for two—mos and myc [37]. In the mouse, the localization of myc to a specific chromosome has also been established [35]. These data are summarized in Table II.

CHROMOSOME CHANGE REVISITED

The cytogenetic model for lymphomagenesis presented in the preceding chapter [6] was based on a correlation between specificity of breakpoints in translocation seen in Burkitt's lymphomas (BL) in humans and plasmacytoma in mice on the one hand, and on the other, the genes located at the breakpoints. In the human tumors, these translocations engage chromosome No. 8 with chromosome No. 14 in the usual form, and with chromosome No. 2 or No. 22 in the variant forms [6,7,29, 32]. In the murine tumors, chromosome No. 15 is involved in translocation with chromosomes No. 6 or No. 12 [39]. In the human, determinants for immunoglobulin heavy chains, $\varkappa$ light chains, and λ light chains have been localized, respectively, to 14q32, 2p12→cen, and 22q11 [28,34,36], and these are also the breakpoints in translocations that affect chromosomes No. 14 and 2 are in bands 14q32 and 2p12 while the break in No. 22 is in band 22q11; the breakpoint in the No. 8 chromosome is in band 8q24 in all tumors [29,32]. The expression of immunoglobulin light chains by tumor cells has been shown to be related to the translocations. Thus, tumors with t(8;22) express λ chains while those with t(2;8) express $\varkappa$

light chains [32]. Tumors with translocation affecting Nos. 8 and 14 produce both ϰ and λ chains [32]. In the mouse, determinants for immunoglobulin heavy chains have been localized to chromosome No. 12 in the region proximal to band F2. Chromosome No. 6 is known to carry the ϰ light chain determinants although their exact location on this chromosome is unknown [29]. The breakpoint in mouse chromosome No. 12 is in band 12F2, while the breakpoint in chromosome No. 15 in all tumors with translocations is in band D3. In parallel with human tumors, murine tumors with t(12;15) produce both ϰ and λ light chains, while those with t(6;15) produce only ϰ light chains [29]. The hypothesis that emerged from these data states that the human No. 8 and the murine No. 15 chromosomes, which do not contain target-cell-function-related genes, contain genes that control target-cell development and proliferation (i.e., oncogenes), and the transposition of the oncogene into transcriptional control of an active gene, by translocation, may lead to the activation of the oncogene and neoplastic proliferation of the target cells [29]. The similarity of this model to the "promoter-insertion" mechanism of ALV-induced lymphoid leukemogenesis in chickens [26], as discussed earlier, is obvious.

As already stated, c-*myc* and c-*mos* have been mapped to chromosome No. 8 [15,37,54] (Figs. 1, 2). The former, at band 8q24, identified the oncogene of BL, because the *myc* gene has been found to be translocated to the 14q32 region in the t(8;14) [15 and our unpublished results]. As would be predicted from the data discussed above, the *myc* gene has been found to be located at band D3 of the murine No. 15 chromosome, and it is also translocated to the No. 12 chromosome in the t(12;15) in the murine plasmacytomas [35].

How general is this model to lymphomagenesis and other forms of carcinogenesis? A review of the cytogenetics of lymphoma permits certain conclusions. The breakpoints on the No. 8 chromosome in one case of BL with a variant translocation involving chromosomes No. 1 and 8 and in one case of non-Hodgkin's lymphoma have been reported to be in band 8q21 rather than 8q24 [19,56]; as stated, c-*mos* maps close to this region, at 8q22 [37]. In lymphoid neoplasia in general, simple or complex translocations between the No. 14 chromosome (breakpoint most often in 14q32) and one of at least 12 so-called donor chromosomes (Nos. 1, 2, 4, 6, 8, 10, 11, 13, 14, 15, 18, and 20) generate the 14q+ marker chromosome [50]. Some specificity also is recognized in these translocations, with t(8;14) and t(14;18) occurring in diffuse histiocytic and nodular lymphomas, respectively [60]. A parallel situation occurs in translocations associated with Philadelphia (Ph¹) chromosome-positive chronic myelogenous leukemia (CML). In the majority (>90%) of patients with this disorder, the translocation is between chromosomes 9 and 22 with break-

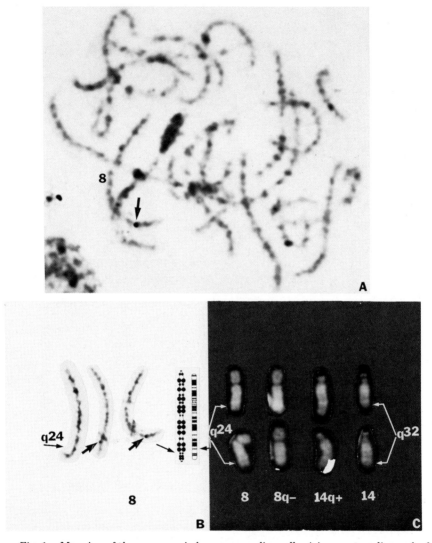

Fig. 1. Mapping of the *myc* gene in human germ-line cells. A is an autoradiograph of a pachytene cell from a normal testis showing a single grain (arrow) at the chromomere corresponding to band 8q24 [see *37* for details of methods]. In B, three No. 8 pachytene bivalents are illustrated. The one on the left is from an unhybridized preparation, while the remaining two are from autoradiographs of hybridized preparations; grains at band 8q24 indicated by thick arrows. The pachytene idiogram and the high-resolution band map of somatic metaphase chromosome No. 8 are included for comparison. In C, partial karyotypes of two cells with the t(8;14) from a non-Hodgkin's lymphoma are presented with breakpoints indicated (arrows in the nontranslocated Nos. 8 and 14). (Studies were performed in collaboration with Drs. W.S. Hayward and B.G. Neel.)

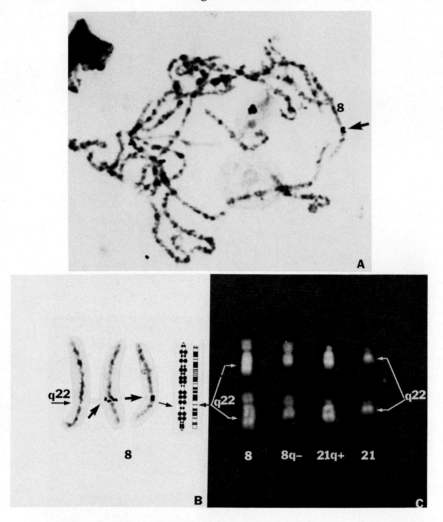

Fig. 2. Mapping of the *mos* gene in human germ-line cells. A is an autoradiograph of a pachytene cell showing two grains (arrow) at the chromomere corresponding to band 8q22 [see *37* for details of methods]. In B, three No. 8 pachytene bivalents are illustrated. The one on the left is from an unhybridized preparation, while the remaining two are from autoradiographs of hybridized preparations; grains at band 8q22 indicated by thick arrows. The pachytene idiogram and the high-resolution band map of somatic metaphase band map No. 8 chromosome are included for comparison. In C, partial karyotypes of two cells with the t(8;21) from an acute nonlymphocytic leukemia are presented, with breakpoints indicated by arrows in the nontranslocated Nos. 8 and 21. (Studies were performed in collaboration with Drs. W.S. Hayward and B.G. Neel.)

points in bands 9q32 and 22q11, respectively [6]. The oncogene c-*abl*, normally situated on the No. 9 chromosome, has been shown to be translocated to the Ph[1] chromosome in this translocation [18]; it may become activated there when brought under the influence of the immunoglobulin λ light chain gene promoters. In the so-called variant translocations that occur in a minority (< 10%) of CML patients [6], the No. 22 chromosome participates in simple or complex translocations with chromosomal regions other than 9q32 [6]. In some of these exceptional cases, cellular oncogenes other than c-*abl* probably become activated by the translocation. For instance, in the case of the Ph[1] chromosome generated by the t(14;22)(q32;q11) described in Figure 4 of the preceding chapter [6] the rearrangement may have brought c-*sis* under the transcriptional control of immunoglobulin heavy chain genes; c-*sis* has been localized to 22q11→ter [14]. These data suggest that any oncogene may be activated by translocation-mediated transposition to the transcriptional control of an active immunoglobulin or other target-cell-function-related gene, and lead to neoplastic proliferation, the corollary being that some or all of the segments from the donor chromosomes contain oncogenes at breakpoints. Simple or complex translocation that has generated marker chromosomes is a commonly found abnormality in tumor cells. Some translocations specifically associated with tumor cell types are presented in Table III. Identification of oncogenes situated at the breakpoints in

TABLE III. SPECIFIC TRANSLOCATIONS ASSOCIATED WITH HUMAN NEOPLASMS*

Translocation	Neoplasm
t(9;22)(q34;q11)	CML, AML, ALL
t(8;21)(q22;q22)	M2-ANLL
t(15;17)(q22;q21)	APL
t(9;11)(p21;q23)	AMoL
t(4;11)(q21;q23)	ALL
t(8;14)(q24;q32)	BL
t(2;8)(p12;q24)	BL-variant
t(8;22)(q24;q11)	BL-variant
t(6;14)(q21;q24)	SCCO
t(3;8)(p25;q21)	SGT

*Data from references [49,50]. p and q refer to short and long arms of chromosomes, respectively. Translocations described by the Paris nomenclature [43]. CML, chronic myelogenous leukemia; AML, acute myelogenous leukemia; M2-ANLL, acute non-lymphatic leukemia of the M2 subset(1); APL, acute promyelocytic leukemia; AMoL, acute monoblastic leukemia; ALL, acute lymphoblastic leukemia; BL, Burkitt's lymphoma; SCCO, serous cystadenocarcinoma of ovary; SGT, salivary gland tumor.

TABLE IV. NONRANDOM GAINS AND LOSSES OF CHROMOSOMES
ASSOCIATED WITH HUMAN NEOPLASMS*

Aberration	Neoplasm†
Monosomy 5	RAEB, ANLL
Monosomy 7	RAEB, ANLL
Monosomy 22	MN
Trisomy 8	ANLL, BP of CML, AC
Trisomy 21	ALL

*Data from references [48-50].
†RAEB, refractory anemia with excess blasts; ANLL, acute non-lymphoblastic leukemia; MN, meningioma; BP, blastic phase; CML, chronic myelogenous leukemia; AC, colonic adenocarcinoma; ALL, acute lymphoblastic leukemia.

translocations can be expected to be instructive in the genetic analysis of neoplastic proliferation.

As discussed in the preceding chapter, translocation is one of two major types of specific chromosome mutations that are seen in tumor cells, the other being aneuploidy [6] (Table IV). The No. 8 chromosome is among those that exhibit nonrandom trisomy in tumor cells. In the case of aneuploidy the mode of action (or activation) of the oncogene must clearly be different from that brought about by translocation, discussed above. The simplest hypothesis here would be one of gene dosage—that even the slight elevation of the oncogene product is due to trisomy being sufficient to disrupt some regulatory function related to development or proliferation of target cells controlled by the gene [29]. Whether this is so in human tumors is unknown, although it is a matter that can be verified by studying the levels of *myc* and *mos* gene expression in tumors in which the No. 8 chromosome is trisomic (Table IV). Cytogenetic analyses of somatic cell hybrids between AKR mouse-derived T-cell lymphoma cells, which are trisomic for the No. 15 chromosome, and normal mouse fibroblasts suggest that the tumor-derived No. 15 chromosome is qualitatively different from its normal homolog [29]; in segregants that were highly tumorigenic, the tumor-derived No. 15 chromosome was found to be overrepresented (five to six copies vs. the expected three) and the normal parent-derived No. 15 chromosome underrepresented (one copy vs. the expected two). In contrast, in segregants that had low tumorigenic ability, two each of tumor-derived and normal-derived No. 15 chromosomes were present [29]. The hypothesis from this observation is that the tumor-derived No. 15 chromosome earlier has undergone a genetic change, such as mutation or permanent activation of the oncogene that controls lymphocyte growth and development; duplication of the already altered chromosome then permits escape from suppressive influence of a *trans*-acting regulatory effect of the normal homolog, leading to an increased level of oncogene product and of neoplastic growth [29].

A GENETIC VIEW OF CARCINOGENESIS

Most of the human cancer burden has been attributed to environmental causes [30]. In experimental carcinogenesis, neoplastic transformation has been shown to be a gradual process, initiation and promotion conventionally being recognized as two major steps. Initiation apparently is a mutational change, occurring at the DNA rather than the chromosomal level [59], although the generality of mutational initiation has been questioned [5]. Promotional events, which may be of many kinds including genetic kinds, are considered to convert the initiated cell into a neoplastic one. Abundant cytogenetic evidence shows that neoplastic cells are genetically unstable, and that the consequent variation sets in motion the origin of and selection for stem lines [6].

The activation of cellular oncogenes can be predicted to turn out to be the common denominator in many, though not all, forms of carcinogenesis in vertebrates. However, it is not known whether this activation involves single or multiple genes before a target cell is converted to a tumor cell. Also unknown is whether there are any hierarchical or cascading relationships between oncogene activations in the development of a given tumor. In the classical ALV-induced lymphoid leukemia in the chicken, although activation of the *myc* gene has been identified, evidence for the involvement of a second transforming gene has been presented [10,11]. Probably, in the majority of human tumors, more than one gene will be found to be involved. A recent study presented evidence for differentiation-stage-specific transforming genes in human and murine lymphoid neoplasia [31]. The mechanisms for oncogene activation themselves must be of more than one kind; among genetic kinds, one which may involve a point mutation and two which involve chromosome change have been discussed here.

An important issue in this connection is raised by the questions: At what stage in neoplastic development does chromosome change occur and what, if any, are the antecedent and consequent cellular events? Formally, chromosome-change-mediated oncogene activation may be viewed as an initiating event that occurs in normal target cells. The events antecedent to the necessary chromosome change would be spontaneous (but rare) chromosome breaks, randomly distributed throughout the genome, or chromosome breaks that occur frequently in the cells of certain rare inherited genetic disorders [22], or, chromosome breaks induced by some clastogenic carcinogen. The considerable amount of evidence conflicting with this view has been discussed in the preceding chapter [6].

Alternatively, the chromosome abnormalities may arise in carcinogen-initiated cells as part of their own genetic variation and become fixed by virtue of selective properties stemming from the chromosome mutation which then would be promotional in nature. The Epstein-Barr virus (EBV)-related BL, as discussed previously [6], is a good example of this model. The mechanism of transformation induced by this virus is virtually unknown, although a virus-specific 48K protein subunit complexed with a cellular 53K protein subunit has been demonstrated in Epstein-Barr nuclear antigen (EBNA)-positive B cells transformed by EBV; the 53K cellular protein itself has been found to be present in a wide variety of transformed cells as well as in spontaneous and induced tumors [13,29, 33]. The initiating event in this case, therefore, may be the activation of genetic machinery for the production of the cellular 53K protein subunit and its complexing with the viral 48K subunit, without involving apparent chromosome change. A cell so initiated may be termed preneoplastic [29]. Genetic instability so characteristic of neoplastic cells probably begins at this stage, and when translocations involving the right chromosomes occur, *myc* or other oncogenes become activated, converting the preneoplastic cell to a neoplastic cell. Persons with either primary or acquired immunodeficiency disorders are prone to infection with EBV, cytomegalovirus, and other opportunistic herpes-type DNA viruses, and BL, along with other forms of neoplasia, frequently occurs in them. These neoplastic proliferations have been shown to commence often as polyclonal expansions of lymphoid cells in which typical monoclonal tumors carrying the appropriate chromosome markers may arise [44].

SUMMARY

The place of chromosome change in neoplastic transformation has been examined in light of recent research on retroviral oncogenes and their cellular counterparts. These phylogenetically ubiquitous and highly conserved DNA sequences probably determine normal vertebrate cellular functions such as proliferation, differentiation, and development. Their abnormal activation and expression leads to neoplastic transformation of normal cells. Genetic as well as nongenetic mechanisms of oncogene activation probably exist, and the former may range from gene to chromosome mutation. Because tumorigenesis is a gradual and evolutionary process it is suggested here that the development of a given malignancy may involve activation of multiple oncogenes.

(Note: We prepared this review with the main aim of making the discussion presented in the preceding chapter current [6], realizing full well that the almost explosive rate of accumulation of new information in this area may render some of the data outdated by the time this book is published.).

Acknowledgments. We are grateful to Drs. Dirk Bootsma, Carlo Croce, and Philip Leder for making available to us preprints of their manuscripts in press [14,18,35,54] and for permitting us to cite their work in this review.

LITERATURE CITED

1. BENNETT J.M., CATOVSKY D., DANIEL M.-T., FLANDRIN G., GALTON D.A.G., GARLNICK R., SULTAN C.: Proposals for the classification of the acute leukemias. Br. J. Hematol. 33:451-458, 1976.
2. BISHOP J.M.: Enemies within: The genesis of retrovirus oncogenes. Cell 23:5-6, 1981.
3. BISHOP J.M.: Oncogenes. Sci. Am. 246:81-93, 1982.
4. BLAIR D.G., OSKARSSON M., WOOD T.G., McCLEMENTS W.L., FISCHINGER P.J., VANDE WOUDE G.: Activation of the transforming potential of a normal cell sequence: A molecular model for oncogenesis. Science 212:941-943, 1981.
5. CAIRNS J.: The origin of human cancers. Nature 289:353-357, 1981.
6. CHAGANTI R.S.K.: This volume.
7. CHAGANTI R.S.K., JHANWAR S.C., ARLIN Z., KOZINER B., AMBATI A., ANDREEFF M., CLARKSON B.D.: A non-T, non-B acute lymphoblastic leukemia (L3) with t(8;22) and two 14q+ chromosomes. Cancer Genet. Cytogenet. (in press).
8. CHANG E.H., FURTH M.E., SCOLNICK E.M., LOWY D.R.: Tumorigenic transformation of mammalian cells induced by a normal human gene homologous to the oncogene of Harvey murine sarcoma virus. Nature 297:479-483, 1982.
9. COOPER G.M.: Cellular transforming genes. Science 218:801-806, 1982.
10. COOPER G.M., NEIMAN P.E.: Transforming genes of neoplasms induced by avian lymphoid leukosis viruses. Nature 287:656-659, 1980.
11. COOPER G.M., NEIMAN P.E.: Two distinct candidate transforming genes of lymphoid leukosis virus-induced neoplasms. Nature 292:857-859, 1981.
12. COOPER G.M., OKENQUIST S., SILVERMAN L.: Transforming activity of DNA of chemically transformed and normal cells. Nature 284:418-421, 1980.
13. CRAWFORD L.: The origins of p53 in relation to transformation. Adv. Viral Oncol. 2: 3-31, 1982.
14. DALLA-FAVERA R., GALLO R.C., GIALLONGO A., CROCE C.M.: Chromosomal localization of the human homolog (c-sis) of the simian sarcoma virus onc gene. Science 218: 686-687, 1982.
15. DALLA-FAVERA R., BERGNI M., ERIKSON J., PATTERSON D., GALLO R.C., CROCE C.M.: Human c-myc onc gene is located on the region of chromosome 8 that is translocated in Burkitt lymphoma cells. Proc. Natl. Acad. Sci. USA 79:7824-7827, 1982.
16. DALLA-FAVERA R., FRANCHINI G., MARTINOTTI S., WONG-STAAL F., GALLO R.C., CROCE C.M.: Chromosomal assignment of the human homologues of feline sarcoma virus and avian myeloblastosis virus onc genes. Proc. Natl. Acad. Sci. USA 79: 7414-7417, 1982.
17. DE FEO D., GONDA M.A., YOUNG H.A., CHANG E.H., LOWY D.R., SCOLNICK E.M.,

ELLIS R.W.: Analysis of divergent rat genomic clones homologous to the transforming gene of Harvey murine sarcoma virus. *Proc. Natl. Acad. Sci. USA* 78:3328-3332, 1981.
18. DE KLEIN A., VAN KESSEL A.G., GROSVELD G., BARTRAM C.R., HAGEMEIJER A., BOOTSMA D., SPURR N.K., HEISTERKAMP N., GROFFEN J., STEPHENSON J.R.: A cellular oncogene (c-*abl*) is translocated to the Philadelphia chromosome in chronic myelocytic leukemia. *Nature* 300:765-767, 1982.
19. DEMEOCQ F., BERNARD A., BOUMSELL L., BEZOU M.J., TURCHINI M.F., MALPUECH G.: Leucemie aigue lymphoblastique de type Burkitt. *Arch. Fr. Pediatr.* 38:355-357, 1981.
20. DER C.J., KRONTIRIS T.G., COOPER G.M.: Transforming genes of human bladder and lung carcinoma cell lines are homologous to the *ras* genes of Harvey and Kirsten sarcoma viruses. *Proc. Natl. Acad. Sci. USA* 79:3637-3640, 1982.
21. ERIKSON R.L.: Avian sarcoma viruses: Molecular biology. *In* Klein G. (ed.): "Viral Oncology." New York: Raven Press, pp. 39-53, 1980.
22. GERMAN J.: Genes which increase chromosome instability in somatic cells and predispose to cancer. *Prog. Med. Genet.* 8:61-101, 1972.
23. GOLDFARB M., SHIMIZU K., PERUCHO M., WIGLER M.: Isolation and preliminary characterization of a human transforming gene from T24 bladder carcinoma cells. *Nature* 296:404-409, 1982.
24. HAGEMEIJER A., HAHLEN K., SIZOO W., ABELS J.: Translocation (9;11)(p21;q23) in three cases of acute monoblastic leukemia. *Cancer Genet. Cytogenet.* 5:95-105, 1982.
25. HAYWARD W.S., NEEL B.G.: Retroviral gene expression. *Curr. Top. Microbiol. Immunol.* 91:218-276, 1981.
26. HAYWARD W.S., NEEL B.G., ASTRIN S.M.: Activation of a cellular *onc* gene by promoter insertion in ALV-induced lymphoid leukosis. *Nature* 290:475-480, 1981.
27. HEISTERKAMP N., GROFFEN J., STEPHENSON J.R., SPURR N.K., GOODFELLOW P.M., SOLOMON E., CARRITT B., BODMER W.F.: Chromosomal localization of human cellular homologues of two viral oncogenes. *Nature* 299:747-749, 1982.
28. KIRSCH I.R., MORTON C.C., NAKAHARA K., LEDER, P.: Human immunoglobulin heavy chain genes map to a region of translocation in malignant B lymphocytes. *Science* 216:301-303, 1982.
29. KLEIN G.: The role of gene dosage and genetic transpositions in carcinogenesis. *Nature* 294:313-318, 1981.
30. KNUDSON A.G.: Genetics and cancer. *In* Burchenal J.H., Oettgen H.F. (eds.): "Cancer Achievements, Challenges, and Prospects for the 1980s." New York: Grune and Stratton, pp. 381-396, 1981.
31. LANE M.A., SAINTEN A., COOPER G.M.: Stage-specific transforming genes of human and mouse B- and T-lymphocyte neoplasms. *Cell* 28:873-880, 1982.
32. LENOIR G.M., PREUD'HOMME J.L., BERNHEIM A., BERGER R.: Correlation between immunoglobulin light chain expression and variant translocation in Burkitt's lymphoma. *Nature* 289:474-476, 1982.
33. LUKA J., JORNVALL H.: The DNA-binding protein p53 in Epstein-Barr virus transformed cells. Properties of this protein, other p53 forms, and the Epstein-Barr virus nuclear antigen. *Adv. Viral Oncol.* 2:155-172, 1982.
34. MALCOM S., BARTON P., MURPHY C., FERGUSON-SMITH M.A., BENTLY D.L., RABBITS T.H.: Localization of human immunoglobulin $\varkappa$ light chain variable region gene to the short arm of chromosome 2 by *in situ* hybridization. *Proc. Natl. Acad. Sci. USA* 79:4957-4961, 1982.

35. MARCU K.B., HARRIS L.J., STANTON L.W., ERIKSON J., WATT R., CROCE C.M.: Transcriptionally active c-*myc* oncogene is contained within NIARD, a DNA sequence associated with chromosome translocations in B cell neoplasia. *Proc. Natl. Acad. Sci. USA* (in press) 1983.
36. MCBRIDE O.W., SWAN D., LEDER P., HEITER P., HOLLIS G.: Chromosomal location of human immunoglobulin light chain constant region genes. Human Gene Mapping 5 (1981): Fifth International Workshop on Human Gene Mapping. *Cytogenet. Cell Genet.* 32:297-298, 1982.
37. NEEL B.G., JHANWAR S.C., CHAGANTI R.S.K., HAYWARD W.S.: Two human c-*onc* genes are located on the long arm of chromosome 8. *Proc. Natl. Acad. Sci. USA* 79: 7842-7846, 1982.
38. NESBITT M.N., FRANKE U.: A system of nomenclature for band patterns of mouse chromosomes. *Chromosoma* 41:145-158, 1973.
39. OHNO S., BABONTIS M., WEINER F., SPIRA J., KLEIN G., POTTER M.: Non-random chromosome changes involving the distal end of chromosome 15 and the Ig-gene carrying chromosomes (Nos. 12 and 6) in pristane-induced mouse plasmacytomas. *Cell* 18:1001-1007, 1979.
40. OSKARSSON M., MCCLEMENTS W.L., BLAIR D.G., MAIZEL J.V., VANDE WOUDE G.: Properties of a normal mouse cell DNA sequence (*sarc*) homologous to the *src* sequence of Moloney sarcoma virus. *Science* 207:1222-1224, 1980.
41. PAPAGEORGE A.G., SCOLNICK E.M., DHAR R., LOWY D.R., CHANG E.H.: Mechanism of activation of a human oncogene. *Nature* 300:143-149, 1982.
42. PARADA L.F., TABIN C.J., SHIH C., WEINBERG R.A.: Human EJ bladder carcinoma oncogene is homologue of Harvey sarcoma virus *ras* gene. *Nature* 297:474-478, 1982.
43. PARIS CONFERENCE (1971): Standardization in Human Cytogenetics. "Birth Defects: Original Article Series." New York: National Foundation, VIII:7, 1972.
44. PENN I.: Depressed immunity and the development of cancer. *Clin. Exp. Immunol.* 46:459-474, 1981.
45. PRAKASH K., MCBRIDE O.W., SWAN D.C., DEVARE S.G., TRONICK S.R., AARONSON S.A.: Molecular cloning and chromosomal mapping of a human locus related to the transforming gene of Moloney murine sarcoma virus. *Proc. Natl. Acad. Sci. USA* 79:5210-5214, 1982.
46. PULCIANI S., SANTOS E., LAUVER A.V., LONG L.K., ROBBINS K.C., BARBACID M.: Oncogenes in human tumor cell lines: Molecular cloning of a transforming gene from human bladder carcinoma cells. *Proc. Natl. Acad. Sci. USA* 79:2845-2849, 1982.
47. REDDY E.P., REYNOLDS R.K., SANTOS E., BARBACID M.: A point mutation is responsible for the acquisition of transforming properties by the T24 human bladder carcinoma oncogene. *Nature* 300:149-152, 1982.
48. REICHMANN A., MARTIN P., LEVIN B.: Chromosomal banding patterns in human large bowel cancer. *Int. J. Cancer* 28:431-440, 1981.
49. SANDBERG A.A.: "The Chromosomes in Human Cancer and Leukemia." New York: Elsevier, 1980.
50. SANDBERG A.A., WAKE N.: Chromosomal changes in primary and metastatic tumors and in lymphoma: Their nonrandomness and significance. *In* Arrighi F.E., Rao P.N., Stubblefield E (eds.): "Genes, Chromosomes, and Neoplasia." New York: Raven Press, pp. 297-333, 1981.
51. SANTOS E., TRONICK S.R., AARONSON S.A., PULCIANI S., BARBACID M.: T24 human bladder carcinoma oncogene is an activated form of the normal human homologue of BALB- and Harvey-MSV transforming genes. *Nature* 298:343-347, 1982.

52. SWAN D.C., MCBRIDE O.W., ROBBINS K.C., KEITHLEY D.A., REDDY E.P., AARONSON S.A.: Chromosomal mapping of the simian sarcoma virus *onc* gene analogue in human cells. *Proc. Natl. Acad. Sci. USA* 79:4691-4695, 1982.
53. SWAN D., OSKARSSON M., KEITHLEY D., RUDDLE F.H., D'EUSTACHIO P., VANDE WOUDE G.F.: Chromosomal localization of the Moloney sarcoma virus mouse cellular (c-*mos*) sequence. *J. Virol.* 44:752-754, 1982.
54. TAUB R., KIRSCH I., MORTON C., LENOIR G., SWAN D., TRONICK S., AARONSON S., LEDER P.: Translocation of the c-*myc* gene into the immunoglobulin heavy chain locus in human Burkitt's lymphoma and murine plasmacytoma cells. *Proc. Natl. Acad. Sci. USA* 79:7837-7841, 1982.
55. TEMIN H.M.: Origin of retroviruses from cellular movable genetic elements. *Cell* 21: 599-600, 1980.
56. Third International Workshop on Chromosomes in Leukemia, 1980. *Cancer Genet. Cytogenet.* 4:95-142, 1981.
57. WEINBERG R.A.: Oncogenes of spontaneous and chemically induced tumors. *Adv. Cancer Res.* 36:149-163, 1982.
58. WEINBERG R.A.: Fewer and fewer oncogenes. *Cell* 30:3-4, 1982.
59. WEINSTEIN I.B., YAMASAKI H., WIGLER M., LEE L.-S., FISHER P., JEFFREY A., GRUNBERGER D.: Molecular and cellular events associated with the action of initiating carcinogens and tumor promoters. *In* Griffin A.C., Shaw C.R. (eds.): "Carcinogens: Identification and Mechanisms of Action." New York: Raven Press, pp. 399-418, 1979.
60. YUNIS J.J., OKEN M.M., KAPLAN M.E., ENSURD K.M., HOWE R.R., THEOLOGIDES A.: Distinctive chromosomal abnormalities in histologic subtypes of non-Hodgkin's lymphoma. *N. Engl. J. Med.* 307:1231-1236, 1982.

TUMOR PROGRESSION AND CLONAL EVOLUTION: THE ROLE OF GENETIC INSTABILITY

PETER C. NOWELL

INTRODUCTION

Current discussions of the natural history of cancer often include the view that most tumors are unicellular in origin (i.e., "clones") and that the clinical and biological phenomenon of tumor progression results from "clonal evolution," the sequential appearance within the neoplastic clone of subpopulations that are more and more genetically aberrant. In this chapter, I will review these concepts and the supporting evidence as well as some current ideas concerning the mechanisms underlying clonal evolution: specifically, the view that genetic instability within the tumor, usually acquired, leads to a greater frequency of mutational events and provides the basis for generation of increasingly abnormal neoplastic sublines. Also, I will consider briefly the evidence for such genetic instability and possible explanations for its existence.

This chapter necessarily will introduce a number of topics that are considered elsewhere in this volume in more detail, including the "chromosome breakage syndromes" as models for some types of genetic lability which may be important in the somatic evolution of tumor cell populations. It is hoped that this discussion will provide a conceptual framework concerning what is and is not known about tumor development, and that this framework, although simplified, will help to place in perspective the more specific aspects of genetic instability and chromosome breakage with which this volume primarily is concerned.

THE CLONAL NATURE OF TUMORS

A large body of evidence indicates that most tumors are unicellular in origin. A simplified formulation would suggest that a carcinogen (e.g., radiation, a chemical, a virus) interacts with the genome of a single cell to confer on that cell a heritable, selective growth advantage. When that cell divides, its progeny escape, to some degree, from local growth regulation and begin clonal expansion as a neoplasm.

This working definition of neoplasia does not consider the nature of the initial event within the first altered cell, and a detailed discussion is not appropriate here. It should be recognized, however, that several fundamental questions about this primary phenomenon remain unanswered. Although the heritable change induced by the carcinogen acts like a mutation, the specific alterations in gene structure or function that occur remain undefined. (It is clear that visible chromosome change is not necessarily required, as some early tumors have a normal karyotype.) The specific gene products that are critically altered in the initial neoplastic cell are also unknown, although experimental data suggest increasingly that molecules on the cell surface, perhaps receptors for local growth regulators, are involved. A third major area of ignorance concerning the earliest stage of neoplasia, however, involves the nature of such local regulators (chalones, growth factors). Cytogenetic studies are of little direct assistance in solving these mysteries surrounding the initial events in tumorigenesis; however, the variety of nonrandom chromosomal changes that are being recognized in many tumors through banding techniques may indicate specific sites in the genome where important "cancer genes" are located [19,50].

Evidence That Tumors Are Clones

Cytogenetic, biochemical, and immunological data provide evidence of the unicellular origin of tumors. Chromosome abnormalities are often the same throughout all the cells of a given tumor. Where there is a spectrum of cytogenetic changes, these are often demonstrably related. Even in highly malignant solid neoplasms with much karyotypic variation, characteristic marker chromosomes, which are recognizable in all cells of a given tumor, indicate that the entire neoplastic population is derived from a single cell of origin [19,39].

Studies of the distribution of the enzyme glucose-6-phosphate-dehydrogenase (G6PD) in many tumors [13] and of the monoclonal characteristics of the immunoglobulin produced by a number of lymphoproliferative neoplasms also support this clonal concept. It should be stressed that the term "clone" is being used solely to indicate the unicellular origin of the tumor and not to indicate a homogeneity of the neoplastic population. Through somatic mutation, a malignancy in its later stages may come to consist of a large number of genetically different subpopulations that are derived from the original clone, and it is therefore not surprising that many tumors prove to be heterogeneous for whatever phenotypic characteristic is being examined. This clonal hypothesis does not rule out the possibility that the application of a carcinogen to an organ or an area of skin can produce many potentially neoplastic cells; it simply indicates that as a tumor evolves from such an area, the progeny of one or, at most, a very few cells ultimately overgrow the site and account for the macroscopic neoplasm.

TUMOR PROGRESSION AND CLONAL EVOLUTION

It has long been recognized that there is a tendency for many tumors to become progressively more aggressive in their behavior and more "malignant" in their characteristics, although the time course may be quite variable. The stepwise nature of this clinical and biological phenomenon was recognized by Foulds [16], and others suggested that it might reflect the sequential appearance within the tumor of increasingly genetically altered subpopulations [9,30]. It was suggested further that this "clonal evolution" [6,23,37] might result from enhanced genetic lability within the tumor cell population, and thus increase the probability of further genetic errors (often recognizable cytogenetically) and their subsequent selection.

Most of the common human cancers, by the time they are observed clinically, represent the end stage of multiple steps in tumor development. Much of the current emphasis in basic cancer research is on the very earliest events in tumorigenesis, which ultimately may prove to be small quantitative changes in gene function that are perhaps even reversible. Relatively little attention is being directed to the subsequent alterations—and their underlying causes—that eventually lead to the fully developed clinical cancer. It is here that more extensive genetic changes, including chromosome breakage and rearrangements, may be of greatest importance.

Biological and Clinical Aspects of Tumor Progression

The changes in cellular morphology and behavior that occur during tumor progression have been described extensively [16]. Of greatest clinical significance is the acquisition by the neoplastic cells of the capacity to invade locally and to metastasize; this is still the fundamental definition of malignancy. In addition, with time, there is a tendency for tumor cell populations to increase their proliferative capacity and to show further evidence of escape from local growth control mechanisms. Usually this appears to reflect not a shortening of the cell-cycle time but an increase in the "growth fraction," the proportion of cells within the neoplastic clone that continues to proliferate actively instead of progressing to terminal differentiation or cell death. Further escape from growth regulation may also be related, in some neoplasms, to altered response to circulating hormones, perhaps through loss of specific receptors.

Concurrently, as they become more malignant, it is common for tumors to show morphological and metabolic alterations that are generally interpreted as "loss of differentiation." Organelles and metabolic functions necessary for specialized activities of the cell tend to decrease or disappear. Whether this reflects an actual "block" in differentiation or again, simply an increased growth fraction, with a higher proportion of cells remaining in an "undifferentiated" proliferating state, may vary in different cancers.

In addition, certain products which appear to help the malignant cells to invade and metastasize may be elaborated, such as tumor angiogenic factor [15] and various proteolytic enzymes [28]. Decreased antigenicity and acquisition of drug resistance may also appear within the neoplastic cell population; these characteristics have obvious selective advantages, but the significance of certain other tumor cell products such as fetal antigens and inappropriate hormones is more difficult to interpret [7,43].

There may be considerable variability in the time period in which these characteristics of tumor progression become apparent. In some instances, the properties of far advanced malignancy may be established before the neoplasm reaches macroscopic size; in other instances, well-differentiated, slow-growing tumors may persist for years before undergoing a relatively abrupt shift to more aggressive behavior. It is also common, particularly with rapidly growing cancers, to find heterogeneity within the tumor regardless of which parameter is measured, so that some cells in the neoplastic population may appear much further advanced in tumor progression than others.

Clonal Evolution as the Basis for Tumor Progression

It has been suggested by a number of workers [6,9,23,30,37] that the clinical and biological events described above as "tumor progression" represent both the effects of genetic instability in the neoplastic cells and the sequential selection of variant subpopulations produced as a result of that genetic instability. A model of this concept of "clonal evolution" is illustrated in Figure 1. In this formulation, most variants that arise in the tumor cell population do not survive; but those few mutants that have a selective growth advantage expand to become predominant subpopulations within the neoplasm and demonstrate the characteristics which we recognize as tumor progression. The continued presence of multiple subpopulations within the tumor, as illustrated in Figure 1, provides the basis for the heterogeneity that is typically observed.

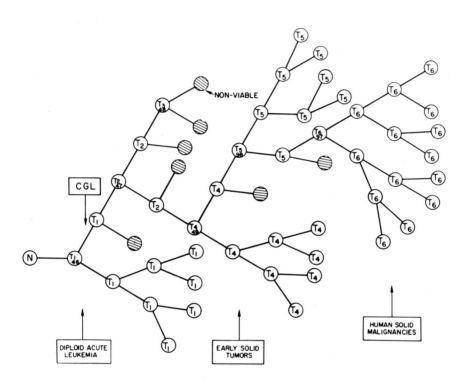

Fig. 1. Model of clonal evolution in neoplasia. Carcinogen-induced change in progenitor normal cell (N) produces a diploid tumor cell (T_1, 46 chromosomes), with growth advantage permitting clonal expansion to begin. Genetic instability of T_1 cells leads to production of variants (illustrated by changes in chromosome number, T_2 to T_6). Most variants die, due to metabolic or immunologic disadvantage (hatched circles); occasionally one has an additional selective advantage (for example, T_2, 47 chromosomes), and its progeny become the predominant subpopulation until an even more favorable variant appears (for example, T_4). The stepwise sequence in each tumor differs (being partially determined by environmental pressures on selection) and results in a different, aneuploid karyotype predominating in each fully developed malignancy (T_6). Earlier subpopulations (for example, T_1, T_4, T_5) may persist sufficiently to contribute to heterogeneity within the advanced tumor. Biological characteristics of tumor progression (for example, morphological and metabolic loss of differentiation, invasion and metastasis, resistance to therapy) parallel the stages of genetic evolution. Human tumors with minimal chromosome change (diploid acute leukemia, chronic granulocytic leukemia) are considered to be early in clonal evolution; human solid cancers, typically highly aneuploid, are viewed as late in the developmental process (from Nowell, [39]).

Evidence for the Clonal Evolution Model

Most of the data that support the clonal evolution concept have been derived from chromosome studies; this may have caused excessive emphasis on the role of cytogenetic abnormalities, as compared to submicroscopic alterations, in discussing the types of genetic change involved in this process. In general, advanced malignancies show more extensive chromosomal aberrations than do earlier stages of neoplasia [39,50]. For most of the common human cancers, however, only single studies of a given tumor have been carried out. It has generally not been possible to determine, through sequential studies, whether biological progression of the neoplasm to more "malignant" characteristics was associated with the emergence of new predominant subpopulations of tumor cells which contain additional genetic alterations that are recognizable cytogenetically.

Such serial studies have been done, however, in certain of the human hematopoietic tumors, as well as in a few experimental malignancies in animals. Best documented are the findings in chronic granulocytic leukemia (CGL), in which both the early and late stages usually are accompanied by chromosomal abnormalities. Typically, the early stage of CGL is a clinically benign disease characterized by a neoplastic clone that expands relatively slowly and consists mostly of differentiated myeloid cells. At this time, the neoplastic cells almost always contain the Philadelphia chromosome (Ph); a translocation from the long arm of chromosome 22, usually to the long arm of chromosome 9) as the only cytogenetic change. After several years, the clinical picture often changes dramatically. The well-differentiated cells are replaced by a rapidly expanding population of undifferentiated myeloblasts, which crowd the blood and bone marrow and lead to the death of the patient in so-called "blast crisis."

In this accelerated phase of CGL, a subpopulation of cells with karyotypic abnormalities in addition to the Ph chromosome often becomes predominant. These additional changes are not consistent from case to case, but frequently involve one or more of several specific alterations, such as a second Ph, iso17q, or trisomy 8 [51]. It thus is possible to speculate that in CGL the disease is initiated by a visible genetic change in a marrow stem cell (the Ph chromosome), which alters its response to local growth regulation, and that tumor progression results from a second mutation in a cell of the original clone, which is often recognizable cytogenetically, and which allows a more aggressive subpopulation to develop and overwhelm the patient. (Recent evidence suggests that, in some instances, the neoplastic clone in CGL may originate with a stem cell in which there is no visible chromosome change and that the Ph chromosome may represent the first step in further tumor progression [14].

The specific genes and gene products involved in these stages are not known; but it is possible that the second cytogenetic abnormality within the clone might further alter membrane receptors for local growth regulators so that nearly all of the cells with the additional chromosome change remain blasts and very few undergo terminal differentiation. This would lead to a rapid expansion of the stem cell pool and the clinical picture of the blast crisis.

Similar sequential patterns of the karyotypic alteration associated with clinical progression have been reported in other human leukemic and preleukemic disorders [39,50], but many cases do not follow the protracted time course that is necessary for repeated investigations. We have recently studied over a 5-yr period a patient with the rare T-cell variant of chronic lymphocytic leukemia (CLL) whose tumor cells showed cytogenetic evolution in parallel with changes in the biological and clinical characteristics of her disease. Initially, her neoplastic lymphocytes had a normal karyotype and typical T-cell characteristics (rosette formation with sheep erythrocytes and response to T-cell mitogens). Over several years, although her benign clinical disease remained unchanged and required no treatment, a pseudodiploid subpopulation, with several translocations including $3q^+$ and $14q^+$ marker chromosomes (Fig. 2A), gradually replaced the diploid leukemic cells, and there was a concurrent decrease in the previously observed T-cell characteristics. This evidence of biological progression ("loss of differentiation") was not accompanied by any change in the patient's clinical course [41].

After several more years, however, the patient abruptly developed large abdominal masses of poorly differentiated "histiocytic" lymphoma, which lacked membrane markers to suggest the cell of origin. These cells had approximately 70 chromosomes, with several of the same abnormal markers ($3q^+$, $14q^+$) previously observed in the pseudodiploid clone (Fig. 2B), clearly indicating the derivation of the lymphoma from the original T-cell leukemia [40]. In this patient the final stage of tumor progression (clinically termed Richter syndrome [40]), apparently resulted from the growth within the neoplastic clone of a new subpopulation with major additional genetic alterations and the biological characteristics of a rapidly expanding lymphoma. It is interesting that in this patient, there was cytogenetic evidence of clonal evolution that produced early biological changes in the cells without clinical significance, and finally, extensive alterations in growth characteristics that proved fatal.

Some chromosomal data have also been obtained in experimental animals that support the concept that clonal evolution underlies tumor progression. In a number of primary sarcomas induced in mice and rats by the Rous sarcoma virus, the karyotype was normal initially [31,35]. Serial

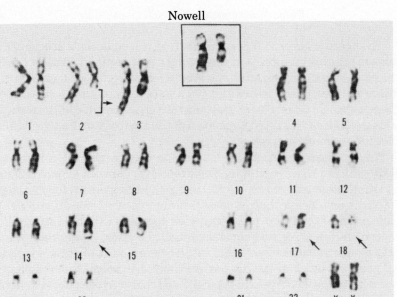
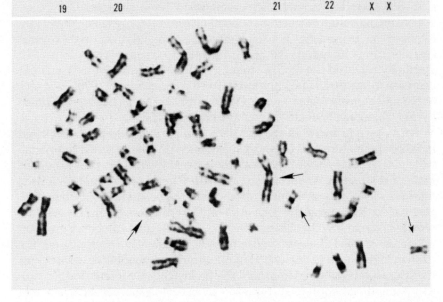

Fig. 2. A. Representative karyotype from a neoplastic lymphocyte clone in a patient with chronic T-cell leukemia. Abnormalities include a translocation from the long arm of a No. 2 chromosome to a No. 3, producing a 3q⁺ marker chromosome; translocation from an 18 to the long arm of a 14, producing a 14q⁺; and an extra band in the long arm of a No. 17. The insert illustrates normal and abnormal No. 3 chromosomes from another cell (From Nowell et al. [41]). B. Metaphase with 70 chromosomes from an abdominal "histiocytic" lymphoma which subsequently developed in the same patient. Large arrows point to the same 3q⁺ and 14q⁺ marker chromosomes as illustrated in A. They indicate that the lymphoma resulted from clonal evolution of the original T-cell neoplasm. The small arrows indicate two i17q markers not present in the leukemic phase of the disease (from Nowell et al. [40]).

biopsies of individual tumors as they gradually acquired more malignant properties, such as increased growth rate and reduced collagen production, revealed the sequential appearance of new subpopulations (identified cytogenetically), which overgrew and replaced the original diploid tumor cells. In the studies of Rous sarcomas in rats, which involved both primary lesions and transplantable tumors derived from them, banding data have indicated a relatively consistent pattern of stepwise karyotypic change: first, the addition of chromosome 7, then 13, and then 12 [26,35]. These further cytogenetic alterations apparently provided additional selective growth advantages over earlier tumor cells, both diploid and aneuploid.

Thus, through the use of cytogenetic techniques, it has been possible to identify the sequential appearance of more and more genetically aberrant subpopulations in a number of human and animal tumors. As these subpopulations parallel the existence of new and more aggressive characteristics of the neoplastic cells, the various findings cited above thus support the concept of tumor progression illustrated in Figure 1.

INCREASED GENETIC LABILITY OF TUMOR CELL POPULATIONS

One component of the hypothesis just discussed is the view that clonal evolution within tumors results from enhanced mutability of the neoplastic cells. When tumors are examined histologically, obvious mitotic abnormalities are often seen. In fact, such observations led von Hansemann [60] and Boveri [4] to the earliest theories concerning an important role of chromosomal alterations in the development of cancer. Such histologic studies of cancer also suggest that the genetic instability represented by mitotic abnormalities may become more pronounced as a neoplasm evolves. In advanced malignancies a wide range of mitotic aberrations are commonly observed with each cell generation, as compared with relatively few variants in early benign lesions [46].

Experimental data indicating that tumor cell populations are more labile genetically than comparable normal cells are being reported with increased frequency, and are considered in more detail elsewhere in this volume [36,52]. Both *in vivo* and *in vitro*, there is some evidence that neoplastic cells may be more susceptible to chromosome breakage, nondisjunction and ploidy changes, sister-chromatid exchange (SCE), and other genetic alterations than are comparable normal cells [8,47,48,55,57,61]. Nichols [36] has demonstrated that some SV40-transformed cell lines show increased mutation rates as compared to controls. There are even limited experimental data which indicate that this enhanced mutability increases with tumor progression.

Taken together, the findings suggest that there is indeed a greater frequency of mitotic errors and other genetic changes in many neoplastic cell populations. Further, the results indicate that the high level of mitotic activity in some tumors accounts only partially for this difference, and that, in addition, each cell division carries an increased risk of genetic variation. More data are needed, however, to determine how consistently this is the case, and to explore in detail the underlying mechanisms that may account for it.

Possible Mechanisms for Increased Genetic Instability in Tumor Cells

Table I summarizes a variety of mechanisms that could be responsible for the apparent increased mutability of tumor cell populations and provide the basis for clonal evolution and tumor progression. The categories listed are by no means mutually exclusive; at various stages of tumor development and in different neoplasms, different mechanisms could operate. I will discuss each category briefly and attempt to indicate how these considerations may be relevant to the more detailed coverage in other chapters.

Inherited defects. In a small segment of the population, increased genetic instability in neoplastic cells may result not from an acquired alteration but instead may reflect an inherited gene defect present in all cells of the body. The so-called chromosome breakage syndromes, such as Bloom's syndrome (BS), Fanconi's anemia (FA), ataxia-telangiectasia (AT), and xeroderma pigmentosum (XP) [18,21,54], are considered in detail elsewhere in this volume. These individuals appear to have an inherited defect in DNA repair, or in some other aspect of DNA "housekeeping," although the details have not been worked out completely in all instances. As a somewhat oversimplified generalization, it can be suggested that the genetic lability that results from the inherited gene defect leads to persistent chromosome aberrations, to cytogenetically abnormal clones, and, ultimately, to the increased incidence of neoplasia that characterizes these syndromes. The genetic instability underlying the original emergence of neoplastic clones in these individuals would, of course, persist in their tumors and contribute to subsequent clonal evolution and clinical progression.

Inherited gene defects with similar effects may be present in individuals who are without recognizable clinical syndromes. There are children, for instance, with unexplained familial disorders of blood formation, who do not fit any of the recognized chromosome-breakage syndromes [27, 38]. These children may exhibit cytogenetically abnormal clones among circulating lymphocytes or in the bone marrow, and some ultimately progress to leukemia. It has not yet been demonstrated that such patients

TABLE I. POSSIBLE MECHANISMS OF GENETIC INSTABILITY IN TUMOR CELL POPULATIONS

A. Inherited defects
  1. "Chromosome breakage" syndromes
  2. Subclinical gene defects
  3. Constitutional chromosome abnormalities
B. Acquired Defects
  1. Gene mutations
    a. DNA repair
    b. DNA replication
    c. Mitotic apparatus
  2. Chromosome alterations
    a. Aneuploidy
    b. Translocations and "transposable elements"
    c. SCE; HSR and DMs
C. Extracellular factors
    a. Viruses
    b. Radiation and mutagenic chemicals
    c. Nutritional deficiencies

carry a gene defect that influences karyotypic stability, but the parallels with known chromosome-breakage syndromes suggest that some of these unexplained disorders may represent "subclinical" inherited defects in DNA housekeeping.

It has also been recognized that cells from individuals with constitutional chromosome abnormalities, particularly Down's syndrome, show evidence of increased susceptibility to cytogenetic damage and rearrangement when exposed to clastogenic agents *in vitro* [53]. It is not clear whether the specific chromosomal alteration in the patient's cells has, itself, a destabilizing effect, or whether both the constitutional abnormality and the increased fragility reflect an inherited defect in chromosomal stability analogous to that discussed in the preceding paragraphs. In at least some families the latter possibility seems to be supported by the presence of different constitutional chromosome alterations in family members as well as a general familial increase in cancer incidence [25,33].

Acquired defects. For the vast majority of patients with cancer it is assumed that there is no constitutional abnormality in genetic stability, and that the increased lability within the neoplastic clone is the result of an acquired alteration. Many kinds of acquired defects have been proposed as the source of the instability, with some supporting evidence in studies of different tumors.

Single gene mutations of various types could destabilize the genome. Such "mutator" genes might lead to abnormalities in DNA repair similar to those in the chromosome-breakage syndromes or other defects in repair mechanisms. Alternatively, such a gene might involve the DNA synthetic apparatus; as a result, more error-prone pathways would be utilized within the tumor cells and thereby increase the probability of subsequent mutations. The product of a mutator gene might even involve the mitotic apparatus itself and lead, for example, to defective polymerization of microtubules, instability of the mitotic spindle, and an increased probability of nondisjunction and other chromosomal rearrangements. Although these various possibilities have not been explored widely, there are limited data to indicate the action of each of these various types of mutator genes in certain neoplastic cell populations [6,22,29,58].

In addition to these specific mutations, it is possible that *chromosomal alterations,* once they are established within the tumor cell population, may themselves contribute to the continuing (and perhaps increasing) genetic instability within the neoplastic cells. Aneuploidy, for instance, could result from one of the mutagenic mutations described and so might appear during evolution of the tumor. Once present, it could readily contribute to the production of further genetic errors. Aneuploid cells are more susceptible than are normal cells to further chromosomal rearrangements and may also be more easily damaged by clastogenic agents [53].

Even balanced translocations in a cell may increase the probability of mitotic errors; and recently the related concept of "transposable elements," a phenomenon originally described in maize, has been reintroduced as a possible important consideration in neoplasia [32,52]. According to this concept, certain DNA segments may move about within the genome and exert various kinds of destabilizing effects on other segments, which are adjacent to their sites of insertion.

Several additional types of chromosomal alterations, recognizable in tumor cells, might increase genetic instability in the neoplastic clone in addition to producing specific and immediate alterations in gene products. It has been postulated, for example, that abnormal sister-chromatid exchanges might play a significant role in tumor development by allowing expression of critical recessive genes [49]; thus, if a somatic alteration in the neoplastic cells produced increased SCEs within the population, the probability of further significant genetic changes would be enhanced also. Similarly, the phenomena of homogeneous staining regions (HSR) and related double minutes (DMs) could have a dual effect. The HSR, as observed in various tumors, appears to represent an area of gene amplification within a chromosome, and the increased gene product from this site could have an important role in the growth of certain neoplasms [3]. If, as has been suggested, the HSR can break down into separate DMs that are ultimately lost from the cell, the HSR could also contribute to genetic instability in the system [1].

Various acquired specific mutations and chromosomal alterations thus may underlie the observed genetic lability in different tumor cell populations, by operating through various forms of gene duplication, deletion, position effects, and imbalance. It should be recognized that different types of genetic change could occur at different times during tumor development, even within the same neoplasm. I have noted already that chromosomal alterations might result from a gene mutation acquired earlier, and then, in turn, contribute to a cascade of increasing instability. It is also possible that, in some instances, the acquired mutagenic mutation might actually occur *before* the first cell acquires a permanent selective growth advantage, and becomes "neoplastic" (according to our working definition). In connection with chemical carcinogenesis in the rat liver, it has been pointed out that most of the earliest nodules in the exposed animal subsequently regress spontaneously [12], and apparently have only a temporary growth advantage within an hepatic microenvironment that is badly damaged by the cytotoxic effects of the carcinogen. However, within the proliferating cells of these "preneoplastic" nodules, there does appear to be an increased probability of mutation, which leads to the outgrowth of a true neoplasm from an occasional nodule. Clearly, both the acquisition of a heritable growth advantage and the acquisition of genetic instability may be very important early events in tumor development, and the specific conditions of carcinogenesis in a particular individual may determine which occurs first.

Extracellular factors. In addition to the inherited and acquired mechanisms that may underlie genetic instability in different neoplasms, it is also possible that extracellular factors contribute to the sequential mutational events involved in clonal evolution. If, for instance, an oncogenic *virus* is the causative agent, its incorporation into the genome of the cell not only may trigger the initial transformation event but also have a continuing destabilizing effect on adjacent segments of the host-cell genome. Some theories of viral oncogenesis, such as Temin's "provirus" concept [59], propose that viral elements function, in some respects, like the "transposable elements" mentioned above. There are also circumstances in which the persistence of intact virus within the tumor cells could have a continuing, direct, damaging effect on the genome [36].

Similarly, in an individual the continued presence of a long-lived carcinogenic *chemical* or *radioisotope*, or repeated doses of clastogenic materials through occupational or other exposure could lead to sequential mutations within the tumor [*11*]. In this way, the mutagenic therapeutic agents used in cancer treatment (radiation, chemotherapy) might contribute significantly to later stages of clonal evolution and tumor progression in some patients.

It has even been suggested that *nutritional* changes within a neoplasm play a role in its genetic instability. Deficiencies of single, essential amino acids and of oxygen have been shown to increase the frequency of nondisjunction in cell culture. With the reduced circulation in many areas of rapidly growing tumors, such phenomena well might help to explain the many mitotic abnormalities observed in aggressive malignancies [*17,24,46*].

Inasmuch as both intracellular and extracellular mechanisms may be important in the observed genetic lability of tumor cell populations, it is worth noting two additional and related examples to indicate how several such factors might operate in the natural history of specific human neoplasms. In both the Burkitt lymphoma and the T-cell leukemias that develop in patients with AT, the neoplastic clone is characterized very often by an abnormality of the terminal portion of chromosome 14 (Fig. 2A). This has been suggested as the site of a specific gene, perhaps involving immunoglobulin synthesis, which, when altered, somehow provides a selective growth advantage for lymphoid cells [*23,42*]. In the case of the Burkitt tumor, in a formulation somewhat different from that suggested by Klein [*23*], one can envision that infection with the Epstein-Barr virus could be the source of genetic instability in the patient's lymphoid population, even though evidence for this effect is lacking *in vitro*. Malarial infection could be a nonspecific stimulus for the cells to proliferate; and the combination of instability and proliferation could result in an occasional cell acquiring the specific alteration in 14q that leads to emergence of a frankly neoplastic clone.

In AT, the instability in the system would result from the inherited defect in DNA repair, and the associated immune dysfunction in these patients could contribute to unbalanced proliferation of various lymphoid subsets. Again, the combination of instability and increased proliferation could lead to chromosomal rearrangements, to clones with various degrees of selective advantage, and ultimately to involvement of the terminal portion of 14q (with or without additional cytogenetic changes) and a clone that expanded rapidly enough to be considered a true neoplasm [*20,38*].

Such a hypothetical sequence of events may be oversimplified, but it does permit phenomena that appear to be important in tumor development to be integrated into already existing concepts of "initiation" and "promotion" [2,12]. Thus, characteristics of the "initiated" cell could include both genetic instability and, or, heritable growth advantage, and "promotion" would be the proliferative stimulus that allows the effects of these characteristics to be expressed in an actively dividing population.

ROLE OF HOST FACTORS IN CLONAL EVOLUTION

The discussion thus far has focused on alterations within the neoplastic cells, but obviously factors in the host environment will provide the selective pressures that determine which mutant cells grow as predominant subpopulations during tumor progression.

Despite continuing debate on the efficacy and even the existence of "immune surveillance," it seems likely that the host immune system represents one type of selective pressure on evolving neoplasms, particularly in the early stages. Currently evidence is emerging that nonimmunological cytotoxic effects directed against tumors by such cells as the macrophage and "natural killer" (NK) cells also may be of major significance [56].

There is much to be learned about the local regulatory substances (chalones, growth factors) to which the neoplastic population responds abnormally. We know little about the importance of increases or decreases in the levels of these factors. Clearly, damage to the microenvironment of an early neoplasm may seriously disrupt local regulatory mechanisms and play a significant role in tumor development. In considering chemical carcinogenesis in the rat liver, Farber [12] has stressed the cytotoxic effects of many of the agents used as well as the selective advantage to emerging tumor cells of resistance to this cytotoxicity. Similar considerations may be important in the pathogenesis of certain human neoplasms. It seems probable that chemicals such as benzene and the alkylating agents used in cancer therapy cause leukemia in humans, not only by direct mutagenic effects on marrow cells but also through damage to the local microenvironment, which allows potentially leukemic clones to expand [38]. Any circumstance that alters the local environment and allows competition among subpopulations of a developing neoplasm may have a significant influence on the continuing evolution of the tumor.

In the later stages of human neoplasia other host factors may also be important. The environmental pressures generated by the general health and nutrition of the patient, his exposure to infectious agents, and even the therapeutic efforts of the physician may all serve to accelerate the appearance of new sublines within the tumor [16,45]. Obviously, additional knowledge is needed of both the external influences on mutant cell populations once they are produced and the types and causes of the genetic changes themselves. Both extracellular and intracellular elements affect the natural history of cancer in a particular patient, including the variable latent periods that may occur before the original appearance of the tumor or between initial therapy and subsequent relapse.

CLINICAL IMPLICATIONS OF CLONAL EVOLUTION AND GENETIC INSTABILITY IN TUMORS

The general concept of clonal evolution based on genetic lability within neoplastic cell populations presents a somewhat discouraging prospect from the standpoint of clinical control of cancer. For many years, the hope has been to find a consistent metabolic alteration in tumor cells which could be exploited therapeutically. It seems increasingly likely that if such a metabolic change is the first step in many neoplasms, it may represent only a relatively minor quantitative shift in the expression of some critical gene product; and identification of that first alteration will be extremely difficult because of the many evolutionary steps (see Fig. 1) between the initial change and the fully developed, clinical malignancy. These sequential stages are multiple and to some degree random, for they reflect the particular environmental pressures that influence the development of each tumor.

Under these circumstances, it may be necessary to consider each advanced malignancy as an individual therapeutic problem. If so, immunotherapy seems to be the most hopeful approach to the specific destruction of residual tumor cells, following after the maximum effect of nonspecific modalities such as surgery, radiation, and current chemotherapy. However, our present knowledge is not adequate to allow confident manipulation of the immune system to the benefit of the patient [56].

This approach ultimately may prove useful, but we must still recognize the handicap to therapy that is imposed by the genetic lability of the tumor cell population. As variants are produced in increasing frequency during tumor progression, the neoplasm will continue to have a marked capacity for generating mutant sublines, which may be resistant to whatever treatment the physician may introduce. Unfortunately for the patient, the same capacity for variation and selection that permits the evolution of a malignant population from the original aberrant cell also provides the opportunity for the tumor to adapt successfully to the inimical environment of therapy.

If complete eradication of the tumor cell population continues to be a therapeutic problem, one might also consider the potential for reversing the neoplastic process, as several authors have suggested [5,10,34,44]. Is it likely that a "cure" can be produced by providing an environment that forces the tumor cell population to cease unbalanced proliferation and to move into a state of controlled differentiation? A few such circumstances have been demonstrated experimentally, both *in vivo* and *in vitro*, usually with tumors having a diploid or near-diploid karyotype and some characteristics of "embryonic rests" (e.g., neuroblastoma, teratocarcinoma), which presumably are still somewhat responsive to normal developmental influences [5,10,34,44]. This approach is of particular interest for neoplasms such as the diploid acute leukemias illustrated on the left side of Figure 1; but, for the highly aneuploid common malignancies indicated on the right side of Figure 1, the possibility of generating *in vivo* the conditions necessary to force normal patterns of differentiation in these neoplastic populations seems much less likely. The predominant sublines in these tumors have been selected through many steps for proliferative capacity and lack of response to growth controls. However, even in tumors with major chromosomal abnormalities, banding studies often indicate that the components of the normal genome are present still, although rearranged; thus, their potential reversibility to normal growth characteristics cannot be excluded categorically. Certainly, these lines of investigation should be continued, as well as increased efforts to understand and control the mechanisms of genetic instability that permit early benign diploid tumors to evolve into the highly aneuploid malignancies which are the typical clinical presentation of human cancer.

SUMMARY

1. Considerable evidence exists indicating that most tumors are clones (i.e., unicellular in origin), and that the phenomenon of tumor progression results from acquired genetic variation within the original clone allowing sequential selection of more aggressive subpopulations.

2. There is increasing evidence also that tumor cell populations are, to some degree, more unstable genetically than comparable normal cells. This lability might result from inherited gene defects, acquired activation of "mutator" genes, chromosomal rearrangements, or continued effects of clastogenic agents in the tumor environment.

3. Genetic alterations in neoplastic cell populations, the result of this instability, are frequently recognizable cytogenetically, particularly in the later stages of tumor development; and studies of patients with the chromosome–breakage syndromes may help to elucidate the mechanisms underlying some of these changes when they occur in the tumor cells of individuals without inherited gene defects.

4. From a clinical standpoint this concept of clonal evolution and tumor progression is discouraging because each advanced human malignancy may be highly individual karyotypically and biologically when first encountered by the physician. Therefore, each patient's cancer may require individual specific therapy, and even this may be thwarted by the emergence of genetically variant sublines that are resistant to treatment.

LITERATURE CITED

1. BALABAN-MALENBAUM, G., GILBERT, F.: Double minute chromosomes and the homogeneously staining regions in chromosomes of a human neuroblastoma cell line. *Science* 198:739-742, 1978.
2. BERENBLUM, I.: Sequential aspects of chemical carcinogenesis in skin. *In* Cancer: A Comprehensive Treatise (Becker, F.F. ed.), New York: Plenum Press, 323-344, 1975.
3. BIEDLER, J., SPENGLER B.: Metaphase chromosome anomaly: Association with drug resistance and cell-specific products. *Science* 141:185-187, 1976.
4. BOVERI, T.: *In* Zur Frage der Entstehung maligner Tumoren, Jena: Gustave Fischer Verlag, 1914.
5. BRAUN, A.C.: The Cancer Problem. New York: Columbia University Press, 1969.
6. CAIRNS, J.: Mutation, selection, and the natural history of cancer. *Nature* 255: 197-200, 1975.
7. COGGIN, J.H., JR., ANDERSON, N.G.: Cancer, differentiation, and embryonic antigens: Some central problems. *Adv. Cancer Res.* 19:105-113, 1974.
8. DANES, B.S.: Increased *in vitro* tetraploidy: Tissue specific within the heritable colorectal cancer syndromes with polyposis coli. *Cancer* 41:2330-2334, 1978.
9. DE GROUCHY, J., DE NAVA, C.: A chromosomal theory of carcinogenesis. *Ann. Intern. Med.* 69:381-391, 1968.
10. DIBERARDINO, M.A., KING, T.J.: Transplantation of nuclei from the frog renal adenocarcinoma. II. Chromosomal and histologic analysis of tumor nuclear-transplant embryos. *Dev. Biol.* 11:217-222, 1965.
11. EVANS, H.: (This volume).
12. FARBER, E., CAMERON, C.: The sequential analysis of cancer development. *Adv. Cancer Res.* 31:125-225, 1980.
13. FIALKOW, P.: Clonal origin of human tumors. *Annu. Rev. Med.* 30:135-143, 1979.
14. FILAKOW, P., DENMAN, A., JACOBSON, R., LEVENTHAL, M.: Chronic leukemia: Origin of some lymphocytes from leukemic stem cells. *J. Clin. Invest.* 62:815-822, 1978.
15. FOLKMAN, J., KLAGSBRUN, M.: Tumor angiogenesis: Effect on tumor growth and immunity. *In* Symposium on Fundamental Aspects of Neoplasia (Gottlieb, A., ed.), New York: Springer-Verlag; 331-340, 1975.
16. FOULDS, L.: Neoplastic Development, Vol. 2. New York: Academic Press, 1975.
17. FREED, J.J., SCHATZ, S.A.: Chromosome aberrations in cultured cells deprived of single essential amino acids. *Exp. Cell Res.* 53:393-398, 1969.
18. GERMAN, J.: The association of chromosome instability, defective DNA repair, and cancer in some rare human genetic diseass. *In* Human Genetics (Armendares, S., Lisker, R., eds.), Amsterdam: Excerpta Medica, 64-68, 1977.

19. HARNDEN, D.: The relationship between induced chromosome aberrations and chromosome abnormality in tumor cells. *In* Human Genetics (Armendares, S., Lisker, R., eds), Amsterdam: Excerpta Medica, 355-366, 1977.
20. HECHT, F., KAISER-MCCAW, B.: (This volume).
21. HECHT, F., MCCAW, B.: Chromosome instability syndromes. *Prog. Cancer Res. Ther.* 3:105-123, 1977.
22. HESTON, L.: Alzheimer's diseases, trisomy 21, and myeloproliferative disorders: Associations suggesting a genetic diathesis. *Science* 196:322-323, 1977.
23. KLEIN, G.: Lymphoma development in mice and humans: Diversity of initiation is followed by convergent cytogenetic evolution. *Proc. Natl. Acad. Sci. USA* 76: 2442-2446, 1979.
24. KOLLER PC: The role of chromosomes in cancer biology. New York: Springer-Verlag, 1972.
25. LAW, I.P., HOLLINSHEAD, A.C., WHANG-PENG, J., DEAN, J.H., OLDHAM, R.K., HERBERMAN, R.B., RHODE, M.C.: Familial occurrence of colon and uterine carcinoma and of lymphoproliferative malignancies. *Cancer* 39:1229-1236, 1977.
26. LEVAN, G., MITELMAN, F.: G-banding in Rous rat sarcomas during serial transfer: Significant chromosome aberrations and incidence of stromal mitoses. *Hereditas* 84:1-14, 1976.
27. LI, F.P., POTTER, N.U., BUCHANAN, G.R., VAWTER, G., WHANG-PENG, J., ROSEN, R.B.: A family with acute leukemia, hypoplastic anemia and cerebellar ataxia: Association with bone marrow C-monosomy. *Am. J. Med.* 65:933-939, 1978.
28. LIOTTA, L.A., TRYGGVASON, K., GARBISA, S., FOLTZ, C., SHAFIE, S.: Metastatic propensity correlates with tumor cell degradation of basement membrane collagen. *Nature* 284:67-68, 1980.
29. LOEB, L., BATTULA, N., SPRINGGATE, C., SEAL, G.: On mutagenic DNA polymerase and malignancy. *In* Fundamental Aspects of Malignancy. Berlin: Springer-Verlag, 243-256, 1975.
30. MAKINO, S.: Further evidence favoring the concept of the stem cell in ascites tumors of rats. *Ann. N.Y. Acad. Sci.* 64:818-823, 1956.
31. MARK, J.: Rous sarcoma in mice: The chromosomal progression in primary tumors. *Eur. J. Cancer* 5:307-318, 1969.
32. MCCLINTOCK, B.: The origin and behavior of mutable loci in maize. *Proc. Natl. Acad. Sci. USA* 36:344-355, 1950.
33. MILLER, O.J., BREG, W.R., SCHMICKEL, R.D., TRETTER, W.: A family with a XXXXY male, a leukemic male, and two 21 trisomic mongoloid females. *Lancet* ii:78-79, 1961.
34. MINTZ, B.: Genetic mosaicism and *in vivo* analyses of neoplasia and differentiation. *Annu. Symp. Cancer Res.* 30:27-53, 1978.
35. MITELMAN, F.: The chromosomes of fifty primary Rous rat sarcomas. *Hereditas* 69: 155-162, 1971.
36. NICHOLS, W.: (This volume).
37. NOWELL, P.: The clonal evolution of tumor cell populations. *Science* 194:23-28, 1976.
38. NOWELL, P.: Preleukemia. Cytogenetic clues in some confusing disorders. *Am. J. Pathol.* 89:459-476, 1977.
39. NOWELL, P.: Cytogenetics. *In* Cancer: A Comprehensive Treatise, Vol. 1: Etiology, Second Ed. (Becker, F., ed.), New York: Plenum, 3-46, 1982.
40. NOWELL, P., FINAN, J., GLOVER, D., GUERRY, D.: Cytogenetic evidence for the clonal nature of Richter's syndrome. *Blood* 58:183-186, 1981.

41. NOWELL, P., ROWLANDS, D., DANIELE, R., BERGER, B., GUERRY, D.: Changes in membrane markers and chromosome patterns in chronic T cell leukemia. *Clin. Immunol. Immunopathol.* 12:323–330, 1979.
42. NOWELL, P., SHANKEY, T., FINAN, J., GUERRY, D., BESA, E.: Proliferation, differentiation, and cytogenetics of chronic leukemic B lymphocytes cultured with mitomycin-treated normal cells. *Blood* 57:444–451, 1981.
43. ODELL, W.D.: Humoral manifestations of nonendocrine neoplasms: Ectopic hormone production. *In* Textbook of Endocrinology (Williams, R., ed.), Philadelphia: Saunders, 1105–1115, 1974.
44. O'HARA, M.: Teratomas, neoplasia, and differentiation: A biological overview. I. The natural history of teratomas. *Invest. Cell. Pathol.* 1:39–63, 1978.
45. OHNO, S.: Genetic implication of karyological instability of malignant somatic cells. *Physiol. Rev.* 51:496–526, 1971.
46. OKSALA, T., THERMAN, E.: Mitotic abnormalities and cancer. *In* Chromosomes and Cancer (German, J., ed.), New York: Wiley, 239–263, 1974.
47. OTTER, M., PALMER, C., BAEHNER, R.: Elevated sister chromatid exchange rate in childhood lymphoblastic leukemia. *Proc. Am. Assoc. Cancer Res.* 19:202, 1978.
48. PARSHAD, R., SANFORD, K.K., TARONE, R.E., JONES, G.M., BAECK, A.E.: Increased susceptibility of mouse cells to fluorescent light-induced chromosome damage after long-term culture and malignant transformation. *Cancer Res.* 39:929–933, 1979.
49. PASSARGE, E., BARTRAM, C.: Somatic recombination as possible prelude to malignant transformation. *In* Birth Defects: Original Article Series. New York: The National Foundation, 177–180, 1976.
50. ROWLEY, J.D.: Chromosome abnormalities in cancer. *Cancer Genet. Cytogenet.* 2: 175–198, 1980.
51. ROWLEY, J.D.: Ph-positive leukaemia, including chronic myelogenous leukemia. *Clin. Haematol.* 9:55–86, 1980.
52. SAGER, R.: (This volume).
53. SEABRIGHT, M.: Patterns of induced aberrations in humans with abnormal autosomal complements. *Chrom. Today* 5:293–297, 1976.
54. SETLOW, R.: Repair deficient human disorders and cancer. *Nature* 271:713–717, 1978.
55. SHIRAISHI, Y., SANDBERG, A.A.: Effects of various chemical agents on sister chromatid exchanges, chromosome aberrations, and DNA repair in normal and abnormal human lymphoid cells. *J. Natl. Cancer Inst.* 62:27–33, 1979.
56. SIEGEL, B., COHEN, S.: Tumor immunity. *Fed. Proc.* 37:2212–2214, 1978.
57. SOKOVA, O., VOLGAREVA, G., POGOSIANTZ, H.: Effect of fluorafur on chromosomes of normal and malignant Djungarian hamster cells. *Genetika* 12:156–159, 1976.
58. STRAUSS, B.: (This volume).
59. TEMIN, H.: The DNA provirus hypothesis. *Science* 192:1075–1080, 1976.
60. VON HANSEMANN, D.: Uber asymmetrische Zellteilung in Epithelkrebsen und deren biologische Bedeutung. *Virchow's Arch. Pathol. Anat. Physiol.* 119:298–307, 1890.
61. WEINER, F., DALIANIS, T., KLEIN, G., HARRIS, H.: Cytogenetic studies on the mechanism of formation of isoantigenic variants in somatic cell hybrids. *J. Natl. Cancer Inst.* 52:1779–1785, 1974.

Subject Index

Acquired agammaglobulinemia, 105
Acquired genetic defects, and genetic instability in tumor cells, 423-424
Actinic keratoses, in xeroderma pigmentosum, 66
Age
  of death in Werner's syndrome, 86
  at diagnosis of Fanconi's anemia, 43, 44, 45
  at diagnosis of Werner's syndrome, 85-86
  and nondisjunction of X chromosomes, 259
  at onset of cancer in ataxia-telangiectasia, 107-111
  at onset of cancer in Bloom's syndrome, 101, 103
  and radiation-induced chromosome aberrations, 261
  and sun sensitivity in xeroderma pigmentosum, 64
  and telangiectatic erythema of Bloom's syndrome, 12-13
Alphafetoprotein levels, in ataxia-telangiectasia, 29, 30
Amniotic fluid cells, and prenatal diagnosis of Bloom's syndrome, 18
Anabolic steroids. *See* Androgen therapy

Androgen therapy in Fanconi's anemia, 51-53, 56
  and cancer, 118-120
Angiosarcomas and vinyl chloride monomer exposure, 263
Antenatal testing, and Bloom's syndrome, 18
Antibody response, and chromosomal rearrangements, 36
AP sites in DNA, 282-286
Aplastic anemia, *See* Childhood aplastic anemia
Amegakaryocytic thrombocytopenia, and development of Fanconi's anemia, 43
Aryl hydrocarbon hydroxylase (AHH) inducibility, and response to cigarette smoke, 8, 271
Ascites tumor, xvi, xxx
Ashkenazim, and Bloom's syndrome, 19
Autoimmune mechanisms, and ataxia-telangiectasia, 35
Ataxia-telangiectasia, 23-37, 193-201
  cancer incidence in, 8, 23-24
  case reports of cytogenetic studies of, 193-197
  chromosome aberrations in, 142-148
  and chromosome 14, 198-200
  and chromosome 14q translocations, 29

Ataxia-telangiectasia (cont.)
  cytogenetics of chromosome breakage in, 147–148
  frequency, 23
  and genetic instability of tumor cells, 426
  hypersensitivity to x-rays, 382
  immune deficiency, chromosome instability, and cancer in, 193–201
  immunologic aberrations in, 24–25
  incidence and types of cancer in, 105–112
  model for cancer susceptibility in, 35–37
  mutagen sensitivity in, 211
  mutant T-lymphocyte clones in, 378–381
  neurologic findings in, 24
  pathogenesis models, 29–35
  and radiation sensitivity, 25–28
  serum alphafetoprotein levels in, 24
  sister-chromatid exchanges in, 178–179
  stages in chromosome changes in, 198
  symptoms, 23–24
  wasted mouse model for, 30–31
  and x-ray hypersensitivity, 240–244
  See also Chromosome-breakage syndromes
Atherosclerosis, in Werner's disease, 87
Atomic energy workers, and radiation-induced chromosome aberrations, 261
Atrophy of skin, in xeroderma pigmentosum, 66, 67–68
Autosomal dominant gene transmission, and cancer-proneness, 98

Autosomal recessive heredity
  of Bloom's syndrome, 19
  classification of chromosome-breakage syndromes by, 206
  and Fanconi's anemia, 44
  and pathogenesis of ataxia-telangiectasia, 29–31

Bacteria
  bypass of lesions during DNA synthesis in, 301–307
  insertion sequences, 341–342
  mutagenesis in, 281
Basal cell carcinomas in xeroderma pigmentosum, 66–67
Basal cell nevus syndrome, 206
  cellular characteristics of, 216
  and dominant genes, 98
Benzo(a)pyrene (BP), and carcinogenicity of cigarette smoke, 267–271
Bleomycin-induced chromosome breaks in ataxia telangiectasia cells, 178
Blood t-lymphocytes
  chromosome aberrations in ataxia-telangiectasia, 142–147
  in Fanconi's anemia, 50
  ionizing radiation-induced chromosome aberrations in 259–261
  sister-chromatid exchange study in Bloom's syndrome, 139–140
Bloom's syndrome, 11–20
  basis for chromosome instability in, 140–142
  cancer incidence in, 7–8, 16–17, 347, 349–350
  chromosome aberrations in, 136–139

and chromosome-mutation
theory of
carcinogenesis, 352–355
clinical features, 12–16
compared with xeroderma
pigmentosum, 121
cytogenetic features of, 17
diagnosis, 18
discovery of, 11–12
DNA replication and sister-
chromatid exchanges
in, 239
genetics of, 19
in vitro studies of molecular
defect in, 19–20
management, 18–19
mutagen sensitivity in, 212
mutations and sister-chromatid
exchanges in,
carcinogenesis and,
236–237
possible causes of cancer
proneness in, 350–352
prenatal diagnosis, 18
recessive inheritance of, 98
sister-chromatid exchange in,
139–142, 175–177
types of neoplasms in,
100–104, 350–351
See also Chromosome
breakage syndromes
Bloom's Syndrome Registry,
11, 103
Bone marrow, in Fanconi's
anemia, 47–48, 51
Bone marrow, cryopreservation
and Bloom's syndrome, 19
Bone marrow failure, and
antileukemic drugs in
Bloom's syndrome, 17
See also Aplastic anemia
Bone marrow transplantation, in
Fanconi's anemia, 57
Boveri's hypothesis, 333–336
Brain. See Cerebellum; Cerebral
cortex

Breakpoint
in chromosome, xvii
Bruton's agammaglobulinemia,
105
BrdU method and sister-
chromatid exchange
studies, 139, 169, 170,
171, 172
Burkitt's lymphoma, and
chromosome
translocations, xvii, 36,
147

Cancer
in Ataxia-telangiectasia, 8,
105–112, 193–201
in Bloom's syndrome, 7–8,
16–17, 100–104
chromosome-breakage
syndromes and patterns
of, 97–127, 135
and chromosome damage, 253
clonal evolution model of, see
Clonal evolution model
of cancer
in Fanconi's anemia, 53–58,
113–120
as preventable disease, 3
and primary
immunodeficiency
105–106
production in animals, 5–7
role of diet in, 4
single cell origin of, xv–xvii
and viral interactions with
mammalian cell
genomes, 317–328
in Werner's syndrome,
123, 124
in xeroderma pigmentosum,
69, 120–122
See also Cancer and genomic
rearrangements;
Carcinogenesis;
Carcinogens;
Chromosome mutation

Cancer (cont.)
    theory of carcinogenesis; Skin cancer, Somatic mutation theory of cancer
Cancer and genomic rearrangements, 333
    Boveri's hypothesis, 333–336
    and genomic rearrangements as causal agents, 336–340
    transposition as mechanism for, 340–342
    and tumorigenicity as multistage process, 342–344
Cancer-proneness
    cellular sensitivity to mutagens and carcinogens in, *see* Mutagen hypersensitivity
    and chromosome mutation theories of cancer, 347–356
    and dominant genes, 97–98
    identification of, patterns of neoplasia and, 97
    and recessive genes, 98–99
Cancer research
    and carcinogenesis in animals, 5–7
    interaction of epidemiology and laboratory research, 3–4
    and selection of animal models, 7
Canine venereal tumor, xxx
Carcinogenesis
    chromosome-mutation theory of, 352–355, 359–386
    and DNA damage and chromosome aberrations, 361
    and genetic defects, *see* Genetic defects
    and lipid oxidation in cell membranes, 8
    mutagenicity in animals and, 5–7
    and mutations and sister-chromatid exchanges in xeroderma pigmentosum and Bloom's disease, 236–237
    and nucleotide sequences for diet-associated cancers, 4
    virus-induced, 6
    See also Carcinogens
Carcinogens
    and agent-specific-induced chromosome changes, 364–365
    and cancer production in animals, 5
    cellular sensitivity to, *see* Mutagen hypersensitivity
    in cigarette smoke, 267–271
    effects on chromosomes, 253–276
    -induced tumorigenesis, 360–361
    metabolism of, genetic diseases and, 8
    See also Chemical mutagens and carcinogens
Catalase, and blood lymphocyte chromosome studies in Werner's syndrome, 88
Cataracts, in Werner's syndrome, 85, 86
Cell motility, and microtubular defect hypothesis of pathogenesis of ataxia-telangiectasia, 33
Cell transformation
    virus-induced, 321–324, 327–328
Cerebellar ataxia. *See* Ataxia-telangiectasia
Cerebellum

neuronal loss in xeroderma
pigmentosum, 68–69
Purkinje cells, in ataxia-
telangiectasia, 24, 34
Cerebral cortex, neuronal loss in
xeroderma pigmentosum,
68–69
Chediak-Higashi syndrome, 207
cellular characteristics of, 214
CHEF/16 and 18 cells, and
genomic rearrangement
studies, 340
Chemical mutagens and
carcinogens
cigarette smoke, 263, 265,
267–276
and DNA damage, 255–256
and genetic instability of
tumor cells, 425–426
-induced tumors, and agent-
specific chromosome
changes, 364–365
-induced chromosome
aberrations in humans,
261–263, 265–276
vinyl chloride monomer,
263, 266
Chemotherapy and Fanconi's
anemia with leukemia, 55
Childhood aplastic anemia and
Fanconi's anemia, 43
Chinese hamster embryo cells
and chromosome changes
associated with
tumorigenicity, 337–338
cigarette smoke studies in,
267–269
Chinese hamster ovary (CHO)
cells
fusion with Bloom's syndrome
cells, 176–177
in sister-chromatid exchange
studies, 174
Choreoathetosis, in ataxia-
telangiectasia, 24
Chromosome aberrations

in ataxia-telangiectasia,
142–148
in Bloom's syndrome 7–8,
136–139
carcinogenic rays and
chemicals and, 253–276
in cigarette smokers, 264,
265, 267–276
and clonal evolution model of
cancer, 414
as crucial events in cancer,
347–356, 359–386
from DNA synthesis, 307–309
in Fanconi's anemia, 148–151
and genetic instability of
tumor cells, 424–425
and genomic rearrangements
in maize, 339
ionizing radiation-induced *in
vitro* and *in vivo*,
259–261
molecular alterations in DNA
and, 281–309
radiation-induced in nuclear
dockyard workers, 259,
261, 264, 265
and tumorigenicity, 337–338
virus-induced, 317–321
in Werner's syndrome, 88–89,
153–155
in xeroderma pigmentosum,
151–153
See also Chromosome
breakage syndromes;
Premature chromosome
condensation
Chromosome breakage
syndromes
autosomal dominant disorders
without known
karyotypic instability,
207, 210, 216–217
autosomal recessive, 206,
208–209, 211, 212,
213, 214
and chromosome mutation

Chromosome breakage syndromes (cont.)
    theory of cancer, xiv–xxi, 373–384
    classification of, 206–207, 210–211
    and clonal proliferation, 378–382
    common features of, 203
    cytogenetics of, *see* Cytogenetics of chromosome-breakage syndromes
    data sources, 98–99
    disorders with specific chromosomal abnormality and predisposition to cancer, 210–211, 216–217
    mutagen and carcinogen hypersensitivity in, 203–223
    and neoplasia-associated traits, 375–384
    patterns of neoplasia associated with, 97–127
    regulation of responses to DNA damage in, 235–245
    spontaneous characteristics of, 210
    and tissue-specific breakages, 205
    types of neoplasms in, 375
    undiscovered, 204–205
    with unstable karyotypes and various modes of inheritance, 206, 216–217
    See also Ataxia-telangiectasia; Bloom's syndrome; Fanconi's anemia; Werner's syndrome; Xeroderma pigmentosum
Chromosome deletion, and characterization of Philadelphia chromosome, xvii, xix, 35
Chromosome 14, in premalignant and malignant clones in ataxia-telangiectasia, 198–200
Chromosome 14q, translocation in ataxia-telangiectasia, 29
Chromosome instability
    and ataxia-telangiectasia, 8, 35–37
    in Bloom's syndrome, 11, 17, 140–142
    and cancer and immune deficiency in ataxia-telangiectasia patients, 193–201
    and cancer proneness of ataxia-telangiectasia, 112
    and cancer risk, 321
    in Fanconi's anemia, 44, 114–115
    and sister-chromatid exchange studies, 169
    stages in ataxia-telangiectasia, 198
    and Werner's syndrome, 85
Chromosome mapping techniques, and two-hit theory of viral transformation, 328
Chromosome mutation theory of carcinogenesis, xv–xx, 352–355, 359–386
    and agent-specific chromosome changes, 364–365
    and chromosome-breakage syndromes, 373–384
    and chromosome changes in neoplastic cells, 360–372
    historic background, 359–360
    and normal karyotypes in tumor cells, 370–372
    and oncogenes, 397–408
    and somatic cell hybridization studies, 372–373

and specificity of chromosome
changes, 361–364
and target-cell-specific
chromosome changes,
365–370
Chromosome pulverization. *See*
Premature chromosome
condensation
Chromosomes
breakage and
rearrangement, 343
breakpoint specificity, xvii,
xix
in Fanconi's anemia, 48–51
hypotonic technique, xix
and oncogenes, 401–407
specific translocations in
neoplastic cells,
362–363
spontaneous changes in human
somatic cells, 258–259
suppression of tumorigenicity
by, 372–373
in tumor cells, 333
See also Philadelphia
chromosome, Sister-
chromatid exchange,
Chromosome 4, xix, xx
Chromosome 7, xix, xx
and under Chromosome
Chronic myelogenous leukemia,
Philadelphia chromosome
as marker of, 35, 359–360
Cigarette smoke, and
chromosome aberrations,
264, 265, 267–276
clonal evolution model of cancer,
413–430
clinical implications of,
428–429
role of host factors in,
427–428
Clonal proliferation, and cancer
in chromosome breakage
syndromes, 378–382
Clones, marked by chromosome
14 rearrangement in ataxia-
telangiectasia, 198–200

benign, xvi–xx
mutant, xv–xxi
neoplastic, xxx
Cockayne's syndrome, 207
cellular characteristics of, 214
and xeroderma
pigmentosum, 70
Complementation analysis, and
genetic heterogeneity, 99
Computed tomography scans, in
xeroderma
pigmentosum, 69
Con-A capping of lymphocytes,
and ataxia-
telangiectasia, 33
Conditional chromosomal
breakage syndromes, 205
Congenital malformations
in Bloom's syndrome, 16
in Fanconi's anemia, 45,
46–47
See also Genetic defects
Consanguinity. *See* Parental
consanguinity
Controlling elements, and
chromosome breakage and
rearrangements, 343
Controlling elements in maize
genetic control of time and
frequency of
transposition and, 339
and reversible suppression of
gene expression,
339–340
Corn. *See* Maize; *Zea mays*
Cyclic AMP levels and
microtubular defect
hypothesis of pathogenesis
of ataxia-telangiectasia, 33
Cyclic GMP levels and
microtubular defect
hypothesis of pathogenesis
of ataxia-telangiectasia, 33
Cyclophosphamide therapy in
ataxia-telangiectasia, 25
Cytogenetic analysis of tumor
cells, 360. *See also*
Cytogenetics of

Cytogenetic analysis (cont.)
  chromosome breakage
    syndromes
Cytogenetics of chromosome
  breakage syndromes,
    135–157
  ataxia-telangiectasia, 142–148,
    193–197
  Bloom's syndrome, 136–142
  Fanconi's anemia, 148–151
  Werner's syndrome, 153–156
  xeroderma pigmentosum,
    151–153

De Sanctis-Cacchione syndrome,
  and xeroderma
    pigmentosum, 68
Deafness, in Fanconi's anemia,
  46–47
Delayed hypersensitivity skin
  tests, in ataxia-
    telangiectasia, 25
Developmental defects, and
  cancer in chromosome
    breakage syndromes, 382
Diabetes mellitus, and Werner's
  disease, 87
Diet, cancer and, 4, 6
  See also Overfeeding
DNA
  in Bloom's syndrome, 19
  and cellular repair
    mechanisms, 286–292
  insertion of sequences between
    pair of bacteria IS
    elements, 341–342
  loss of purine or pyrimidine
    base from, 282–286
  molecular alterations in,
    mutation and
    chromosome aberrations
    and, 281–309
  reaction with mutagens,
    282–286

-related enzyme defect in
  ataxia-telangiectasia,
    31–32
removal of O^6-methylguanine
  adduct from, 288–292
and sister chromatid-
  exchange-inducing
    agents,
    172–174
viral, insertion into host
  DNA, 319
DNA crosslinks
  hypersensitivity to induction
    of, in Fanconi's
    anemia, 179
  sister-chromatid exchange
    induction by, 174
DNA damage
  in ataxia-telangiectasia,
    200–201
  and cancer in animals, 5
  cancer induction and, 342–344
  chemically and x-ray-induced
    compared, 255–258
  and cigarette smokers,
    265, 267
  and DNA repair mechanisms,
    292–297
  during DNA replication, and
    sister-chromatid
    exchanges, 237–240
  excision in xeroderma
    pigmentosum, 237
  regulation of responses to,
    mutagen
    hypersensitivity and
    chromosome-breakage
    syndromes and,
    235–245
  and sister-chromatid
    exchanges, 173, 184
  and stage of cell cycle,
    254–255
  and tumorigenicity, 342–343

in xeroderma pigmentosum, 7, 135, 152-153, 177-178
x-rays and, 254
DNA ligase deficiency, in Fanconi's anemia, 150
DNA metabolism, and congenital defects with high cancer incidence, 7-8
DNA N-glycosylases and mutagens, 285-286
DNA repair, 286-292
   in ataxia-telangiectasia, 30
   in chromosome fragility syndromes, 321
   and classification of xeroderma pigmentosum, 218
   and enzyme defects in ataxia-telangiectasia, 31-32
   in Fanconi's anemia, 51
   human diversity in, 211-212
   and mutagen hypersensitivity, 221-222
   and radiation therapy in cancer patients with ataxia-telangiectasia, 25-28
   of ultraviolet-induced damage, and xeroderma pigmentosum, 63, 64, 70-78
   and xeroderma pigmentosum, 120-122, 153, 287-288, 290
DNA replication
   DNA damage during, and sister-chromatid exchanges, 237-240
   and sister-chromatid exchange, 169, 171
   and unremoved adducts, 292-297
   in Werner's syndrome, 89-90
DNA replication fork
   in Bloom's syndrome, 176

and sister-chromatid exchange induction, 174
DNA synthesis
   and bypass of lesions, 301-307
   and mutagenesis in bacteria, 281
   mutagens and, 297-301
   and mutations and chromosome aberrations, 307-309
   and RNA tumor viruses, 335
Dominant genes, and cancer-proneness, 97-98
Down's syndrome
   cancer risk in, 210
   cellular characteristics, 217
   chromosome breakage in, 205
Drosophila, transposable elements in, 335
Drosophila sperm, chromosome-damaging properties of x-rays and alkylating agents in, 256
Dwarfism, in Cockayne's syndrome, 207
Dyskeratosis congenita, 98, 206
   cellular characteristics, 216

Ear abnormalities, in Fanconi's anemia, 47
Embryonic cell transplantation, production of teratocarcinomas in animals and, 5
Enzyme defects, DNA-related, and ataxia-telangiectasia, 31-32
Enzymes, heat-labile, in Werner's syndrome, 90-91
Epidemiology of cancer, and identification of causative factors, 4

Epstein-Barr virus,
    transformation of human
    lymphocytes, 319
Eye abnormalities, in Fanconi's
    anemia, 46

Face, in Bloom's syndrome,
    12, 18
Familial defects. *See* Genetic
    defects
Familial retinoblastoma, cellular
    characteristics of, 216
Fanconi's anemia, 43–58
    acute leukemia and, 54–55
    age at diagnosis, 43
    androgen treatment in, 51–53,
        113–120
    and cancer incidence and type,
        113–120
    chromosome abnormalities in,
        48–51
    clinical presentation, 45
    clonal proliferation in, 382
    complications, 53–58
    cytogenetics of chromosome
        breakage in, 148–151
    diagnosis of, 44
    frequency, 43
    and hepatic disease and
        malignancies, 53–54,
        113–120
    inheritance pattern of, 44
    laboratory studies in, 47–51
    mortality in, 51–52
    mutagen sensitivity in, 213
    physical findings in, 46–47
    and regulation of DNA
        repair, 244
    sister-chromatid exchanges in,
        179–183
    and susceptibility to SV–40
        induced chromosome
        breakage, 324
    treatment, 51–53, 113–120
    See also Chromosome-
        breakage syndromes
Fibroblasts
    and chromosome aberrations

    in ataxia-telangiectasia,
        xviii, 142–147
    and chromosome aberrations
        in Fanconi's anemia,
        148–149
    sister-chromatid exchange in,
        Bloom's syndrome and,
        17, 140–142
    in Werner's syndrome, 88–91
Food intake. *See* Overfeeding
Founder effect, and Bloom's
    syndrome in
    Ashkenazim, 19
Free radicals, and chromosome
    breakage in Werner's
    syndrome, 88–89

Gardner's syndrome
    cellular characteristics of, 216
    and dominant genes, 97, 98
Gastrointestinal infection, in
    Bloom's syndrome, 13, 16
Genes
    cancer and, *see* Cancer and
        genomic rearrangements
    differences in radiosensitivity
        of, 212, 214
    virus-induced transfer of, 321
    See also Autosomal recessive
        genes; Dominant genes;
        Genetic disorders;
        Recessive genes
Genetic complementation, and
    classification of xeroderma
    pigmentosum, 70–77
Genetic defects, and susceptibility
    to cancer, 7–9
    See also Genetic disorders
Genetic disorders
    cancer prone, 336
    and DNA damage and cancer
        induction, 342
    and genetic instability in
        tumor cells, 422–423
    See also Ataxia-telangiectasia;
        Bloom's syndrome;
        Fanconi's anemia;

Werner's syndrome;
  Xeroderma
  pigmentosum
Genetically determined immune
  deficiencies
  and cancer, 125, 126, 127
  and cancer in ataxia-
    telangiectasia, 105–112
Genomes, transposable elements
    in, 335
  See also Cancer and genomic
    rearrangements
Glucose-6-phosphate
  dehydrogenase, xvi
Growth deficiency
  in Bloom's syndrome, 12, 13,
    14–15, 18
  in Fanconi's anemia, 46
  in progeria, 207
  in Werner's syndrome, 85, 86
  in xeroderma pigmentosum,
    68, 69
Growth potential, of Werner's
  syndrome cells in vitro,
  155
Growth spurt before puberty, and
  Bloom's syndrome, 13

Hair, premature graying of, in
  Werner's syndrome, 86
Hair loss, in Werner's
  syndrome, 86
Heterogeneity, of ataxia-
  telangiectasia, 29–30
Heterozygotes
  and Bloom's syndrome, 19
  Fanconi's anemia, 44
  of xeroderma pigmentosum,
    78
Hodgkin's disease, and primary
  immunodeficiency
  disorders, 377–378
Hokkaido Imperial University,
    xxvii
Huntington's chorea, cellular
  characteristics of, 217

Hutchinson-Gilford syndrome.
  See Progeria
Hydrocarbons, and mutability of
  xeroderma pigmentosum
  cells, 219
Hypogonadism, in Werner's
  syndrome, 87

IgA deficiency, in ataxia-
  telangiectasia, 25, 29
IgG levels, in ataxia-
  telangiectasia, 25
IgM levels, in ataxia-
  telangiectasia, 25
Immune surveillance concept,
  376–377
Immune system, and clonal
  evolution model of
  cancer, 427
Immunodeficiency
  and cancer and chromosome
    instability in ataxia-
    telangiectasia patients,
    193–201
  and cancer incidence in ataxia-
    telangiectasia, 32–33
  and cancer-proneness in
    Bloom's syndrome, 352
  and cancer-proneness of
    chromosome breakage
    syndromes, 376–378
  genetically determined, *see*
    Genetically-determined
    immunodeficiency
  types of, 105
  and xeroderma
    pigmentosum, 69
Immunodeficiency Cancer
  Registry, 200
Immunofunction, in Werner's
  syndrome, 91
Immunoglobulins
  absence of, 105
  in ataxia-telangiectasia, 25
  in Bloom's syndrome, 16
Immunoregulation, and
  pathogenesis of ataxia-
  telangiectasia, 32–33

Infertility, in Bloom's syndrome, 16
Initiation, in carcinogen-induced tumorigenesis, 360
Insertion sequences of bacteria, 341
Ionizing radiation, xix. *See also* Radiation
Interferon, 3

Lesch-Nyhan cells, SV40-induced revertants of, 326
Leukemia
  in ataxia-telangiectasia patients, chromosome instability and immune deficiency in, 197
  and Bloom's syndrome, 16–17, 19, 103, 347, 349–350
  chromosome changes in, 333–334
  and Fanconi's anemia, 54–55, 113–114, 149–150
  target-cell-related chromosome changes in, 367–368
  x-ray induced, xvi
  See also Chronic myelogenous leukemia
Lipid oxidation, and carcinogenesis, 8
Liver cancer, in Fanconi's anemia, 53–54
Liver disease, and Fanconi's anemia, 53–54
Lupus erythematosus, differentiation from Blooms' syndrome, 11
Lymph nodes, in ataxia-telangiectasia, 24
Lymphocytes
  chemically-induced chromosome damage in, 256–258
  chromosome studies in cigarette smokers, 264, 265, 267–276

Con-A capping of, ataxia-telangiectasia and, 33
sister-chromatid exchange in, Bloom's syndrome and, xix, 17
  and tests of chemicals, 263
  in Werner's syndrome, 88–89
  See also PHA-stimulated lymphocytes
Lymphocytotoxic assays in ataxia-telangiectasia, 24–25
Lymphosarcoma, in ataxia-telangiectasia, 25
Lyon hypothesis, xvi

Maize, transposable elements in, 338–340
Malignant melanoma, in xeroderma pigmentosum, 67
Menarche in Bloom's syndrome, 16
Menses, in Werner's syndrome, 87
Mental development, in Bloom's syndrome, 16
Mental retardation, in Fanconi's anemia, 46
Mice
  induced tumors in, and target-cell specific chromosome changes, 265–267
  model for ataxia telangiectasia, 30–31
  virus-induced cell transformation in, 323
Mitomycin C, and sister-chromatid exchange formation in fibroblasts from Fanconi's anemia patients, 179–183
Microophthalmia, in Fanconi's anemia, 46
Microtubular defects, and pathogenetic models of ataxia telangiectasia, 33–34

Mole,
  cytogenetics of, xxx
Mortality
  in ataxia-telangiectasia, 24
  in Fanconi's anemia, 51–52, 57–58
  and neurological abnormalities in xeroderma pigmentosum, 69
  in Werner's syndrome, 86
Mosaic, xviii
Muscle atrophy
  in ataxia telangiectasia, 24
  in Werner's syndrome, 86–87
Mutagen hypersensitivity
  in ataxia-telangiectasia, 211
  in Bloom's syndrome, 212
  in chromosome-breakage and other cancer-prone syndromes, 203–223
  and classification of cancer-prone syndromes, 204, 205
  and DNA repair defects and cancer, 221–222
  and DNA repair deficiencies, 219–220
  in excision-deficient xeroderma pigmentosum, 208–209
  in Fanconi's anemia, 213
  and human diversity, 211–212, 214–215
  and regulation of responses to DNA damage, 235–245
  research potentials of, 203
  and regulatory signals for DNA repair and regulation, 240–244
  and undiscovered chromosome breakage syndromes, 204–205
  and use of transformed cell lines, 214–215
  xeroderma pigmentosum as model of, 215, 218–219
Mutagenesis
  in bacteria, 281
  and sister-chromatid exchange, 174–175
Mutagenic-carcinogens, 5–7
  SCE-inducing, 171–174
Mutagens
  and cellular repair mechanisms, 286–292
  DNA reaction with, 282–286
  and DNA repair mechanisms, 292–297
  and DNA synthesis and bypass of lesions, 301–307
  and inhibition of DNA synthesis, 297–301
Mutation
  and clonal evolution model of tumor progression, 421–427
  molecular alterations in DNA and, 281–309
  and carcinogenesis, 8
  from DNA synthesis, 307–309
  spontaneous, in Bloom's disease and xeroderma pigmentosum, 236–237
  virus-induced, 320, 324–326

N-acetoxy-N-2-acetylaminofluorene
  and adduct removal from DNA, 293–297
  and DNA lesions, mutations and chromosome aberrations and, 307–309
  DNA reaction with, 282–286
N-methyl-N'-nitro-N-nitrosoguanidine
  DNA reaction with, 282–286
  and O^6 methylguanine or 3-methyladenine removal from DNA, 290–291, 293–297
α-naphtho-flavone (ANF) and blockade of cigarette smoke-induced chromosome aberrations, 268–270

Neuroendocrine dysfunctions, and ataxia-telangiectasia, 35
Neurofibromatosis, and dominant genes, 98
Neurological abnormalities, and xeroderma pigmentosum, 63, 64, 68–69
Nuclear dockyard workers, gamma radiation-induced chromosome aberrations in, 259, 261, 264, 265
Nutritional changes, and genetic instability of tumor cells, 426
Nutritional deficiency inborn, and ataxia-telangiectasia pathogenesis, 34

O^6-methylguanine, removal from DNA, mutagens and, 288–297
Occupational radiation, chromosome aberrations in nuclear dockyard workers, 259, 261, 264, 265
Ocular abnormalities, in xeroderma pigmentosum, 67–68
Oncogenes, xvii, 335–336, 398–399
and chromosome mutation theory of cancer, 354–355, 397–408
mapping of, 400–401
from transformed cells, 399–400
viral, 397–399
Osteoporosis, in Werner's syndrome, 87
Overfeeding, cancer caused by, 6
Oxygen metabolism
and chromosome instability in Werner's syndrome, 155–156
in Fanconi's anemia, 150

Pancytopenia, in Fanconi's anemia, 50
Parental consanguinity
and ataxia–telangiectasia as autosomal recessive genetic disorder, 29
and Bloom's syndrome, 19
and cytoskeletal hypothesis of ataxia-telangiectasia, 34
and Fanconi's anemia, 44
in Werner's syndrome, 86
and xeroderma pigmentosum, 64
Patent ductus arteriosis and Fanconi's anemia, 47
Peripheral blood lymphocytes. *See* Lymphocytes
Peripheral vascular disease, in Werner's syndrome, 87
PHA-stimulated lymphocytes
and chromosome aberrations in Fanconi's anemia, 148
and chromosome studies in Werner's syndrome, 153–154, 155–156
Phenotypic expression
of Fanconi's anemia, 44
of xeroderma pigmentosum, 63
Philadelphia chromosome, as marker of chronic myelogenous leukemia, xvii, xix, 35, 359
Photophobia, in xeroderma pigmentosum, 67
Phytohemagglutinin-stimulated blood lymphocyte chromosomes, in cytogenetic studies of ataxia-telangiectasia patients, 194–197
plant chromosomes, roentgen or radium rays and breakage of, 253

Poly (ADP-ribose) polymerase synthesis and increased radiation damage in ataxia-telangiectasia, 28
Polycyclic aromatic hydrocarbons, and carcinogenicity of cigarette smoke, 267–271
Pregnancy, in Bloom's syndrome, 16
Premature aging. *See* Progeria; Werner's syndrome
Premature chromosome condensation, virus-induced, 317–318, 320
Prenatal diagnosis of Bloom's syndrome, 18
Primary immunodeficiency disorders, 377–378
Progeria, 207
  cellular characteristics of, 214
Promotion, in carcinogen-induced tumorigenesis, 360
Provirus, xxii–xxiii
Purkinje cells in cerebellum, in ataxia-telangiectasia, 24, 34
Pyrimidine dimers in DNA, and ultraviolet light-produced cancers, 5

Quadriradial configuration (Qr)
  in Bloom's syndrome, 136, 138–139, 175–176
  in Fanconi's anemia, 149

Radiation
  and genetic instability of tumor cells, 425–426
  -induced chromosome aberrations, 259, 261, 264, 265 *See also* X-rays
  and induction of chromosome aberrations, 259–261
  and tumor induction, 336
Radiation therapy, and ataxia-telangiectasia and cancer, 25–28
Radiosensitivity, human diversity in, 212, 214
Rauscher leukemia virus, and chromosome aberrations, 318, 321
Recessive genes, and cancer-prone syndromes, 98–99, 135
Recombination DNA technology, and future studies of sister-chromatid exchanges, 184
Renal abnormalities in Fanconi's anemia, 46
Retinoblastoma genes, 97
Retinoblastomas, and chromosome 13 deletion, 210–211, 217
Respiratory tract infections, in Bloom's syndrome, 13
Retroviruses, and oncogenes, 335–336, 397–399
RNA tumor viruses. *See* Retroviruses
Rothmund-Thomson syndrome, 207
  cellular characteristics of, 214
  differentiation from Werner's syndrome, 85
Rous sarcoma virus
  -induced specific chromosome changes in tumor tissue, 364
  and two-hit hypothesis of viral transformation, 327–328

Sarcomas, in xeroderma pigmentosum, 67
Scleroderma, differentiation from Werner's syndrome, 85
Sea urchin eggs, abnormal chromosomes and abnormal development in, 253

Serum alphafetoprotein, in ataxia-telangiectasia, 24
Sex differences
  in age of diagnosis of Fanconi's anemia, 43
  in nondisjunction of X chromosomes, 258–259
  in skin lesions in Bloom's syndrome, 12–13
  in xeroderma pigmentosum incidence, 64
Sexual development, immature, in xeroderma pigmentosum, 68, 69
Sister-chromatid exchanges, 169–184
  in ataxia telangiectasia, 178–179
  basic features of, 169–171
  biological significance of, 184
  in Bloom's syndrome, 17, 19–20, 139–142, 175–177, 236–237
  in chromosome-fragility diseases, 175–183
  in cigarette smokers, 265, 267–275
  defined, 169
  detection, 169
  and diagnosis of Bloom's syndrome, 18
  DNA-replication fork and, 174
  in Fanconi's anemia, 179–183
  frequency in human tissue, 171
  induction by exogenous agents, 171–174
  mechanism of production, 237–240
  molecular analysis of formation of, 183–184
  mutagenesis and, 174–175
  virus-induced increase in, 318, 321
  in Werner's syndrome, 89
  in xeroderma pigmentosum, 153, 177–178, 236–237
Skin. *See* Skin atrophy, Skin changes; Sun sensitivity
Skin atrophy, in progeria, 207
Skin cancer, in xeroderma pigmentosum, 7, 66–67
Skin changes
  in Fanconi's anemia, 46
  in xeroderma pigmentosum, 63, 65–67
Skin lesions
  in ataxia-telangiectasia, 23
  in Bloom's syndrome, 12–13
  in Werner's syndrome, 85, 87
  in xeroderma pigmentosum variant group, 77–78
Slow virus infection, and ataxia-telangiectasia pathogenesis, 34–35
Somatic cell hybridization, genetic analysis of neoplastic growth by, 372–373
Somatic crossing-over, and chromosome-mutation theory of carcinogenesis, 352–353
Somatic mutation theory of cancer
  in chromosome-breakage syndromes, 383–384
  and mutagenicity of chemical carcinogens, 219
  and spontaneous mutation rate in chromosome-breakage syndromes, 204–205
  and virus-induced cell transformation, 321–324
Somatic recombination, and cancer proneness in Bloom's syndrome, 351

SOS repair system in bacteria, 281
Spm system of maize, 339–340
Squamous cell carcinomas
  in Fanconi's anemia, 56
  in xeroderma pigmentosum, 66–67
src genes, 335–336
Stemline concept, xvi, xxx
Steroid hormones, and sister-chromatid exhange induction, 172
Stress erythropoiesis, in Fanconi's anemia, 48
Sun sensitivity
  in Cockayne's syndrome, 207
  in Rothmund-Thomson syndrome, 207
  and telangiectatic erythema in Bloom's syndrome, 12
  in xeroderma pigmentosum, 63, 64
Superperoxide dismutase, and blood lymphocyte chromosome studies in Werner's syndrome, 88–89
SV-40 virus
  cell tranformation studies, 322–323
  and chromosome aberrations, 318, 319
  and gene mutations, 324–326
  hypertransformability of Fanconi's anemia cells by, 115
  -induced chromosome breakage, in Fanconi's anemia, 324
  and sister-chromatid exchange in xeroderma pigmentosum cells, 177–178
  transformed xeroderma pigmentosum cells, DNA-damaging agents and, 153

T-cells, in ataxia-telangiectasia, 24, 29
T-helper-cells, in ataxia-telangiectasia, 24
T-lymphocytes
  in ataxia-telangiectasia, 24–25
  in Fanconi's anemia, 151
Telangiectases. *See* Ataxia-telangiectasia
Telangiectatic erythema, in Bloom's syndrome, 12, 18
Teratocarcinomas, and transplantation of embryonic animal cells, 5
Testes, undescended, in Fanconi's anemia, 46
Testicular atrophy, in Werner's syndrome, 87
Testicular hypogonadism, in Bloom's syndrome, 16
Tetradecanoyl-phorbol-acetate, and sister-chromatid exchange induction, 172
Thumb anomalies, in Fanconi's anemia, 45, 46
Thymus, in ataxia-telangiectasia, 24
T-lymphocyte clones, in ataxia-telangiectasia, 378–381
Tonsils, in ataxia-telangiectasia, 24
Tr chromatid-interchange configurations, in Fanconi's anemia, 149
Transformation processes, oncogenes and, 397–408
Translocations
  associated with human neoplasma, 405
  in ataxia-telangiectasia, 145
Transposable elements, as mechanism of genomic rearrangement, 338–342
Trisomy 13, 210

Trisomy 18, 210
Trisomy 21. *See* Down's syndrome
Tritiated thymidine, and sister-chromatid exchange studies, 169, 171
Tumor cells
  abnormal mitoses in, 359
  chromosome abnormalities in, 359–360
  chromosome constitution of, 333, 353–355
  genetic instability of, 421–427
  normal karyotypes in, 370–372
  oncogenes from, 399–400
  specific translocations in, 362–363
Tumor viruses, and chromosome damage, 342–343
Tumors
  and clonal evolution, *see* Clonal evolution model of cancer
  and ocular abnormalities in xeroderma pigmentosum, 68
  potential reversibility of, 428–429
  See also Cancer
Tumors
  specificity of chromosome changes in, 361–364

Ultrasonography, and fetal size determinations, diagnosis of Bloom's syndrome and, 18
Ultraviolet irradiation, and cancer in xeroderma pigmentosum, 120–122
Ultraviolet light
  damaging effects of, 70–78, 253

  and DNA repair mechanisms, 293
  -induced mutations and sister-chromatid exchanges in xeroderma pigmentosum, 219, 283
  and repair of DNA damage in xeroderma pigmentosum, 177
  and sister-chromatid exchange-inducing agents, 173
Urine, acid glycosaminoglycans, in Werner's syndrome, 90

V gene translocation, and lymphoma-inducing gene(s), 36
Variable immunodeficiency. *See* Acquired agammaglobulinemia
Variegated translocation mosaicism, in Werner's syndrome, 88
Vinyl chloride monomer (VCM), and induced DNA damage and chromosome aberrations, 263, 266
Virus-induced genetic changes, 317–328
  and cell transformation studies, 321–324
  mutations, 324–326
  two-hit hypothesis of cellular transformation, 327–328
  types of, 317–321
Viruses
  and carcinogenesis in animals, xxii–xxiii, 6
  and genetic instability of tumor cells, 425
  role in cancer, xxiii
  See also Slow virus infection; virus-induced genetic changes

Voice
  in Bloom's syndrome, 16
  in Werner's syndrome, 86

Werner's syndrome, 85–91
  cancer in, 123, 124
  chromosome abnormalities in, 88–89
  clinical characteristics, 85–91
  cytogenetics of chromosome breakage in, 153–156
  growth potential of fibroblasts in, 89–90
  hair graying and loss in, 86
  incidence, 86
  premature aging in, 90–91
  vascular disease in, 87
  See also Chromosome-breakage syndromes
Wiskott-Aldrich syndrome, 98, 105

X chromosomes, nondisjunction, 258–259
Xeroderma pigmentosum, 63–78
  acute sun sensitivity in, 64
  cancer and, 69, 120–122
  chronic cutaneous changes in, 65–67
  clinical manifestations, 63
  and Cockayne's syndrome, 70
  and complementation groups classifications, 70–77, 218
  as conditional chromosomal-breakage syndrome, 205
  cytogenetics of chromosome breakage in, 151–153
  diversity of characteristics of, 211–212
  and DNA damage and replication, 237, 244, 287–288, 290, 292–293
  genetic subgroups of, 63
  heterozygotes, 78
  hypersensitivity to DNA-damage in, 135, 235–236
  as model for mutagen hypersensitivity studies, 215, 218–219
  mutagen sensitivity in excision-deficient, 208–209
  mutations and sister-chromatid exchanges in, carcinogenesis and, 236–237
  neurological abnormalities in, 68–69
  ocular abnormalities in, 67–68
  prevalence, 63–64
  recessive inheritance of, 98
  sister-chromatid exchange in, 153, 177–178
  skin cancer and, 7
  variants, 77–78, 239
  See also Chromosome-breakage syndromes
Xeroderma Pigmentosum Registry, 121
X-rays
  and ataxia-telangiectasia cells, 147, 178, 200–201, 240–244
  and controlling elements in maize, 338
  and DNA damage, 254, 256–258

Yeast, transposable elements in, 335

Zea mays, transposable elements in, 335